D0900179

ALLE·ZEIT·WACH
1842

The Eukaryotic Ribosome

Edited by Heinz Bielka

With Contributions by

Heinz Bielka Joachim Stahl Ulrich-Axel Bommer
Heinz Welfle Franz Noll Peter Westermann

With 36 Figures and 29 Tables

Springer-Verlag
Berlin Heidelberg New York 1982

AKM Prof. Dr. Heinz Bielka
Dr. rer. nat. Ulrich-Axel Bommer
Dr. med. Franz Noll
Dr. agr. Joachim Stahl
Dr. sc. nat. Heinz Welfle
Dr. sc. nat. Peter Westermann

Zentralinstitut für Molekularbiologie
der Akademie der Wissenschaften der DDR
Lindenberger Weg 80
DDR - 1115 Berlin

Sole distribution rights for all non-socialist countries
granted to Springer-Verlag Berlin Heidelberg New York

ISBN 3-540-11059-3 Springer-Verlag Berlin Heidelberg New York
ISBN 0-387-11059-3 Springer-Verlag New York Heidelberg Berlin

Printed in the German Democratic Republic

2131/3140-543210

Contents

Abbreviations and glossary

AA-tRNA	Aminoacyl-tRNA
A site	the site on the ribosome at which AA-tRNA during elongation and peptidyl-tRNA after the peptidyltransferase reaction (before translocation) are bound
1-D	one-dimensional (electrophoresis)
2-D	two-dimensional (electrophoresis)
dsRNA	double stranded RNA
e	eukaryotic
EDTA	ethylenediamine tetraacetic acid
EF	elongation factor(s)
ER	endoplasmic reticulum
hnRNA	high molecular weight nuclear RNA
IF	initiation factor(s)
Initiator-tRNA	formylatable methionyl-tRNA, the first AA-tRNA bound to the P site of the small ribosomal subunit during initiation of protein synthesis; see also Met-$tRNA_f$
Met-$tRNA_f$	formylatable methionyl initiator-tRNA (not formylated in eukaryotes, in contrast to prokaryotes)
mRNA	messenger-RNA; occasionally also named template-RNA
NEM	N-ethylmaleimide
P site	the site on the ribosome at which initiator-tRNA during initiation and peptidyl-tRNA during elongation before the transferase reaction are bound
rRNA	ribosomal RNA
S	the Svedberg-unit to express the sedimentation coefficient. $1\ S = 10^{-13}\ s/rad^2$
$s_{20,w}$	sedimentation coefficient at 20 °C in water
SDS	sodium dodecylsulfate
tRNA	transfer-RNA; occasionally also named acceptor-RNA, adapter-RNA, or soluble RNA

I. Introduction

Ribosomes are the subcellular organelles at which the mRNA directed linkage of amino acids to peptide chains takes place.

Crick [1] and Orgel [2] suggested that the primitive apparatus for protein biosynthesis might have consisted originally only of RNA and that the addition of proteins during evolution refined this machinery with regard to efficiency and fidelity of the translation of genetic messages.

In prokaryotes (*E. coli*) about two-thirds of the mass of each ribosomal subunit is RNA and only one-third is protein; altogether, the *E. coli* ribosome contains 55 proteins. Ribosomes from eukaryotes are bigger and more complicated than those from prokaryotes. Ribosomes from rat liver, for example, consist of at least 70 to 75 different proteins, and their total mass amounts to roughly 50 per cent of the particle.

Many molecular biologists share the opinion that the ribosome will be the first organelle of the cell the structure and function of which will be elucidated at the molecular level. Both subunits of the *E. coli* ribosome can be entirely reconstituted in vitro from the various macromolecular components. Analysis of this assembly process afforded already some important insights into the molecular mechanism of interactions between RNAs and the proteins for the structural organization of the ribosomal subunits and their function in the translation process. Furthermore, numerous physical and chemical properties of the ribosomal RNAs and proteins, especially of *E. coli*, and their localization in the ribosomal particles have been elucidated. The main results of these studies have been described in the last decade in various publications [3—37].

Although ribosomes were originally discovered in eukaryotic cells (see Chapter II), we know at present more about prokaryotic ribosomes than about their eukaryotic counterparts. This is due mainly to the following facts: (1) as already mentioned, the prokaryotic ribosome is less complex

Contents

than the eukaryotic one, at least as far as the number of the macromolecular constituents is concerned; (2) in bacteria numerous ribosomal mutants are available the analysis of which essentially contributed to the understanding of the functional meaning of ribosomal proteins; (3) much more laboratories are working with bacterial ribosomes than with ribosomes from eukaryotes.

Although the structural and functional architecture of eukaryotic ribosomes is most likely similar ore even the same as that of E. *coli* ribosomes, the number and properties of the chemical components (RNA, proteins) and hence various physical and chemical properties of the ribosomal particles differ in detail. It is precisely this difference that determines the importance of studies on eukaryotic ribosomes, which can perhaps be made clear by asking the question: why is the eukaryotic ribosome, as a result of evolution, organized in a more complex way, consisting of about 70 to 75 different proteins, although the smaller *E. coli* ribosome with only 55 proteins can basically do the same job? Perhaps for the realization of genetic programs, aside from the control at the transcription level, also translation regulation may play an important role in the more complex eukaryotic cell, which therefore requires a correspondingly higher organized apparatus. This level of regulation in gene expression might be of biological significance for growth and differentiation of eukaryotic cells.

During the last few years more and more laboratories have turned to experiments with eukaryotic ribosomes. Thus, these investigations have now reached such a remarkable level which justifies the attempt to summarize and interpret the main lines of present knowledge. Eukaryotic cells are equipped with two complete systems for protein biosynthesis, namely one in the cytoplasm and one in the mitochondria (plant cells possess of a further translation system in the plastides). The protein synthesis machineries of these cell organelles are in many aspects similar to the ribosomes and the other components of translation in bacteria (prokaryote-type of translation machinery) and differ significantly from the cytoplasmic ribosomes.

The reader of this book should be aware of its aims and limitations; therefore we deem it necessary to make the following editorial comments.

The individual chapters deal preferentially with the cytoplasmic or so-called 80 S type ribosomes of eukaryotic cells; for comparison only a few properties of ribosomes and their components of mitochondria and plastides will be mentioned in the given chapters.

Also, the book deals chiefly with the structure and function of the eukaryotic ribosome and its constituents. Properties of non-ribosomal components interacting with the ribosome during the translation process, such as tRNA and mRNA, are summarized only; the reader interested in these topics is referred to more comprehensive reviews which are included in the reference list of Chapter XI.

A separate chapter on antibiotics and some other effectors interacting with ribosomes and the translation process, planned originally, was omitted because of excellent review articles and books that have appeared recently [29—31].

The topics of this book are restricted mainly to the description of experimental results and conclusions, while experimental approaches and techniques have not been considered.

Despite efforts toward a comprehensive and objective representation, the choice and treatment of the vast material available cannot be completely free of subjective traits. We are also aware that this book cannot cover all the valuable contributions of all authors investigating the structure and function of eukaryotic ribosomes, despite their importance in this field. Therefore, the reader is referred to earlier publications, especially to the books of PETERMANN [3], of SPIRIN and GAVRILOVA [4], of NOMURA et al. [5] which cover the original literature on ribosomes up to the late 1960s and early 1970s, respectively, and to more recent reviews [25—46].

Except for Chapters II and III, the literature mainly of the last 10 to 15 years has been included. For technical reasons the original literature could be considered only up to fall of 1979; essential results published in 1980 and early 1981 will be summarized as "Notes added in proof" at the end.

As to the designation of ribosomal proteins the proposed uniform nomenclature, published in 1979 [47] will be used throughout.

Finally, the editor and the authors of this book would welcome all critical and helpful suggestions by the readers.

Berlin—Buch, 1982

Heinz BIELKA

References

[1] CRICK, F. H. C.: The origin of the genetic code. J. Mol. Biol. **38**, 367—379 (1968)

[2] ORGEL, L. E.: Evolution of the genetic apparatus. J. Mol. Biol. **38**, 381—393 (1968)

[3] PETERMANN, M. L.: The physical and chemical properties of ribosomes. Elsevier Publ. Comp., New York, 1964.

[4] SPIRIN, A. S., and L. P. GAVRILOVA: The ribosome. Springer, New York, 1969

[5] NOMURA, M., A. TISSIÈRES, and P. LENGYEL (Eds.): Ribosomes. Cold Spring Harbor Laboratory (1974)

[6] TRAUB, P.: Structure, function and in vitro reconstitution of *E. coli* ribosomes. Current Topics Microbiol. **62**, 1—93 (1970)

[7] KURLAND, C. G.: Structure and function of the bacterial ribosome. Ann. Rev. Biochem. **41**, 377—408 (1972)

[8] GARRETT, R. A., and H. G. WITTMANN: Structure and function of the ribosome. Endeavour **32**, 8—14 (1973)

[9] NOMURA, M.: Assembly of bacterial somes. Science **179**, 864—873 (1973)

[10] PONGS, O., K. H. NIERHAUS, V. ERDMANN, and H. G. WITTMANN: Active sites in *E. coli* ribosomes. FEBS Lett. **40**, S28 to S37 (1974)

[11] WITTMANN, H. G.: Purification and identification of *Escherichia coli* ribosomal proteins. In ref. [5] 93—114 (1974)

[12] WITTMANN, H. G., and B. WITTMANN-LIEBOLD: Chemical structure of bacterial ribosomal proteins. In ref. [5] 115—140 (1974)

13] FELLNER, P.: Structure of the 16S and 23S ribosomal RNAs. In ref. [5] 169—191 (1974)

[14] NOMURA, M., and W. A. HELD: Reconstitution of ribosomes: Studies of ribosome structure, function and assembly. In ref. [5] 193—223 (1974)

[15] ZIMMERMANN, R. A.: RNA-protein interactions in the ribosome. In ref. [5] 225—269 (1974)

[16] TRAUT, R. R., R. L. HEIMARK, T. T. SUN, J. W. B. HERSHEY, and A. BOLLEN: Protein topography of ribosomal subunits from *Escherichia coli.* In ref. [5] 271—308 (1974)

[17] JASKINAS, S. R., M. NOMURA, and J. DAVIES: Genetics of bacterial ribosomes. In ref. [5] 333—368 (1974)

[18] SCHLESSINGER, D.: Ribosome formation in *Escherichia coli.* In ref. [5] 393—416 (1974)

[19] WITTMANN, H. G.: Structure, function and evolution of ribosomes. Europ. J. Biochem. **61**, 1—13 (1976)

[20] BRIMACOMBE, R., K. H. NIERHAUS, R. A. GARRETT, and H. G. WITTMANN: The ribosome of *Escherichia coli.* Progr. Nucleic Acid Res. Mol. Biol. **18**, 1—44 (1976)

[21] KÜCHLER, E.: Chemische Methoden zur Untersuchung der Ribosomenstruktur. Angew. Chemie **88**, 555—564 (1976)

[22] KURLAND, C. G.: Structure and function of the bacterial ribosome. Ann. Rev. Biochem. **46**, 173—200 (1977)

[23] BRIMACOMBE, R., G. STÖFFLER, and H. G. WITTMANN: Ribosome structure. Ann. Rev. Biochem. **47**, 217—249 (1978)

[24] SPIRIN, A. S.: Energetics of the ribosome. Progr. Nucleic Acid Res. Mol. Biol. **21**, 39—62 (1978)

[25] WITTMANN, H. G.: A comparison of ribosomes from prokaryotes and eukaryotes. Symp. Soc. Gen. Microbiol. **20**, 55—76 (1970)

[26] NANNINGA, N.: Structural aspects of ribosomes. Internat. Rev. Cytol. **35**, 135—188 (1973)

[27] VAN HOLDE, K. E., and W. E. HILL: General physical properties of ribosomes. In ref. [5] 53—91 (1974)

[28] DAMASCHUN, G., J. J. MÜLLER, and H. BIELKA: Scattering studies of ribosomes and ribosomal components. Methods Enzymology **LIX**, 706—750 (1978)

[29] VASQUEZ, D.: Inhibitors of protein synthesis. FEBS Lett. **40**, S63—S84 (1974)

[30] VASQUEZ, D.: Translation inhibitors. Internat. Rev. Biochem. **18**, 169—232 (1978)

[31] VASQUEZ, D.: Inhibitors of protein biosynthesis. Springer, Berlin—Heidelberg—New York, 1979

[32] HADJIOLOV, A. A., and N. NIKOLAEV: Maturation of ribosomal ribonucleic acids and the biogenesis of ribosomes. Progr. Biophys. Mol. Biol. **31**, 95—144 (1976)

[33] COX, R. A., and E. GODWIN: Ribosome structure and function. Internat. Rev. Biochem. **7**, 179—253 (1975)

[34] COX, R. A.: Structure and function of prokaryotic and eukaryotic ribosomes. Progr. Biophys. Mol. Biol. **32**, 193—231 (1977)

[35] LUCAS-LENARD, J., and F. LIPMANN: Protein synthesis. Ann. Rev. Biochem. **40**, 409—448 (1971)

[36] HASELKORN, R., and L. B. ROTHMAN-DENES: Protein synthesis. Ann. Rev. Biochem. **42**, 397—438 (1973)

[37] GRUNBERG-MANAGO, M., and F. GROS: Initiation mechanism of protein synthesis. Progr. Nucleic Acid Res. Mol. Biol. **20**, 209—284 (1977)

[38] BATTANER, E., and D. VASQUEZ: Inhibitors of protein synthesis by ribosomes of the 80-S type. Biochim. Biophys. Acta **254**, 316—330 (1971)

[39] SMELLIE, R. M. S. (Ed.): The structure and function of eukaryotic ribosomes. The Biochem. Soc., London (1973)

[40] TRAUGH, J. A., and R. R. TRAUT: Recent advances in the preparation of mammalian ribosomes and analysis of their protein composition. Methods Cell Biology **7**, 67—103 (1973)

[41] WOOL, I. G., and G. STÖFFLER: Structure and function of eukaryotic ribosomes. In ref. [5], 417—460 (1974)

[42] WARNER, J. R.: The assembly of ribosomes in eukaryotes. In ref. [5], 461—488 (1974)

[43] REEDER, R. H.: Ribosomes from eukaryotes: Genetics. In ref. [5], 489—518 (1974)

[44] WEISSBACH, H., and S. OCHOA: Soluble factors required for eukaryotic protein synthesis. Ann. Rev. Biochem. **45**, 191—216 (1976)

[45] BIELKA, H., and J. STAHL: Structure and function of eukaryotic ribosomes. Internat. Rev. Biochem. **18**, 79—166 (1978)

[46] WOOL, I. G.: The structure and function of eukaryotic ribosomes. Ann. Rev. Biochem. **48**, 719—754 (1979)

[47] MCCONKEY, E. H., H. BIELKA, J. GORDON, S. M. LASTICK, A. LIN, K. OGATA, J.-P. REBOUD, J. A. TRAUGH, R. R. TRAUT, J. R. WARNER, H. WELFLE, and I. G. WOOL: Proposed uniform nomenclature for mammalian ribosomal proteins. Molec. Gen. Genetics **169**, 1—6 (1979)

II. A short historical survey

Contents

At the end of the last century, GARNIER [1] and BENSLY [2] described in the cytoplasm of cells filamentous elements characterized by a high affinity for basic dyes; these structures were called ergastoplasm by GARNIER.

BRACHET [3, 4] and CASPERSSON [5, 6], using biochemical and cytochemical methods, respectively, demonstrated that this basophilia was caused by a high content of RNA located in the cytoplasm and, furthermore, that a direct relationship exists between the amount of RNA of cells and their rate of protein synthesis. Therefore, both authors concluded that the RNA containing elements of the basophilic structures of a cell are involved in the biosynthesis of proteins.

By differential centrifugation of homogenates of tumor tissues CLAUDE [7—9] isolated a postmitochondrial particulate fraction which he called "small granules" and later "microsomes". This fraction consisted of pieces of the phospholipid containing membranes of the endoplasmic reticulum and of free and membrane-bound nucleoprotein particles. In the following years this subcellular fraction has attracted increasing interest of cell morphologists and biochemists.

In 1955/56 PALADE and coworkers [10, 11] in their electron microscopic studies described RNA rich particles of about 10 to 20 nm diameter and high electron density in thin sections of whole cells and in the isolated microsomal fraction. Depending on the tissue the particles were found either attached to the membranes of the endoplasmic reticulum or free in the cytoplasm (see Chapter III). These studies supported the assumptions of CLAUDE that the nucleoprotein particulate fraction, obtained by differential centrifugation of tissue homogenates, in fact represented native organelles of the cell.

The ribonucleoprotein particles as part of the ergastoplasm or microsomes of eukaryotic tissues (animals, plants, yeast) have been extensively analyzed by ultracentrifugal studies in the 1950s by PETERMANN et al. [12, 13], by Tso et al. [14] and by CHAO and SCHACHMAN [15, 16]. These authors observed discrete peaks of the "macromolecular nucleoprotein" in the analytical ultracentrifuge with sedimentation coefficients in the range of 40 S, 60 S and 80 S. The particles of these fractions were shown to contain about 40% RNA.

SCHACHMAN's group [17] analyzed extracts of various bacterial species and found ribosomal particles corresponding to 50 S and 30 S. The existence of 50 S and 30 S ribosomal subunits of 70 S monomeric ribosomes from *E. coli* was also demonstrated in 1958 by TISSIÈRES and WATSON [18, 19] and later on by HALL and SLATER [20] and by HUXLEY and ZUBAY [21]. The latter authors [20, 21] demonstrated *E. coli* ribosomal particles and their subunits by electron microscopy. From the size, the chemical composition and the sedimentation data, molecular weights in the order of 10^6 were calculated for this particular fraction.

Depending on the various techniques used for their detection, isolation and chemical characterization, the granular nucleoproteins of the cell were called "ribonucleoprotein particles", "small granules", "small dense granules", "ultramicrosomes", "200 Å granules", "Palade-granules" or "Palade particles".

In 1958 a meeting on "ribonucleoprotein particles and protein synthesis" was held by the Biophysical Society at the Massachusets Institute of Technology (USA) under the guidance of R. B. ROBERTS [22] at which he introduced the name "Ribosomes" for the ribonucleoprotein particles, discovered and described before by CLAUDE and PALADE, respectively.

In 1950 HULTIN [23] and BORSOOK and his coworkers [24] using labeled amino acids observed that microsomal components are involved in the biosynthesis of proteins. Both by in vivo and in vitro experiments, it could be demonstrated that the microsomes were the first and highes labeled subcellular fraction. Ribosomes as the proper site of protein synthesis were found by ZAMECNIK's group in 1955 [25]. These authors demonstrated that upon disintegration of the microsomes by dosium deoxycholate, radioactively labeled proteins were present mainly in the insoluble fractions, that means, on the ribosomes. In 1959 SCHACHTSCHABEL and ZILLIG [26] and MCQUILLEN et al. [27] and later on other authors [28—30] could show that also in bacteria ribosomes are the sites of protein synthesis.

During the 1953—58s, using various subcellular fractions and other ingredients, ZAMECNIK and his coworkers [31—39] found in cell-free incorporation experiments with radioactively labeled amino acids the following components essential for protein synthesis: ribosomes (microsomes), enzymes from the $105000 \times g$ supernatant, "soluble RNA" (= tRNA), ATP and GTP. The formation of aminoacyl-AMP and aminoacyl-tRNA as intermediates has been demonstrated by HOAGLAND et al. [37, 38]. The enzymes from the $105000 \times g$ supernatant were characterized as aminoacyl-tRNA synthetases. BERG and OFENGAND [40] purified in 1958 a synthetase and they demonstrated that the enzyme in the presence of ATP causes activation of the amino acid and its linking to tRNA. From their experiments, HOAGLAND et al. [38] suggested an adapter hypothesis on the basis of nucleotide base pairing between the aminoacyl-tRNA and a template RNA. At the same time, CRICK [41] and CRICK et al. [42] proposed a similar hypothesis, involving a triplet code. A triplet code had already been suggested before by DOUNCE [43] in 1952.

Evidence for a template RNA at which the synthesis of specific proteins takes places came from studies of VOLKIN and ASTRACHAN [44] in 1956 and lateron from similar experiments of NOMURA et al. [45] studying the synthesis of phage specific proteins in E. *coli* cells infected with T 2-phages. The presence of specific RNA templates also in uninfected cells was demonstrated by GROS et al. [46]. In 1961 JACOB and MONOD [47] introduced for these metabolically unstable template RNAs the name "messenger RNA" (mRNA). The DNA-template origin of the mRNA was shown in 1961 by HALL and SPIEGELMAN [48] using the DNA × RNA hybridization technique. Also studying the synthesis of phage specific proteins in T 2 infected *E. coli* cells, BRENNER et al. [49] were able to show that the phage specific mRNA is attached to ribosomes. In 1962/63, polysomes as "clusters" of ribosomes connected by mRNA were demonstrated and analyzed by electron microscopy and ultracentrifugation for bacteria [50—53] and various animal cells [54—63]. In further experiments it has been shown that polysomes are the site of protein synthesis [30, 58—63] and models of the function of polysomes in protein synthesis have been proposed by several authors [58, 62, 63].

In 1958 SIEKEVITZ and PALADE [64] suggested that "digestive proteins" are produced mainly by "ribonucleoprotein particles" attached to the membranes of the endoplasmic reticulum. This hypothesis was supported lateron by experiments in which SIEKEVITZ and PALADE [65] could show that chymotrypsinogen is synthesized by membrane-bound ribosomes. Similar findings were made lateron also for other secretory proteins [66, 67]. The molecular mechanism by which polysomes producing "export proteins" are attached to membranes were elucidated mainly by BLOBEL et al. [68] (see Chapters X and XI).

The studies mainly of NIRENBERG and MATTHAEI [69] and of OCHOA's group [70, 71] (for reviews see [72, 73]), using defined polynucleotides as templates in cell free ribosomal amino acid incorporation systems resulted in the analysis of the genetic amino acid code. Using defined trinucleotides [74] and ribonucleotides containing repeating trinucleotide sequences [75], NIRENBERG et al. [74] and KHORANA's group [75] definitely revealed the triplet nucleotide nature of the amino acid code.

Studies on the mechanism of protein biosynthesis at the ribosomes published in 1960 by BISHOP et al. [76] demonstrated that proteins are synthesized by a stepwise linkage of amino acids from the amino terminal to the carboxyl terminal groups. Experiments of numerous authors (for reviews see [77—83]) have shown in the following time that the biosynthesis of proteins at the ribosomes, through which the mRNA moves along during elongation in steps of three nucleotides, is a complex process mainly consisting of three main events, namely "initiation", "elongation", and "termination". Each of these processess in turn is complex in itself, involving different components and mechanism (see Chapter XI).

Analysis of the chemical components of ribosomes was started in 1959 by LITTAUER and EISENBERG [89] and by KURLAND [84] studying the RNA of *E. coli* ribosomes and by WALLER and HARRIS [86] concerning the proteins. KURLAND [84] clearly demonstrated that the 30 S subunit of *E. coli* ribosomes contains one molecule of 16 S RNA and the large 50 S subunit one molecule of 23 S RNA. In 1963, ROSSET and MONIER [85] discovered the 5 S RNA as a further constituent of the 50 S subunit of *E. coli* ribosomes.

The studies of WALLER and HARRIS [86] demonstrated that the ribosomes of *E. coli* are formed by a large number of various proteins. From the work of WITTMANN and his group [87] the heterogeneity of ribosomal proteins of *E. coli* was clearly demonstrated.

The high molecular weight RNA components of the large and the small ribosomal subunit of eukaryotic cells were analyzed as 28 S and 18 S RNA by HALL and DOTY [88]. The 5.8 S RNA of eukaryotic ribosomes which is in many ways analogoues to the 5 S RNA of *E. coli* ribosomes was described in 1967 by FORGET and WEISSMAN [90] and an eukaryotic specific 5 S ribosomal RNA in 1968 by BROWN and WEBER [91] (see Chapter VI). The molecular heterogeneity of the proteins of eukaryotic ribosomes was clearly demonstrated by two-dimensional electrophoresis in polyacrylamide gels at the early 1970 s by our group [92] and by SHERTON and WOOL [93].

The isolation and functional characterization of eukaryotic specific initiation factors of protein synthesis in animal cells mainly by STAEHELIN's [94, 95] and ANDERSON's [96] group has shown a higher complexity of the translation process in eukaryotics (see Chapter XI) in comparison to bacteria.

Over the last decade important progress has been made in elucidating the structure and function of ribosomes, mainly those of *E. coli*, but also the investigation of eukaryotic ribosomes has now reached a remarkable level which will be outlined in the following chapters.

References

[1] GARNIER, CH.: Les filaments basaux des cellules glandulaires. Bibliographie anat. **5**, 278—289 (1897)

[2] BENSLY, R. R.: The structure of the mammalian gastric glands. Quart. J. Microscop. Sci. **41**, 361—389 (1899)

[3] BRACHET, J.: La localication des acides pentosenucléiques dans les tissus animaux et les oeufs d'amphibiens en voie de développement. Arch. de Biol. **53**, 207—257 (1942)

[4] BRACHET, J.: Biochemical cytology. Acad. Press, New York, 1957

[5] CASPERSSON, T.: Studien über den Eiweißumsatz der Zelle. Naturwiss. **29**, 33—43 (1941)

[6] CASPERSSON, T.: Cell growth and cell function. W. W. Norton, New York, 1950

[7] CLAUDE, A.: Particulate components of normal and tumor cells. Science **91**, 77—78 (1940)

[8] CLAUDE, A.: Particulate components of cytoplasm. Cold Spring Harbor Symp. Quant. Biol. **9**, 263—270 (1941)

[9] CLAUDE, A.: Fractionation of mammalian cells by differential centrifugation. II. Experimental procedures and results. J. Exptl. Med. **84**, 61—89 (1946)

[10] PALADE, G. E.: A small particulate component of the cytoplasm. J. Biophys. Biochem. Cytol. **1**, 59—68 (1955)

[11] PALADE, G. E., and P. SIEKEVITZ: Liver microsomes. An integrated morphological and biochemical study. J. Biophys. Biochem. Cytol. **2**, 171—198 (1956)

[12] PETERMANN, M. L., and M. G. HAMILTON: The purification and properties of cytoplasmic ribonucleoprotein from rat liver. J. Biol. Chem **224**, 725—736 (1957)

[13] PETERMANN, M. L.: The physical and chemical properties of ribosomes. Elsevier Publ. Comp., Amsterdam, London, New York 1964

[14] TSO, P., J. BONNER, and J. VINOGRAD: Microsomal nucleoprotein particles from pea seadlings. J. Biophys. Biochem. Cytol. **2**, 451—465 (1956)

[15] CHAO, F. C., and H. K. SCHACHMAN: The isolation and characterization of a macromolecular component from yeast. Arch. Biochem. Biophys. **61**, 220—230 (1956)

[16] CHAO, F. C.: Dissociation of macromolecular ribonucleoprotein of yeast. Arch. Biochem. Biophys. **70**, 426—431 (1957)

[17] SCHACHMAN, H. K., A. B. PARDEE, and R. Y. STANIER: Studies on the macromolecular organization of microbial cells. Arch. Biochem. Biophys. **38**, 245—260 (1952)

[18] TISSIÈRES, A., and J. D. WATSON: Ribonucleoprotein particles from *E. coli*. Nature **182**, 778—780 (1958)

[19] TISSIÈRES, A., J. D. WATSON, D. SCHLESSINGER, and B. R. HOLLINGWORTH: Ribonucleoprotein particles from *Escherichia coli*. J. Mol. Biol. **1**, 221—233 (1959)

[20] HALL, C. E., and H. S. SLATER: Electron microscopy of ribonucleoprotein particles from *E. coli*. J. Mol. Biol. **1**, 329—332 (1959)

[21] HUXLEY, H. E., and G. ZUBAY: Electron microscope observations on the structure of microsomal particles from *E. coli*. J. Mol. Biol. **2**, 10—18 (1960)

[22] ROBERTS, R. B. (Ed.): Microsomal particles and protein synthesis. 1st Symp. Biophys. Soc. Pergamon Press, New York, Oxford (1958)

[23] HULTIN, T.: Incorporation in vivo of ^{15}N

labeled glycine into liver fractions of newly hatched chicks. Exptl. Cell Res. **1**, 376 to 381 (1950)

[24] Borsook, H., C. L. Deasy, A. J. Hagen-Smit, G. Keighley, and P. H. Lowy: Metabolism of C^{14}-labeled glycine, L-histidine, L-leucine, and L-lysine. J. Biol. Chem. **187**, 839—848 (1950)

[25] Littlefield, J. W., E. B. Keller, J. Gross, and P. C. Zamecnik: Studies on cytoplasmic ribonucleoprotein particles from the liver of the rat. J. Biol. Chem. **217**, 111—123 (1955)

[26] Schachtschabel, D., and W. Zillig: Untersuchungen zur Biosynthese der Proteine. Hoppe-Seyler's Z. Physiol. Chem. **314**, 262—275 (1959)

[27] McQuillen, K., R. B. Roberts, and R. J. Britten: Synthesis of nascent protein by ribosomes in *E. coli*. Proc. Natl. Acad. Sci. U.S. **45**, 1437—1447 (1959)

[28] Lamberg, M., and P. C. Zamecnik: Amino acid incorporation by extracts of *E. coli*. Biochim. Biophys. Acta **42**, 206—211 (1960)

[29] Tissières, A., D. Schlessinger, and F Gros: Amino acid incorporation into pro. teins by *E. coli* ribosomes. Proc. Natl-Acad. Sci. U.S. **46**, 1450—1463 (1960)

[30] Risebrough, R. W., A. Tissières, and J. D. Watson: Messenger-RNA attached to active ribosomes. Proc. Natl. Acad. Sci. U.S. **48**, 430—436 (1962)

[31] Zamecnik, P. C.: Incorporation of radioactivity from D,L-leucine-1-^{14}C into proteins of rat liver homogenates. Fed. Proc. **12**, 295 (1953)

[32] Keller, E. B., P. C. Zamecnik, and R. B. Loftfield: The role of microsomes in the incorporation of amino acids into proteins. J. Histochem. Cytochem. **2**, 378—386 (1954)

[33] Zamecnik, P. C., and E. B. Keller: Relation between phosphate energy donors and incorporation of labeled amino acids into proteins. J. Biol. Chem. **209**, 337—354 (1954)

[34] Zamecnik, P. C., E. B. Keller, J. W. Littlefield, M. B. Hoagland, and R. B. Loftfield: Mechanism of incorporation of labeled amino acids into protein. J. Cell Comp. Physiol. **47**, 81—102 (1956)

[35] Keller, E. B., and P. C. Zamecnik: The effect of guanosine diphosphate and triphosphate on the incorporation of labeled amino acids into proteins. J. Biol. Chem. **221**, 45—59 (1956)

[36] Zamecnik, P. C., M. L. Stephenson, J. F. Scott, and M. B. Hoagland: Incorporation of ^{14}C-ATP into soluble RNA isolated from the 105000×g supernatant of rat liver. Fed. Proc. **16**, 275 (1957)

[37] Hoagland, M. B., P. C. Zamecnik, and M. L. Stephenson: Intermediate reactions in protein biosynthesis. Biochim. Biophys. Acta **24**, 215—216 (1957)

[38] Hoagland, M. B., M. L. Stephenson, J. F. Scott, L. I. Hecht, and P. C. Zamecnik: A soluble ribonucleic acid intermedaite in protein synthesis. J. Biol. Chem. **231**, 241—257 (1958)

[39] Zamecnik, P. C., M. L. Stephenson, and L. I. Hecht: Intermediate reactions in amino acid incorporation. Proc. Natl. Acad. Sci. U.S. **44**, 73—78 (1958)

[40] Berg, P., and E. J. Ofengand: An enzymatic mechanism for linking amino acids to tRNA. Proc. Natl. Acad. Sci. U. S. **44**, 78—88 (1958)

[41] Crick, F. H. C.: On protein synthesis. Soc. Exptl. Biol. Symp. London **12**, 138—163 (1958)

[42] Crick, F. H. C., L. Barnett, S. Brenner, and R. J. Watts-Tobin: General nature of the genetic code for proteins. Nature **192**, 1227—1232 (1961)

[43] Dounce, A. L.: Duplicating mechanism for peptide chain and nucleic acid synthesis. Enzymologia **15**, 251—258 (1952)

[44] Volkin, E., and L. Astrachan: Phosphorus incorporation in *Escherichia coli* ribonucleic acid after infection with bacteriophage T2. Virology **2**, 149—161 (1956)

[45] Nomura, M., B. D. Hall, and S. Spiegelman: Characterization of RNA synthesized in *Escherichia coli* after bacteriophage T2 infection. J. Mol. Biol. **2**, 306 to 326 (1960)

[46] Gros, F., H. Hiatt, W. Gilbert, C. G. Kurland, R. W. Risebrough, and J. D. Watson: Unstable ribonucleic acid revealed by pulse labelling of *Escherichia coli*. Nature **190**, 581—585 (1961)

[47] Jacob, F., and J. Monod: Genetic regulatory mechanism in the synthesis of proteins. J. Mol. Biol. **3**, 318—356 (1961)

[48] Hall, B. D., and S. Spiegelman: Sequence complementarity of T2-DNA and T2-

specific RNA. Proc. Natl. Acad. Sci. U.S. **47**, 137–146 (1961)

[49] Brenner, S., F. Jacob, and M. Meselson: An unstable intermediate carrying information from genes to ribosomes for protein synthesis. Nature **190**, 576–581 (1961)

[50] Spyrides, G. J., and F. Lipmann: Polypeptide synthesis with sucrose gradient fractions of *E. coli* ribosomes. Proc. Natl. Acad. Sci. U.S. **48**, 1977–1983 (1962)

[51] Gilbert, W.: Polypeptide synthesis in *E. coli*. I. Ribosomes and the active complex. J. Mol. Biol. **6**, 374–388 (1963)

[52] Schaechter, M.: Bacterial ribosomes and their participation in protein synthesis in vivo. J. Mol. Biol. **7**, 561–568 (1963)

[53] Barondes, S. H., and M. W. Nirenberg: Fate of a synthetic polynucleotide directing cell-free protein synthesis. II. Association with ribosomes. Science **138**, 813–817 (1962)

[54] Slayter, H. S., J. R. Warner, A. Rich, and C. E. Hall: The visualization of polyribosomal structure. J. Mol. Biol. **7**, 652 to 657 (1963)

[55] Warner, J. R., A. Rich, and C. E. Hall: Electron microscope studies on ribosomal clusters synthesizing hemoglobin. Science **138**, 1399–1403 (1962)

[56] Marks, P. A., E. R. Burks, and D. Schlessinger: Protein synthesis in erythroid cells: I. Reticulocyte ribosomes active in stimulation of amino acid incorporation. Proc. Natl. Acad. Sci. U.S. **48**, 2163 to 2171 (1962)

[57] Rich, A., J. R. Warner, and H. M. Goodman: The structure and function of polyribosomes. Cold Spring Harbor Symp. Quant. Biol. **28**, 269–285 (1963)

[58] Gierer, A.: Function of aggregated reticulocyte ribosomes in protein synthesis. J. Mol. Biol. **6**, 148–157 (1963)

[59] Warner, J. R., P. M. Knopf, and A. Rich: A multiple ribosomal structure in protein synthesis. Proc. Natl. Acad. Sci. U.S. **49**, 122–129 (1963)

[60] Penman, S., K. Scherrer, Y. Becker, and J. E. Darnell: Polyribosomes in normal and poliovirus-infected HeLa cells and their relationship to messenger-RNA. Proc. Natl. Acad. Sci. U.S. **49**, 654–662 (1963)

[61] Wettstein, F. O., T. Staehelin, and H. Noll: Ribosomal aggregates engaged in protein synthesis: Characterization of the ergosome. Nature **197**, 430–435 (1963)

[62] Noll, H., T. Staehelin, and F. O. Wettstein: Ribosomal aggregates engaged in protein synthesis: Ergosome breakdown and messenger-ribonucleic acid transport. Nature **198**, 632–638 (1963)

[63] Goodman, H. M., and A. Rich: Mechanism of polyribosome action during protein synthesis. Nature **199**, 318–322 (1963)

[64] Siekevitz, P., and G. E. Palade: A cytochemical study on the pancreas of the guinea pig. I. Isolation and enzymatic activities of cell fractions. J. Biophys. Biochem. Cytol. **4**, 203–217 (1958)

[65] Siekevitz, P., and G. E. Palade: A cytochemical study on the pancreas of the guinea pig. V. In vivo-incorporation of leucine-1-C^{14} into chymotrypsinogen of various cell fractions. J. Biophys. Biochem. Cytol. **7**, 619–630 (1960)

[66] Campbell, P. N.: Functions of polyribosomes attached to membranes of animal cells. FEBS Lett. **7**, 1–7 (1970)

[67] Tata, J. R.: Ribosomal segregation as a possible function for the attachment of ribosomes to membranes. Sub-Cell. Biochem. **1**, 83–89 (1971)

[68] Blobel, G., and B. Dobberstein: Transfer of proteins across membranes. I. Presence of proteolytically and unprocessed nascent immunglobulin light chains on membrane-bound ribosomes of murine myeloma. J. Cell. Biol. **67**, 835–851 (1975)

[69] Nirenberg, M. W., and J. H. Matthaei: The dependence of cell-free protein synthesis in *E. coli* upon naturally occuring or synthetic polynucleotides. Proc. Natl. Acad. Sci. U.S. **47**, 1588–1602 (1961)

[70] Lengyel, P., J. F. Speyer, and S. Ochoa: Synthetic polynucleotides and the amino acid code. Proc. Natl. Acad. Sci. U.A. **47**, 1936–1942 (1961)

[71] Speyer, J. F., P. Lengyel, C. Basilio, and S. Ochoa: Synthetic polynucleotides and the amino acid code. Proc. Natl. Acad. Sci. U.S. **48**, 441–448 (1962)

[72] Ochoa, S.: Chemical basis of heredity, the genetic code. Experientia **20**, 1–12 (1964)

[73] Bernfield, M. R., and M. W. Nirenberg: RNA codeworts and protein synthesis. Science **147**, 479–484 (1965)

[74] Nirenberg, M. W., M. R. Bernfield, R.

Brimacombe, J. Trupin, F. Rottmann, and C. O' Neal: RNA codewords and protein synthesis. VII. On the general nature of the RNA code. Proc. Natl. Acad. Sci. U.S. **53**, 1161–1168 (1965)

[75] Morgan, A. R., R. D. Wells, and H. G. Khorana: Studies on polynucleotides. LIX. Further codon assignments from amino acid incorporation directed by ribopolynucleotides containing repeating trinucleotide sequences. Proc. Natl. Acad. Sci. U.S. **56**, 1899–1906 (1966)

[76] Bishop, J., J. Leaky, and R. Schweet: Formation of the peptide chain of hemoglobin. Proc. Natl. Acad. Sci. U.S. **46**, 1030–1038 (1960)

[77] Moldave, K.: Nucleic acids and protein synthesis. Ann. Rev. Biochem. **34**, 419 to 448 (1965)

[78] Lengyel, P., and D. Soll: Mechanism of protein synthesis. Bact. Rev. **33**, 264–301 (1969)

[79] Lucas-Lenard, J., and F. Lipmann: Protein biosynthesis. Ann. Rev. Biochem. **40**, 409–448 (1971)

[80] Haselkorn, R., and L. B. Rothman-Denes: Protein synthesis. Ann. Rev. Biochem. **42**, 397–438 (1973)

[81] Weissbach, H., and S. Ochoa: Soluble factors required for eukaryotic protein synthesis. Ann. Rev. Biochem. **45**, 191–216 (1976)

[82] Grunberg-Manago, M., and F. Gros: Initiation mechanism of protein synthesis. Progr. Nucleic Acid. Res. Mol. Biol. **20**, 209–284 (1977)

[83] Bielka, H., and J. Stahl: Structure and function of eukaryotic ribosomes. Internat. Rev. Biochem. **18**, 79–166 (1978)

[84] Kurland, C. G.: Molecular characterization of ribonucleic acid from *E. coli* ribosomes. I. Isolation and molecular weights. J. Mol. Biol. **2**, 83–91 (1960)

[85] Rosset, R., and R. Monier: A propos de la présence d'acide ribonucléique de faible poids moléculaire dans les ribosomes d'*E. coli*. Biochim. Biophys. Acta **68**, 653–656 (1963)

[86] Waller, J. P., and J. I. Harris: Studies on the composition of the proteins from *Escherichia coli* ribosomes. Proc. Natl. Acad. Sci. U.S. **47**, 18–23 (1961)

[87] Kaltschmidt, E., and H. G. Wittmann: Ribosomal proteins. XII. Number of proteins in small and large ribosomal subunits of *E. coli* as determined by two-dimensional gel electrophoresis. Proc. Natl. Acad. Sci. U.S. **67**, 1276–1282 (1970)

[88] Hall, B., and P. Doty: The preparation and physical chemical properties of ribonucleic acid from microsomal particles. J. Mol. Biol. **1**, 111–126 (1959)

[89] Littauer, U. Z., and H. Eisenberg: Ribonucleic acid from *Escherichia coli*. Preparation, characterization and physical properties. Biochim. Biophys. Acta **32**, 320–337 (1959)

[90] Forget, B. G., and S. M. Weissman: Low molecular weight RNA components from KB cells. Nature **213**, 878–882 (1967)

[91] Brown, D. D., and C. S. Weber: Gene linkage by RNA × DNA hydbridization. I. Unique DNA sequences homologous to 4S, 5S and rRNA. J. Mol. Biol. **34**, 661–680 (1968)

[92] Welfle, H., J. Stahl, and H. Bielka: Studies on proteins of animal ribosomes. XIII. Enumeration of ribosomal proteins of rat liver. FEBS-Lett. **26**, 228–232 (1972)

[93] Sherton, C. C., and I. G. Wool: Determination of the number of proteins in liver ribosomes and ribosomal subunits by two-dimensional polyacrylamide gel electrophoresis. J. Biol. Chem. **247**, 4460–4467 (1972)

[94] Schreier, M. H., B. Erni, and T. Staehelin: Initiation of mamalian protein synthesis. I. Purification and characterization of seven initiation factors. J. Mol. Biol. **116**, 727–753 (1977)

[95] Trachsel, H., B. Erni, M. H. Schreier, and T. Staehelin: Initiation of mamalian protein synthesis. II. The assembly of the initiation complex with purified initiation factors. J. Mol. Biol. **116**, 755–767 (1977)

[96] Adams, S. L., B. Safer, W. F. Anderson, and W. C. Merrick: Eukaryotic initiation complex formation. Evidence for two distinct pathways. J. Biol. Chem. **250**, 9083–9089 (1975)

III. Ribosomes within the cell

The organization, localization and number of ribosomes in cells can be evaluated mainly by two methods, namely by electron microscopic analysis of ultrathin sections of tissues and by ultracentrifugation of tissue homogenates in sucrose gradients.

High resolution studies of the cell morphology were made possible after the ultrathin sectioning technique had been developed by PORTER [1] and by SJÖSTRAND [2]. As mentioned in Chapter II, PALADE et al. [3, 4] (for further references see Chapter II) in their electron microscopic studies described a small particulate component in "microsomes" and in the cytoplasm of cells of various tissues. These studies revealed already that the ribonucleoprotein particles, called later "Ribosomes" by ROBERTS (see Chapter II), are present in the cells of all animal tissues with the exception of the adult erythrocyt and that their amount varies considerably from one cell type to another. Furthermore, it was shown that ribosomes are present both free in the cytoplasm as well as attached to the membranes of the endoplasmic reticulum (so-called membrane-bound ribosomes). Besides single particles, ribosomes were found organized in "short rows, within which the granules occur at regular intervals" and in "linear series, spaced at more or less intervals" [3]. These structures were then analyzed in more detail by several authors and called polysomes or polyribosomes (see Chapter II).

Recently it has been demonstrated for HeLa cells that free polysomes in the cytoplasm are associated via their mRNAs to the nonfilamentous portion of a cytoskeletal structure [5].

The amount of ribosomes and polysomes as well as their distribution and organization within the cell depend on the type of tissue, the functional state of cells and their stage of development and differentiation.

Contents

1. Proportion of ribosomes in tissues

The lowest number of ribosomes are found in granulocytes of the blood, in cells of seminal epithelia, in spermatocytes and in spermatides, the highest concentration in embryonic cells, in proliferating cells and in cells of adult tissues, active in protein synthesis, such as those of glandular organs (pancreas, mammary gland, salivary gland) [3]. Variations in the concentration of the total amount of ribosomes linked to differentiation of spermatocytes, granulocytes and lymphocytes were already demonstrated by the studies of PALADE in 1955 [6]. A decrease of the amount of ribosomes and polysomes during maturation of reticulocytes was shown by various authors [7—10]; an average of $3.2 \cdot 10^4$ per cell has been estimated [80]. Starvation of animals [11—14] and of cells in vitro [14] also results in a decreased number of ribosomes and polysomes, respectively, while recovery from starvation leads to an increase of ribosomes. Increased amounts of ribosomal particles were also found in proliferating liver cells after partial hepatectomy [15—17].

Furthermore, the number of ribosomes per cell is influenced by hormones. Administration of ACTH, e.g., results in an increased amount of ribosomes in the adrenal cortex [18—20]. Also cortisone [21, 22] and estradiol [23] lead to elevated levels of ribosomes in the liver of rats and male chicken, respectively, just as testosterone in the kidney [24], while castration decreases the number of polysomes and ribosomes in the kidney of mice [24]. Increased concentrations of ribosomes were found also after administration of cysteine in liver cells [25].

The total amount of ribosomes in the normal rat liver is about 1 to $5 \cdot 10^6$ particles per cell [155].

Like the total amount of ribosomes, also the ratio of ribosomes (monomeric ribosomes) to polysomes depends on various biological conditions. The percentage of single ribosomes of the so-called run-off type in the free polysomal and ribosomal fraction is about 23% in the liver, 38% in the spleen and 52% in the pancreas of the rat [156]; this proportion was found not to be influenced by fasting overnight. In hepatomas the amount of polysomes, especially of the heavier ones is significantly reduced, while the concentration of monosomes and dimeric ribosomes is increased [26—34].

During reticulocyte maturation a decrease in the polysome content [7,35], as well as a shift in the distribution of polysomes from larger aggregates to lighter ones and free monosomes has been observed [8, 9, 35, 36].

Starvation leads to a reversible shift to lighter polysomes and to an increase of free monosomes [37, 38, 39] while addition of nutrients to starved cells results in a rapid conversion of monomeric ribosomes to polysomes [39]. Also during liver regeneration after partial hepatectomy, the pool of free ribosomal monosomes is rapidly decreased in favour of an increased amount of polysomes [37, 40]. Furthermore, addition of epidermal growth factor to cultured chick embryo epidermis leads to an increased formation of polysomes from monomeric ribosomes [41].

Studies with *Strongylocentrotus purpuratus* [42] and *Ilyanassa absoleta* [43], respectively, have clearly shown that unfertilized eggs have only a small population of polysomes. Fertilization induces an increase in polysomes the amount of which further increases during the following development; simultaneously the polysome profiles are shifted toward heavier classes. Also in neurospora conidia a stricking increase in the content of polysomes during the germination process has been observed [44].

The ratio of monomeric ribosomes to polysomes is furthermore influenced by hormones [45]. Administration of hydrocortisone to adrenalectomized rats [21, 22] and of estradiol to male chicken [23] elevate the number of polysomes and lead to a marked shift to heavier polysomes in the liver just as thyrotropin in the dog thyroid [46]. A quantitative evaluation by density gradient centrifugation and electron microscopy [47] revealed in the normal sheep thyroid 20% monosomes, 35% light, and 45% heavy polysomes.

Cyclic alterations in the content and shape of polysomes were also observed during the cell cycle [48—52]. A significant disaggregation of polysomes to single ribosomes occurs during cell division [50—52], especially in metaphase cells of HeLa and synchronized Chines hamster cells [48, 49]. Reformation of polysomes from single ribosomes takes place during transition from metaphase to interphase cells [49, 50, 52]. On the other hand, in studies with the slime mold *Physarum polycephalum* it was found that both the size and distribution of polysomal preparations were essentially constant through the mitotic cycle [53].

The proportion of single ribosomes to polysomes depends furthermore on nutritional conditions of animals [54—57, 157]. In the liver and pancreas of fasted animals or of animals fed an incomplete amino acid diet the percentage of single ribosomes is increased, while during and after refeeding an increased rate of ribosomal aggregation to polysomes with a concomitant decrease in monomers has been observed.

2. Free and membrane-bound ribosomes

Most eukaryotic cells have two populations of ribosomes according to whether they are bound to the membranes of the endoplasmic reticulum or exist free in the cytoplasm.

As a rule membrane-bound ribosomes or polysomes, respectively, are found preferentially in highly differentiated cells, especially of tissues producing mainly secretory proteins, such as the acinar cells of the pancreas, the mammary gland, the salivary gland, special cells of the hypophysis and plasma cells of the lymphatic tissue [3]. On the other hand, in rapidly proliferating cells, such as in embryonic, germinating and tumor tissues and in granulocytes and spermatocytes most of the ribosomal particles and polysomes exist free in the cytoplasm [3].

In the pancreas mainly membrane-bound ribosomes are present [3, 58—61] (Fig. 1). The quantitative analysis of the proportion of membrane-bound ribosomes to free ribosomesin the rat pancreas revealed that the amount of membrane-bound ribosomes increases during the 21st day of development to the 4th day after birth from 45% to 75%, while the amount of free polysomes decreases correspondingly [61].

Predominantly membrane-attached ribosomes were found also in glandular cells of the gastric mucosa [62, 63], in osteoblasts of young rats [64] and in mesenchymal cells producing collagen [65].

In the normal liver of adult animals free and membrane-bound ribosomal populations are present [59, 66—71, 155] (Fig. 2). The ratio of membrane-bound to free ribosomal particles in the liver of adult rats is about 75%/25% [69, 70, 71, 155]. By molecular nucleic acid hybridization it was found that close to 98% of the mRNA coding for albumin are organized in membrane-bound polysomes [72].

In regenerating liver [17], in starved cultured cells [39], in hepatomas [26, 32, 33, 69, 72, 73] as well as in other tumors [75—77] the proportion of free ribosomes and polysomes is elevated, while the number of membrane-bound polysomes is more or less reduced, depending on the degree of differentiation of the tissues. Recovery from starvation leads to a rapid conversion of monomeric ribosomes to polysomes and their attachment to membranes [39]. An increase in the amount of membrane-

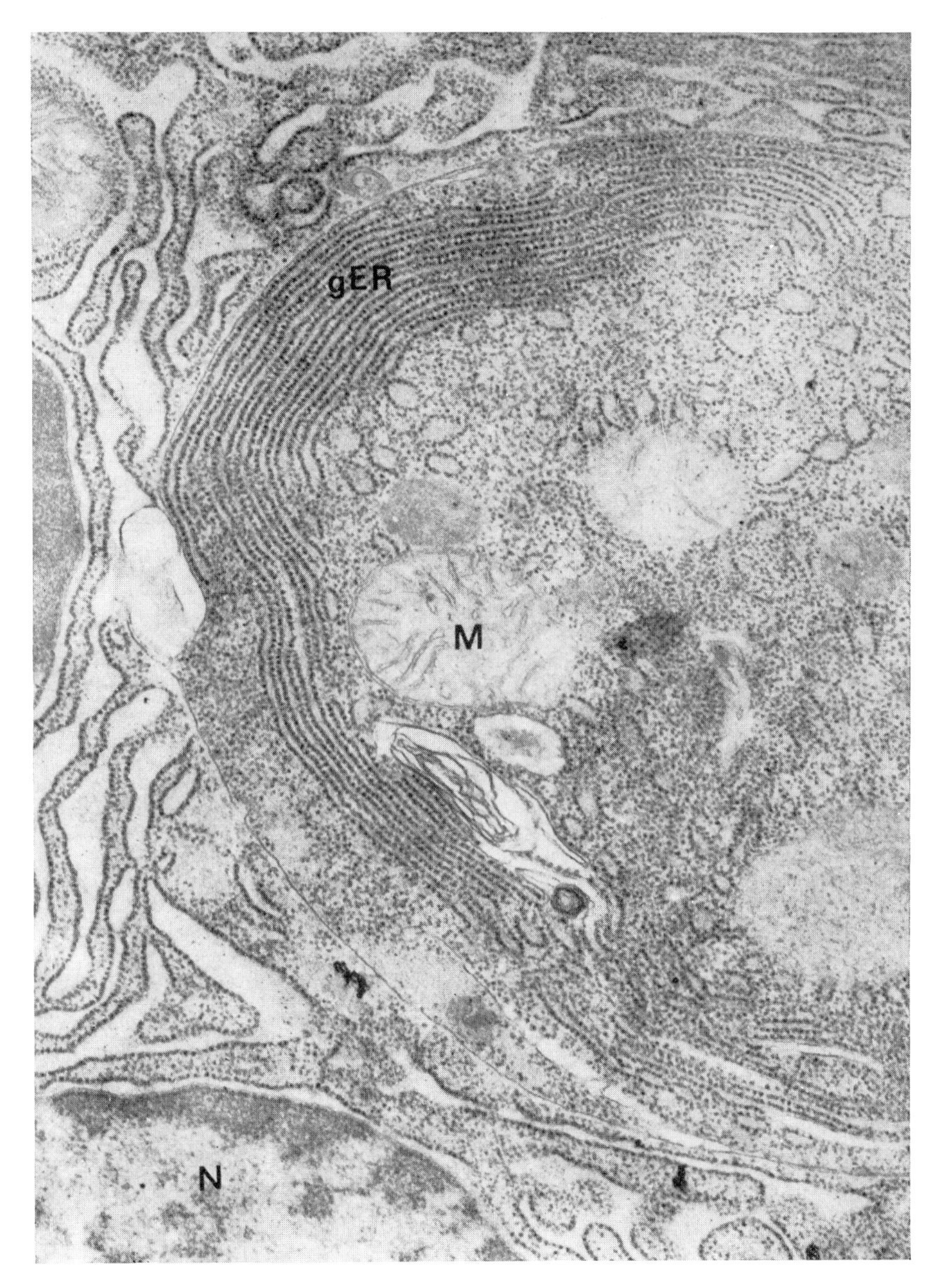

Fig. 1a. Electron micrograph of a thin section of the rat pancreas, showing mainly membrane-bound ribosomes (granular endoplasmic reticulum, gER) and polysomes arranged in rosette-like structures. M: mitochondrion, N: nucleus. × 10000.

Original, courtesy Prof. Dr. D. Bierwolf (Berlin)

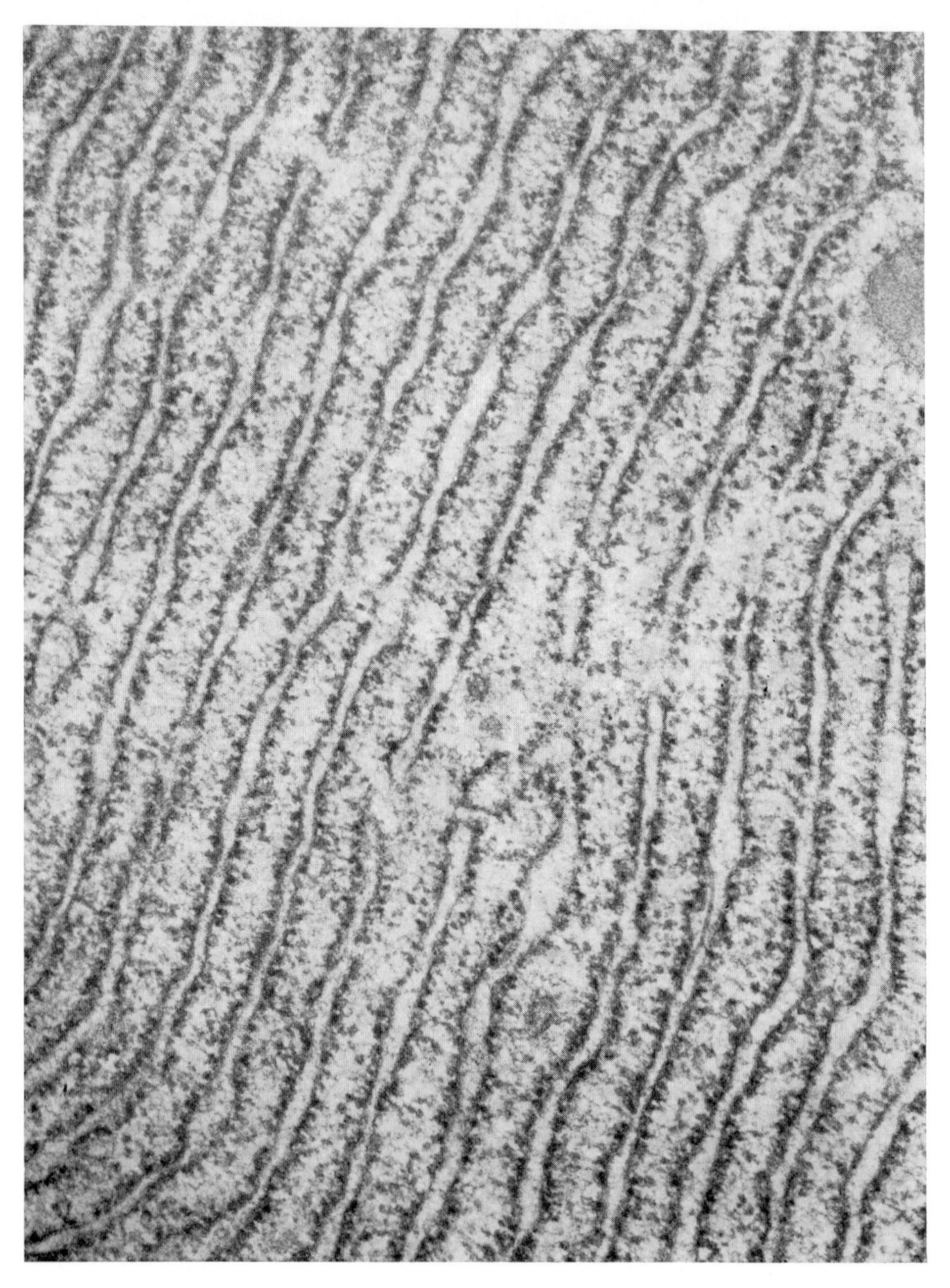

Fig. 1b. Higher magnification (× 75000) of an electron micrograph of a thin section of the rat pancreas. Edge-on-view of membran-bound ribosomes of the granular endoplasmic reticulum.

Original, courtesy Prof. Dr. H. DAVID (Berlin)

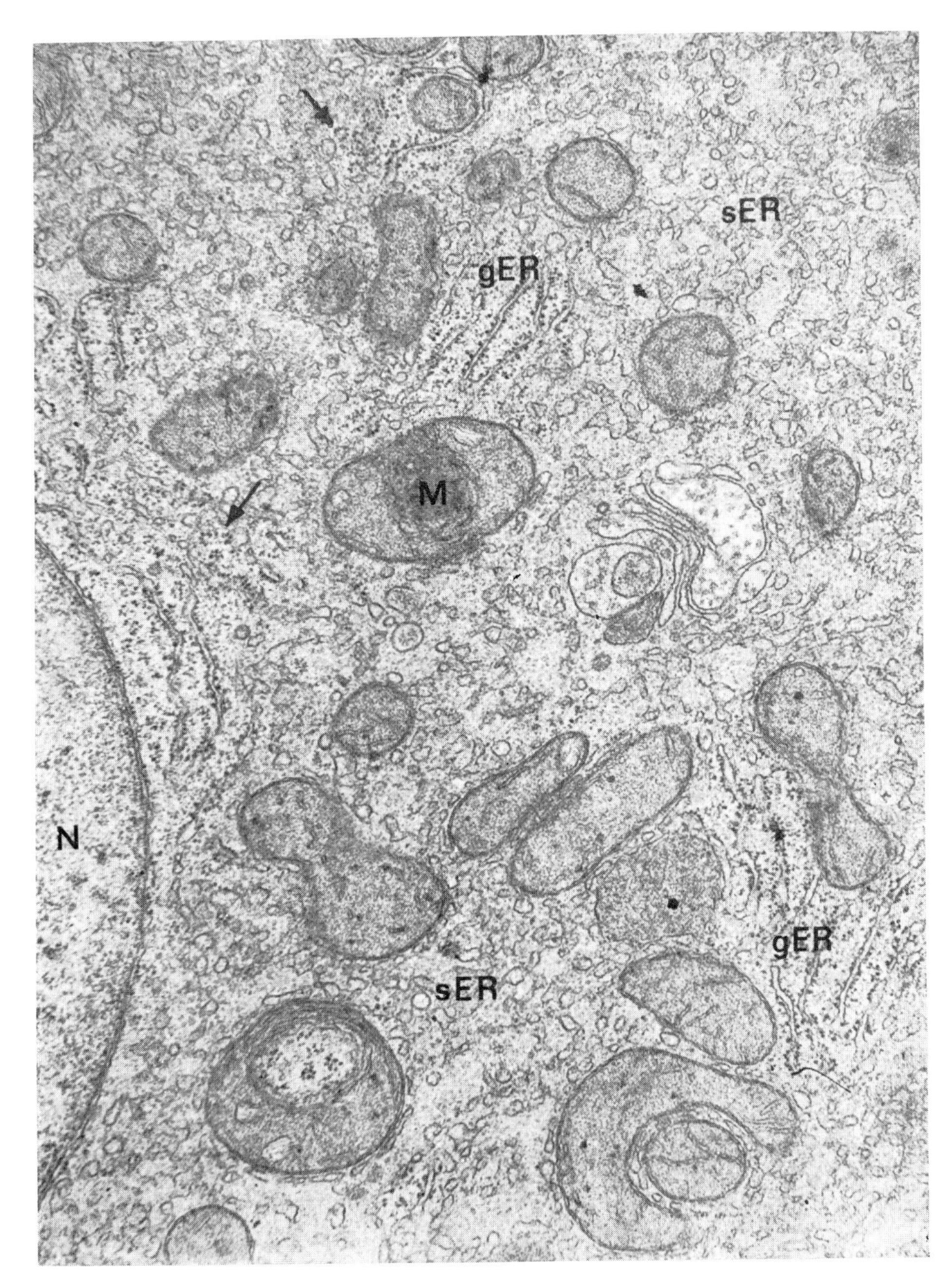

Fig. 2. Electron micrograph of a thin section of a rat liver cell with agranular (smooth, sER) and granular (gER) portions of the endoplasmic reticulum and free ribosomes and polysomes (↗), respectively. M: mitochondrion, N: nucleus, $\times 25000$.
Original, courtesy Prof. Dr. H. David (Berlin)

bound ribosomes was also found during the development of liver tissue [78] and after treatment of adrenalectomized rats with cortisol [21].

In mouse kidney which secrets only a little protein, only 25% of the ribosomes are bound to membranes [79], and in reticulocytes about 22% of the total rRNA only was found associated to membranes [80]. In cultured HeLa cells 10 to 20% of the ribosomes are associated to the membranes of the endoplasmic reticulum [81].

The brain tissue contains mostly free ribosomal particles [82—85]. In the Nissle-bodies of matured neurons, however, most of the ribosomes are attached to the membrane systems of these cells [84, 86].

In regenerating cells of *Amblystoma* larvae [87] and in proliferation cells of hydra [88], ribosomes were mainly found free in the cytoplasm, while the cells of the differentiating tissues of these species contained predominantly ribosomes attached to membranes.

Also in plant cells the amount and distribution of ribosomes depends significantly on the development and age of the plants [89—93]. The content of ribosomes in young leaves is 5 to 10 times that of old leaves. In young plant cells free ribosomes are particularly abundant; as growth and differentiation progresses, the proportion of ribosomes attached to membranes increases. In cotyledones and acinar cells of plants mainly membrane-bound polysomes are present while, on the other hand, hair cells contain predominantly free polysomes (Fig. 3).

In yeast cells the relative amount of free ribosomes shows maxima at each duplication of the initial cell mass. In the stationary phase the ratio of free to membrane-bound ribosomes is about 60% and 40% respectively [94].

Electron microscopic and biochemical studies revealed that polysomes are attached to membranes by the large subunit of the ribosomes (for further details and references see Chapter X).

The analysis of proteins synthesized on free and on membrane-bound polysomes resulted in the hypothesis that the two populations of polysomes synthesize different sets of proteins. According to this hypothesis secretory proteins are produced preferentially by polysomes attached to the membranes of the endoplasmic reticulum [95—103] and proteins retained by the cell mainly by so-called free, not membrane-bound polysomes [100—103] (for more details see Chapters X and XI).

3. Ribosomes in nucleoli

As described in Chapter I, in eukaryotic cells ribosomes are found, besides in mitochondria and plastides, mainly in the cytoplasmic compartment (free and membrane-bound). Ribosomal particles were furthermore found to occur also in the nucleoli of cells [86, 104—111]. The presence of ribosomes particles within the nucleolar region of the cell nucleus is connected with the site and mechanism of their biogenesis (see Chapter VII).

4. Structure of polysomes in the cell

Already the studies of PALADE [3] in 1955 have revealed that ribosomes occur within the cell in the form of "parallel double rows, loops, spirals, circles and rosettes".

Later on such helical [93, 112—119] and spiral-like structures [65, 93, 112, 120 to 124] of polysomes were found in cells of numerous animal and plant tissues (Figs. 1 to 4). Helical arrangements of polysomes are found mainly in free polysomes while rosettes and spiral-like structures seem to exist predominantly in membrane-bound polysomes.

As to the structure of isolated polysomes see Chapter V.

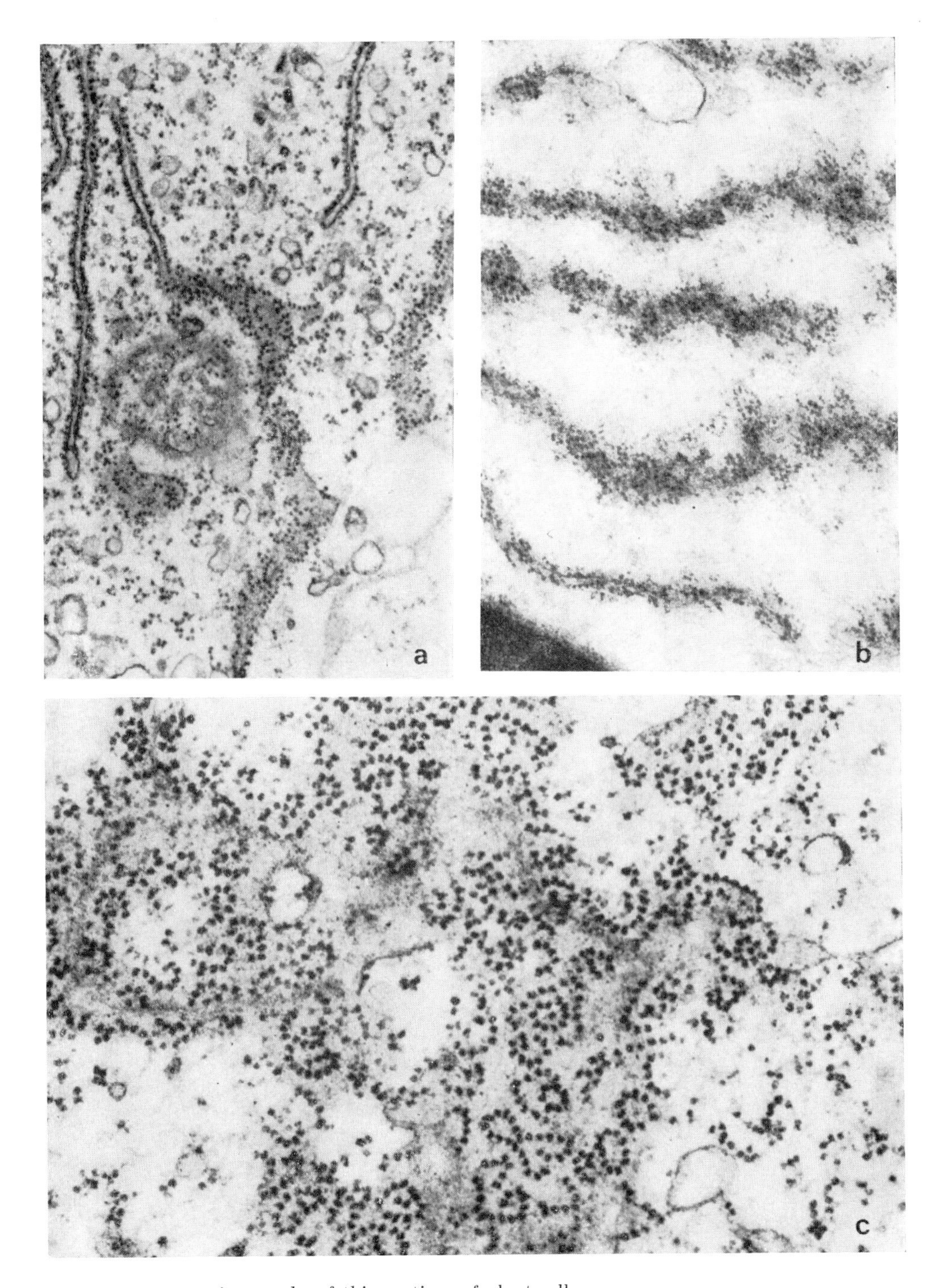

Fig. 3. Electron micrographs of thin sections of plant cells.

a) and c) Free and membrane-bound ribosomes in root cells of radish, mainly arranged in rosette-like polysomes. a) $\times$ 33000, c) $\times$ 57000.

From H. T. Bonnet and E. H. Newcomb [122]. With permission of The Rockefeller University Press (New York).

b) Membrane-bound polysomes arranged in rosette-like structures in seadlings of *Vicia faba*. $\times$ 60000.

Original, courtesy Dr. D. Neumann (Halle)

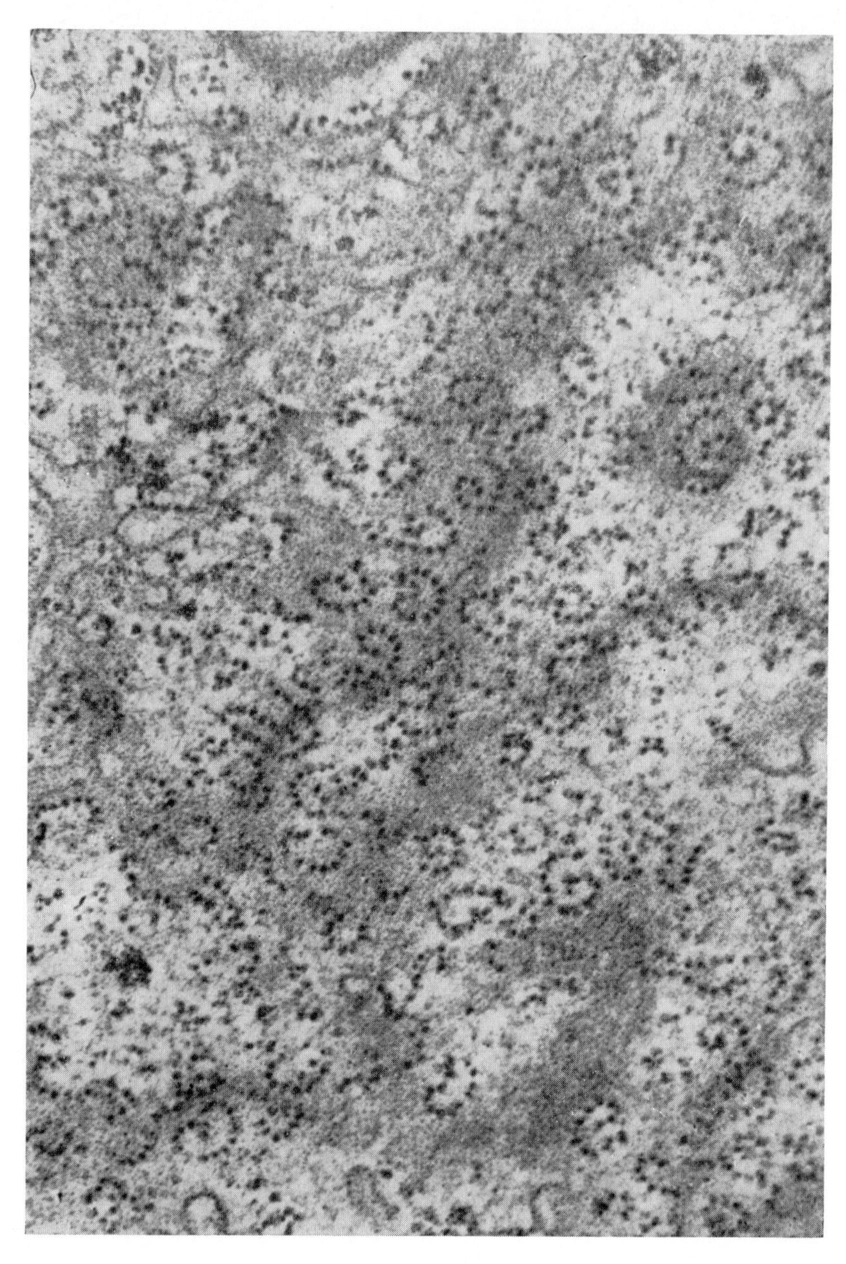

Fig. 4a. Electron micrograph of a thin section of the rat liver, showing polysomes in various arrangements. × 90000.

Original, courtesy Prof. Dr. H. DAVID (Berlin)

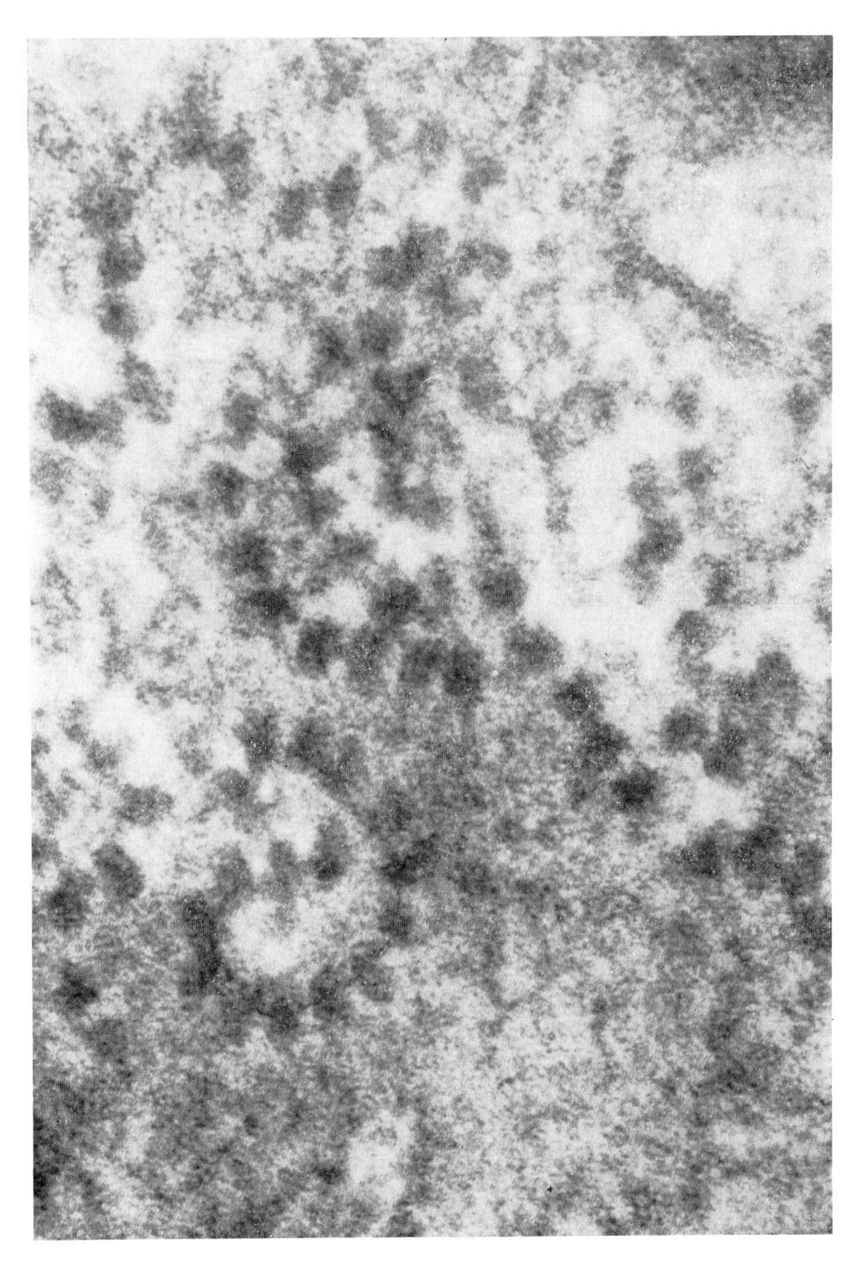

Fig. 4b. Higher magnification ($\times$ 400000) of polysome arrangements in a rat liver cell. Note the localization of the mRNA molecule between the small and the large ribosomal subunits, especially in the bottom left-hand part of the micrograph.
Original, courtesy Prof. Dr. H. DAVID (Berlin)

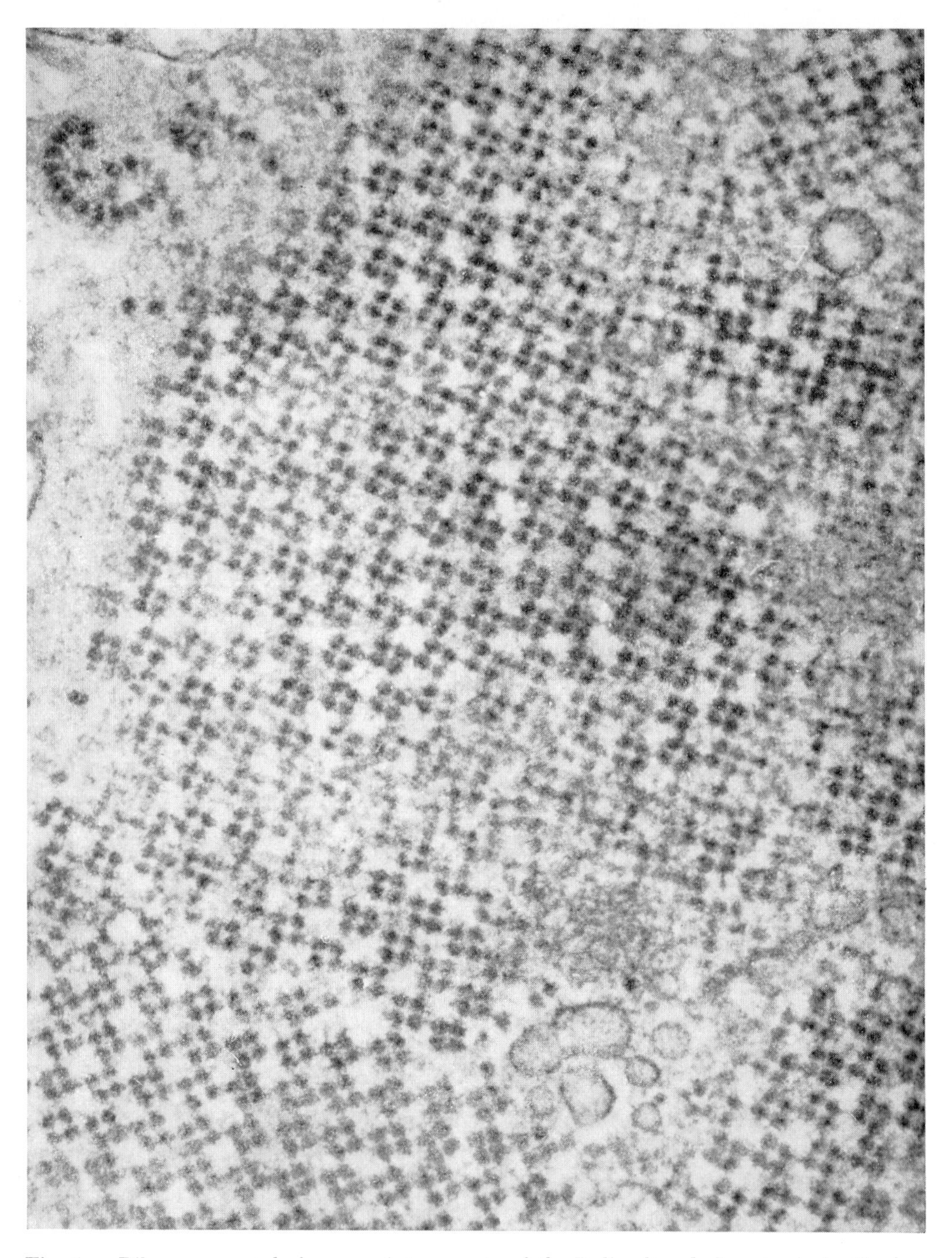

Fig. 5a. Ribosome crystals from a winter oocyte of the Italian lizard, *Lacerta sicula*. En face view, showing the ribosomes arranged as tetramers on a P 4 crystal lattice (unit cell dimension 59.5 nm). × 120000.

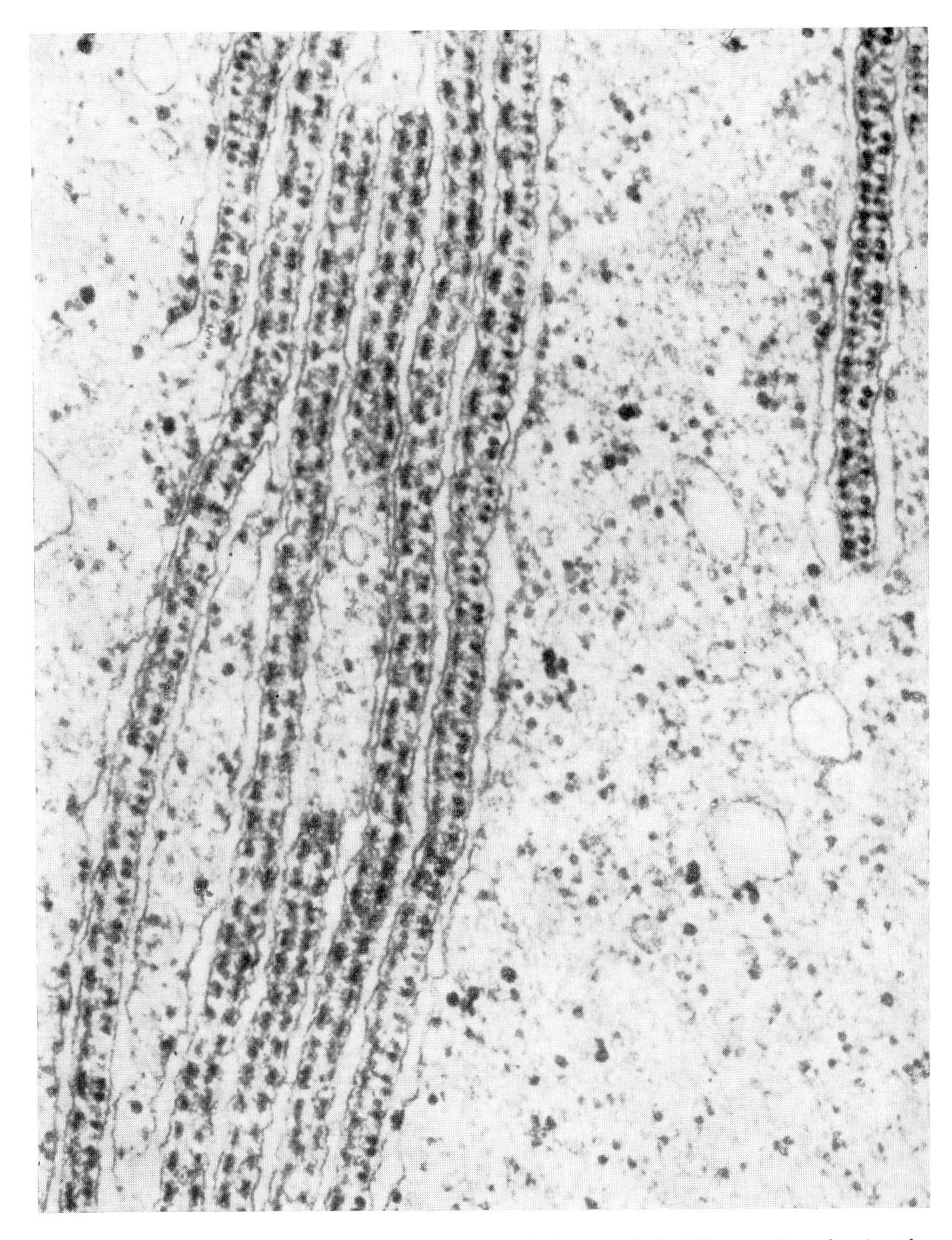

Fig. 5b. Ribosome crystals from a winter oocyte of *Lacerta sicula*. Edge-on-view showing the crystals to be organized as two-layer sheets, bounded on either side by endoplasmic reticulum membranes. × 90000.

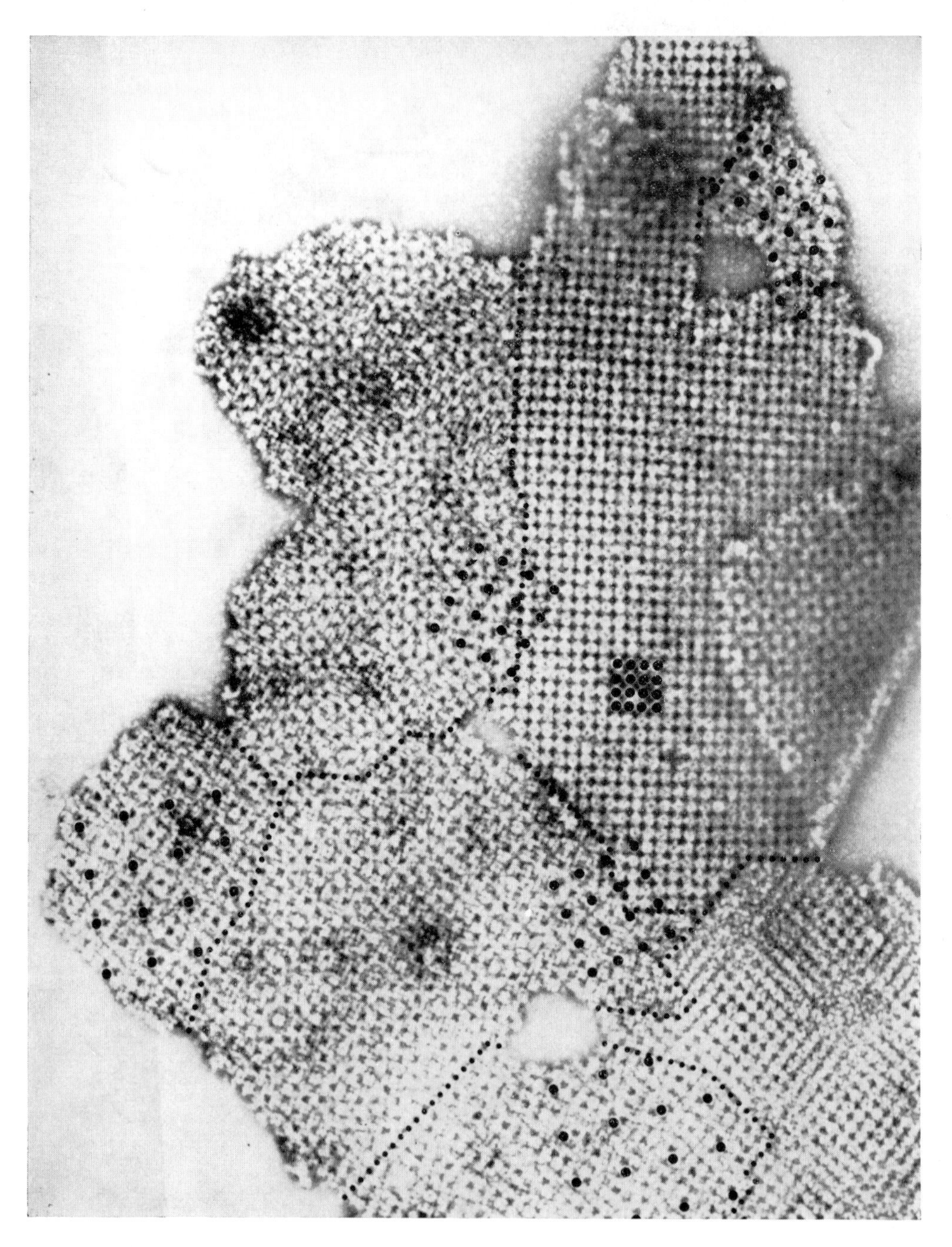

Fig. 5c. Isolated crystalline sheet (from N. UNWIN and C. TADDEI, J. Mol. Biol. **114**, 491, 1977). The different patterns arise from superposition of the two, variably oriented, P 4 crystal layers. × 35000.

Originals, courtesy Dr. N. UNWIN (Cambridge)

5. Ribosomal crystals

In several tissues (liver, oocytes, ganglia, bone marrow, feather cells, muscle, embryonic cells) of various species of animals (rats, guinea pigs, chicken, lizards, amoeba) and in fertilized eggs of a fern, highly organized complexes of ribosomal particles, also called chromatoid bodies by several authors, have been observed in cells under special conditions.

Crystal-like ribosomal structures appear chiefly in two types, namely as helices packed in a hexagonal array [124—130] or in the form of sheets composed of two layers [131—154] (Fig. 5).

Helical arrangements of ribosomes in crystal-like structures were found mainly in cells of various species of Entamoeba [125, 126, 128, 130].

In the sheet crystals, the ribosomes are organized in each of the two layers as tetramers [133, 134, 136, 141—145, 148, 149, 152] on a P 4 space group lattice [134, 141, 152, 153]. Tetramer ribosomal aggregates have been observed in chicken embryo tissues under hypothermia [134, 136, 138 to 142, 144, 145, 148, 154], in hypothermized chicken bone marrow cells [151], in dying neurons of the embryonic chick spinal cord [149] and in oocytes and follicular cells of lizards during the winter hibernation period [131, 135, 146, 147, 152] or in the summer months after cold treatment [147]. Ribosome cristals have been isolated by centrifugation from the postnuclear supernatant [148].

The physiological significance of the formation of crystal-like ribosomal structures is not yet definitely known. When hypothermized eggs of chicken are rewarmed, the tetrameric ribosomes of the sheet crystals are disassembled into monomers (80 S ribosomes) [145]. Crystal-like arrangements of ribosomes in immature chicken bone marrow cells after hypothermic treatment could not be observed in fully differentiated cells [151]. Furthermore, it has been observed that ribosomes arranged in crystal-like sheets in oocytes and follicular cells of lizards undergo transformation into free ribosomes during spring [146].

The tetramers of the sheet crystals consist of 80 S ribosomes, but not of polysomes, because they can be dissociated into 80 S monomers and into 60 S and 40 S ribosomal subunits by low Mg^{++} concentrations without puromycin [142, 144]. The ribosomes within the tetrameric crystals do not carry nascent polypeptide chains [144] or tRNA [148] and have no endogenous mRNA activity [144, 147, 148] but can be stimulated by poly(U) [141, 144, 147]. From these observations it has been concluded that ribosomes organized in crystal-like structures may be a storage form of inactive ribosomes [146].

6. References

[1] Porter, K., and F. Kallman: The properties and effects of osmium tetroxide as a tissue fixative with special reference to its use for electron microscopy. Exptl. Cell. Res. 4, 127—141 (1953)

[2] Sjöstrand, F. S.: Systems of double membranes in the cytoplasm of certain tissue cells. Nature 171, 31—32 (1953)

[3] Palade, G. E.: A small particulate component of the cytoplasm. J. Biophys. Biochem. Cytol. 1, 59—68 (1955)

[4] Palade, G. E., and P. Siekevitz: Liver microsomes. An integrated morphological and biochemical study. J. Biophys. Biochem. Cytol. 2, 171—198 (1956)

[5] Lenk, R., L. Ransom, Y. Kaufmann, and S. Penman: A cytoskeletal structure with associated polyribosomes obtained from HeLa cells. Cell 10, 67—78 (1977)

[6] Palade, G. E., Studies on the endoplasmic reticulum. II. Simple dispositions in cells in situ. J. Biophys. Biochem. Cytol. 1, 567—582 (1955)

[7] Marks, P. A., R. A. Rifkind, and D. Danon: Polyribosomes and protein synthe-

sis during reticulocyte maturation in vitro. Proc. Natl. Acad. Sci. U.S. **50**, 336—342 (1963)

[8] Danon, D., Zehavi-Willner, and G. R. Berman: Alteration in polyribosomes of reticulocytes maturing in vivo. Proc. Natl. Acad. Sci. U.S. **54**, 873—879 (1965)

[9] Nijhof, W., and P. K. Wierenga: The shift in the distribution of polysome-classes in the course of the maturation process of reticulocytes. Biochim. Biophys. Acta **361**, 367—371 (1974)

[10] Grasso, J. A., A. L. Sullivan, and Sai-Chung Chan: Studies of the endoplasmic reticulum and plasma membrane-bound ribosomes in erythropoietic cells. J. Cell Sci. **31**, 165—178 (1978)

[11] Petermann, M. L., and M. G. Hamilton: The influence of age, sex, pregnancy, starvation, and other factors on the cytoplasmic ribonucleoproteins of rat liver. J. Biophys. Biochem. Cytol. **4**, 771—776 (1958)

[12] Ramsey, J. C., and W. J. Steele: Effect of starvation on the distribution of free and membrane-bound ribosomes in rat liver and on the content of phospholipid and glycogen in purified ribosomes. Biochim. Biophys. Acta **447**, 312—318 (1976)

[13] Henshaw, E. C., C. A. Hirsch, B. E. Morton, and H. H. Hiatt: Control of protein synthesis in mammalian tissues through changes in ribosome activity. J. Biol. Chem. **246**, 436—446 (1971)

[14] Lynch, A. G., and N. W. Klein: Polysome activity in relation to growth and protein starvation in brains and hearts of cultured early chick embryos. Biochim. Biophys. Acta **519**, 194—203 (1978)

[15] Lieberman, I., and P. Kane: Synthesis of ribosomes in the liver after partial hepatectomy. J. Biol. Chem. **240**, 1737 to 1741 (1965)

[16] Zweig, M., and J. W. Grisham: Free and bound hepatic polyribosomes after partial hepatectomy: Pool size and sedimentation patterns. Biochim. Biophys. Acta **246**, 70—80 (1971)

[17] Loeb, J. N., and L. L. Yeung: Free and membrane-bound ribosomes in regenerating rat liver. Biochim. Biophys. Acta **520**, 623—629 (1978)

[18] Fiala, S., E. E. Sproul, and A. E. Fiala: The action of ACTH on nucleic acids and subcellular elements in the adrenal cortex. J. Biophys. Biochem. Cytol. **2**, 115—126 (1956).

[19] Bransome, E. D., and W. J. Reddy: Studies of adrenal nucleic acids: The influence of ACTA, unilateral adrenalectomy and growth hormone upon adrenal RNA and DNA in the dog. Endocrinology **69**, 997—1008 (1961)

[20] Farese, R. V.: Further studies on the stimulation of adrenal protein synthesis by ACTH: An effect on microsomes and ribosomes. Endocrinology **74**, 579—585 (1964)

[21] Cox, R. F., and A. P. Mathias: Cytoplasmic effects of cortisol in liver. Biochem. J. **115**, 777—787 (1969)

[22] Kulkarni, S. B., M. S. Netrawali, D. S. Pradhan, and A. Sreenivasan: Action of hydrocortison at a translational level in the liver. Mol. Cell Endocrin. **4**, 195—203 (1976)

[23] Bast, R. E., S. A. Garfield, L. Gehrke, and J. Ilan: Coordinate of ribosome content and polysome formation during estradiol stimulation of vitellogenin synthesis in immature male chick livers. Proc. Natl. Acad. Sci. U.A. **74**, 3133—3137 (1977)

[24] Kochakian, C. D., M. Nishida, and T. Hirone: Ribosomes of mouse kidney: Regulation by androgens. Am. J. Physiol. **217**, 383—391 (1969)

[25] Emmelot, P., I. J. Mizrahi, R. Naccarato, and E. L. Benedetti: Changes in functions and structure of the endoplasmic reticulum of rat liver cells after administration of cysteine. J. Cell. Biol. **12**, 177—180 (1962)

[26] Webb, T. E., G. Blobel, and V. R. Potter: Polyribosomes in rat tissue: I. A study of in vivo patterns in liver and hepatomas. Cancer Res. **24**, 1229—1237 (1964)

[27] Mizrahi, I. J., and P. Emmelot: On the mode of action by which the carcinogen dimethylnitrosamine inhibits protein synthesis in the liver. Biochim. Biophys. Acta **91**, 362—364 (1964)

[28] Mizrahi, I. J., and G. C. DeVrie: Instability of polyribosomes derived from

rats pretreated with the hepatocarcinogen dimethylnitrosamine. Biochem. Biophys. Res. Commun. **21**, 555—561 (1965)

[29] Utsunomiya, T., and J. S. Roth: Studies on the function of intracellular ribonucleases. IV. Some observations on the properties of ribosomes and polysomes from rat liver and hepatoma. J. Cell Biol. **29**, 387—393 (1966)

[30] Webb, T. E., and V. R. Potter: Polyribosomes in rat tissues. IV. On the abnormal dimer peak in hepatomas. Cancer Res. **24**, 1022—1025 (1966)

[31] Mansbridge, J. N., and A. Korner: The polysomes and the messenger ribonucleic acid content of hepatoma 223. Biochim. Biophys. Acta **119**, 92—98 (1966)

[32] Stahl, J., and H. Bielka: Lokalisation und Organisation von Ribosomen in Leber und Hepatomgewebe. Acta Biol. Med. Germ. **19**, 249—258 (1967)

[33] Kwan, S.-H., T. E. Webb, and H. P. Morris: Diversity and nature of ribosomal pools in hepatoma 7800 and host liver. Biochem. J. **109**, 617—623 (1968)

[34] Webb, T. E., and H. P. Morris: Properties of the inactive ribosomal components in rat liver and hepatoma. Biochem. J. **115**, 575—582 (1969)

[35] Rifkind, R. A., D. Danon, and P. A. Marks: Alterations in polyribosomes during cell erythroid maturation. J. Cell Biol. **22**, 599—611 (1964)

[36] Marbaix, G., A. Burny, G. Huez, B. Lebleu, and J. Temmerman: Evolution of the polysome distribution during in vivo reticulocyte maturation. Europ. J. Biochem. **13**, 322—325 (1970)

[37] Webb, T. E., G. Blobel, and V. R. Potter: Polyribosomes in rat tissues. III. The response of the polyribosome pattern of rat liver to physiologic stress. Cancer Res. **26**, 253—257 (1966)

[38] Wunner, W. H., J. Bell, and H. N. Munro: The effect of feeding with a tryptophan-free amino acid mixture on rat liver polysomes and ribosomal ribonucleic acid. Biochem. J. **101**, 417—428 (1966)

[39] Lee, S. Y., V. Krsmanovic, and G. Brawerman: Attachment of ribosomes to membranes during polysome formation in mouse sarcoma 180 cells. J. Cell Biol. **49**, 683—691 (1971)

[40] Rizzo, A. J., and T. E. Webb: Concurrent changes in the concentration of monomeric ribosomes and the rate of ribosome synthesis in rat liver. Biochim. Biophys. Acta **169**, 163—174 (1968)

[41] Cohen, S., and M. Stastny: Epidermal growth factor. III. The stimulation of polysome formation in chick embryo epidermis. Biochim. Biophys. Acta **166**, 427—437 (1968)

[42] Infante, A. A., and M. Nemer: Accumulation of newly synthesized RNA templates in a unique class of polyribosomes during embryogenesis. Proc. Natl. Acad. Sci. U.S. **58**, 681—688 (1967)

[43] Mirkes, P. E.: Polysomes and protein synthesis during development of *Ilyanassa obsoleta*. Exptl. Cell Res. **74**, 503—508 (1972)

[44] Mirkes, P. E.: Polysomes, ribonucleic acid, and protein synthesis during germination of *Neurospora crassa* conidia. J. Bacteriol. **117**, 196—202 (1974)

[45] Tata, J. R.: Hormonal regulation of growth and protein synthesis. Nature **219**, 331—337 (1968)

[46] Lecocq, R. E., and J. E. Dumont: In vivo and in vitro effects of thyrotropin on ribosomal pattern of dog thyroid. Biochim. Biophys. Acta **299**, 303—311 (1973)

[47] Lecocq, R. E., F. Cantraine, E. Keyhani, A. Claude, C. Delcroix, and J. E. Dumont: Quantitative evaluation of polysomes and ribosomes by density gradient centrifugation and electron microscopy. Anal. Biochem. **43**, 71—79 (1971)

[48] Scharff, M. D., and E. Robbins: Polyribosome disaggregation during metaphase. Science **151**, 992—995 (1966)

[49] Steward, D.-L., J. R. Shaeffer, and R. M. Humphrey: Breakdown and assembly of polyribosomes in synchronized Chinese hamster cells. Science **161**, 791 to 793 (1968)

[50] Hodge, L. D., E. Robbins, and M. D. Scharff: Persistence of messenger RNA through mitosis in HeLa cells. J. Cell Biol. **40**, 497—507 (1969)

[51] Fan, H., and S. Penman: Regulation of protein synthesis in mammalian cells. II. Inhibition of protein synthesis at the level of initiation during mitosis. J. Mol. Biol. **50**, 655—670 (1970)

[52] Eremenko, Z., and P. Volpe: Polysome translation state during the cell cycle. Europ. J. Biochem. **52**, 203—210 (1975)

[53] Brewer, E. N.: Polysome profils, amino acid incorporation in vitro, and polysome aggregation following disaggregation by heat shock through the mitotic cycle in *Physarum polycephalum*. Biochim. Biophys. Acta **277**, 639—645 (1972)

[54] Fleck, A., J. Shepherd, and H. N. Munro: Protein synthesis in rat liver: Influence of amino acids in diet on microsomes and polysomes. Science **150**, 628—629 (1965)

[55] Staehelin, T., E. Verney, and H. Sidransky: The influence of nutritional change on polyribosomes of the liver. Biochim. Biophys. Acta **145**, 105—119 (1967)

[56] Van Venrooij, W. J., and C. Poort: Rate of protein snythesis and polyribosome formation in the frog pancreas after fasting and feeding. Biochim. Biophys. Acta **247**, 468—470 (1971)

[57] Ip, C. C. Y., and A. E. Harper: Effect of threonine supplementation on hepatic polysome patterns and protein synthesis of rats fed a threonine-deficient diet. Biochim. Biophys. Acta **331**, 251—263 (1973)

[58] Sjöstrand, F. S., and V. Hanzon: Membrane structures of cytoplasm and mitochondria in exocrine cells of mouse pancreas by high resolution electron microscopy. Exptl. Cell Res. **7**, 393—414 (1954)

[59] Bernhard, W.: Ultrastructural aspects of nucleo-cytoplasmic relationship. Exptl. Cell Res. Suppl. **6**, 17—50 (1958)

[60] Sjöstrand, F. S., and L. G. Elfvin: The granular structure of mitochondrial membranes and of cytomembranes as demonstrated in frozen-dried tissue. J. Ultrastruct. Res. **10**, 263—292 (1964)

[61] Hack, P., and H. David: Quantitative Untersuchungen zum Verhältnis freier und membrangebundener Ribosomen in der exokrinen Pankreaszelle während der Entwicklung. Acta Biol. Med. Germ. **22**, 685—686 (1969)

[62] Helander, H. F.: Ultrastructure of fundus glands of the mouse gastric mucosa. J. Ultrastruct. Res. Suppl. **4**, 1—123 (1962)

[63] Ito, S., and R. J. Winchester: The fine structure of the gastric mucosa in the rat. J. Cell Biol. **16**, 541—577 (1963)

[64] Cameron, D. A.: The fine structure of osteoblasts in the metaphysis of the tibia of the young rat. J. Biophys. Biochem. Cytol. **9**, 583—595 (1961)

[65] Leeson, C. R., and T. A. Leeson: An unusual arrangement of ribosomes in mesenchymal cells. J. Cell. Biol. **24**, 324—328 (1965)

[66] Bernhard, W., A. Gautier, and C. Rouiller: La notion de "microsomes" et le problème de la basophilie cytoplasmique. Arch. Anat. Microscop. Morphol. Exptl. **43**, 236—275 (1954)

[67] Palade, G. E., and P. Siekevitz: Liver microsomes: An integrated morphological and biochemical study. J. Biophys. Biochem. Cytol. **2**, 171—198 (1956)

[68] Wilson, S. H., and M. B. Hoagland: Studies on the physiology of rat liver polyribosomes: Quantitation and intracellular distribution of ribosomes. Proc. Natl. Acad. Sci. U.S. **54**, 600—607 (1965)

[69] Webb, T. E., G. Blobel, V. R. Potter, and H. P. Morris: Polyribosomes in rat tissue. II. The polyribosome distribution in the minimal deviation hepatomas. Cancer Res. **25**, 1219—1224 (1965)

[70] Morton, B., C. Nwizu, E. C. Henshaw, C. A. Hirsch, and H. H. Hiatt: The isolation of large polysomes in high yield from unfractionated tissue homogenates. Biochim. Biophys. Acta **395**, 28—40 (1975)

[71] Ramsey, J. C., and W. J. Steele: A procedure for the quantitative recovery of homogeneous populations of undegraded free and bound polyomes from rat liver. Biochem. **15**, 1704—1712 (1976)

[72] Yap, S. H., R. K. Strair, and D. A. Shafritz: Distribution of rat liver albumin mRNA in membrane bound and free polysomes as determined by molecular hybridization. Proc. Natl. Acad. Sci. U.S. **74**, 5397—5401 (1977)

[73] Howatson, A. F., and A. W. Ham: Electron microscope study of sections of two rat liver tumors. Cancer Res. **15**, 62—69 (1955)

[74] Novikoff, A. B.: A transplantable rat liver tumor induced by 4-dimethylamino-

azobenzene. Cancer Res. **17**, 1010—1027 (1957)

[75] DALTON, A. J., and M. D. FELIX: The electron microscopy of normal and malignant cells. Ann. New York Acad. Sci. **63**, 1115—1140 (1955)

[76] BERNHARD, W.: Electron microscopy of tumour cells and tumor viruses. A review. Cancer Res. **18**, 491—509 (1958)

[77] SMETANA, K., and H. BUSCH: On the ultrastructure of the Walker 256 carcinosarcoma. Cancer Res. **23**, 1600—1603 (1963)

[78] CHU, M.-L. H., H. M. JERNIGAN, and M. FRIED: Polymorphism in fowl serum albumin. VII. Distribution and activity of free and membrane-bound polysomes in developing fowl liver. Biochim. Biophys. Acta **518**, 298—307 (1978)

[79] PRIESTLEY, G. C., M. L. PRUYN, and R. A. MALT: Glycoprotein synthesis by membrane-bound ribosomes and smooth membranes in kidney. Biochim. Biophys. Acta **190**, 154—160 (1969)

[80] BURKA, E. R.: The distribution of RNA and ribosomes in reticulocytes. Biochim. Biophys. Acta **166**, 672—680 (1698)

[81] ATTARDI, B., B. CRAVIOTO, and G. ATTARDI: Membrane-bound ribosomes in HeLa cells. I. Their proportion to total cell ribosomes and their association with messenger RNA. J. Mol. Biol. **44**, 47—70 (1969)

[82] HANZON, V., and G. TOSCHI: Electron microscopy on microsomal fractions from rat liver. Exptl. Cell Res. **16**, 256—271 (1956)

[83] TOSCHI, G.: A biochemical study of brain microsomes. Exptl. Cell Res. **16**, 232—255 (1956)

[84] ESCHNER, K., and P. GLEES: Free and membrane-bound ribosomes in maturing neurones of the chick and their possible functional significance. Experientia **19**, 301—303 (1963)

[85] MERITS, I., J. C. CAIN, E. J. RDZOK, and F. N. MINARD: Distribution between free and membrane-bound ribosomes in rat brain. Experientia **25**, 739—740 (1969)

[86] PALAY, S. L., and G. E. PALADE: The fine structure of neurons. J. Biophys. Biochem. Cytol **1**, 69—88 (1955)

[87] HAY, E. D.: The fine structure of blastema cells in differentiating cartilage cells in regenerating limbs of *Amblystoma* larvae. J. Biophys. Biochem. Cytol. **4**, 583—592 (1958)

[88] SLAUTTERBACK, D. B., and D. W. FAWCETT: The development of the cnidoblasts of hydra. An electron microscope study of cell differentiation. J. Biophys. Biochem. Cytol. **5**, 441—452 (1959)

[89] SETTERFIELD, G.: Structure and composition of plant-cell organelles in relation to growth and development. Can. J. Botany **39**, 469—489 (1961)

[90] TS'O, P. O. P.: The ribosomes — Ribonucleoprotein particles. Ann. Rev. Plant Physiol. **13**, 45—80 (1962)

[91] PAYNE, P. I., and D. BOULTER: Free and membrane bound ribosomes of the cotyledons of *Vicia faba*. Planta **84**, 263—271 (1969)

[92] LARKINS, B. A., and E. DAVIES: Polyribosomes from peas. V. An attempt to characterize the total free and membrane-bound polysomal population. Plant Physiol **55**, 749—756 (1975)

[93] GUNNING, B. E. S., and M. W. STEER: Biologie der Pflanzenzelle. Ein Bildatlas. Gustav Fischer Verlag, Stuttgart—New York, 1977

[94] SCHNEIDER, E., E. R. LOCHMANN, and H. LOTHER: Distribution of membrane-bound and free ribosomes in growing yeast. Biochim. Biophys. Acta **432**, 92—97 (1976)

[95] HENDLER, R. W., A. J. DALTON, and G. G. GLENNER: A cytological study of the albumin-secreting cells on the hen oviduct. J. Biophys. Biochem. Cytol. **3**, 325—330 (1957)

[96] SIEKEVITZ, P., and G. E. PALADE: A cytochemical study on the pancreas of the guinea pig. V. In vivo-incorporation of leucine-1-C^{14} into chymotrypsinogen of various cell fractions. J. Biophys. Biochem. Cytol. **7**, 619—630 (1960)

[97] TAKAGI, M., and K. OGATA: Direct evidence for albumin biosynthesis by membrane bound polysomes in rat liver. Biochem. Biophys. Res. Commun. **33**, 55—60 (1968)

[98] TATA, J. R.: Ribosomal segregation as a possible function for the attachment of ribosomes to membranes. Sub-Cell. Biochem. **1**, 83—89 (1971)

[99] Hallinan, T., C. N. Murty, and J. H. Grant: The exclusive function of reticulum bound ribosomes in glycoprotein biosynthesis. Life Science **7**, 225—232 (1968)

[100] Palade, G. E.: Intracellular aspects of the process of protein synthesis. Science **189**, 347—358 (1975)

[101] Hicks, S. I., J. W. Drysdale, and M. N. Munro: Preferential synthesis of ferritin and albumin by different populations of liver polysomes. Science **164**, 584—585 (1969)

[102] Campbell, P. N., The function of polyribosomes attached to membranes in animal cells. Biochem. J. **117**, 57p—58p (1970)

[103] Campbell, P. N.: Functions of polyribosomes attached to membranes of animal cells. FEBS-Lett. **7**, 1—7 (1970)

[104] Bernhard, W.: Ultrastructural aspects of nucleo-cytoplasmic relationship. Exptl. Cell Res., Suppl. **6**, 17—50 (1958)

[105] Georgiev, G. P., and J. S. Chentsov: On the structural organization of nucleolo-chromosomal ribonucleoproteins. Exptl. Cell Res. **27**, 570—572 (1962)

[106] Birnstiel, M. L., M. I. H. Chipchase, and B. B. Hyde: The nucleolus, a source of ribosomes. Biochim. Biophys. Acta **76**, 454—462 (1963)

[107] Allfrey, V. G.: Nuclear ribosomes, messenger-RNA and protein synthesis. Exptl. Cell Res., Suppl. **9**, 183—212 (1963)

[108] Chipchase, M. I. H., and M. L. Birnstiel: On the nature of nucleolar RNA. Proc. Natl. Acad. Sci. U.S. **50**, 1101—1107 (1963)

[109] Bernhard, W., and N. Granboulan: The fine structure of the cancer cell nucleus. Exptl. Cell Res., Suppl. **9**, 19—53 (1963)

[110] Swift, H.: Cytochemical studies on nuclear fine structure. Exptl. Cell Res., Suppl. **9**, 54—67 (1963)

[111] Wang, T.-Y.: Physico-chemical and metabolic properties of nuclear ribosomes. Exptl. Cell Res., Suppl. **9**, 213—219 (1963)

[112] Shelton, E., and E. L. Kuff: Substructure and configuration of ribosomes isolated from mammalian cells. J. Mol. Biol. **22**, 23—31 (1966)

[113] Waddington, C. H., and M. M. Perry: Helical arrangement of ribosomes in differentiating muscle cells. Exptl. Cell Res. **30**, 599—600 (1963)

[114] Behnke, O.: Helical arrangement of ribosomes in the cytoplasm of differentiating cells of the small intestine of rat foetuses. Exptl. Cell Res. **30**, 597—598 (1963)

[115] Cedergren, B., and I. Harary: In vitro studies on single beating rat heart cells. VI. Electron microscopic studies of single cells. J. Ultrastruct. Res. **11**, 428—442 (1964)

[116] Ecklin, P.: An apparent helical arrangement of ribosomes in developing pollen mother cells of *Ipomoea purpurea*. J. Cell Biol. **24**, 150—153 (1965)

[117] Weiss, P., and N. B. Grover: Helical array of polyribosomes. Proc. Natl. Acad. Sci. U.S. **59**, 763—768 (1968)

[118] Krishan, A., and D. Hsu: Observations on the association of helical polyribosomes and filaments with vincristine-induced crystals in Earle's L-cell fibroblasts. J. Cell. Biol. **43**, 553—563 (1969)

[119] Rosenbaum, R. M., and M. Wittner: Ultrastructure of bacterized and axenic trophozoites of *Entamoeba histolytica* with particular reference to helical bodies. J. Cell Biol. **45**, 367—382 (1970)

[120] Hinz, R. W., G. Barski, and W. Bernhard: An electron microscopic study of the development of the encephalomyocarditis virus propagated in vitro. Exptl. Cell Res. **26**, 571—586 (1962)

[121] Benedetti, E. L., H. Bloemendal, and W. S. Bont: Polyribosomes isolés à partir du foie de rat. C. R. Acad. Sci. (Paris) **259**, 1353—1356 (1964)

[122] Bonnet, H. T., and E. H. Newcomb: Polyribosomes and cisternal accumulations in root cells of radish. J. Cell Biol. **27**, 423—432 (1965)

[123] Parry, G., C. Davie, and D. J. Williams: Topography and mobility of ribosomes on the surface if isolated endoplasmic reticulum of rat liver. J. Mol. Biol. **109**, 589—592 (1977)

[124] Dallner, G., P. Siekevitz, and G. E. Palade: Biogenesis of endoplasmic reticulum membranes. I. Structural and chemical differentiation in developing rat hepatocyte. J. Cell Biol. **30**, 73—96 (1966)

[125] SIDDIQUI, W. A., and M. A. RUDZINSKA: A helical structure in ribonucleoprotein bodies of *Entamoeba invadens*. Nature **200**, 74—75 (1963)

[126] BARKER, D. C., and G. SVIHLA: Localization of cytoplasmic nucleic acid during growth and encystment of *Entamoeba invadens*. J. Cell Biol. **20**, 389—398 (1964)

[127] MORGAN, R. S., and B. G. UZMANN: Nature of the packing of ribosomes within chromatoid bodies. Science **152**, 214—216 (1966)

[128] MORGAN, R. S., H. S. SLAYTER, and D. L. WELLER: Isolation of ribosomes from cysts of *Entamoeba invadens*. J. Cell Biol. **36**, 45—51 (1968)

[129] LAKE, J. A., and H. S. SLAYTER: Three-dimensional structure of the chromatoid body of *Entamoeba invadens*. Nature **227**, 1032—1037 (1970)

[130] LAKE, J. A., and H. S. SLAYTER: Three-dimensional structure of the chromatoid body helix of *Entamoeba invadens*. J. Mol. Biol. **66**, 271—282 (1972)

[131] PORTE, A., and I. P. ZAHND: Structure fine du follicule ovarien de *Lacerta stirpium*. C. R. Séanc. Soc. Biol. **155**, 1058 to 1061 (1961)

[132] CRAIN, S. M., H. BENITEZ, and A. E. VATTER: Some cytologic effects of salivary nerve-growth factor on tissue cultures of peripherical ganglia. Ann. N.Y. Acad. Sci. **118**, 206—231 (1964)

[133] BELL, E., T. HUMPHREYS, H. S. SLATER, and C. E. HALL: Configuration of inactive and active polysomes of the developing down feather. Science **148**, 1739—1741 (1965)

[134] BEYERS, B.: Ribosome crystallization induced in chick embryo tissues by hypothermia. J. Cell Biol. **30**, C1-C6 (1966)

[135] GHIARA, G., and C. TADDEI: Dati citologici e ultrastrutturali su di un particolare tipo di costituenti basofili del citoplasma di cellule follicolari e di ovociti ovarici di rettili. Boll. Soc. Ital. Biol. Sper. **42**, 784—788 (1966)

[136] BEYERS, B.: Structure and formation of ribosome crystals in hypothermic chick embryo cells. J. Mol. Biol. **26**, 155—167 (1967)

[137] MARALDI, N. M., and M. BARBIERI: Ribosome crystallization. I. Study on electron microscopy of ribosome crystallization during chick embryo development. J. Submicr. Cytol. **1**, 159—170 (1969)

[138] BARBIERE, M., L. SIMONELLI, P. SIMONI, and N. M. MARALDI: Ribosome crystallization. II. Ultrastructural study on nuclear and cytoplasmic ribosome crystallization in hypothermic cell cultures. J. Submicr. Cytol. **2**, 33—49 (1907)

[139] MARALDI, N. M., L. SIMONELLI, P. PETTAZONI, and M. BARBIERI: Ribosome crystallization. III. Ribosome and protein crystallization in hypothermic cell cultures treated with vinblastine sulfate. J. Submicr. Cytol. **2**, 51—67 (1970)

[140] BARBIERI, M., P. PETTAZONI, F. BERSANI, and N. M. MARALDI: Isolation of ribosome microcrystals. J. Mol. Biol. **54**, 121—124 (1970)

[141] BYERS, B.: Chick embryo ribosome crystals: Analysis of bonding and functional activity in vitro. Proc. Natl. Acad. Sci. U.S. **68**, 440—444 (1971)

[142] CAREY, N. H., and G. S. READ: The arrangement of ribosomes in tetramers from hypothermic chick embryos. Biochem. J. **121**, 511—519 (1971)

[143] DUCKETT, J. G.: Pentagonal arrays of ribosomes in fertilized eggs of *Pteridium aquilinum*. J. Ultrastruct. Res. **38**, 390—397 (1972)

[144] MORIMOTO, T., G. BLOBEL, and D. D. SABATINI: Ribosome crystallization in chick embryos. I. Isolation, characterization and in vitro activity of ribosome tetramers. J. Cell Biol. **52**, 338—354 (1972)

[145] MORIMOTO, T., G. BLOBEL, and D. D. SABATINI: Ribosome crystallization in chikken embryos. II. Conditions for the formation of ribosome tetramers in vivo. J. Cell Biol. **52**, 355—366 (1972)

[146] TADDEI, C.: Ribosome arrangement during oogenesis of *Lacerta sicula* RAF. Exptl. Cell Res. **70**, 285—292 (1972)

[147] TADDEI, C., R. GAMBINO, S. METAFORA, and A. MONROY: Possible role of ribosomal bodies in the control of protein synthesis in pre-vitellogenic oocytes of the lizard *Lacerta sicula* RAF. Exptl. Cell Res. **78**, 159—167 (1973)

[148] DONDI, P. G., and D. C. BARKER: Some properties of ribosome crystals isolated

from hypothermically treated chick embryos. J. Cell Sci. **14**, 301–317 (1974)

[149] O'Connor, T. M., and C. R. Wyttenbach: Cell death in the embryonic chick spinal cord. J. Cell Biol. **60**, 448–459 (1974)

[150] Löwe, H.: Bildung von kristallartigen Ribosomenaggregaten in glatten Muskelzellen des Meerschweinchenileums nach Inkubation in vitro. Acta Biol. Med. Germ. **34**, 1283–1287 (1975)

[151] Brunelli, M. A., M. Marini, P. Pettazoni, and G. Bubola: Ribosomal crystallization in hypothermized chicken bone marrow. J. Ultrastruct. Res. **60**, 140–147 (1977)

[152] Unwin, P. N. T., and C. Taddei: Pakking of ribosomes in crystals from the lizard *Lacerta sicula*. J. Mol. Biol. **114**, 491–506 (1977)

[153] Barbieri, M.: The primitive ribosome model. J. Theoret. Biol. **47**, 269–280 (1974)

[154] Dondi, P. G.: Analysis, by two-dimensional gel electrophoresis, of ribosome crystal proteins. Biochem. J. **149**, 475 to 476 (1975)

[155] Blobel, G., and V. R. Potter: Studies on free and membrane-bound ribosomes in rat liver. I. Distribution as related to total cellular RNA. J. Mol. Biol. **26**, 279–292 (1967)

[156] DeVries, M., and W. S. Bont: Run-off ribosomes in liver, spleen and pancreas of the rat. Mol. Biol. Rep. **2**, 177–179 (1975)

[157] Sidransky, H., E. Verney, and C. N. Murty: Stability of messenger ribonucleic acid of free and membrane-bound polyribosomes of the livers of rats treated with ethionine. Lab. Invest. **40**, 92–98 (1979)

IV. Some general properties of ribosomes[1)]

Contents

Ribosomes are composed of two subunits of different size, particle mass, shape, chemical composition, and functions.

1. Physical characteristics

The large subunit of eukaryotic ribosomes is characterized by a sedimentation coefficient $S_{20,w}$ of about 60S, the small one by a value close to 40 S; the whole ribosome has a sedimentation coefficient of about 80 S. The 80S monomers dissociate into the subunits during the translation cycle (see Chapter XI) and in vitro depending on the ionic conditions (see Chapter VIII). The sedimentation coefficients are approximative values, which vary depending on the origin of the ribosomes, e.g. on the species of animals or plants. This becomes obvious especially from the molecular masses of the monomers and the subunits for which examples are given in Table 1.

The entire molecular mass of the small ribosomal subunit of rat liver as measured by physical methods was found to be in the order of $1.4 \cdot 10^6$ dalton. This value agrees exactly with the sum of the molecular weight of the 18 S RNA ($0.7 \cdot 10^6$) and the total mass of the proteins ($0.7 \cdot 10^6$ dalton).

For the large rat liver ribosomal subunit a total mass of about $2.9 \cdot 10^6$ dalton has been found by direct measurements with physical methods. This value agrees reasonably with the sum of the molecular masses of the three species of RNA and those of all proteins, which is in the order of $2.8 \cdot 10^6$ dalton.

The mass of cytoplasmic ribosomes from various species varies from $3.9 \cdot 10^6$ dalton in plants, $4.1 \cdot 10^6$ in sea urchins, and $4.3 \cdot 10^6$ in birds to $4.5 \cdot 10^6$ dalton in mammals. Although the mass of the small subunit of all eukaryotic ribosomes is consistently close to about $1.5 \cdot 10^6$ dalton, the large subunits vary in their

---

1) For further comprehensive reviews covering earlier publications see [1, 2, 3]

Table 1. Some properties of 80 S type eukaryotic ribosomes

Source of ribosomal particles	Sedimentation coefficient ($s_{20,w}$) [S]	Total mass (dalton $\cdot 10^{-6}$	Buoyant density in CsCl ($g \cdot cm^{-3}$)	Partial specific volume ($cm^3 \cdot g^{-1}$)	RNA content (%)	Protein content (%)
Rat liver						
Monosomes	79.4 [4, 5]		1.61 [5]		50.9 [4]	48.6 [5]
		4.55 [8]	1.57 [8]	0.635 [8]	49.1 [8]	
		4.35 [11]	1.57 [12]		50.3 [12]	
		4.09 [13]				
Large subunit	59.1 [7]	3.0 [7]	1.62 [6]		57.2 [11]	45 [7]
		2.97 [11]	1.61 [7]		59.0 [12]	
		2.71 [13]	1.61 [11]		57.0 [13]	
			1.60 [12]			
			1.575 [13]			
Small subunit	40.9 [7]	1.5 [7]	1.49 [6]		50.8 [11]	55 [7]
		1.4 [11]	1.55 [7]		44.3 [12]	
		1.38 [13]	1.578 [11]		47.0 [13]	
			1.515 [12]			
			1.635 [13]			
Mouse Liver						
Monosomes		4.56 [8]				
Large subunit		3.01 [8]	1.610 [8]		56.8 [8]	
Small subunit		1.53 [8]	1.556 [8]		46.6 [8]	
HeLa cells						
Monosomes			1.56 [6]			
Large subunit			1.60 [6]			
Small subunit			1.52 [6]			
L cells						
Monosomes			1.55 [14]			
Large subunit			1.57 [14]			
Small subunit			1.49 [14]			
Chicken liver						
Monosomes		4.25 [8]	1.583 [8]		51.8 [8]	
Large subunit		2.78 [8]	1.610 [8]		56.8 [8]	
Small subunit		1.53 [8]	1.551 [8]		45.6 [8]	
Chicken muscle	78.1* [10]	5.2 [10]	1.61* [10]			
	81.1** [10]		1.60**[10]			
Sea urchin						
Monosomes	74 [9]	4.09 [8]	1.575 [9]			
Large subunit	56.2 [9]	2.59 [8]	1.595 [8]		54.1 [8]	
			1.61 [9]			
Small subunit	35.8 [9]	1.50 [8]	1.550 [8]		45.4 [8]	
			1.55 [9]			

Continuet table 1

Source of ribosomal particles	Sedimentation coefficient ($s_{20,w}$) [S]	Total mass (dalton $\cdot 10^{-6}$	Buoyant density in CsCl ($g \cdot cm^{-3}$)	Partial specific volume ($cm^3 \cdot g^{-1}$)	RNA content (%)	Protein content (%)
Artemia salina						
Monosomes	81 [16]	3.8 [16]	1.570 [15]	0.63 [16]	49.5 [15]	50.5 [15]
Paramecium		3.45 [17]			57 [17]	
Neurospora crassa						
Monosomes	77—78 [18, 19]	4.05 [19]	1.575 [19]		50 [19]	
Large subunit	56 [19]	2.50 [19]	1.590 [19]		53 [19]	
Small subunit	36 [19]	1.55 [19]	1.550 [19]		45 [19]	
Yeast						
Monosomes	80.4 [20]	3.95 [20]		0.67 [20]		
	82 [21]	3.60 [21]		0.63 [21]	53 [21]	
Large subunit	60.7 [21]	2.57 [21]		0.635 [21]	50 [21]	
Small subunit	37.8 [21]	1.01 [21]		0.620 [21]	57.5 [21]	
Pea seedlings						
Monosomes	78 [19]	3.93 [19]	1.581 [19]		51 [19]	
		3.84 [11]				
		3.97 [8]				
Large subunit	56 [19]	2.48 [19]	1.591 [19]		53 [19]	
		2.44 [8]	1.591 [8]		53.3 [8]	
		2.45 [11]	1.59 [11]		53 [11]	
Small subunit	36 [19]	1.53 [19]	1.552 [19]		46 [19]	
		1.52 [8]	1.552 [8]		45.8 [8]	
		1.39 [11]	1.575 [11]		50 [11]	

* Free single ribosomes
** Polysomal derived ribosomes

masses from $2.4 \cdot 10^6$ in plants, $2.6 \cdot 10^6$ in sea urchins, and $2.8 \cdot 10^6$ in birds to $3.0 \cdot 10^6$ dalton in mammals. Thus, the increase in the particle mass of the eukaryotic ribosomes is mainly the result of differences in the mass of the large subunit. This increase in the mass of the large subunit is predominantly due to an increase of the molecular weight of its large RNA moiety (see Chapter VI).

Because the content of RNA and protein in both subunits varies depending on their origin the buoyant densities are also different (Table 1).

Some properties of ribosomes from mitochondria and chloroplasts are given in Table 2.

2. Chemical characteristics

Ribosomes consist mainly of RNA and protein. The RNA content of the small subunit is close to 45%, that of the large one in the order of 50—55% (Table 1).

The small ribosomal subunit contains one species of high molecular weight RNA

Table 2. Some properties of ribosomes from mitochondria and chloroplasts

Source of ribosomal particles	Sedimentation coefficient ($s_{20,w}$) [S]	Total mass (dalton · 10^{-6})	Buoyant density in CsCl ($g \cdot cm^{-3}$)	Sedimentation coefficient of RNA ($s_{20,w}$) [S]	Total mass of RNA (dalton · 10^{-6})
Mitochondria					
Rat liver					
Monosomes	55	3.2	1.45		
Large subunit	39—40	2.0		16	0.65
Small subunit	28—30	1.2		13	0.36
Beef liver					
Monosomes	56.3	2.83	1.42		
Large subunit	44.9	1.65	1.43		
Small subunit	30.1	1.10	1.43		
Yeast (Saccharomyces)					
Monosomes	72—77	4.05	1.48		
Large subunit	50—54	2.47		21—25	1.25
Small subunit	32—36	1.69		16—17	0.68
Neurospora					
Monosomes	73				
Large subunit	50—52			23—25	1.28
Small subunit	37—39			16—17	0.72
Chloroplasts					
Chlamydomonas					
Monosomes	68				
Large subunit	43			22	1.1
Small subunit	28			16	0.55
Pea					
Monosomes	67		1.57		
Large subunit	46				1.07
Small subunit	30				0.56

From data found in refs. [22—38]

(18 S type), the large subunit also one species of high molecular weight RNA (28 S type) and furthermore two types of low molecular weight RNAs: one molecule of 5 S RNA and one 5.8S RNA molecule.

The eukaryotic 80S ribosome is composed of at least 70 different proteins, about 40 in the large and 30 in the small subunit, as it has been shown mainly for rat liver ribosomes.

For details about properties of the RNA and protein moieties see Chapter VI.

The gross morphology of ribosomal particles is described in Chapter V, and their functional properties will be presented in Chapters IX and XI.

3. References

[1] Ts'o, P. O. P.: The ribosomes-Ribonucleoprotein particles. Ann. Rev. Plant Physiol. **13**, 45—80 (1962)

[2] Spirin, A. S., and L. P. Gavrilova: The ribosome. Springer, Berlin, Heidelberg, New York, 1969

[3] Petermann, M. L.: The physical and chemical properties of ribosomes. Elsevier Publ. Comp. Amsterdam, London, New York, 1964

[4] Bielka, H., H. Welfle, M. Böttger, and W. Förster: Strukturveränderungen und Dissoziation von Leberribosomen in Abhängigkeit von der Mg^{++}-Konzentration. Europ. J. Biochem. **5**, 183—190 (1968)

[5] Grummt, F., and H. Bielka: Stepwise dissociation of rat liver ribosomes into core particles and split proteins. Biochim. Biophys. Acta **199**, 540—542 (1970)

[6] Ovchinnikov, L. P., and A. S. Spirin: Ribonucleoprotein particles in cytoplasmic extracts of animal cells. Naturwiss. **57**, 514—521 (1970)

[7] Hamilton, M. G., A. Pavlovec, and M. L. Petermann: Molecular weight, bouyant density, and composition of active subunits of rat liver ribosomes. Biochem. **10**, 3424—3427 (1971)

[8] Cammarano, P., A. Romeo, M. Gentile, A. Felsani, and C. Gualerzi: Size heterogeneity in the large ribosomal subunit and observations of the small subunits in eukaryote evolution. Biochim. Biophys. Acta **281**, 597—624 (1972)

[9] Infante, A. A., and M. Nemer: Heterogeneous ribonucleoprotein particles in the cytoplasm of sea urchin embryos. J. Mol. Biol. **32**, 543—565 (1968)

[10] Vournakis, J., and A. Rich: Size change in eukaryotic ribosomes. Proc. Natl. Acad. Sci. U.S. **68**, 3021—3025 (1971)

[11] Cammarano, P., S. Pons, A. Romeo, M. Galdieri, and C. Gualerzi: Characterization of unfolded and compact ribosomal subunits from plants and their relationship to those of lower and higher animals: Evidence for physiochemical heterogeneity among eukaryotic ribosomes. Biochim. Biophys. Acta **281**, 571—596 (1972)

[12] Sherton, C. C., and I. G. Wool: The extraction of proteins from eukaryotic ribosomes and ribosomal subunits. Mol. Gen. Genetics **135**, 97—112 (1974)

[13] Sacchi, A., U. Ferrini, P. Londei, P. Cammarano, and N. M. Maraldi: Mitochondrial and cytoplasmic ribosomes from mammalian tissues. Biochem. J. **168**, 245—259 (1977)

[14] Perry, R. P., and D. E. Kelley: Buoyant densities of cytoplasmic ribonucleoprotein particles of mammalian cells. Distinctive character of ribosome subunits and the rapid labelled components. J. Mol. Biol. **16**, 255—268 (1966)

[15] Nieuwenhuysen, P., E. de Herdt, and J. Clauwaert: Ribonucleic acid and protein content of eukaryotic ribosomes isolated from *Artemia salina*. Anal. Biochem. **88**, 532—538 (1978)

[16] Nieuwenhuysen, P., and J. Clauwaert: Physical properties of *Artemia salina* ribosomes. Biochem. **17**, 4260—4265 (1978)

[17] Reisner, A. H., J. Rowe, and H. M. Macindoe: Structural studies on the ribosomes of paramecium: Evidence for a "primitive animal ribosome". J. Mol. Biol. **32**, 587—610 (1968)

[18] Küntzel, H.: Mitochondrial and cytoplasmic ribosomes from *Neurospora crassa*. J. Mol. Biol. **40**, 315—320 (1969)

[19] Cammarano, P., A. Felsani, A. Romeo, and F. M. Alberghina: Particle weights of active ribosomal subunits from *Neurospora crassa*. Biochim. Biophys. Acta **308**, 404—411 (1973)

[20] Cotter, I., R. I. McPhie, and W. B. Gratzer: Internal organization of the ribosome. Nature **216**, 864—868 (1967)

[21] Mazelis, A. G., and M. L. Petermann: Physical-chemical properties of stable yeast ribosomes and ribosomal subunits. Biochim. Biophys. Acta **312**, 111—121 (1973)

[22] Stegeman, W. J., C. S. Cooper, and C. J. Avers: Physical characterization of ribosomes from purified mitochondria of yeast. Biochem. Biophys. Res. Commun. **39**, 69—76 (1970)

[23] Borst, P., and L. A. Grivell: Mitochondrial ribosomes. FEBS-Lett. **13**, 73—88 (1971)

[24] O'Brien, T. W.: The general occurrence of 55S ribosomes in mammalian liver mitochondria. J. Biol. Chem. **246**, 3409—3417 (1971)

[25] Vignais, P. V., B. J. Stevens, J. Huet, and J. André: Mitoribosomes from *Candida utilis*. Morphological, physical and chemical characterization of the monomer

form and of its subunits. J. Cell Biol. **54**, 468–492 (1972)

[26] Aaij, C., N. Nanninga, and P. Borst: The structure of ribosome-like particles from rat-liver mitochondria. Biochim. Biophys. Acta **277**, 140–148 (1972)

[27] Sacchi, A., F. Cerbone, P. Cammarano, and U. Ferrini: Physicochemical characterization of ribosome-like (55S) particles from rat liver mitochondria. Biochim. Biophys. Acta **308**, 390–403 (1973)

[28] Reijnders, L., P. Sloff, and P. Borst: The molecular weights of the mitochondrial ribosomal RNAs of *Saccharomyces carlsbergensis*. Europ. J. Biochem. **35**, 266–269 (1973)

[29] Heizmann, P.: Maturation of chloroplast rRNA in *Euglena gracilis*. Biochem. Biophys. Res. Commun. **56**, 112–118 (1974)

[30] Hamilton, M. G., nd T. W. O'Brien: Ultracentrifugal characterization of the mitochondrial ribosome and subribosomal particles of bovine liver. Molecular size and composition. Biochem. **13**, 5400–5403 (1974)

[31] Van den Borgert, C., and H. de Vries: The mitochondrial ribosomes of *Neurospora crassa*. Biochim. Biophys. Acta **442**, 227–238 (1976)

[32] Hoober, J. K., and G. Blobel: Characterization of the chloroplastic and cytoplasmic ribosomes of *Chlamydomonas reinhardii*. J. Mol. Biol. **41**, 121–138 (1969)

[38] Chi, J. C. H., and Y. Suyama: Comparative studies on mitochondrial and cytoplasmic ribosomes of *Tetrahymena pyriformis*. J. Mol. Biol. **53**, 531–556 (1970)

[34] Ashwell, M. A., and T. S. Work: The functional characterization of ribosomes from rat liver mitochondria. Biochem. Biophys. Res. Comm. **39**, 204–211 (1970)

[35] Greco, M., P. Cantatore, G. Pepe, and C. Saccone: Isolation and characterization of rat-liver mitochondrial ribosomes highly activ in poly(U)-directed poly(phe) synthesis. Europ. J. Biochem. **37**, 171–177 (1973)

[36] Datema, R., E. Agsteribbe, and A. M. Kroon: The mitochondrial ribosomes of *Neurospora crassa*. I. On the occurrence of 80-S ribosomes. Biochim. Biophys. Acta **335**, 386–395 (1974)

[37] Faye, G.: Mitochondrial ribosomal subunits from *Saccharomyces cerevisiae* rho$^-$-mutants. FEBS Lett. **69**, 167–170 (1976)

[38] Michel, R., G. Hallermayer, M. A. Harmey, F. Miller, and W. Neupert: The 73S ribosome of *Neurospora crassa* is the native mitochondrial ribosome. Biochim. Biophys. Acta **478**, 316–330 (1977)

V. Morphology of ribosomes and polysomes

The gross morphology of ribosomal particles from eukaryotes has been studied by electron microscopy [1—64] and more recently also by small-angle X-ray scattering [65—69]. More detailed information about the organization of the surface of ribosomal particles can be obtained by immune electron microscopy. This technique provides data about the location of the proteins, at least of their antigenic determinants, organized on the surface of the ribosomal particles [41, 43, 51, 52 55, 62, 64] (for details see Chapter IX Section 2e).

Contents

1. Electron microscopy

Electron microscopic studies have been performed using various techniques such as negative and positive staining mostly by uranyl acetate, the stain of choice, and shadow-casting using platinum or tungsten.

The appearance of the ribosomal particles in the electron micrographs are independent of whether the ribosomal samples were fixed by formaldehyde, by glutaraldehyde, or not [17, 36]. Ribosomal particles mounted with the conventional adsorption technique and negatively stained can be affected in their overall dimensions. Images of ribosomes deposited between two thin carbon layers and dried within this "sandwich", however, are far more reproducible and usually larger [58].

Since the first detailed gross morphology of ribosomes of *E. coli* has been described [2], several reports on the ultrastructural features of these organelles obtained from eukaryotic cells have been published (for reviews see [33, 38, 42, 54]). Electron microscopic studies of 80S ribosomes have been mostly concerned with monomers and their subunits [9, 10, 16—18, 20—23, 25—31, 33, 35, 36, 38, 40, 42, 44, 45, 47—50, 55—57, 59—61, 64], polysomes [1, 3—8, 10, 11, 13—16, 19, 24, 32, 40, 42] and also with ribosomal crystals (see Chapter III).

Investigations on the ultrastructure of cytoplasmic ribosomes include particles from rat liver [10, 13, 15, 26, 27, 30, 32 to 36, 38, 40, 42, 44, 45, 47, 48, 54, 55, 59, 61, 64], mouse liver [22, 25], rabbit reticulocytes [3–6, 16, 63], tumor cells [16, 24, 32], *Locusta* [50], *Artemia* [60], *Plasmodia* [29], yeast cells [23, 31, 37], *Geotrichum* [56, 57], *Tetrahymena* [49], *Neurospora* [39], and different plants [9, 11, 17, 18, 20, 21, 28]. Furthermore, results of electron microscopic studies of ribosomes of mitochondria [34, 37, 39, 49, 50], and chloroplasts [17, 18, 20, 21] have been described.

A particular feature of ribosomes is the lack of symmetry. Ribosomes are asymmetric in their shape, in the distribution of their protein and RNA moieties and also in the mutual position of their unequal large and small subunits. They do not posses any repetitive arrangement of their components in a regular quarternary structure like viruses of comparable size [58].

In the following mainly the morphology of rat liver ribosome will be described in detail as representative for the structure of at least most eukaryotic cytoplasmic ribosomes [54].

a) The small ribosomal subunit

A typical general view of small subunits prepared from 80 S ribosomes by puromycin treatment in high KCl is shown in Fig. 6, representing projections of subunits after negative staining with uranyl acetate in different positions on the supporting film in mainly two prolate image types: (1) Images with a slightly bent profile (about 17×25 nm) divided by two contrast lines into three parts called A and C images. (2) Images with a more elongated profile (about 13×23 nm) with one contrast line dividing the images into two unequal parts, the head and the larger body, called type B images.

The images with bent profiles exist in two, more or less frequently observable, nearly enantiomorphic forms referred to as type A (Fig. 6b) and type C (Fig. 6d). Type C images, which are much less frequent, represent projections of 40 S subunits rotated through an angle of 180° about their long axis in comparison to particles in position A. It is assumed that type B images (Fig. 6c) represent 40 S subunits turned through an angle of 90° about their long axis in comparison to type A and type C images [64]. Type B images show a protuberance at the body directed to the stronger contrast line between the body and the head. It is assumed that this protuberance is involved in the contact between the 40 S and the 60 S subunit inside the 80 S ribosome. This protuberance is in particular demonstrable in preparations of 80 S ribosomes at low magnesium concentrations (10^{-4} M) at which dissociation can be made visible in electron micrographs [35].

Native small subunits [63] show a three-lobed image resulting from the apposition of an additional mass of $16 \times 10 \times 6$ nm on one side of the subunit. The site of attachment of this component is located on a prominence extending from the central part of the small subunit and is separated by a cleft from the head of the subunit. Since removal of the eukaryotic initiation factor 3 (eIF-3) from these native small ribosomal subunits by treatment with high concentrations of salt results in subunit profiles indistinguishable in electron micrographs from those of the so-called derived subunits, the native subunits with the three-lobed images have been identified as small subunits complexed with eIF-3.

b) The large ribosomal subunit

The general view of 60 S subunits (Fig. 7a) shows three main profile types: (1) Nearly equilateral triangular profiles ($\sim 24 \times 20$ nm) called lateral views (Fig. 7c). (2) Asymmetric profiles (~ 25

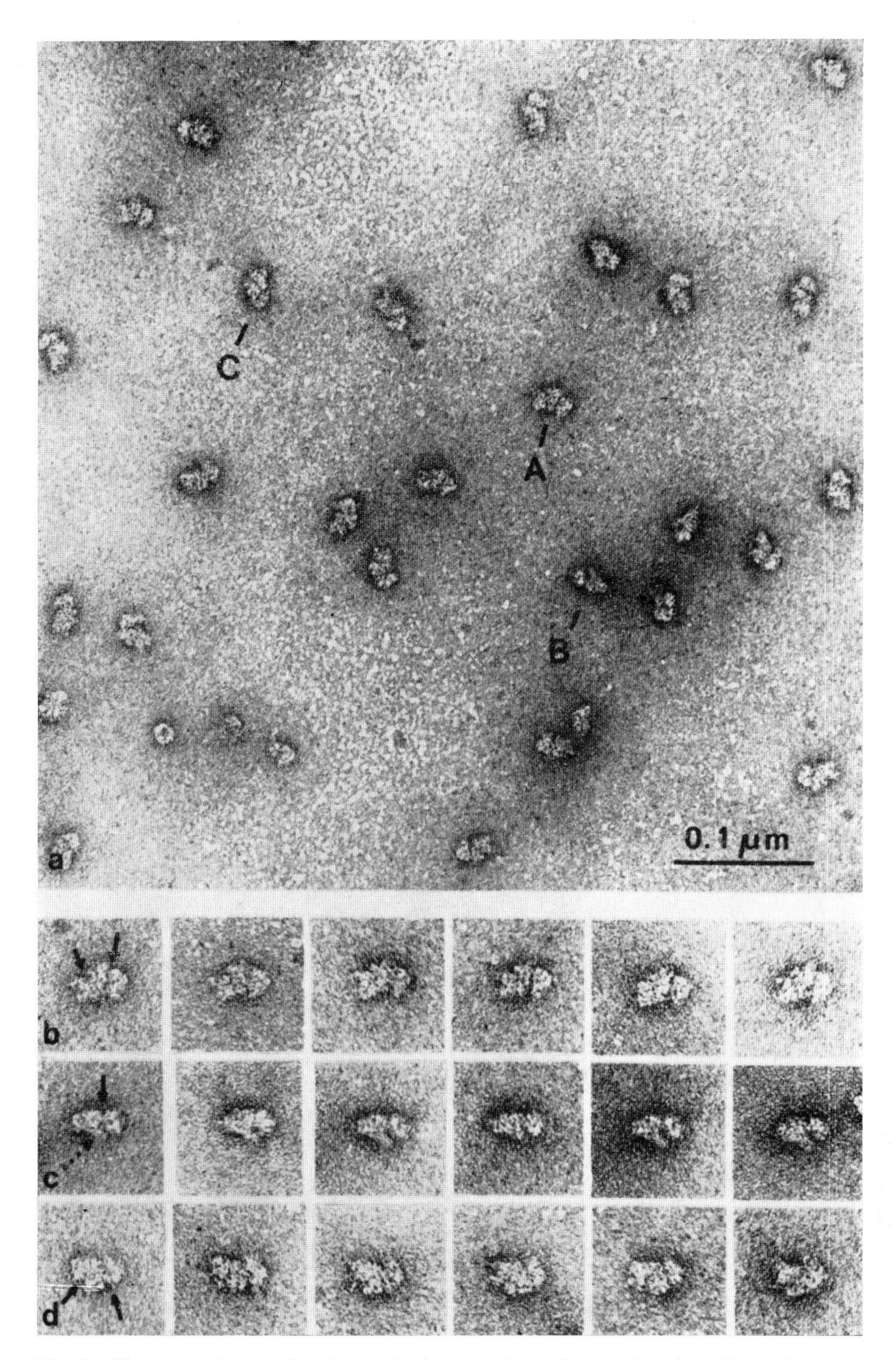

Fig. 6. Electron micrographs of negatively stained small subunits of rat liver ribosomes.
a) General view (× 200000); b) — d) selected images (× 300000); type A images (b), type B images (c), and type C images (d).
From G. LUTSCH et al. [64].

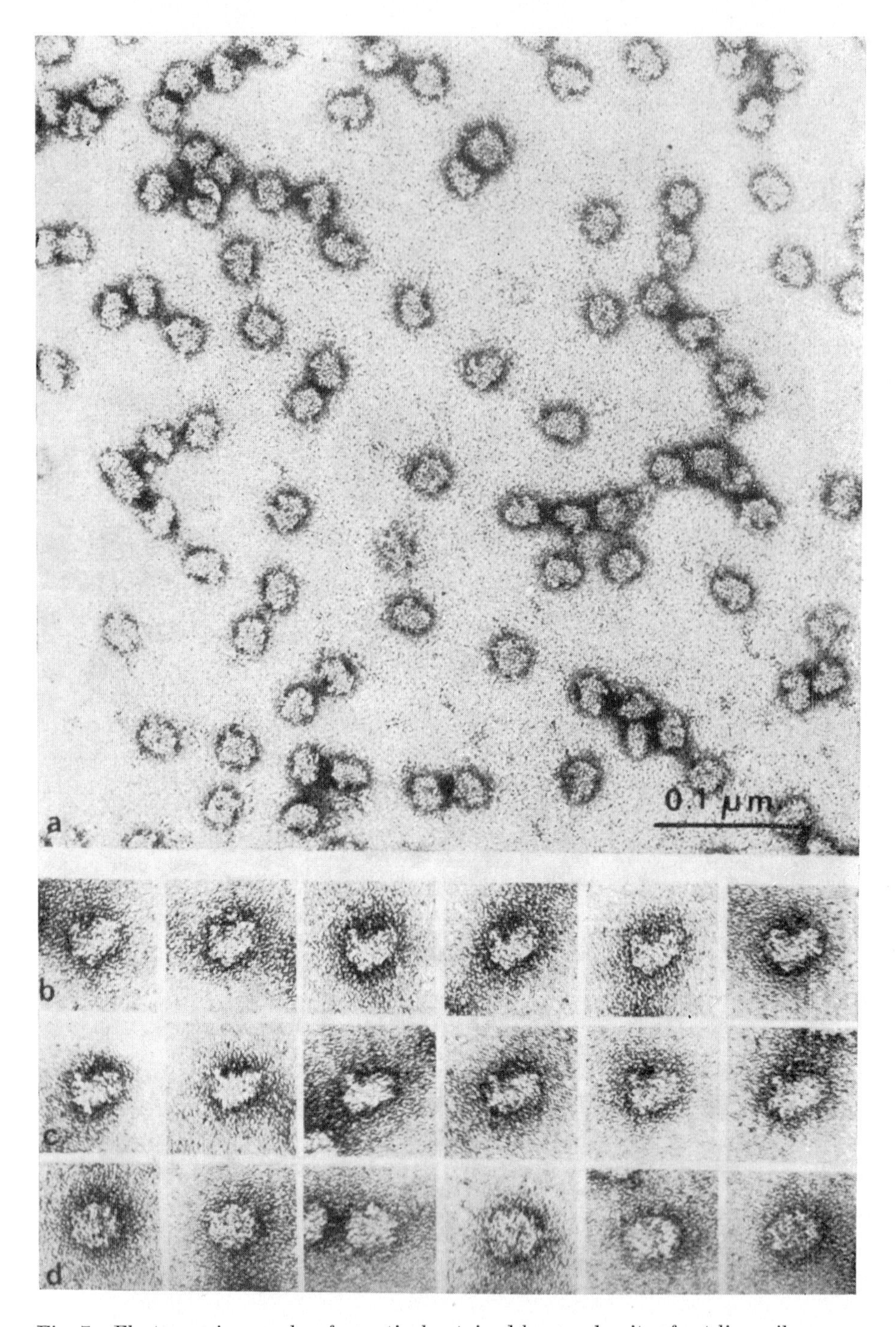

Fig. 7. Electron micrographs of negatively stained large subunits of rat liver ribosomes.
a) General view ($\times$ 200000); b)—d) selected images ($\times$ 300000); frontal views (b), lateral views (c), and top views (d).
From G. LUTSCH et al. [36] and unpublished results.

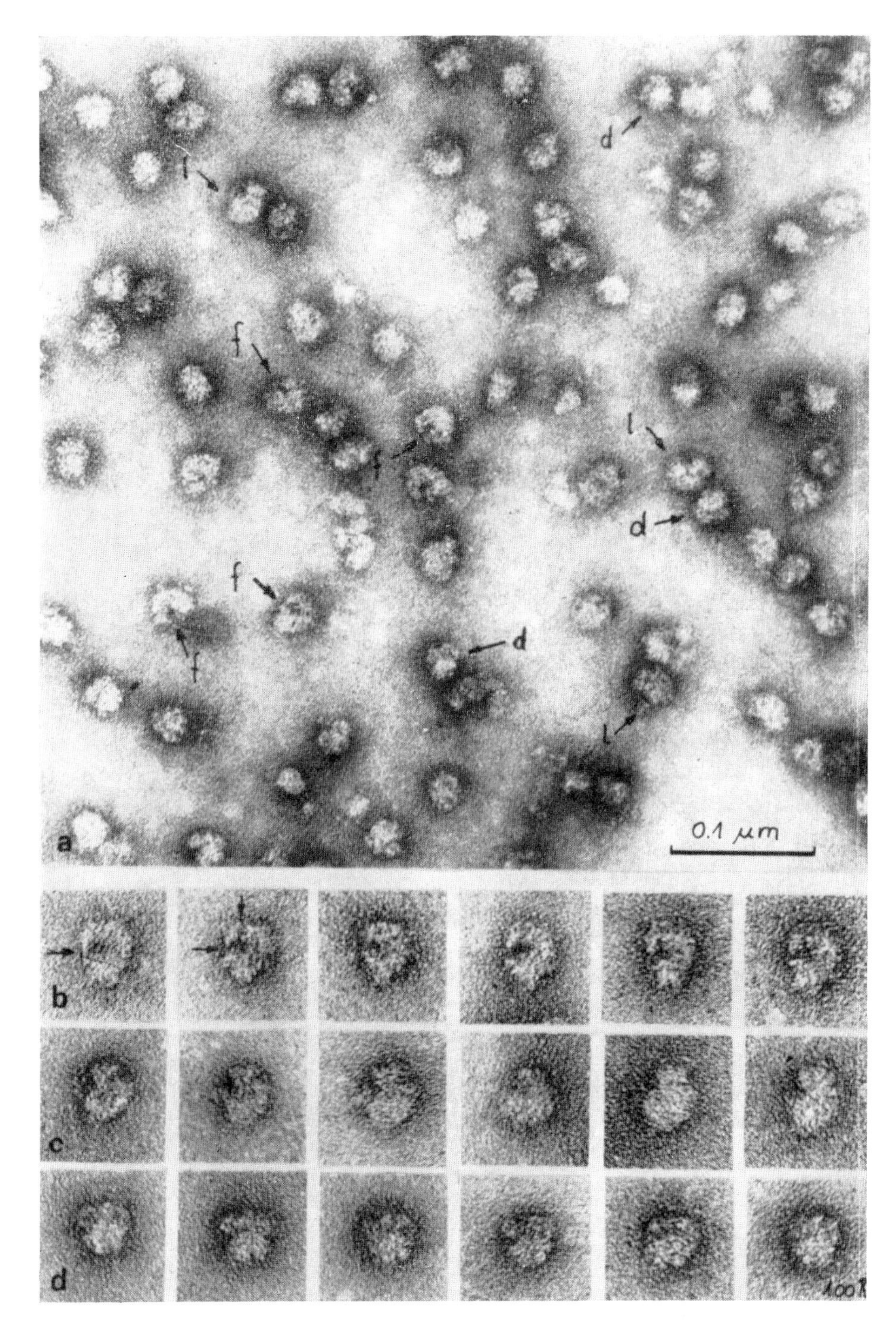

Fig. 8. Electron micrographs of negatively stained monomeric ribosomes from rat liver.
a) General view (× 200000); b)—d) selected images (× 300000); frontal views (b), lateral views (c), and dorsal views (d).
From G. Lutsch et al. [36] and unpublished results.

×20 nm) called frontal views (Fig. 7b), which have a skiff-like shape with a blunted and a pointed end. They are approximately triangular with one flattened or concave side (23 nm long), in which a dense notch (3—4 nm) towards the blunted end of the skiff can be observed (Fig. 7b). The tip of the triangle is excentrically located, pointing to the right, and the notch is on the left side of the image. On the lower tip of the triangular images two lateral incisions can be seen (Fig. 7b) which represent possibly a ring-like groove around the tip of the 60 S subunit (cf. Fig. 9a). The notch in these frontal views of large subunits is thought as an end view of a laterally positioned groove transferring the flattened surface of the subunit [25, 30, 33, 36]. (3) Rounded profiles with a diameter of approximately 23 nm called top views (Fig. 7d). They are likely to represent projections of large subunits laying on their interface which is involved in the contact with the small subunit within the 80 S ribosome [42]. Therefore this profil type cannot be seen in images of large subunit dimers or monomeric ribosomes.

c) *The monomeric ribosome*

After negative staining three main image types depending on the position of the ribosome on the supporting film (Fig. 8a) can be observed: (1) Images with a longish triangular profile (frontal view), (2) images with rounded profiles (dorsal view), and (3) images consisting of a small rectangular profile located laterally on a flattened side of the triangular large subunit profile (lateral view).

Frontal views (Fig. 8b) of images of monomeric ribosomes (~ 29 nm high and 24 nm wide) are characterized by an excentrically located region of high staining intensity with a diameter of 4—5 nm. In most cases this dense spot is located towards the left of the ribosome images, if they are oriented with the small subunit profile horizontally and to the top (left-featured frontal views). The elongated slightly curved small subunit profile (~ 23 nm long and 13 nm wide) is joined through its concave side of the flattened side of the triangular skiff-shaped large subunit profile. Furthermore, the partition of the small subunit, which defines the two unequally sized parts, is visible. The groove on the flattened side of the large subunit profile (in frontal views located at the pointed end on the left side) and the groove on the concave side of small subunit profiles between the head and body of the small subunit form a channel in between the subunits. The enantiomorphic form can be observed in rare cases only.

Lateral views (Fig. 8c) of monomeric ribosomes are characterized by two segments of different size and shape with a thin stained band between both. The rectangular profile of the small subunit with a hight of ~ 12 nm and a width of ~ 14 nm is interpreted as an end view of the prolate shaped subunit. The large subunit profile in lateral views is of type (1) (Fig. 7c). In most cases the small subunit is positioned at the right side of the large subunit (right-lateral views).

Dorsal views (Fig. 8d) of monomeric ribosomes are characterized by a rounded profile (~ 26 nm high and ~ 25 nm wide)

⟶

Fig. 9. B) Models of the rat liver ribosome. From Y. Nonomura et al. [30]. With permission of the authors. Copyright by Academic Press Inc. (London) Ltd.

a) Plasticine model which, when lying in opposite positions over the supporting film, explain the main image types. The channel on the surface of the large subunit has not been emphasized in the model.

b) X-ray radiographics of the model demonstrated in a). 1. Left-featured frontal type; 2. right-lateral type of image.

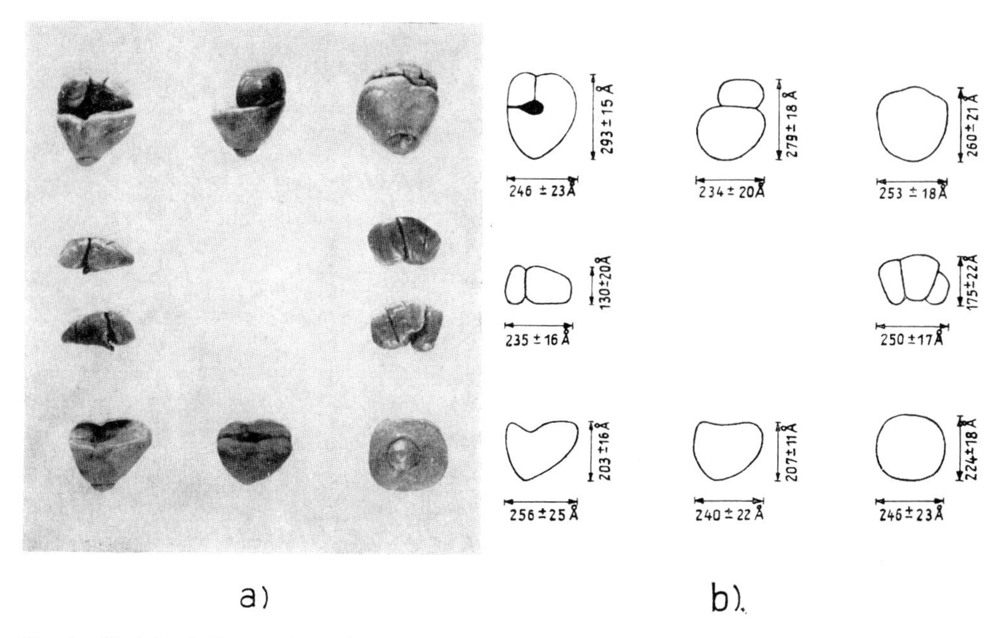

Fig. 9. Models of ribosomal particles.

A) Models of monomeric ribosomes, and small and large subunits, respectively, of rat liver ribosomes. From G. LUTSCH et al. [36].

a) Plasticine models. Upper row: Monomeric ribosomes, middle row: small subunits, lower row: large subunits. b) Dimensions of the various ribosomal image types.

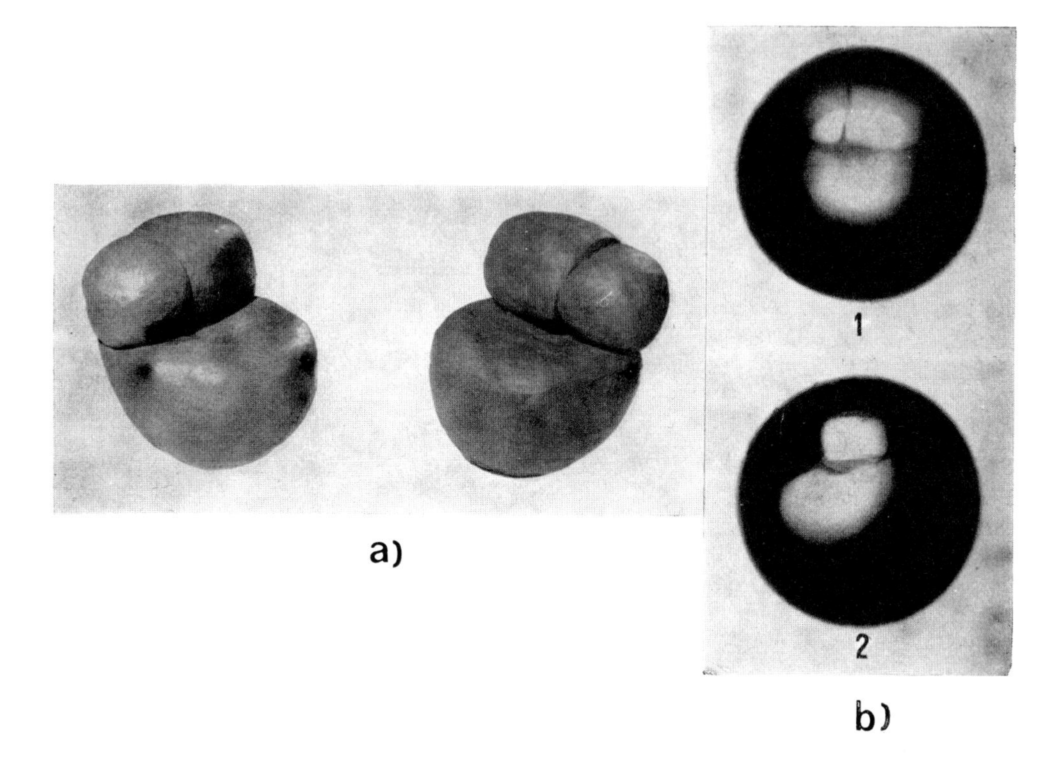

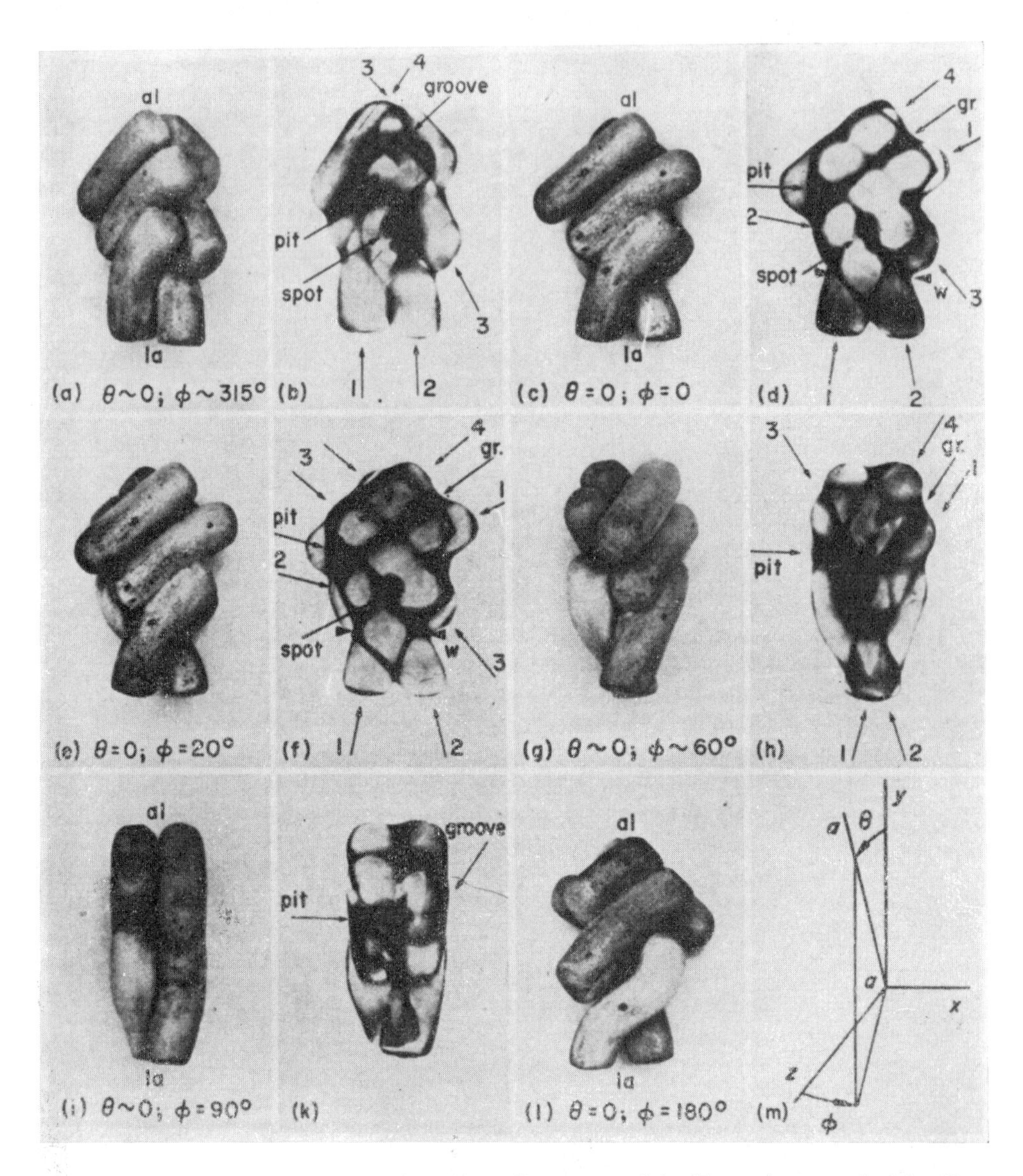

Fg. 9. C) Model of small subunit of rat liver ribosomes and its X-ray photographs (a)—(l) an co-ordinate system (m). From N. A. KISELEV et al. [61]. With permission of the authors Copyright by Academic Press Inc. (London) Ltd.

and some incisions on one side. Additionally, a thin dense line transversing horizontally the dorsal images can be observed.

d) *Three-dimensional models*

Three-dimensional models of rat liver cytoplasmic ribosomes [30, 36, 44, 61] as well as of mitochondrial ribosomes of yeast [37] have been deduced from two-dimensional electron microscopic data. In general the proposed model of rat liver ribosomes, demonstrated in Fig. 9A, is in. good agreement with another model of ribosomes from the same source [30] (Fig. 9B). Regarding the ribosomal overall shape these models are also in conformity with the three-dimensional structure of the *E. coli* ribosome except in size [51–53, 60]. Thus, the protuberance as a distinct feature of rat liver small subunits has its counterpart in the small ribosomal subunit of *E. coli* in form of a "platform" [53] or a "bulge" [46], whereas other authors propose instead of such structures two "lobes" of unequal size protruding from the body of the small subunit of the *E. coli* ribosome [51]. The models for rat liver ribosomes [30, 36] agree better with the asymmetric model [52, 53] than with the more symmetric one for *E. coli* ribosome [51].

Another model for rat liver ribosomes has been proposed [44], which differs from those mentioned before (Fig. 9C), concerning the gross and fine morphology. In Fig. 9C this model is presented for the small subunits of rat liver ribosomes, revealing four elongated domains in some places further subdivided into 5 nm diameter strands, which are interpreted as parts of a longer ribonucleoprotein strand [61].

e) *Polysomes*

The structure of polysomes or polyribosomes has been studied in cell sections [6–8, 11, 13, 15, 19, 22] (see also Chapter III) as well as in isolated specimens [3–6, 10, 16, 30, 32, 33, 38, 40]. Within the polysomal structure single ribosomes are connected by the mRNA strand which has a thickness of 1–1.5 nm in positively stained and 1.5–3 nm in negatively stained probes [4, 5, 10, 33, 42]. These differences in the diameters suggest that the mRNA molecule between the ribosomes has a secondary structure and/or is covered by protein [30, 33].

In polysomes the same main image types of ribosomes as described for isolated monomeric ribosomes are present (Fig. 10). However, lateral views are more frequently seen in electron micrographs of free rat liver polysomes than in preparations of isolated monomers, probably resulting from a restriction in free rotation of the particles caused by the mRNA strand [30]. From electron microscopic observations it can be concluded that the mRNA molecule transverses the ribosome in a channel at the level of the plane between both subunits in a direction perpendicular to the long axis of the small subunit. This channel between both subunits is represented by the dense spot seen in frontal views of ribosomes (see Fig. 8b).

Polysomes are of variable size; up to 70 or even more ribosomes have been counted in a presumed single polysome [12]. Small polysomes of four to nine monomers are seen as linear arrays or in rosette-like configuration [3, 4, 16, 40]. In such circular structures the 40 S subunit is oriented towards the center of the rosettes. Larger polysomes are arranged in shapes suggestive of flattened helices [16] or as clusters or strands, in which the ribosomes are visible as beads on a thread-like structure [3, 4, 13, 40].

Models of free polysomes, in which the ribosomes are arranged in a helix, have been proposed [13, 16, 19].

The configuration of membrane-bound polysomes and of ribosomal crystals are described in more detail in Chapter III.

2. Small-angle X-ray scattering

X-rays cannot be focused like beams of electrons. Therefore, X-rays yield instead of an image of ribosomes only a diffraction pattern, a record of direction and intensities of the waves deflected by their interaction with the ribosomes [65]. The advantage of examining ribosomes in solution by scattering methods compared with electron microscopy is that the formation of artefacts formed during drying by dehydratation of the specimens [58] is redu-

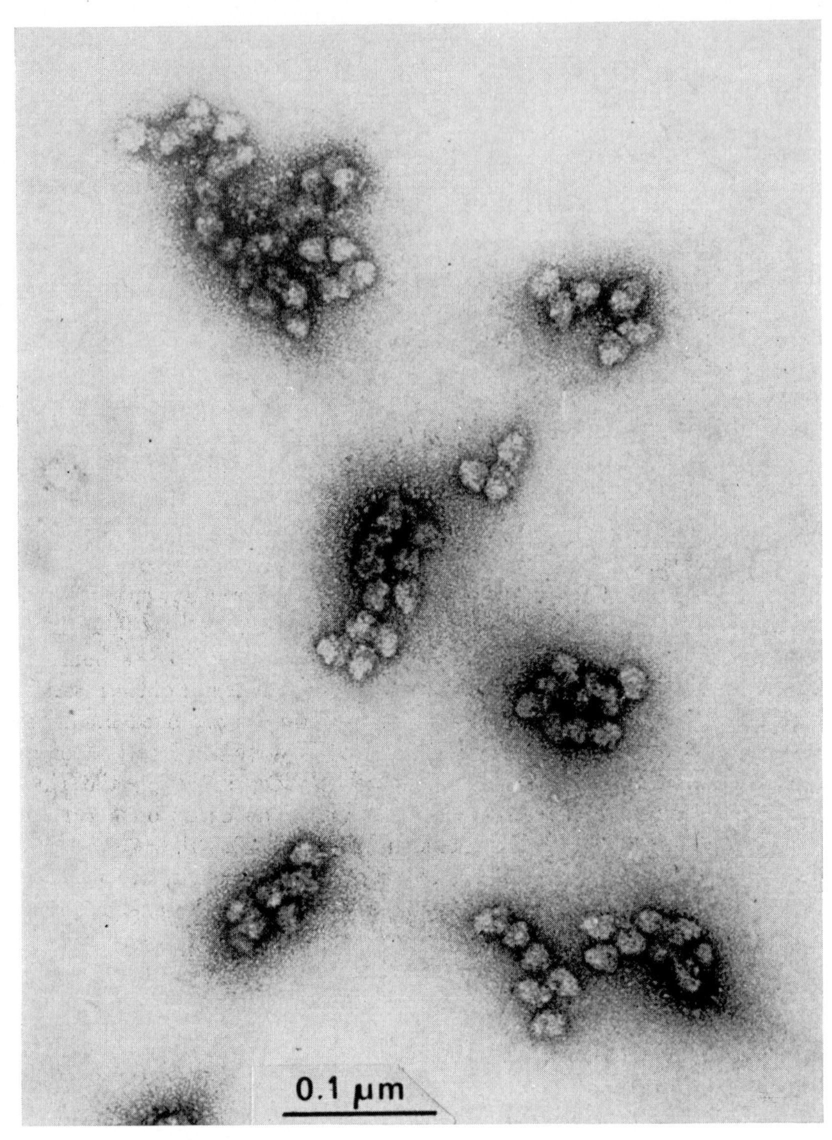

Fig. 10. Electron micrographs of polysomes of rat liver. G. Lutsch and F. Noll, unpublished.

ced and that the complicated spatial structure is not projected to a plane (two-dimensional image) [68].

a) The monomeric ribosome

The dimensions of ribosomes in solution derived from scattering data are larger by 10–20% when compared with electron microscopic data. The differences are caused by solvatation of the ribosomes. The greatest diameter of the ribosome was determined to be 37 nm while for 60 S subunits a mean value of 28 nm has been determined [66]. The mass centres of both subunits are spaced apart by 28 nm [66], this means that they must be positioned excentrically within the ribosome. The dimensions of the ribosomes increase with decreasing Mg^{++} concentrations by deconvolution which has been observed also in electron micrographs [35]. While the increase in the greatest diameter of the 80 S particles is 25%, the increase in the diameter of the 60 S subunit is 10% only.

Supposing two allosteric conformations of the small subunit, the oblate (O-) and the prolate (P-) conformation, a transition of the O- to the P-conformation can explain the increase in the dimensions of 25% (= 9 nm). The cyclic transition P—O—P—O—P... is the basis of a proposed dynamic ribosome model, called "rack-and-roll model", which tries to explain the translocation of the ribosome along the mRNA during the elongation process [67]. From scattering data obtained at low Mg^{++} concentrations it was concluded that the ribosome is organized like a core-shell model, in which the core has a higher protein portion than the rRNA-rich shell [66, 67].

b) Polysomes

The data from small-angle X-ray scattering [68] show that rat liver polysomes in solution have no stretched structures, but are more compact. The distances between the mass centres of neighbouring ribosomes is close to 35 nm. The free space between two ribosomes is about 7 nm, assuming the ribosome diameter near the channel for the mRNA to be 28 nm. The greatest diameter of a polysome with eight ribosomes is 130 nm. A helical arrangement of the ribosomes bound on the mRNA is the most probable structure of polysomes in solution. The pitch of the helix with six ribosomes per turn in a polysome with eight ribosomes equals 33 nm. This polysomal helix requires a bent of the mRNA molecule of about 110° at each ribosome which agrees with the proposed allosteric prolate conformation of the small subunit [67].

3. References

[1] Palade, G. E.: A small particulate component of the cytoplasm. J. Biophys. Biochem. Cytol. **1**, 59—68 (1955)

[2] Huxley, H. E., and G. J. Zubay: Electron microscopic observations of the structure of microsomal particles from *Escherichia coli*. J. Mol. Biol. **2**, 10—18 (1960)

[3] Warner, J. R., A. Rich, and C. E. Hall: Electron microscope studies of ribosomal clusters synthesizing hemoglobin. Science **138**, 1399—1403 (1962)

[4] Warner, J. R., P. M. Knopf, and A. Rich: A multiple ribosomal structure in protein synthesis. Proc. Natl. Acad. Sci. U.S. **49**, 122—129 (1963)

[5] Slayter, H. S., J. R. Warner, A. Rich, and C. E. Hall: The visualization of polyribosomal structure. J. Mol. Biol. **7**, 625 to 657 (1963)

[6] Mathias, A. P., R. Williamson, H. E. Huxley, and S. Page: Occurrence and function of polysomes in rabbit reticulocytes. J. Mol. Biol. **9**, 154—167 (1964)

[7] Benedetti, E. L., H. Bloemendal, and W. S. Bont: Polyribosomes isolés à partir du foie de rat. C. R. Acad. Sci. **259**, 1353—1356 (1964)

[8] Cedergren, B., and J. Harary: In vitro studies on single beating rat heart cells. VI. Electron microscopic studies of single cells. J. Ultrastruct. Res. **11**, 428 to 442 (1964)

[9] Bayley, S. T.: Physical studies on ribosomes from pea seadlings. J. Mol. Biol. **8**, 231—238 (1964)

[10] Dass, C. M. S., and S. T. Bayley: A structural study of rat liver ribosomes. J. Cell. Biol. **25**, 9—22 (1965)

[11] Bonnett, H. T., and E. H. Newcomb: Polyribosomes and cisternal accumulations in root cells of radish. J. Cell Biol. **27**, 423—432 (1965)

[12] Campbell, P. N.: The biosynthesis of proteins. Progr. Biophys. Mol. Biol. **15**, 3—38 (1965)

[13] Pfuderer, P., P. Cammarano, D. R. Holladay, and G. D. Novelli: A helical polysome model. Biochim. Biophys. Acta **109**, 595—606 (1965)

[14] Bell, E., T. Humphreys, H. S. Slayter, and C. E. Hall: Configuration of inactive and active polysomes of the developing down feather. Science **148**, 1739—1741 (1965)

[15] Dallner, G., P. Siekevitz, and G. E. Palade: Biogenesis of endoplasmic reti-

culum membranes. J. Cell Biol. **30**, 73–96 (1966)

[16] SHELTON, E., and E. L. KUFF: Substructure and configuration of ribosomes isolated from mammalian cells. J. Mol. Biol. **22**, 23–31 (1966)

[17] MILLER, A., U. KARLSON, and N. K. BOARDMAN: Electron microscopy of ribosomes isolated from tobacco leaves. J. Mol. Biol. **17**, 487–489 (1966)

[18] ODINTSOVA, M. S., V. I. BRUSKOV, and E. V. GOLUBEVA: Comparative study on ribosomes of chloroplasts and cytoplasm of some plants. Biochimija **32**, 1047–1059 (1967)

[19] WEISS, P., and N. B. GROVER: Helical array of polyribosomes. Proc. Natl. Acad. Sci. U.S. **59**, 763–768 (1968)

[20] BRUSKOV, V. I., and M. S. ODINTSOVA: Comparative electron microscopic studies on chloroplast and cytoplasmic ribosomes. J. Mol. Biol. **32**, 471–473 (1968)

[21] BRUSKOV, V. I., and N. A. KISELEV: Electron microscopy investigation of the structure of cytoplasmic ribosomes of bean leaves. J. Mol. Biol. **38**, 443–445 (1968)

[22] FLORENDO, N. T.: Ribosome substructure in intact mouse liver cells. J. Cell Biol. **41**, 335–339 (1969)

[23] GHOSH, N.: Studies on the conversion of 60S yeast ribosomes into a 50S component. Biochem. J. **115**, 1005–1007 (1969)

[24] SUZUKI, I., H. KAMEI, and M. TAKAHASHI: Ultrastructural study on the ribosome helix and its change after antigenic stimulazion in P3-J-cells. Exp. Molec. Path. **11**, 28–37 (1969)

[25] FLORENDO, N. T., and L. B. FOSTER: Studies on the fine structure of large subunits of ribosomes from mouse liver. J. Ultrastruct. Res. **30**, 1–6 (1970)

[26] HAGA, J. Y., M. G. HAMILTON, and M. L. PETERMANN: Electron microscopic observations on the large subunit of the rat liver ribosomes. J. Cell Biol. **47**, 211–221 (1970)

[27] BENDER, R., S. H. BELLMAN, and R. GORDON: Art and the ribosome: A preliminary report on the three-dimensional structure of individual ribosomes determined by an algebraic reconstitution technique. J. theor. Biol. **29**, 483–487 (1970)

[28] AMELUNXEN, F., and E. SPIESS: Investigations on the structure of ribosomes. A contribution to the conformation of 80S ribosomes from *Pisum sativum*. Cytobiol. **4**, 293–306 (1971)

[29] AIKAWA, M., and R. T. COOK: Ribosomes of the malarial parasit *Plasmodium knowlesi*. II. Ultrastructural features. Comp. Biochem. Biophys. **39**, 913–917 (1971)

[30] NONOMURA, Y., G. BLOBEL, and D. D. SABATINI: Structure of liver ribosomes studied by negative staining. J. Mol. Biol. **60**, 303–323 (1971)

[31] FUNAKOSHI, H., and K. ISO: Characteristic patterns observed in electron microscopic photos of negatively stained yeast 80S ribosomes. J. Biochem. **70**, 201–204 (1971)

[32] BERDINSKYKH, N. K., V. P. MANYKKOV, and V. V. KOZAK: Electron-microscopic study of polyribosomes of tumor and normal cells. Dopov. akad. nauk Ukr. RSR Ser. B (russ.) 1971, 934–939

[33] SABATINI, D. D., Y. NONOMURA, T. MORIMOTO, and G. BLOBEL: Structural studies on rat liver and chicken embryo ribosomes. FEBS Symp. **23**, 147–173 (1972)

[34] AAIJ, C., N. NANNINGA, and P. BORST: The structure of ribosome-like particles from rat liver mitochondria. Biochim. Biophys. Acta **277**, 140–148 (1972)

[35] BIELKA, H., K. WAHN, F. NOLL, and G. LUTSCH: Studies on the structure of animal ribosomes. II. Loosening and unlocking of rat liver ribosomes at low Mg^{++} concentrations: An electron microscopic study. Acta Biol. Med. Germ. **29**, 607–619 (1972)

[36] LUTSCH, G., H. BIELKA, K. WAHN, and J. STAHL: Studies on the structure of animal ribosomes. III. Electron microscopic investigations of isolated rat liver ribosomes and their subunits. Acta Biol. Med. Germ. **29**, 851–876 (1972)

[37] VIGNAIS, P. V., B. J. STEVENS, J. HUET, and J. ANDRÉ: Mitoribosomes from *Candida utilis*. Morphological, physical and chemical characterization of the monomer form and of its subunits. J. Cell Biol. **54**, 468–492 (1972)

[38] NANNINGA, N.: Structural aspects of ribosomes. Int. Rev. Cytol. **35**, 135–188 (1973)

[39] KURIYAMA, Y., and D. J. L. LUCK: Membrane-associated ribosomes in mitochon-

dria of *Neurospora crassa*. J. Cell Biol. **59**, 776—784 (1973)

[40] LUTSCH, G., K. WAHN, H. BIELKA, I. JUNGHAHN, J. STAHL, and F. NOLL: Electron microscopic investigations of rat liver ribosomes. In: "Ribosomes and RNA metabolism" (Ed. by J. ZELINKA and J. BALAN). Slovak Acad. Sci., Bratislava, Vol. 1, 1973, p. 81—94

[41] WABL, M. R.: Electron microscopic localization of two proteins on the surface of the 50 S ribosomal subunit of *Escherichia coli* using specific antibody markers. J. Mol. Biol. **84**, 241—247 (1974)

[42] LAKE, J. A., D. D. SABATINI, and Y. NONOMURA: Ribosome structure as studied by electron microscopy. In: "Ribosomes" (Eds. M. NOMURA, A. TISSIÉRES, and P. LENGYEL). Cold Spring Harbor Laboratory, New York 1974, p. 543—557

[43] STÖFFLER, G.: Structure and function of the *Escherichia coli* ribosome: Immunochemical analysis. In: "Ribosome" (Eds. M. NOMURA, A. TISSIÉRES, and P. LENGYEL). Cold Spring Harbor Laboratory, New York **1974**, p. **615**—**667**

[44] KISELEV, N. A., V. YA .STEL'MASHCHUK, M. I. LERMAN, and O. YU. ABAKUMOVA: On the structure of liver ribosomes. J. Mol. Biol. **86**, 577—586 (1974)

[45] KISELEV, N. A., V. YA. STEL'MASHCHUK, V. L. TSUPRUN, M. I. LERMAN, and O. YU. ABAKUMOVA: The structure of liver ribosomes. Acta Biol. Med. Germ. **33**, 795 to 807 (1974)

[46] VASILIEV, V. D.: Morphology of the ribosomal 30 S subparticle according to electron microscopic data. Acta Biol. Med. Germ. **33**, 779—793 (1974)

[47] LUTSCH, G., F. NOLL, and H. BIELKA: Electron microscopic studies on the structure and dissociation of rat liver ribosomes. Acta Biol. Med. Germ. **33**, 813—816 (1974)

[48] MEYER, M., W. S. BONT, M. DE VRIES, and N. NANNINGA: Electron microscopic and sedimentation studies on rat-liver ribosomal subunits. Europ. J. Biochem. **42**, 259—268 (1974)

[49] CURGY, J.-J., G. LEDOIGT, B. J. STEVENS, and J. ANDRÉ: Mitochondrial and cytoplasmic ribosomes from *Tetrahymena pyriformis*. Correlative analysis by gel electrophoresis and electron microscopy. J. Cell Biol. **60**, 628—640 (1974)

[50] KLEINOW, W., W. NEUPERT, and F. MILLER: Electron microscope study of mitochondrial 60 S and cytoplasmic 80 S ribosomes from *Locusta migratoria*. J. Cell Biol. **62**, 860—875 (1974)

[51] TISCHENDORF, G. W., H. ZEICHARDT, and G. STÖFFLER: Architecture of the *E. coli* ribosome as determined by immune electron microscopy. Proc. Natl. Acad. Sci. U.S. **72**, 4820—4824 (1975)

[52] BOUBLIK, M., W. HELLMANN, and H. E. ROTH: Localization of ribosomal proteins L 7/L 12 in the 50 S subunit of *Escherichia coli* ribosomes by electron microscopy. J. Mol. Biol. **107**, 479—490 (1976)

[53] LAKE, J. A.: Ribosome structure determined by electron microscopy of *Escherichia coli* small subunits, large subunits and monomeric ribosomes. J. Mol. Biol. **105**, **131**—**159** (1976)

[54] COX, R. A.: Structure and function of prokaryotic and eukaryotic ribosomes. Prog. Biophys. Mol. Biol. **32**, 103—231 (1977)

[55] LUTSCH, G., F. NOLL, H. THEISE, and H. BIELKA: Localization of ribosomal protein S 2 in rat liver ribosomes by immune electron microscopy. Acta Biol. Med. Germ. **36**, 287—290 (1977)

[56] SHIMIZU, K., J. HOSOI, and I. MIYAZAWA: The identification of rear profile of monomer ribosomes from *Geotrichum candidum* with electron microscopy. J. Electron Microsc. **26**, 153—154 (1977)

[57] SHIMIZU, K., J. HOSOI, K. OHARA, and Y. MIYAZAWA: Electron microscopic study of negatively stained ribosomal subunits from *Geotrichum candidum*. J. Electron Microsc. **27**, 111—117 (1978)

[58] BOUBLIK, M., W. HELLMANN, and A. K. KLEINSCHMIDT: Size and structure of *E. coli* ribosomes by electron microscopy. Cytobiol. **14**, 293—300 (1977)

[59] LUTSCH, G., K. P. PLEISSNER, G. WANGERMANN, and F. NOLL: Studies on structure of animal ribosomes. 8. Application of a digital image processing method to enhancement of electron micrographs of small ribosomal subunits. Acta Biol. Med. Germ. **36**, K 59—K 62 (1977)

[60] BOUBLIK, M., and W. HELLMANN: Comparison of *Artemia selina* and *E. coli* ribosome structure by electron microscopy.

Proc. Natl. Acad. Sci. U.S. **75**, 2829–2833 (1978)

[61] Kiselev, N. A., V. Ya. Stel'mashchuk, G. Lutsch, and F. Noll: Structure of small subparticles of liver ribosomes. J. Mol. Biol. **126**, 109–115 (1978)

[62] Strycharz, W. A., M. Nomura, and J. A. Lake: Ribosomal proteins L 7/L 12 localized at a single region of the large subunit by immune electron microscopy. J. Mol. Biol. **126**, 123–140 (1978)

[63] Emanuilov, I., D. D. Sabatini, J. A. Lake, and Ch. Freienstein: Localization of eukaryotic initiation factor 3 on native small ribosomal subunits. Proc. Natl. Acad. Sci. U.S. **75**, 1389–1393 (1978)

[64] Lutsch, G., F. Noll, H. Theise, G. Enzmann, and H. Bielka: Localization of proteins S 1, S 2, S 16 and S 23 on the surface of small subunits of rat liver ribosomes by immune electron microscope. Molec. Gen. Genet. **176**, 281–291 (1979)

[65] Engelman, D. M., and P. B. Moore: Neutron-scattering studies of the ribosome. Sci. American **235**, 44–54 (1976)

[66] Damaschun, G., J. J. Müller, H. Bielka, and M. Böttger: Studies on the structure of animal ribosomes. V. Studies on the size and structure of rat liver ribosomes by small angle X-ray scattering. Acta Biol. Med. Germ. **33**, 817–823 (1974)

[67] Damaschun, G., J. J. Müller, and H. Bielka: Untersuchungen zur Struktur tierischer Ribosomen. VI. Ein dynamisches Modell der Ribosomenstruktur. Acta Biol. med. Germ. **34**, 229–239 (1975)

[68] Müller, J. J., G. Damaschun, and H. Bielka: Studies on the structure of rat-liver polysomes by small-angle X-ray scattering. Eur. J. Biochem. **90**, 547–553 (1978)

VI. Chemical components

Contents

The eukaryotic 80 S ribosome is composed of at least 75 different proteins and 4 different molecules of RNA. The isolation and detailed analysis of the properties of these compounds is a prerequisite for the understanding of the structural and functional organization of the ribosome.

This chapter deals with the present knowledge of the properties of the protein and RNA moieties of eukaryotic ribosomes[1]).

1. Ribosomal proteins

a) *Electrophoretic separation and number*

Electrophoretic techniques play an important role in the investigation of ribosomal proteins. For years one-dimensional polyacrylamide gel electrophoresis has been a useful tool for the analysis of the protein mixtures extracted from eukaryotic ribosomes [11—37]. By this method the total protein fraction of rat liver 80 S ribosomes was separated into about 24 protein bands [12, 13]. The proteins of the 60 S subunit revealed qualitatively comparable patterns of 24 bands, whereas the protein mixture of the 40 S subunit was separated into about 18 bands [13].

Much more information has been obtained by two-dimensional electrophoresis of ribosomal proteins as demonstrated at first for the *E. coli* ribosome [38, 39]. Later on several modifications of this method have been developed with respect to buffer systems, the presence of urea, SDS or other detergents, the size and construction of the equipment and other parameters [40—53]. This technique is now an indispensable and most effective method for the separation and characterization of ribosomal proteins.

---

[1]) For further reviews on eukaryotic ribosomal proteins and ribonucleic acids see [1—7] and [4, 5, 8—10] respectively.

Table 3. Number of proteins in ribosomal particles

Source of ribosomes	80 S monosomes	60 S subunits	40 S subunits	References
Rat liver	76	34	27	[56]
		39	31	[55]
		40	30	[78]
	69—72	39	30	[57, 64]
			28	[31]
		40	33	[66]
		36	26	[67]
		46	42	[80]
	77	43	34	[70]
Rat muscle	69	38	31	[64]
Mouse liver	70	39	31	[76]
Rabbit liver	75			[54]
		37	33	[74]
Rabbit reticulocytes	62—63	36—37	26	[42]
		40	33	[58]
		46	30	[61]
		39	32	[65]
	70	40	30	[84]
	75	42	33	[72]
Human placenta	64—66	36	28	[59]
Human tonsils		35	34	[63]
HeLa cells		37—38	29	[79]
		40		[88]
		47	35	[77]
Krebs II ascites cells		35—37	28—29	[79]
L cells		45	34	[81]
Chinese hamster cells	72	41	31	[89]
Xenopus ovary		37	26—28	[60]
Artemia salina			32	[326]
Pea seeds		28	24	[35]
	83	44—55	32—40	[94]
		47	32	[90]
Wheat germ			30	[326]
Dictyostelium discoideum		38	32	[32]
Podospora anserina		41	28	[91]
Saccharomyces carlsbergensis		39	30	[95]
		41	30	[96, 102]
Saccharomyces cerevisiae	67	37	30	[97]
		42	34	[98]
		40—41	28—30	[99]
	74	45	28	[101]

Continued table 3

Source of ribosomes	80 S monosomes	60 S subunits	40 S	References
Chlamydomonas		39	26	[104]
		44	31	[105]
	56			[106]
Polytoma mirum	53			[106]
Euglena	73—83	37—42	33—36	[108]
Neurospora		31	21	[111]
Mitochondria				
Neurospora		39	30	[111]
Xenopus	77—84	40	44	[110]
Rat liver		36	30	[113]
		40	30	[111]
Yeast		38	33	[112]
Chloroplasts				
Chlamydomonas		26	22	[104]
		30	20	[107]
		34	25	[105]
Euglena	56—60	30—34	22—24	[108]

Ribosomal proteins from various eukaryotic species have been separated by 2-D electrophoresis [54—113]. The main purpose of these experiments was to determine their number in the mixture of total ribosomal proteins. Table 3 summarizes the number of ribosomal proteins of cytoplasmic 80 S ribosomes and their subunits of different species as well as of ribosomes of mitochondria and plastides.

Most of the investigations have been performed with proteins of mammalian ribosomes, especially of rat liver [31, 55 to 57, 66—70, 76, 78, 80, 81] and rabbit reticulocytes [42, 58, 61, 65, 72].

Proteins of mammalian ribosomes can be separated by 2-D electrophoresis into complex protein patterns consisting of about 70 different components. The protein mixtures isolated from the separated 40 S and 60 S subunits were resolved into about 30 and 40 components, respectively [42, 53—89].

As an example, the 2-D electrophoretic patterns of the basic proteins of the small and the large subunit of rat liver ribosomes are given in Figs. 11 and 12 ([78]; nomenclature according to [82]). The proteins of the 40 S and 60 S subunit were separated into 30 and 40 components, respectively, moving cationically at pH 8.3. These results are in good agreement with the 2-D patterns of rat liver ribosomal proteins published by other authors [57, 64, 67, 80, 82]. Cytoplasmic ribosomes from other eukaryotic species have a similar number of proteins as mammalian ribosomes [90 to 103].

A smaller number of about 50 proteins was found in ribosomes of chloroplasts [104—108] which are comparable in this respect to prokaryotic ribosomes [109], whereas in mitochondrial ribosomes, as in cytoplasmic ribosomes, about 70 proteins have been analyzed [110—113].

Besides the basic proteins, which represent the majority, some acidic proteins were found in eukaryotic ribosomes [55, 57—59, 61—64, 74, 80, 93—97, 114—115, 117, 127] (especially the acidic phospho-

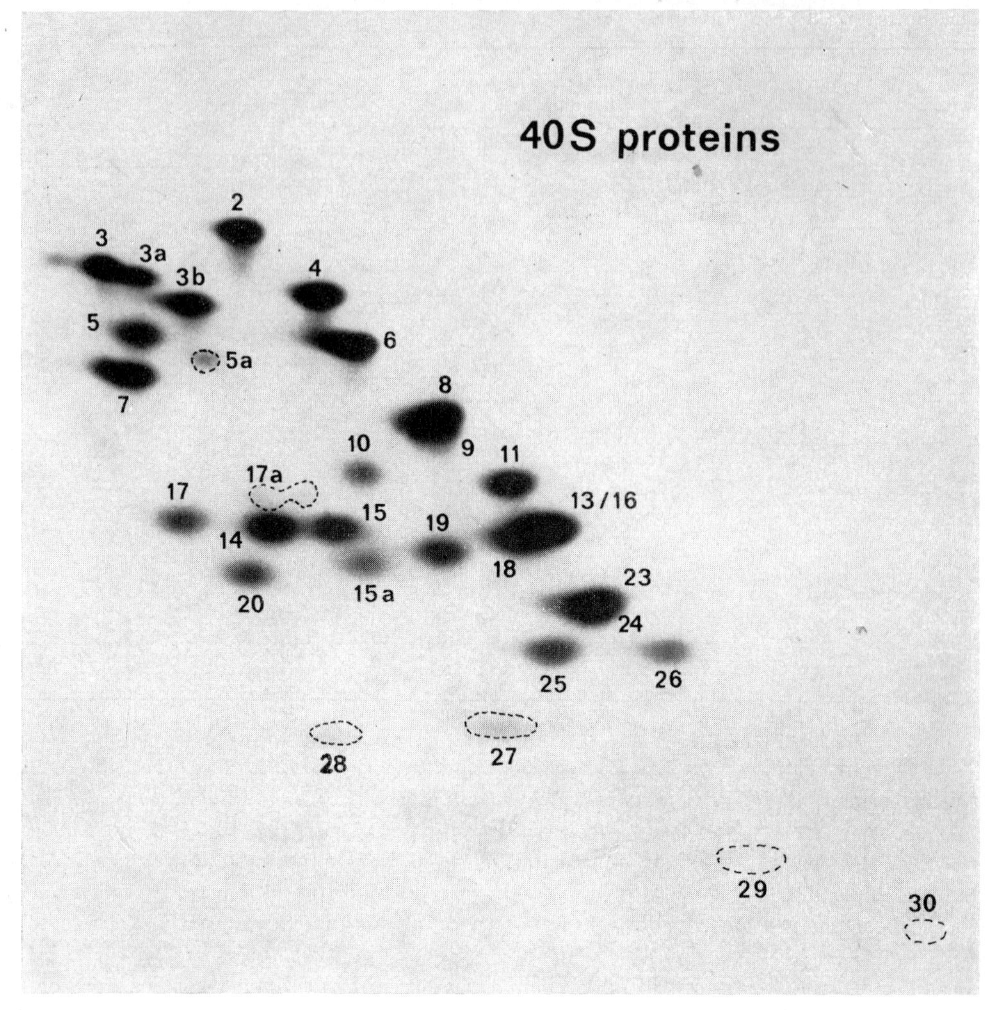

Fig. 11. Two-dimensional electrophoretic pattern of proteins of the small subunit of rat liver ribosomes. From H. WELFLE et al. [78].

First dimension from left to right; second dimension from top to bottom. The spots are designated according to the proposed uniform nomenclature of mammalian ribosomal proteins [82].

proteins of the large ribosomal subunit have recently been investigated thoroughly in many laboratories (see Section 2d of this chapter). Ribosomes from yeast, e.g. are characterized by a rather high number of 7 to 10 acidic proteins [95—102].

In the past, various nomenclatures for mammalian ribosomal proteins have been defined by 2-D electrophoretic patterns obtained in different laboratories [53, 55, 57—59, 61, 63—65, 67, 79, 83, 84]. On the basis of accumulated knowledge a uniform nomenclature was proposed in 1979 [82]. Most of the numbered components were identified according to their positions in the 2-D electrophoretic patterns obtained with the KALTSCHMIDT-WITTMANN system [39], that is electrophoresis in the presence of urea at pH 8.6 in the first, and at pH 4.5 in the second dimension. Unfortunately, it is not yet possible to correlate the new nomenclature with other

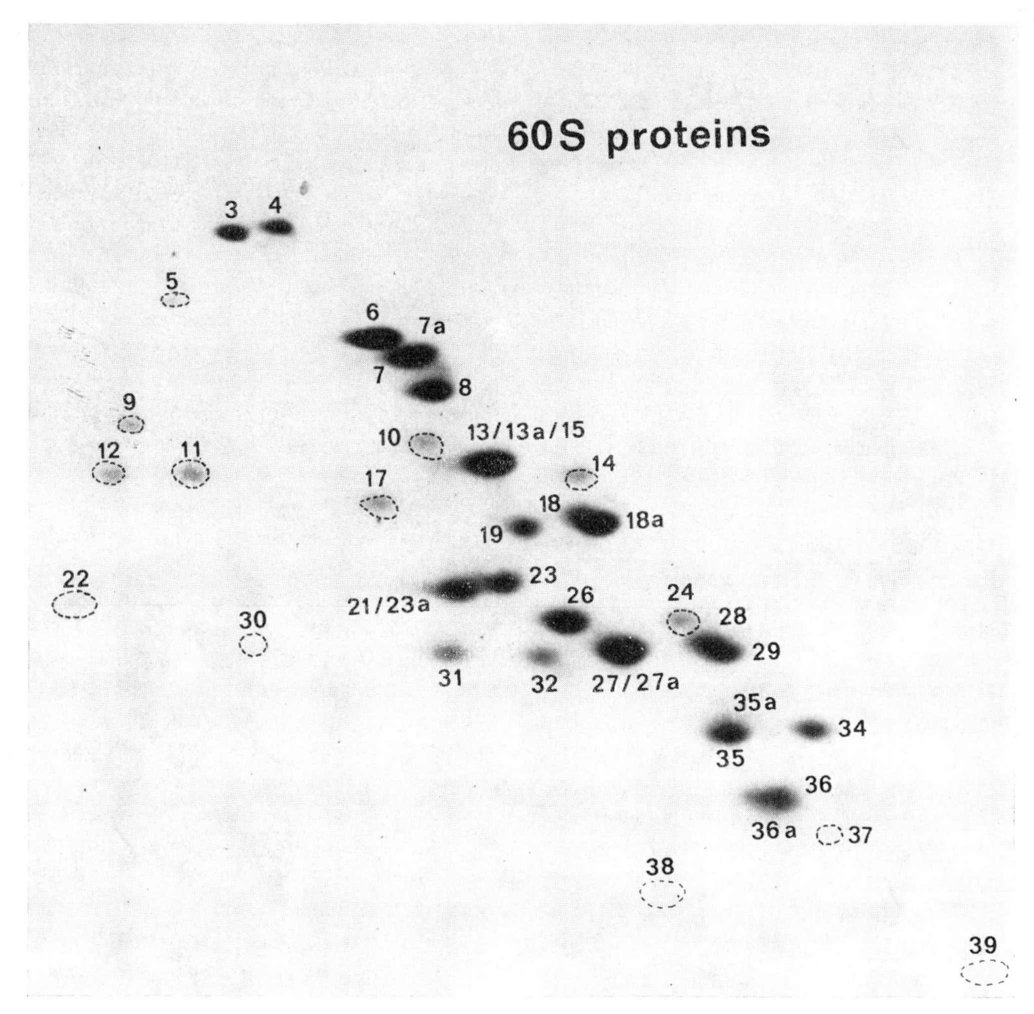

Fig. 12. Two-dimensional electrophoretic pattern of proteins of the large subunit of rat liver ribosomes. From H. WELFLE et al. [78].
For further explanation see legend to Fig. 11.

nomenclatures, which were defined on the basis of electrophoretic systems using low pH in the first, and SDS in the second dimension [83, 84, 88, 89].

In the new nomenclature [82], which will be used throughout in this chapter, 33 and 45 proteins of the small and large ribosomal subunit, respectively, are enumerated. However, despite the high separation power of the 2-D electrophoresis, not all ribosomal proteins can be identified as single spots. Some of the numbered components have very similar electrophoretic mobilities and cannot be separated in the KALTSCHMIDT-WITTMANN system; these components, therefore, overlap in the 2-D patterns and where defined by additional criteria obtained by the combination of different methods. Thus, for instance, proteins L 18a, L 23a, L 27a, and L 36a were identified by chromatographic [116, 117] and electrophoretic procedures [78]. These proteins were found by comparison of the 2-D patterns obtained with different electrophoretic techniques using the KALTSCHMIDT-WITTMANN system, on the one hand, and the pH 8.3/SDS system, on the other [78]. Finally, S 27a of the small subunit [118] as well as L 13 a and L 37a of the large one [116, 117] are included in the new nomenclature, which were found by chromatographic procedures only.

Despite the efforts and progress in the separation and analysis of ribosomal proteins it has to be considered that some of the numbered proteins may be artifacts and that, on the other hand, new proteins will be identified as real ribosomal proteins.

Nevertheless, for most of the numbered proteins in the proposed uniform nomenclature, their character as genuine ribosomal proteins is well established and this nomenclature will be helpful for the future to compare the results obtained in different laboratories.

A general and complicated problem is the lack of unambiguous criteria to establish the real ribosomal nature of a protein in question. Up to now empirical definitions basing on the experimental procedures used for the isolation of ribosomes and ribosomal proteins and their characterization have been applied. Recently, a discrimination between structural ribosomal proteins, exchangeable ribosomal proteins and nonribosomal proteins by selective labeling of polysomal proteins of L cells in the presence of low doses of actinomycin D was suggested [81]. Structural ribosomal proteins were not labeled and form stained nonradioactive spots in the 2-D pattern; exchangeable ribosomal proteins were labeled and form stained radioactive spots; nonribosomal proteins were detectable by autoradiography only.

A final estimation of the number of ribosomal proteins is rather difficult; the results are seriously affected by the experimental conditions. Differences in the estimates of the number of ribosomal proteins obtained by various authors may be attributable to minor differences in the preparation of ribosomes and ribosomal subunits, in the method of extraction of the proteins and in the electrophoretic procedures. Ribosomal proteins may be lost at different steps of the preparation while, on the other hand, additional proteins spots may be found due to the presence of nonribosomal proteins or the formation of different derivation products of ribosomal proteins.

Proteolytic degradations [83, 142], carbamylation by ammonium cyanate [55, 57] or oxidation of SH groups [46, 78] have to be considered as possible artificial modifications of ribosomal proteins during the isolation and characterization procedures. For example, two protein components of the small and three components of the large ribosomal subunit have been found originally in the 2-D patterns of rat liver ribosomal proteins and were enumerated as S 5, S 6 and L 15 [55] as well as L 17 and L 32 [55, 57, 65, 67]. These proteins have been identified later on as derivatives of ribosomal proteins produced by oxidation during the experiments [78]. Nevertheless, proteins L 17 and L 32 were included as L 16 and L 32, respectively, in the new uniform nomenclature [82] and were prepared recently [116, 143] and considered as real ribosomal proteins.

The protein composition of ribosomes can differ depending on the buffers and salt concentrations used for the isolation and purification of ribosomes. Unwashed rat liver ribosomes, e.g., contain additional acidic proteins in comparison to washed ribosomes [13, 115]. Treatment of eukaryotic ribosomes with 0.5 M KCl at 2 to 5 mM $MgCl_2$ is considered as an appropriate procedure to remove nonribosomal proteins associated with the particles [119, 120]. On the other hand, removal of loosely bound structural ribosomal proteins during purification of ribosomes may be another reason of differences in the electrophoretic patterns and of misinterpretations [121]. Furthermore, differences in the phosphorylation of ribosomal proteins depending on the physiological conditions of a given tissue can influence their electrophoretic 2-D patterns [66, 144].

The 2-D patterns of the protein mixtures extracted from the separated subunits depend on the method used for the dissociation of ribosomes [62, 68] (see Chapter VIII). Dissociation of rat liver ribosomes with EDTA removes a 7 S complex consisting of 5 S RNA and protein L 5 from the large subunit [68]. Therefore, protein L 5 is absent in the 2-D pattern of large subunits obtained by treatment with EDTA. Usually most of the experiments were performed with protein mixtures from subunits prepared by the high KCl-puromycin method [122, 123] which are active in in vitro-protein synthesis.

As to their electrophoretic mobility, proteins of identical properties have not been found in the small and large subunit of rat liver ribosomes [55]. Most of the small and large subunit proteins can be clearly distinguished. Proteins S 8 and L 13 have very similar mobilities and were therefore thought to be the same [57, 64]; recently published amino acid compositions of proteins S 8 and L 13, however, demonstrate that both proteins are unique [124, 125].

Variable partition of certain proteins of rat liver and muscle between the two ribosomal subunits has been found and explained by the different methods used to isolate ribosomes from these tissues [64]. Another example is the variable location of probably identical phosphoproteins found on the 60 S subunit of ribosomes isolated from Krebs II ascites cells and on the 40 S subunit of ribosomes isolated from BHK-21/C 13 cells [126]. These proteins were found also in the 60 S subunit isolated from BHK cells as a result of modifications of the method used for the isolation of the ribosomes [127].

For the preparation of ribosomal proteins a number of methods is available, namely extraction with acetic acid [128], acetic acid in the presence of magnesium ions [129, 130], hydrochloric acid [36], sulfuric acid [286], 2-chlorethanol and hydrochloric acid [131], LiCl [132], LiCl and urea [133], guanidine-HCl [134, 135], or treatment of the particles with sodium dodecyl sulphate [136], ribonuclease [137] and trichloro acetic acid [138]. The protein solutions are usually lyophilized or the proteins are precipitated with acetone [36, 139]. Depending on the special problem, one or the other method can be used suitably, but the most efficient procedure seems to be the commonly applied extraction of ribosomal proteins with acetic acid in the presence of 33 mM Mg^{++} [140, 141].

b) Preparation of single ribosomal proteins

First attempts to separate proteins of eukaryotic ribosomes on a preparative scale and to analyze some of their properties were described in 1967 and 1969, respectively, using one-dimensional polyacrylamide gel electrophoresis [19, 145] or carboxymethyl cellulose chromatography [20, 21, 146]. The isolation of pure single proteins of eukaryotic ribosomes is rather difficult due to the high complexity and the degree of similarity in their chemical properties. Prefractionation of the total protein mixture and combination of different methods were found to be essential for the subsequent effective isolation of single proteins.

The first effective step for the prefractionation of ribosomal proteins is the dissociation of the ribosomes into their subunits dividing the proteins into two different groups [56, 57]. Splitting of ribosomes bound to DEAE-Sephadex by LiCl [147] and differential extraction of proteins from ribosomes using 67% acetic acid and increasing concentrations of Mg^{++} [148] were described as other useful prefractionation steps.

Carboxymethyl cellulose chromatography and gel filtration enabled the isolation of 12 relatively pure proteins of the small ribosomal subunit from rat liver

[31] and the estimation of their molecular weights. In the same way other authors [149] have prepared 24 proteins of the small subunit of rat liver ribosomes. The identity and purity of the isolated proteins were checked by one- and two-dimensional gel electrophoresis. The amino acid composition and molecular weights have been estimated and the tryptic patterns analyzed [149, 150], demonstrating that the isolated proteins are unique. Relatively large amounts of 21 proteins of the small subunit of rat liver ribosomes have been prepared by combinations of ion exchange chromatography, gel filtration, preparative polyacrylamide gel electrophoresis in sodium dodecyl sulphate and fractional extraction with perchloric acid [151]. The phosphoprotein S 6 was isolated from the protein mixture of in vitro phosphorylated 40 S subunits of rabbit reticulocyte ribosomes by fractionation on phosphocellulose in urea and gel filtration on Sephadex G-100 [154].

Progress in the isolation of single proteins from both subunits of rat liver ribosomes was achieved more recently by using a relatively simple preparative 2-D electrophoretic procedure in polyacrylamide gels. By this technique 24 proteins of the small, and 24 proteins of the large subunit of rat liver ribosomes were isolated, and their amino acid compositions were determined [152].

On the basis of an improved group fractionation [153] large scale preparation of the proteins of the small [118, 125, 143] of the large subunit [116, 117, 124, 143] of rat liver ribosomes has been developed. 33 proteins of the 40 S and 49 proteins of the 60 S ribosomal subunit have been isolated, and their amino acid compositions and molecular weights were estimated. However, it should be taken into account that the isolation of a pure component from the mixture of ribosomal proteins does not yet establish its status as a genuine ribosomal protein.

c) *Molecular weights*

The molecular weights of ribosomal proteins from a variety of organisms have been estimated. In Table 4 the distribution of the molecular weights and the average values of small and large ribosomal subunit proteins are presented. Most of the published data were obtained by electrophoresis of the mixtures of ribosomal proteins in polyacrylamide gels in the presence of SDS [25, 26, 28, 32, 35, 59, 80, 81, 85, 88, 89, 96, 102, 104—106, 155, 156]. Frequently also SDS electrophoresis of single proteins isolated by chromatographic procedures [31, 116, 118, 124, 125, 149] or extracted from polyacrylamide gels after 2-D electrophoretic separation [33, 65, 67, 74, 79, 83, 102, 157] has been used for the determination of their molecular weights. Molecular weights of isolated proteins have been estimated also by gel filtration [146] and by analytical ultracentrifugation [146, 149, 159].

Table 5 gives the molecular weights of single proteins of rat liver ribosomes as estimated in four laboratories [67, 78, 80, 116—118, 124, 125].

The estimation of molecular weights by electrophoresis in the presence of SDS is a convenient and simple technique but the values are not very precise owing to principal limitations of this method. Furthermore, modifications of the electrophoretic conditions and technical problems may explain the fact that the molecular weights of the same rat liver ribosomal proteins estimated in different laboratories deviate up to 20% or even more (cf. Table 5).

The total mass of basic proteins of the 40 S subunit of rat liver ribosomes is approximately $0.7 \cdot 10^6$ and that of the 60 S subunit is close to $1.05 \cdot 10^6$ [78]. These values are afflicted with a rather high error resulting from experimental errors of the molecular weight estimation and the uncertainty of the exact final number of ribosomal proteins as discussed in Section 1a of this chapter.

Table 4. Molecular weights of proteins of eukaryotic cytoplasmic ribosomes

Source of ribosomes	Distribution of molecular weights		Average molecular weights		Method	References
	60 S subunit proteins	40 S subunit proteins	60 S subunit proteins	40 S subunit proteins		
Rat liver[d]	12000–27700	10400–27400			a	[146]
	9000–27600	12100–31200			b	[146]
		9200–67000			c	[31]
		10000–30500			c	[149]
		9700–30600			b	[149]
		11300–33100			c	[125]
		11200–41500				[118, 125]
	11600–41800				c	[116, 124]
Rat liver[e]	10000–53700	10100–44000	28000	25400	c	[157]
	10000–60000	10000–38000	23900	23000	c	[67]
	13000–75000	11700–42000			c	[83]
	10000–28000				b	[145]
Rat liver[f]	9500–55500				c	[85]
	7000–69100	8400–70600			c	[156]
	10000–57000	10000–40000			c	[80]
	11000–55500	8000–35500	26500	22500	c	[78]
Mouse liver[f]			27000	29000	c	[26]
	9600–56000	7700–39200			c	[85]
Rabbit liver[f]	9700–55700	7800–39800			c	[85]
Rabbit liver[e]	9400–52000	9000–35000	23600	19600	c	[74]
Chicken liver[f]	9600–54000	7600–40000			c	[85]
Toad liver[f]		7800–43000			c	[85]
Rabbit reticulocytes[f]	9000–63000	16000–58500			c	[24]
	8000–57000	6000–39000			c	[28]
	10100–53800	7900–38800			c	[85]
Rabbit reticulocytes[e]	8900–58500	8500–39300	31000	25000	c	[65]
HeLa cells[f]	13000–53000				c	[88]
HeLa cells[e]	14100–43500		22900	20900	c	[79]
Human placenta[f]	13000–53000				c	[59]
Ascites cells[e]	13700–44000	10000–38000	23000	20600	c	[79]
L cells[f]	12300–51200	13000–47000			c	[81]
Chinese hamster cells[f]	10200–58200	9700–46600			c	[89]
Yeast[f]			18800	17500	c	[96]
	10500–60000	10500–40000			c	[101]
Yeast[e]	10000–48400	11800–31000	21800	21300	c	[102]
Dictyostelium[f]	13000–73000	16900–84000			c	[32]
Dictyostelium[e]	12400–48500	9000–42700	22300	20400	c	[33]
Euglena[f]			26000	32000	c	[26]
Chlamydomonas[f]	13000–54000	12500–32000	23000	21000	c	[104]
	10000–35000	10000–50000			c	[105]
Pisum sativum[f]	10500–77500	10000–61500			c	[158]

[a]Gel filtration; [b]Sedimentation equilibrium; [c]SDS electrophoresis; [d]Single proteins isolated by chromatographic procedures; [e]Single proteins isolated by electrophoretic procedures; [f]Total protein mixture

Table 5. Molecular weights of the basic proteins from 40 S and 60 S rat liver ribosomal subunits

40 S subunit proteins	From references				60 S subunit proteins	From references			
	[67]	[78]	[80]	[118, 125, 143]		[67]	[78]	[80]	[116, 117, 124, 143]
S 2	38000	35000	33000	33100	L 3	54000	51000	57000	37800
3	35000	35500	30000	30400	4	60000	55500	57000	41800
3a	35000	35500	31500	32000	5	39000	40000	36000	32500
3b	35000	32000	29000	30400	6	40000	45500	41000	33000
4	32000	31000	29500	29500	7	40000	35500	32000	29200
5	26000	27000	22000	22800	7a	32000	35500	31500	28700
5a	—	—	21000	21500	8	30000	34000	31000	28400
6	37000	34000	31500	31000	9	24000	27500	25000	24700
7	23000	26000	24000	22200	10	25000	30500	27000	24200
8	28000	28500	26000	26800	11	22000	24000	21000	21300
9	25000	24000	23500	24300	12	19000	24500	18500	18700
10	24000	23000	19000	20100	13	27000	29000	27000	26300
11	21000	25000	21000	20700	13a	29000	26000	23500	24600
13	21000	20000	15500	18600	14	28000	30500	27000	25800
14	20000	21000	17000	17300	15	29000	26000	25000	24500
15	22000	20500	19500	19600	16	20000	—	—	18700
15a	17000	16000	12500	15700	17	23000	28000	26000	22100
16	18000	19000	17000	17100	18	26000	26000	24000	24500
17	18000	21000	17500	18000	18a	23000	26000	23500	21300
17a	—	—	19000	—	19	28000	30500	26000	25300
18	23000	19000	18000	18500	20	—	—	—	16200
19	20000	18500	14000	17100	21	22000	28500	21000	20300
20	17000	18500	13000	16500	22	15000	19000	13000	16100
21	10000	12000	<10000	12300	23	22000	24000	17500	15600
23	21000	22000	18500	18800	23a	22000	21000	21000	18000
24	21000	22000	18500	18800	24	19000	25500	23000	—
25	21000	21500	16000	17000	25	14000	18500	13500	17500
26	21000	18000	13500	16500	26	19000	25500	19500	18600
27	14000	12500	<10000	14500	27	16000	21000	18000	17800
27a	—	—	13000	12800	27a	16000	23500	19500	18000
28	11000	10000	<10000	11300	28	17000	23500	18500	17800
29	—	8000	<10000	11200	29	21000	31500	27000	20500
30	—	12000	<10000	—	30	14000	18500	12500	14500
					31	15000	19000	13500	15600
					32	20000	22000	17500	17200
					33	17000	—	—	15600
					34	17000	19500	14000	15800
					35	19000	20000	17000	17500
					35a	13000	16500	13000	13700
					36	16000	16500	13000	14300
					36a	16000	22000	17500	16200
					37	16000	19500	14000	15400
					37a	—	14500	<10000	12800
					38	10000	11000	<10000	11500
					39	—	11000	<10000	11500

From hydrodynamical measurements have been determined not only the molecular weights but also the shape of individual proteins of the small subunit of rat liver ribosomes in solution [159]. Most of the ribosomal proteins studied are elongated with maximal lengths of 7.7 to 19.4 nm (see Table 20, Chapter IX).

d) Amino acid composition and amino acid sequences

The amino acid composition of the total ribosomal protein mixture is characterized by a high content of charged amino acids [19, 24, 108, 158]. In total rat liver ribosomal protein the sum of the basic amino acids Lys, Arg and His ranges from 21.5 to 22.4 mol% [19, 24], while the content of Asp and Glu amounts to 17.2 mol% [19, 24]. Close to 5.5 mol% of the acidic amino acids were described to be amidated [24]. Similar values were found also for other eukaryotic ribosomal proteins. Proteins of cytoplasmic ribosomes of *Euglena gracilis* contain 21.3 mol% and 18.8 mol% basic and acidic amino acids, respectively; the corresponding values for chloroplast ribosomal proteins are 20.6 and 20.9 mol% [108].

First attempts to estimate the amino acid composition of single rat liver ribosomal proteins were performed with electrophoretically [19] and chromatographically [20, 146] isolated fractions. Now results of various estimations of the amino acid composition of most proteins of the small [118, 125, 143, 149, 152] and the large subunit [116, 117, 124, 143, 152] are available. The sum of the basic amino acids varies from 15 to 18 mol% for the proteins of the small subunit of rat liver ribosomes and from 16 to 30 mol% for those of the large subunit; the values for the acidic amino acids are 15 to 24 mol% and 14 to 22 mol%, respectively [152].

The hitherto published values of the amino acid compositions obtained in different laboratories are, for a number of single proteins, not yet in very good agreement and, therefore, further studies are necessary to obtain more precise data.

In general, the amino acid compositions of single proteins are fairly similar, but most of them differ significantly from each other by a few amino acids. Therefore, the data support the conclusion that proteins which are electrophoretically different in urea- and/or SDS containing systems are unique.

Sequence analysis studies have been performed in a few cases only. Partial sequences [162, 163] and, recently, the complete sequence of the acidic ribosomal phosphoproteins from Artemia salina [164], the amino terminal sequence of the acidic "A" protein of wheat ribosomes and the complete sequence of protein 44 from the large subunit of yeast ribosomes [165] have been estimated. Amino terminal sequences of three proteins of the small and nine proteins of the large subunit of rat liver ribosomes have been published [7].

e) Stoichiometry

An approximative determination of the stoichiometry of proteins in ribosomal particles can be made from data about their total mass in the subunits, the number of unique proteins, and their molecular weights. For rat liver small and large ribosomal subunits, total masses of the proteins of about $0.7 \cdot 10^6$ and $1.3 \cdot 10^6$ Dalton, respectively, were estimated by physical and chemical methods. From the number of proteins and their average molecular weight, values of $0.7 \cdot 10^6$ and $1.05 \cdot 10^6$ for the basic proteins of the small and large subunit, respectively, were calculated [78]. Taking into account the presence of some acidic proteins in both subunits, such calculations are in favor of a 1:1 stoichiometry for most of the proteins in both ribosomal subunits of rat liver.

More direct estimations of the copy number of individual proteins were tried by quantitation of the relative masses of single proteins with different methods after electrophoretic separation [33, 65, 86, 102, 166, 167, 168]. In these experiments mostly the amount of dye bound to individual protein fractions was determined and taken as being proportional to the protein mass.

In another approach, ribosomal proteins of hepatoma ascites cells were labeled in vivo with [^{3}H]-lysine. The relative content of ^{3}H-Lys residues in each ribosomal protein was estimated after 2-D electrophoretic separation and compared with the corresponding calculated value, obtained from the amino acid composition and the molecular weight. The results support the one-copy hypothesis for most of the proteins of the small ribosomal subunit of rat liver [167].

On the other hand also repeated and fractional proteins have been found in the small [167] and the large [86] subunit of rat liber ribosomes and in reticulocyte ribosomes [65, 166]. A rather large number of fractional and repeated proteins were described also for ribosomes of *Dictyostelium discoideum* [33] and yeast [102]. Using a radioimmunoassay, 1.8 copies of the acidic phosphoprotein eL 12 were found in the large subunit of *Artemia salina* ribosomes [169].

A number of technical problems is connected with the determination of the stoichiometry of eukaryotic ribosomal proteins, and the estimation of the exact number of copies of each protein in the ribosomal particles needs further experiments. All methods used until now yield only the number of copies in isolated ribosomes which is influenced more or less by the preparation methods of ribosomes and their proteins and may not really reflect the native protein composition of the ribosomes in the cell.

f) *Posttranslational modifications*

Like many other proteins, also ribosomal proteins are subject to various posttranslational modifications.

One of the most interesting events is phosphorylation which therefore was studied in many details (see Section 2 of this chapter).

Other modifications are methylation and acetylation.

Methylation

Methylated amino acids have been detected in ribosomal proteins from *Euglena gracilis* [170], cultured muscle cells [171], HeLa cells [172—175], yeast [165, 175 to 178], and *Blastocladiella emersonii* [179].

Methylation of proteins of the small [172, 177, 178] as well as of the large ribosomal subunit [165, 172, 176, 178] has been detected. ε-N-mono-methyllysine is the main methylated amino acid in ribosomal proteins of *Blastocladiella emersonii* [179]. In HeLa cells N^6,N^6-Dimethylarginine was found to be the major methylated amino acid. Both subunits of HeLa cell ribosomes contain furthermore ε-N-trimethyllysine and in lower amounts also ε-N-dimethyllysine. The 60 S ribosomal subunit contains significantly more ε-N-trimethyllysine than the 40 S ribosomal subunit [172]. Also in yeast the level of methylation is higher in the 60 S ribosomal subunit [177] than in the small one [176].

Two proteins of the small (called S 31 and S 32) and two proteins of the large ribosomal subunit (called L 15 and L 41) were identified to be methylated in yeast [178]. These proteins became associated to the ribosomal precursor particles in a late stage during biogenesis of ribosomes. In HeLa cells at least seven ribosomal proteins are methylated. Methylation of the small subunit proteins varies during the cell cycle and reaches the highest level at the late G 1 phase while, on the other hand, methylation of the 60 S subunit proteins shows less variations during the cell cycle and exhibits the highest level at the early S 1 phage [175].

Acetylation

Rat liver ribosomal proteins were found to become acetylated in vivo within 5 min after administration of [^{3}H]acetate [180, 181, 182]; N-α-acetyllysine and N-ε-acetyllysine were identified as labeled amino acids [180]. Application of aldosterone to adrenalectomized rats reduces the amount of acetylation of ribosomal proteins in the renal cortex, whereas acetylation is increased in the renal medulla in a reversed correlation to protein biosynthesis [183].

Incubation of rabbit reticulocytes with [^{3}H]acetate [184] or of rat liver ribosomes with [^{14}C]acetyl-CoA [181] results in a rapid incorporation of labile acetyl groups into ribosomal proteins. One acetylated protein with a molecular weight of 29000 was identified in the 60 S subunit of reticulocyte ribosomes [185]. In rat liver ribosomes also one protein of the large subunit was found acetylated. This protein, charactericed by a molecular weight of 43000, can be extracted from the 60 S subunit with EDTA as a complex with 5 S RNA and consequently should correspond to L 5 [68] although final identification by 2-D electrophoresis is still lacking.

The functional importance of acetylation of ribosomal proteins is still unknown. Complete inhibition of acetylation by sodium fluoride and partial inhibition by cycloheximide suggests a correlation between the acetylation of ribosomal proteins and the formation of the initiation complex [184]. In this context it is worthwhile to note that the ribosomal 5 S RNA seems to be involved in the formation of the initiation complex in eukaryotes (see Chapter XI). Possibly acetylation of protein L 5, which is complexed with 5 S RNA, influences the properties of the binding region for initiator-tRNA in the eukaryotic ribosome.

g) Comparison of ribosomal proteins of different tissues

Earlier studies on ribosomal proteins from different tissues of the same species of animals have not shown any significant qualitative differences in their electrophoretic behavior in 1-D systems, as e.g. demonstrated for ribosomal proteins of liver and muscle of the rat [12, 24], liver, kidney, and testicles of the rat [14], liver, kidney, brain and heart of rat, rabbit, and chicken [15], brain and liver of mice [30] or liver and skin of chick embryos [11].

Also 2-D electrophoretic analysis have not elucidated significant tissue specificities of ribosomal proteins. Very similar 2-D pattern were obtained for ribosomal proteins from muscle and liver of the rat [64], rabbit reticulocytes and liver [85, 186], brain, kidney, liver and gut of the mouse [187], and kidney, cecal appendix and reticulocytes of the rabbit [54, 188]. Minor differences which were sometimes observed may be due to differences in the preparation of ribosomal particles and their proteins and to specific properties of the various tissues.

2-D electrophoretic patterns of ribosomal proteins have not revealed any significant qualitative differences between liver and hepatoma [189, 190] and other tumors [190]. A few differences have been described in the 2-D patterns of ribosomal proteins from a mouse neuroblastoma and rat forebrain [191] as well as between normal tissue and Ehrlich ascites tumor cells for which an additional acidic high molecular weight ribosomal protein has been described [187]. Similar protein patterns were obtained from polysomes of normal CV-1 green monkey kidney cells and SV 40-transformed cells [192]. Variations in the protein pattern of Ehrlich ascites tumor cell polyribosomes have been reported depending on the growth conditions with respect to three proteins present in higher amounts in polyribosomes from starved cells [193].

Changes in the protein composition of ribosomes depending on the stage of differentiation have not been found for rat liver [188] and for eggs, blastomers, prism stages and gastrulae of the sea urchin [11]. On the other hand stage specific ribosomal proteins have been described for the total population of ribosomes in *Drosophila melanogaster* [194—199]. This finding was not confirmed for *Drosophila virilis* [200]. The qualitative and quantitative changes in the protein composition of *Drosophila melanogaster* ribosomes were observed mainly during the third larval instar [198], which was not considered in the studies with *Drosophila virilis* [200] and thus may explain the discrepancies.

h) *Species specificities and evolution*

Ribosomal proteins from different mammalian species are very similar or even identical as far as analyzed by electrophoretic methods [29, 41, 54, 82, 84, 87, 186, 188, 201—207]. Furthermore, only minor differences were found between ribosomal proteins from animals of various phyla (mammals, birds, reptiles and fishes) [84, 85, 202—205]. From such studies it has been concluded that ribosomal proteins have evolved slowly during evolution of vertebrates [202, 204]. A rather high degree of evolutionary microdivergence was described for chicken and rat liver ribosomal proteins [208].

More recently ribosomal proteins from rat, rabbit, chicken and trout have been carefully compared using several electrophoretic systems [323]. Most of the ribosomal proteins have identical positions in the 2-D patterns, but also clear-cut differences have been found in the electrophoretic behavior of several proteins. The number of different proteins increases with the evolutionary distance between the species compared.

The results of studies on species specificities of ribosomal proteins with immunological methods will be described in Section 3b of this chapter.

Ribosomes from lower species of animals and from plants have similar properties as ribosomes from vertebrates with respect to number and molecular weights of their proteins (see Sections 1a and c of this chapter). However, the individual proteins from lower species of animals and plants differ markedly in their 2-D electrophoretic mobility in comparison to their counterparts of ribosomes from vertebrates; that means that the positions of the protein spots in the 2-D electrophoretic patterns do not coincide [202, 204]. Therefore, a definite correlation of related proteins from ribosomes of vertebrates and invertebrates is not yet possible. One exception are the acidic phosphoproteins eL 7/L 12 found in rat liver and *Artemia salina* ribosomes [127, 162—164, 169] which possibly correspond to the acidic phosphoprotein L 44 from yeast ribosomes [96, 97, 165]. A protein corresponding to the main phosphoprotein S 6 of the small subunit of mammalian ribosomes was found also in the small ribosomal subunit of *Physarum polycephalum* [321].

Ribosomal proteins of prokaryotes and eukaryotes have been compared by coelectrophoresis of ^{125}I-labeled proteins from one species with an excess of unlabeled proteins of the other one. Overlapping spots have been found in the 2-D electrophoretic patterns of ribosomal proteins from *E. coli* on the one hand and *Saccharomyces cerevisiae*, wheat germ and rat liver, on the other [209]. However, the results are too scarce to establish homologies of such proteins and rather reflect limitations of the methods employed.

The electrophoretic comparison of proteins prepared from ribosomes of different strains of *Drosophila* revealed complete identity between *D. melanogaster* and *D. simulans* [211]; between the evolutionary more distant strains *D. melanogaster* and *D. virilis* only two different protein spots were found [332].

A high degree of evolutionary conservation of ribosomal proteins was also found within several species of higher plants, as demonstrated by the very similar 2-D electrophoretic patterns of cytoplasmic ribosomal proteins from seedlings of the dicotyledon pea and the monocotyledon corn [94, 210].

The ribosomal proteins of *Saccharomyces cerevisiae* [97—99, 101] and *Saccharomyces carlsbergensis* [95, 96, 102] have similar properties.

However, comparative investigations performed under strictly controlled identical conditions have not been published until now and therefore it is not yet possible to decide whether strain specific differences really exist.

i) Ribosomal proteins of mitochondria and chloroplasts

Properties of ribosomes from the two cell organelles have been reviewed a few years ago [212]. Characterization, especially of the proteins from mitochondrial ribosomes, has not been accomplished in the meantime because of difficulties mainly in the preparation of sufficient amounts of pure protein fractions.

Proteins from mitochondrial ribosomes were characterized by one-dimensional [22, 213] and by two-dimensional electrophoretic techniques [110—113, 214, 215]. The proteins of mitochondrial ribosomes differ with respect to number, molecular weights, and positions in the electrophoretic 2-D patterns from the proteins of cytoplasmic ribosomes, as demonstrated for *Neurospora crassa* [22, 111, 168], *Saccharomyces cerevisiae* [112] and *Xenopus laevis* [110, 214].

2-D electrophoretic patterns of ribosomal proteins from mitochondria of rat liver [215] revealed 60 intensive and 47 weak spots which, however, do not yet allow safe identification as genuine ribosomal proteins. More recently 35 and 30 proteins were found in the large and small subunit, respectively, of mitochondrial ribosomes from rat liver [113]. Coelectrophoresis of the proteins of mitochondrial ribosomes from bovine liver with radioactively labeled proteins of mitochondrial ribosomes from rat liver demonstrated only 7 and 3 identical proteins in the large and small subunit, respectively. This finding points to a rapid evolutionary divergence of proteins in mammalian mitochondrial ribosomes in contrast to cytoplasmic ribosomes [113].

Also for more closely related organisms such as *Xenopus mulleri* and *Xenopus laevis* 4 and 3 specific proteins, respectively, were found in mitochondrial ribosomes [214]. Some proteins of mitochondrial ribosomes from different strains of yeast have modified electrophoretic mobilities [112]. At least no common proteins could be identified in mitochondrial ribosomes of evolutionary distant species such as the rat and *Neurospora crassa* [111].

Chloroplast ribosomal proteins can be distinguished from those of cytoplasmic ribosomes of pea seedlings [18], spinach leaves [17], and *Chlamydomonas* [23] by one-dimensional electrophoresis. These results are basically confirmed by two-dimensional electrophoretic analysis; chloroplast and cytoplasmic ribosomal proteins from spinach leaves [94], from wheat leaves [216], from *Chlamydomonas reinhardii* [104, 105], and from *Euglena gracilis* [108, 217], respectively, exhibit only small similarities.

Chloroplast ribosomal proteins are similar to the proteins of prokaryotic ribosomes with respect to their lower number, higher amount of acidic protein molecules and their lower molecular weights. Proteins of cytoplasmic ribosomes of plants, however, are more similar in this respect to ribosomal proteins of vertebrates.

k) Ribosomal mutants with altered proteins

Altered proteins have been observed mostly in chloroplast ribosomes of erythromycin [218, 219] and streptomycin-resistant mutants [105, 107, 220] of *Chlamydomonas reinhardii*. Different ribosomal proteins have been found also in a streptomycin-resistant mutant of *Euglena* in comparison to the wild type [221].

The first example for an electrophoretically altered protein of eukaryotic cytoplasmic ribosomes was demonstrated in a cylcoheximide resistant mutant of the fungus *Podospora anserina* [91]. One protein of the large subunit of the mutant ribosomes was found to migrate slightly different in comparison to the wild type. Furthermore, nine Chinese hamster cell mutants resistant against emetine have been analyzed by 2-D electrophoresis. Only one ribosomal protein of the small subunit from the mutant clone Em^r-2 was found to be altered in comparison to the wild type protein [89]. Emetine resistance and the modified ribosomal protein phenotypes are due to mutation linked to the same chromosome in the Chinese hamster genome; most likely they are due to the same mutation [222].

2. Phosphorylation of ribosomal proteins

Phosphorylation of ribosomal proteins was described for the first time in 1970 for proteins of rat liver [223] and rabbit reticulocyte ribosomes, respectively [224], and has been reviewed in 1974 [225] and more recently [6].

In vivo phosphorylation is a common event in eukaryotes, which is not normally found in prokaryotes [226], very probably due to the lack of protein kinases in bacteria [227]. Only after infection of *E. coli* with bacteriophage T 7 some ribosomal proteins are phosphorylated by a protein kinase induced by the phage [227]. In vitro, proteins of prokaryotic ribosomes are accessible to phosphorylation by protein kinases from rabbit skeletal muscle [229] or rabbit reticulocytes [230].

The proteins of eukaryotic ribosomes can be phosphorylated both in vivo and in vitro. Phosphorylation of ribosomal proteins from many sources has been described as summarized in Table 6.

Phosphorylation has become increasingly attractive in the last few years, because it is affected in vivo by a number of various stimuli, which suggests that this process might be involved in the regulation of protein synthesis at the translational level. There is, however, no clear-cut evidence so far that phosphorylation of ribosomal proteins really influences the functional properties of ribosomes.

a) *In vitro phosphorylation and dephosphorylation*

A fair number of isolated proteins can be phosphorylated in vitro whereas in intact ribosomal particles fewer proteins are accessible to phosphorylation (see also Chapter IX, Section 2c). Serine and, to a lesser extent, threonine residues were found to be phosphorylated in ribosomal proteins [223, 224, 232, 236, 242, 263, 266, 276, 277, 304].

In vitro phosphorylation was performed with ribosome-bound protein kinases or with more or less purified enzymes isolated from the cytosol or from salt wash fractions of ribosomes. Different types of protein kinases were found in eukaryotic cells as reviewed recently [322]. Two cAMP dependent and one cAMP independent protein kinases, e.g. were isolated from the cytosol of reticulocytes; these enzymes and two additional protein kinases were found associated with the protein synthesizing complex of reticulocytes [185, 330].

Most of the in vitro experiments have been performed with cAMP regulated protein kinases using ATP as phosphate donor [223, 232, 236, 239, 266, 268, 270, 271, 274—276, 278], but also cAMP independent enzymes [278, 302, 309] and enzymes catalyzing the transfer of the phosphoryl group of GTP to ribosomal proteins [230, 242, 266, 269, 308] have been used.

Enzymatic dephosphorylation of the proteins of rat liver ribosomes phosphorylated before in vitro can be accomplished with rabbit skeletal muscle glycogen synthetase-D phosphatase [255], rat liver histone phosphatase [255] and with ribosomal protein phosphatase from rat liver

Table 6. Survey of various species and tissues in which phosphorylated ribosomal proteins have been found

Species/Tissue	References
Rat liver	[66, 117, 144, 162, 163, 169, 186, 223, 231–258]
Rat tissues	[259–262, 269]
Rabbit reticulocytes	[154, 186, 224, 263–269, 271]
Rabbit tissues	[186, 224, 272, 273]
Gland tissues	[274–278]
Cell cultures	
Baby hamster kidney	[279–282]
Hamster islet cell tumor	[283]
Sarcoma 180 tumor	[284]
Novikoff hepatoma ascites	[285, 286]
Krebs II ascites	[126, 127, 260, 281, 282, 287–289]
Ehrlich ascites tumor	[290, 314]
HeLa	[252, 291–295]
Mouse L cells	[291, 297–299]
Mouse plasmacytoma	[300]
Mouse myeloma	[301]
Mouse fibroblast	[334]
Chicken fibroblasts	[302, 303]
Trout	[304]
Plants	[272, 305, 306]
Yeast	[97, 272, 307–313]
Artemia salina	[162, 163, 169, 315–317]
Tetrahymena pyriformis	[318–320]
Physarum polycephalum	[321]

and Krebs ascites cells [236]. In vitro phosphorylated 40 S subunits of rabbit reticulocyte ribosomes are dephosphorylated by a phosphoprotein phosphatase activity isolated from rabbit reticulocytes [271].

Until now it has not been possible to find ribosomal functions that are significantly altered by in vitro phosphorylation. Inhibition of amino acid incorporation in an in vitro system by protein kinases and cAMP has been described and explained by a possible phosphorylation of the ribosomes [270]. On the other hand, no differences were found between phosphorylated and control ribosomes with respect to their abilities to interact with elongation factors eEF-T and eEF-G [237], with initiation factors [237], and with globin mRNA [255], respectively.

b) In vivo phosphorylation

Unlike in vitro, only one ribosomal protein of the small subunit and two acidic proteins of the large one were found phosphorylated in vivo in normal animal cells. In HeLa cells [292, 295], in L cells after viral infection [297], in ascites cells [288, 289], in mouse myeloma cells [301] and in regenerating rat liver [66] a few other proteins were found phosphorylated. The additionally phosphorylated proteins are S 2 [288, 289, 292, 295, 297], S 3/S 3a [66, 289, 295, 297], S 16 [295], L 6 [289] and L 14 [289, 292, 297]. Recently two further phosphoproteins with molecular weights of 43000 and 37000 were described in ribosomes of HeLa cells [77].

In yeast, besides two proteins corresponding to the basic protein of the small and to the acidic protein of the large ribosomal subunit of animal cells, some additional phosphoproteins were found [97, 101, 307, 309, 310]. Three acidic and two basic phosphoproteins were found in the large and the small ribosomal subunit, respectively, of *Physarum polycephalum* [321]. One protein of the small subunit, obviously corresponding to the basic small subunit protein of animal cells, takes up about 70% of the total activity incorporated into the ribosomal protein fraction [321].

c) Phosphorylation of ribosomal protein S 6

In vivo only one 40 S subunit protein, namely S 6, was found to be phosphorylated in ribosomes of rat liver [239, 241], rabbit reticulocytes [268], baby hamster kidney fibroblasts [280, 282], bovine anterior pituitary gland[276], hamster islet cell tumor [283], ascites cells [288, 290, 314], Hela cells [77, 292], mouse myeloma cells [301], plants [305] and *Tetrahymena pyriformis* [318]. Protein S 6 can take up up to five phosphoryl groups [77, 144, 241, 261, 268].

The degree of phosphorylation of S 6 depends, in particular, on the physiological conditions of the tissues and is influenced by a number of effects.

Phosphorylation of S 6, e.g., was found to be stimulated by cAMP [244, 261, 267, 276, 283] and dibutyryl-cAMP [261, 267, 276]. In other cases, however, phosphory lation of protein S 6 was not found to be enhanced by increased intracellular cAMP levels [144, 296].

Treatment of animals with the hormones glucagon [231, 244, 283], thyroxin [235], and epinephrine [144] as well as inhibition of protein synthesis by puromycin [243] and cycloheximide [243, 295] increases phosphorylation of S 6.

Infection of cell cultures with vaccinia virus [292, 293], vesicular stomatitis virus [297] and mengovirus [290] and changes of the growth conditions of cell cultures [78, 260, 280, 289, 300, 318, 319, 320] also influence phosphorylation of S 6. Phosphorylation of S 6 was also found to be dependent on the mitotic cell cycle [279, 334]. In monolayer cultures of Chinese hamster ovary cells a predominant phosphoprotein was found during the mitotic phase which disappeared from the electrophoretic pattern when the cells entered the G 1 phase [279]. In mouse fibroblasts, however, a large increase in phosphorylation of S 6 was observed 5 min after the induction of quiescent cells to enter G 1 phase [334].

Phosphorylation of protein S 6 is increased in regenerating rat liver [67, 241], during hepatic injury [247, 248] and in the liver of diabetic rats [245]; in the latter case, the increased rate of phosphorylation could be reduced to normal values by insulin [245].

Decreased phosphorylation levels of protein S 6 on the other hand were observed after administration of ethionine to rats [144] and in mouse myeloma cells after hypertonic initiation block [301].

Protein S 6 is phosphorylated preferentially in polysomes in comparison to monosomes of reticulocytes [284], sarcoma cells [284], MPC 11 cells [300], and baby hamster kidney fibroblasts [282]. In certain circumstances S 6 becomes phosphorylated also on monoribosomes, e.g. of baby hamster kidney cells [282] and regenerating liver [241]. A close correlation between phosphorylation of S 6 and the rate of RNA degradation was found in *Tetrahymena pyriformis* suggesting a possible connection between phosphorylation and the catabolism of ribosomes [320].

Despite these numerous facts it has not yet been possible to explain the biological role of phosphorylation of ribosomal protein S 6.

d) Acidic ribosomal phosphoproteins

More recently some acidic ribosomal phosphoproteins have been found in yeast [97, 307—310], rat liver [117, 162, 169, 251, 252], *Artemia salina* [162, 169, 315, 316], reticulocytes [269], muscle [260], ascites cells [126, 127, 260, 281, 287, 288], HeLa cells [252, 294] and other cell cultures [252, 281, 298]. These proteins resemble the *E. coli* ribosomal proteins L 7/L 12 with respect to their electrophoretic behaviour; furthermore, they are also highly acidic, poorly stainable in polyacrylamide gels and have comparable molecular weights (12500 to 16000). Moreover, selective phosphorylation of the *E. coli* proteins L 7/L 12 was observed when incubating *E. coli* 50 S ribosomal subunits with rabbit skeletal muscle protein kinase and γ [^{32}P] GTP [229].

Some discrepancies remain, however, with respect to the number and properties of acidic phosphoproteins in eukaryotic ribosomes resulting from experimental difficulties in their characterization [281]; most data are available for rat liver [117, 169] and *Artemia salina* ribosomes [162, 169, 315, 316].

From rat liver ribosomes two such proteins, P 1 and P 2, were isolated and two to three phosphorylated forms of both have been describes [117].

From ribosomes of *Artemia salina* and rat liver two closely related acidic phosphoproteins, called eL 12 and eL 12′, were prepared from ethanol/salt extracts of ribosomes [169]. Partial sequence data suggest that eL 12 and eL 12′ are derived from distinct but closely related genes [169]. Both proteins were also purified in phosphorylated forms, eL 12-P and eL 12′-P. Using a radioimmuno assay, 1.8 copies of eL 12 and 0.9 of eL 12′ were found in the 80 S *Artemia salina* ribosome. These proteins of *Artemia salina* and rat liver have similar isoelectric points and molecular weights. The isoelectric points [169] of eL 12 and eL 12′ are 4.6 and 4.2, respectively, and they have molecular weights of 13000 and 13500, respectively [169, 315, 316].

The total amino acid compositions of *Artemia salina* [169] and rat liver ribosomal proteins L 12/L 12′ and P 1 and P 2 [117, 169] are similar. A comparison of the alanine rich sequences of the L 12-protein from rat liver and *Artemia salina* shows remarkable similarities [162]. This points to the conservation of these proteins during evolution from invertebrates to mammals. Furthermore, structural homologies seem to exist also to the *E. coli* protein L 7/L 12 [162, 163, 164, 331] and to the analogous protein L 20 of *Halobacterium cutirubrum* [324].

Recently, in vivo incorporation of microinjected acidic protein eL 12 from the large subunit of *Artemia salina* ribosomes into oocyte ribosomes was demonstrated, while *E. coli* ribosomal proteins L 7/L 12 or L 7/L 12/L 10 were not incorporated [317].

The acidic phosphoproteins of *A. salina* are functionally similar to the acidic ribosomal proteins L 7/L 12 of *E. coli* [316].

Functionally similar acidic proteins, called L 40 and L 41, of the large subunit of rat liver ribosomes have been prepared in the same way as the acidic phosphoproteins of *Artemia* and rat liver ribosomes from NH_4Cl/ethanol wash solutions of 60 S subunits [2]. In contrast to the acidic proteins, however, phosphorylation of L 40 and L 41 could not be demonstrated [241]. Furthermore, different molecular weights [2] and different electrophoretic mobilities of L 40 and L 41 and of the acidic phosphoproteins, respectively, were found [153].

Immunochemical cross-reactivities of L 40 and L 41 of rat liver ribosomes with proteins L 7/L 12 of *E. coli* have been described [2, 258] whereas, on the other hand, immunochemical cross-reactivities of *Artemia salina* eL 12 protein [315] and rat liver ribosomal proteins P 1/P 2 [117] with L 7/L 12 proteins of *E. coli* ribosomes could not be demonstrated.

Extraction of 60 S ribosomal subunits from ascites cells with NH_4Cl/ethanol yields an acidic phosphoprotein Lγ in the split fraction [127]. Improved electrophoretic separation of Lγ yielded three spots in the 2-D electrophoresis which were interpreted to represent four different species, namely a 13500 and a 14500 dalton species and their corresponding phosphorylated forms [127].

The extent of in vivo phosphorylation of the acidic phosphoprotein Lγ is influenced by the milieu conditions and was found to be reduced in ascites cells incubated in a medium lacking glucose [260].

As established for the *E. coli* proteins L 7/L 12 [325], also the eukaryotic acidic phosphoproteins seem to be located at the interface of the two ribosomal subunits [127, 252].

Immunological cross-reactions between antisera against phosphoprotein Lγ and *E. coli* L 12 were described [127].

The characterization of acidic proteins of eukaryotic ribosomes is obviously very difficult and the published data seem to be not always reproducible. Despite the sometimes conflicting data the available evidence suggests that the acidic phosphoproteins are possibly the eukaryotic counterparts of the *E. coli* proteins L 7/L 12.

3. Immunochemical properties of ribosomal particles and proteins

a) Applicability of antibodies

Immunological approaches using specific antibodies against ribosomal proteins are extremely effective for elucidating the structure and function of ribosomes as shown for prokaryotic [336–339] and now also for eukaryotic ribosomes and their proteins [7, 340–344].

Antibodies against eukaryotic ribosomes [2, 94, 258, 340, 345–360], ribosomal subunits [2, 94, 258, 340, 359–363], total protein of the ribosomal particles [2, 94, 258, 340, 347, 348, 351, 354, 355, 359, 360] and purified individual ribosomal proteins from rat liver [2, 3, 117, 127, 151, 169, 258, 315, 340–344] have been prepared in rabbits [2, 3, 94, 117, 127, 151, 169, 205, 340–342, 344–360, 364–371], sheep [2, 258, 359, 360, 364], goats [361 to 363], pigs [372] and chicken [341, 342, 344], and were used to compare ribosomes and ribosomal proteins from prokaryotes and eukaryotes [2, 7, 117, 127, 169, 258, 315, 377], from different eukaryotic species [2, 94, 169, 205, 315, 345, 348, 351 to 353, 355, 365–368, 373], from different tissues of the same species [3, 351–353, 355, 367], from normal and tumor cells [350, 354, 361–363, 374], from cytoplasmic ribosomes and those of mitochondria and chloroplasts [94, 358]. Furthermore, antibodies were applied to compare the antigenic properties of the small and large ribosomal subunit [2, 340, 359, 363, 364], the antigenicity of individual ribosomal proteins [2, 3, 117, 127, 151, 169, 258, 315, 340–344] and for the estimation of the pool of free ribosomal proteins in the cytoplasmic fraction [364].

Ribosomes and their proteins of eukaryotic organisms are immunogenic in mammals and birds. The antisera contain predominantly antibodies against the protein moiety, whereas the ribosomal RNA is less immunogenic [2, 94, 359, 369, 375]. On the other hand, it has been shown that anti-tRNA antibodies [346], antibodies against polynucleotides [376] and yeast RNA [374] precipitate ribosomes by reaction with RNA, although no specificities regarding the origin of antigens or immunogens could be proven. Antibodies against ribosomes or ribosomal protein of a given animal species could not be raised in the same species [366, 378]. Whereas the antigen-antibody reaction with ribosomal particles can be measured by the quantitative precipitin reaction, the most valuable method for ribosomal proteins is passive tannic hemagglutination [352, 353, 360, 365]. By this procedure one escapes the difficulty that ribosomal proteins aggregate spontaneously, even with the γ-globulin fraction, in an unspecific way [370, 378]. The double radial immunodiffusion is also a very tricky technique, because of the formation of unspecific precipitation lines in many cases [378, 379]. Radioimmunoassays have been developed for the detection of antibodies against ribosomes in sera of patients with systemic lupus erythematosus (SLE) [380, 381], against ribosomal proteins in rabbit antisera [378] or for the quantitation of ribosomes in tissue homogenates [382].

b) *Antigenicity of 80 S ribosomes and their total proteins*

In order to get insights into the evolution of ribosomal proteins besides electrophoretic techniques (see Section 1 h of this chapter) also immunochemical methods have been used [2, 94, 169, 205, 315, 345, 348, 351—353, 355, 360, 365—368, 370, 373]. Most of the experiments were performed with rabbit antisera against 80 S ribosomes and their total protein mixtures of different species of animals. The immunogenicity of heterologous ribosomes or sibosomal proteins is an allusion to study rtructural differences between ribosomal proteins of the donor and those of the animals producing antibodies. It should be mentioned that neither in rabbits [366] nor in chicken [378] antibodies against their own ribosomal proteins could be raised. Rabbit antisera against rat or chicken ribosomal protein gave no immunological cross-reaction with ribosomal proteins from man, ewe, calf, guinea pig, and mouse [366]. The same results were obtained with rabbit antisera against bovine liver ribosomes which did not reveal cross-reactions with ribosmes from the livers of rat, chicken, mouse, and rabbit (Fig. 13) [352, 353, 355]; vice versa, antisera against rat liver ribosomes do not cross-react with ribosomes or their total protein from bovine liver, bovine kidney, and rabbit liver. However, partial cross-reaction could be detected between rabbit antisera against ribosomal proteins of rat liver and the ribosomal proteins of mouse liver which may be caused by the close taxonomical relationship between both species [353, 355].

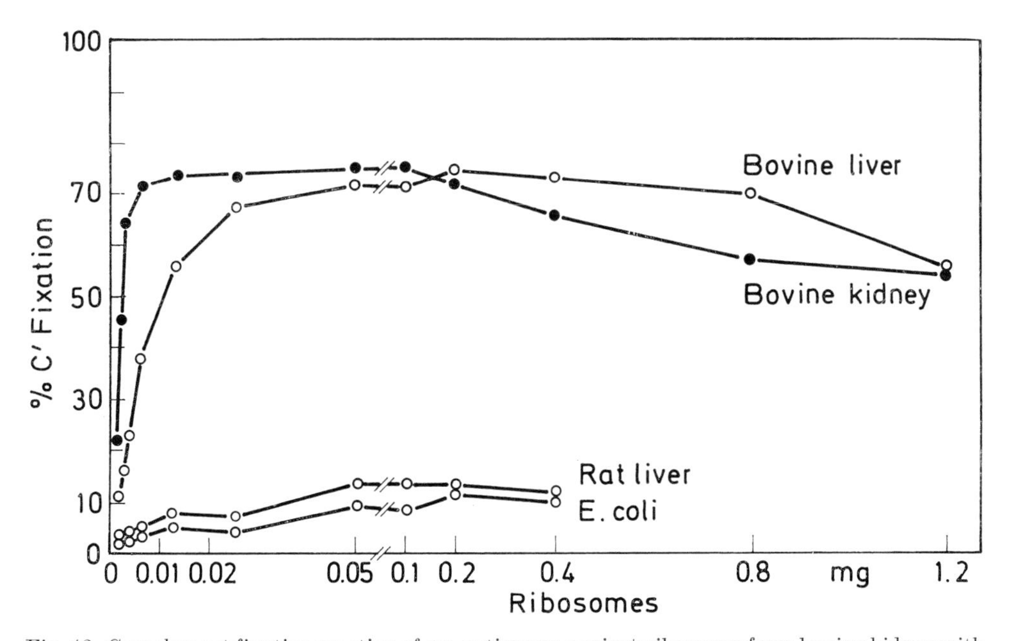

Fig. 13. Complement fixation reaction of an antiserum against ribosomes from bovine kidney with ribosomes from bovine liver, bovine kidney, rat liver, and *E. coli*. From F. Noll and H. Bielka [353].

Another detailed comparison of the antigenicity of ribosomes and ribosomal proteins of rat and chicken liver using rabbit and sheep antisera revealed a weak cross-reaction also between rat and chicken ribosomal proteins [360]. Rabbit antisera against *Neurospora crassa* ribosomes and ribosomal protein did not react with ribosomes from sheep reticulocytes or from *Xenopus leavis* [348]. Immunological experiments using rabbit antisera against 40 S and 60 S subunits of rat liver ribosomes suggested also that there are few, if any, similarities between ribosomal proteins from different species (livers of rat, chicken, bovine and trout) [205].

These immunochemical results indicate that despite of great similarities of the ribosomal protein patterns in the 2-D polyacrylamide gel electrophoresis (see Section 1 h of this chapter) there exist differences in the properties of ribosomal proteins from the analyzed species of vertebrates. These discrepancies could be explained in the following ways: (1) Identical electrophoretic mobilities of proteins do not exclude molecular heterogeneity due to limitations of the 2-D polyacrylamide gel electrophoresis [337]. (2) The size and charge of ribosomal proteins were conserved during evolution, whereas antigenic determinants have been changed by substitution of at least some amino acids. (3) The ribosomal proteins of the immunized animals may have extensive structural homologies with the immunogens and therefore produce antibodies against different sequence regions (the species specific antigenic determinants) only [205, 360].

On the other hand, ribosomes from different tissues of the same animal species exhibit immunological identities as it was shown for rat liver and heart [382], bovine liver and kidney [352, 353, 355] or spleen [351] and *Neurospora crassa* mycelia and conidia [348]. Therefore it can be concluded that obviously no tissue specific antigenic determinants exist among ribosomal proteins as far as tested with total protein mixtures.

Furthermore, comparative immunological experiments have shown that ribosomal proteins from *E. coli*, yeast, plants and animals have no common antigenic determinants [347, 353, 368]. Some authors found partial cross-reactivities between ribosomal protein of E. *coli* and *Neurospora crassa* [347, 348] while others did not [358].

A high degree of evolutionary conservation of ribosomal proteins was found by immunochemical techniques using antisera against ribosomal proteins of pea [94] or of kidney bean [366], and against cytoplasmic ribosomes of several species of higher plants such as bean, pea, lentil, spinach, tobacco and wheat [94, 366].

The immunological comparison of proteins of cytoplasmic ribosomes with those of chloroplasts [94] or mitochondria [94, 358] did not reveal significant sequence homologies. Furthermore, no immunological cross-reactions could be detected between ribosomes of chloroplasts, mitochondria, bacteria and blue-green algae [94].

The question of whether ribosomes of malignant cells do contain the same set of proteins as ribosomes of normal cells or whether ribosomal proteins are altered during the process of malignant transformation was also investigated by immunological methods. The comparison of 80 S ribosomes and their total protein from rat hepatoma and normal rat liver did not exhibit any immunological differences [354] (Fig. 14). Goat antisera against 60 S [361] or 40 S ribosomal subunits [362] of Novikoff hepatoma ascites cells did not discriminate between the corresponding ribosomal subunits from normal and hepatoma cells. Rabbit antisera against ribosomal proteins of the mouse myeloma MOPC 31 C reacted also with MOPC 70 E ribosomes [374].

On the other hand, antigenic differences between ribosomal proteins of normal and

Table 7. Titers of haemagglutination of antisera against single 40 S rat liver ribosomal subunit proteins.
The titers are given as the reciprocals of the antiserum dilution still showing a positive agglutination. 0 = no reaction
(From THEISE et al. [151])

Antigens tested	Antiserum against protein																			
	S 2	S 3	S 3a	S 4	S 5	S 6	S 7	S 8	S 9	S 11	S 13	S 17	S 18	S 19	S 21	S 23/24	S 25	S 28	S 29	S 30
S 2	640	0				0					0			0	0		0			
S 3	0	640		0		0	0			0	10		0	0	0		0	0		0
S 3a		0	80		0											0				
S 4		40		1280				0		0			0					0		
S 5			0		1280											0				
S 6	0	0		0		640		40			0				0	0		0		0
S 7		0					1280				0	0	0	0	0	0	0			
S 8						20		5120			0	0		0			0		0	0
S 9		0							160			0	0					0		
S 11		0								1280			0					0		
S 13	0	0		0		0				80	5120	0	0	0			0	0		
S 17	0										0	640	0	0						
S 18	0	0		0						0	0	0	1280	0				0		0
S 19	0	0		0				0		20	0	0	0	1280	0		0	0		
S 21	0	0				0	0				10			0	2560		0	0		
S 23/24			0		0	40		10								320				
S 25						0		0				0		0	10		320	0		0
S 28		0		0			0				10		0		0		0	2560	0	
S 29		20						20										0	160	0
S 30		0				0		0			0		0				0		0	160

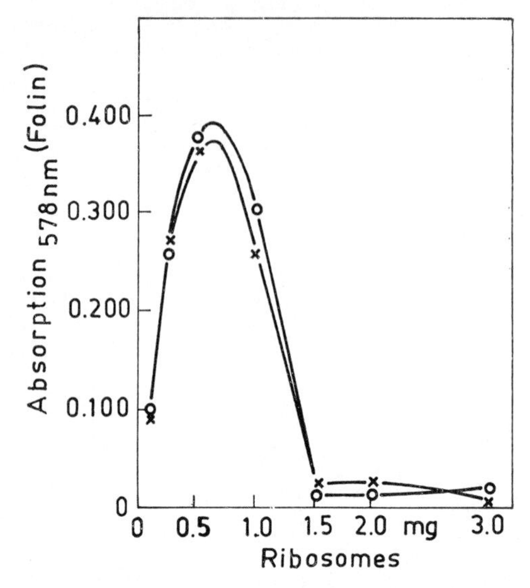

Fig. 14. Quantitative precipitin assay of an antiserum against rat liver ribosomes with ribosomes from rat liver (○) and hepatoma (×).
From F. NOLL and H. BIELKA [354].

malignant tissues have been described. Two extra proteins (molecular weights of 30000 and 65000), responsible for immunological differences between the proteins of 60 S ribosomal subunits of normal liver and Novikoff hepatoma ascites cells were found by absorption of antiserum against tumor 60 S ribosomal subunits with normal 60 S subunits [363]. Differences in the antigenic properties of ribosomes from normal rat liver and the livers of rats fed a carcinogen (N-nitrosomorpholine) were also reported [350].

c) Antigenicity of 40 S and 60 S ribosomal subunits

Ribosomal subunits of rat liver show differences in their immunogenicity and antigenicity [340, 359, 380]. 60 S ribosomal particles as well as the total 60 S subunit proteins are more immunogenic than the 40 S subunits or their total protein mixture in rabbits and sheep, if the antibody titers in the corresponding antisera are compared. Moreover, in anti-80 S ribosome antisera the immunochemical reaction with 60 S subunits is higher than with 40 S subunits [340, 359, 380].

Furthermore, no immunological cross-reactions could be detected between the proteins of the large and the small ribosomal subunits of rat liver [2, 340, 359, 363, 364]. This finding is in accordance with the 2-D polyacrylamide gel electrophoretic data (see Section 1a of this chapter).

d) Immunological properties of purified individual proteins

Antigenic properties of purified individual proteins of the 40 S (Table 7) and the 60 S ribosomal subunits (Table 8) of rat liver ribosomes [3, 151, 340] have been tested by passive haemagglutination. All isolated proteins were found to be immunogenic in rabbits and, as far as tested, also in chicken [344]; however, the response of these animals against the various proteins is different. With antisera against single proteins positive immunological reactions were obtained with the corresponding proteins only, while no significant cross-reactions could be detected. Very weak titers have been found for antisera against some proteins in heterologous tests with

Table 8. Titers of haemagglutination of antisera against single 60 S rat liver ribosomal subunit proteins.
The titers are given as the reciprocals of the antiserum dilution still showing a positive agglutination. 0 = no reaction
(From Noll et al. [340]).

Proteins tested	Antiserum against protein			
	L 28/29	L 35	L 36/36a	L 38
L 19	0	—	0	—
L 28/29	1280	0	0	—
L 33	0	—	—	0
L 34	0	—	—	0
L 35	0	1280	—	—
L 36	0	—	80	—
L 36/36a	0	0	160	—
L 38	0	—	0	320

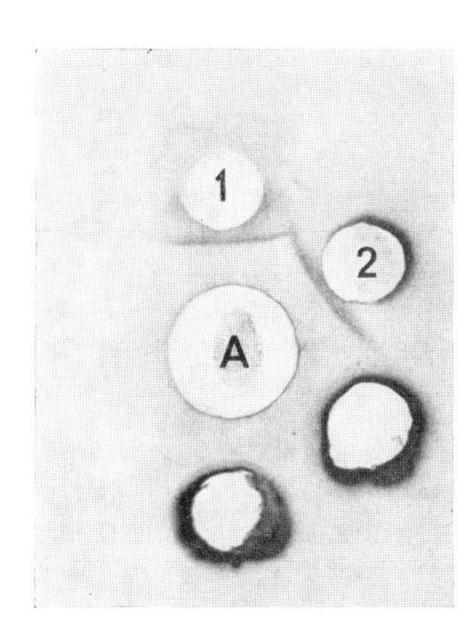

Fig. 15. Radial immunodiffusion of chicken antiserum against total protein of the small ribosomal subunit from rat liver (A) with proteins S 2 (1) and S 3 (2) from rat liver ribosomes. Crossing of the precipitation lines points to structural differences between both proteins. F. Noll, unpublished results.

a few other proteins, but after purification of the antibodies by immune affinity chromatography the antibody preparations were found monospecific [344]. The absence of any partial cross-reactions between the various proteins clearly demonstrated that the isolated proteins of rat liver ribosomes are unique (Fig. 15).

e) Antibodies to ribosomes in sera of patients with SLE

Anti-ribosome antibodies have been detected in sera of patients with systemic lupus erythematosus (SLE) [373, 380, 381, 383—386]. Using various methods it was found that these antibodies react with ribosomes from human and animal sources [373, 383]. Both protein [373, 384, 386] and RNA [373, 385] moieties of the ribosomes form antigenic determinants. The antibodies in SLE sera reacted with both ribosomal subunits, predominantly however with 60 S subunits [380].

4. Ribosomal ribonucleic acids

a) High molecular weight ribosomal ribonucleic acids

Both the large and the small ribosomal subunit of eukaryotic ribosomes contain one moiety of high molecular weight RNA characterized by sedimentation coefficients of 25 to 28 S and of 15 to 18 S, respectively.

Molecular weights

In the past, molecular weights of ribosomal RNAs from a variety of eukaryotes have been determined mainly by physical methods and by electrophoresis in polyacrylamide gels under non-denaturing and denaturing conditions. Some data are given in Tables 9 and 10.

The comparison of the size of the high molecular weight ribosomal RNAs of different species shows that the molecular weight of the 18 S rRNA is nearly constant among various species of animals, while the size of the larger rRNA, the 28 S rRNA type, is higher in more evolved species [389, 417, 418].

The determination of the molecular weights can be seriously affected by the secondary structure of the RNA molecules as demonstrated, e.g. for the mitochondrial rRNA of *Saccharomyces carlsbergensis* [401]. Gel electrophoresis under nondenaturing conditions may lead to considerable errors in the estimation of the molecular weights. The complete abolition of the secondary structure of the polynucleotide chains during gel electrophoresis provides a suitable basis for reliable and rapid measurements of the molecular weights of rRNA [396, 419]. This can be achieved by electrophoresis in the presence of urea at 60 °C [396], in 98% formamide at 30 °C [419] or dry formamide at 45 °C [420]. Other methods for the determination of RNA molecular weights are sedimentation velocity analysis after denaturation with formaldehyde [421] or dimethylsulfoxide [422], sedimentation equi-

Table 9. Molecular weights ($\cdot 10^{-6}$) of cytoplasmic 28 S and 18 S type ribosomal RNAs

Species	RNA from the large subunit	small subunit	References
Fungi	1.36—1.39	0.65—0.70	[387—389]
Fern	1.34	0.72	[389]
Algae	1.3	0.69	[389, 390]
Plants	1.2—1.3	0.65—0.71	[387, 391—394, 395]
Protozoa			
Amoeba	1.57	0.83	[387]
Tetrahymena	1.35	0.7	[387]
	0.59 + 0.59	0.52	[396]
Yeast	1.21	0.72	[396]
Animals			
Lumbricus	1.31	0.66	[387]
Artemia	1.43	0.66	[387]
Galleria	1.4	0.7	[397]
Drosophila	1.4	0.65—0.73	[393—394, 398]
Tenebrio	1.5	0.7	[387]
Sea urchin	1.37	0.65	[387]
Ciona (Tunicate)	1.38	0.65	[387]
Fish	1.4—1.5	0.65	[387, 394]
Frog	1.4—1.5	0.6—0.7	[387, 393, 394]
Reptiles	1 5	0.62	[394]
Chicken	1.6—1.65	0.65	[387, 394]
Marsupial	1.7	0.65	[394]
Mouse	1.65—1.7	0.65	[387, 394]
Rat	1.44	0.66	[396]
Monkey	1.66	0.65	[387]
HeLa cells	1.66	0.65	[387]

librium centrifugation [423—426], determination of the endgroup mass [427], oligonucleotide analysis after degradation with pancreatic ribonuclease [428, 429], and sedimentation and viscosity measurements [430].

Molecular weights can be calculated also from electron microscopic data and are in fairly good agreement with those measured by gel electrophoresis if spacing data are available from reference molecules of the same RNA class [398, 401, 410]; the spacing of the bases in native RNA molecules has been determined to be 0.245 nm [410]. In Table 11 values of the size of ribosomal RNAs are presented which have been calculated from electron microscopic studies of single RNA molecules.

Treatment of the high molecular weight 26 S rRNA of the large ribosomal subunit of insects [404, 431—438] and protozoa [396, 439—444] with heat, dimethylsulfoxide or urea leads to its complete dissociation into two halfs of about equal size. This fragmentation can be explained by the presence of hidden breaks in the rRNA of the large subunit [404, 432, 435] and was also found in the 28 S type rRNA of algae [445], higher plants [446] and in chloroplast 23 S rRNA [406, 447, 448].

Heat induced fragmentation was also observed in the rRNA of old rat liver ribosomes [449, 450]. Since a 140- to 170-nucleotides piece is missing from the center of the mature 26 S RNA two nicks in the rRNA have been postulated [438, 451].

Table 10. Molecular weights (· 10^{-6}) of mitochondrial and chloroplast ribosomal RNAs

Species	RNA from the		References
	large subunit	small subunit	
Mitochondrial ribosomes from			
Yeast	1.29	0.72	[400]
	1.30	0.70	[396]
	1.26	0.68	[401]
Maize	1.25	0.76	[392]
Wheat	1.3	0.75	[395]
Tetrahymena	0.90	0.47	[396, 402]
Neurospora	1.28	0.72	[403]
Smittia	0.66	0.31	[404]
Rat liver	0.95	0.50	[403]
HeLa	0.7	0.4	[405]
Chloroplast ribosomes from			
Chlamydomonas	1.07	0.54	[390]
Phaseolus	1.1	0.56	[391]
Maize, pea and Chlamydomonas	1.1	0.56	[406]
Spinach	1.28	0.54	[407]

Table 11. Size of 28 S and 18 S type ribosomal RNAs estimated by electron microscopy

Species	RNA from the		References
	large subunit [μm]	small subunit [μm]	
Cytoplasmic ribosomes from			
Rat liver		0.57	[408]
	1.70	0.71	[409]
Duck erythroblast	1.22–1.39	0.61–0.65	[410, 411]
BHK cells	1.62	0.65	[412]
HeLa cells	1.16	0.59	[410]
L cells	1.79	0.64	[414]
Xenopus	1.54	0.63	[415]
	1.50	0.68	[409]
Aspergillus	1.10	0.52	[388]
Rudd	1.42	0.64	[409]
Lizard	1.49	0.70	[409]
Chicken	1.62	0.68	[409]
	1.80	0.98	[416]
Pigeon	1.65		[409]
Cat	1.73	0.67	[409]
Monkey	1.70	0.68	[409]

The breaks that interrupt the continuity of the rRNA molecule seem to be located within a central hairpin loop which can be demonstrated by electron microscopy as the most prominent feature in the psoralen cross-linked 26 S rRNA from *Drosophila melanogaster* [437].

Base composition

The base composition of ribosomal RNAs of various fungi, plants and animals are given in Table 12. The two RNA components of the large and small subunit, respectively, are significantly different from each other. While the base compositions of RNA from different tissues of the same species do not differ significantly, rRNAs from the same tissue of different organisms are different in this respect. Compared with lower species, there is an increase in the G + C content, especially of the rRNA of the large subunit in plants as well as in animals of the higher phyla [453].

Ribosomal RNAs are characterized by methylated and other odd bases, as well as by nucleotides methylated in their ribose moiety (2′-O-methylated residues). Nucleotides methylated in the sugar moiety, the base residue or both were described for rRNAs of plants [454—458], yeast [459—463, 469, 496] protozoa [442, 464, 465], *Xenopus* [466], and various mammals [463, 467—495]. Methylation was also observed for chloroplast rRNA [497] and mitochondrial rRNA [400, 498—500]; in the latter, minimal methylation was observed in the RNA of the large ribosomal subunit only while the RNA of the small subunit seems to be unmethylated [400, 415, 501].

Methylation is a posttranscriptional event [454, 459, 470, 479]. Kinetic studies have revealed that methylation of the ribose residues starts at the level of the first rRNA precursor (45 S RNA in HeLa cells; see Chapter VII), whereas the base residues are methylated at later stages of the maturation process of the ribosomal precursor RNA [454, 459, 470, 487, 496]. Most of the early methylations take place on the conserved parts of the 45 S precursor molecule. Extensive homologies between the methylated nucleotide sequences have been found in the rRNA from HeLa cells, hamster cells, mouse L cells, chick embryo fibroblasts and *Xenopus laevis* kidney cells [481].

In HeLa cells, 63 and 38 methyl groups have been found in the 28 S and 18 S rRNA, respectively [481]. Most methylations are found in the ribose residues [470]. For NOVIKOFF hepatoma, 56 2′-O-methyl groups were estimated for the 28 S rRNA and 39 for the 18 S rRNA [477]. The 28 S rRNA of yeast contains a total of 43 methyl groups, the 17 S rRNA a total of 24, 6 of which are attached to the bases on both RNAs [461].

Table 12. Base composition (in per cent) of high molecular weight ribosomal RNAs (From LAVA-SANCHEZ et al. [452] and AMALDI [453])

Species	28 S type rRNA				18 S type rRNA			
	C	A	U	G	C	A	U	G
Fungi	20.6	26.5	24.0	28.9	19.9	25.2	26.0	28.9
Dicotyledons	23.8	23.7	20.0	32.5	22.4	24.9	22.9	30.7
Monocotyledons	28.0	21.2	16.9	33.9	25.0	22.0	20.2	32.8
Insects	19.6	30.8	27.1	22.5	20.3	28.8	27.4	23.5
Echinoderms	26.0	19.9	19.3	34.9	23.8	23.1	23.2	29.9
Amphibians	27.9	19.7	17.4	34.0	24.1	24.1	22.9	28.9
Birds	28.6	18.0	18.8	34.7	25.8	23.3	22.1	28.8
Mammals	29.8	17.2	17.9	35.1	26.3	21.4	21.7	30.6

Methylations occur on both the bases and the ribose moiety of all four nucleotides [461—463, 475, 496].

Besides methylated bases, other odd nucleotides have been found in rRNAs, mainly pseudouridylic acid [462, 465, 471 to 473, 489, 490]. The rRNA of *Crithidia* contains 2′-O-methylpseudouridine and 2′-O-methylinosine [442, 465]. In the 18 S rRNA of Chinese hamster cells [476] and HeLa cells [463] and in the 17 S rRNA of yeast [463, 496], 1-methyl-3-γ-(α-amino-α-carboxypropyl)pseudouridine has been found, and 2′-O-ethylpseudouridine is present in wheat embryo rRNA [455]. In the rRNA of the small subunit of rat liver, chicken liver and yeast N^4-acetylcytidine was identified under conditions known to prevent any destructions of modified labile nucleosides [495]. Furthermore, N^6-methyladenosine and 7-methylguanosine have been found in the 18 S rRNA of rat liver [495].

The biological role of the methylated and the other odd bases in the rRNA is not yet quite clear. Lack of methionine results in the synthesis of submethylated 45 S and 32 S precursor rRNA [502], but no mature rRNA is formed. Submethylated rRNA cannot form stable ribosomal subunits. Therefore, it seems that methylation and other modifications of the nucleotides are prerequisites for the accurate folding of the polynucleotide chain of the precursor RNA which is necessary for the subsequent correct attack of nucleases during processing of the precursor and specific interactions with proteins during ribosome assembly.

Sequences

Until now, only fragments of the high molecular weight ribosomal RNAs of eukaryotes have been sequenced [397, 442, 463, 481, 492, 504—530, 543], but rapid progress can be expected because the methods of DNA sequencing can be applied to rDNAs [531, 532]. cDNA of 28 S rRNA of rat liver has already been prepared [533] and larger quantities of rDNA were obtained by plasmids [534, 535]. Using these techniques sequences of about 20 nucleotides at the 3′ termini of 18 S rRNA of mouse, silk worm, wheat embryo and the slime mold have been determined [528].

Until now most of the work has been done by the analysis of oligonucleotides obtained by enzymatic digestion of 18 S rRNA from different species. The structure of the rRNA from different species was compared by homochromatography fingerprinting of T 1 RNAse or pancreatic RNase derived oligonucleotides. Furthermore, 3′ terminal and 5′ terminal nucleotide sequences as well as sequences of methylated oligonucleotides from different species have been estimated and compared.

The structure of the rRNA species from the free and membrane-bound ribosomal fractions from HeLa cells were found to be very similar using T 1 RNase and combined T 1 RNase and pancreatic RNase fingerprinting procedures [536]. The electrophoretic mobility and composition of 3′ terminal oligonucleotides obtained by T 1 digestion of 18 S rRNA from rat is similar to that of other rRNAs [514]. Comparison of the large T 1 fragments of 18 S rRNA from rat, mouse, hamster, and man by two-dimensional homochromatography showed a high degree of similarity [537]. A small number of point mutations only have been found. The 18 S rRNA of man has more oligonucleotides in common with those of Chinese hamster than rat and mouse [537]. The primary structure of the high molecular weight rRNA of human, chicken and *Xenopus laevis* cells was found very similar; some sequences, however, revealed interspecies differences [526]. Small differences in the fingerprint patterns of oligonucleotides of rRNA from normal and hepatoma tissues [525, 538] and of polysomal rRNA from AH-130 tumor cells and adult liver cells have been described [539] and discussed in terms of transcription of different redundant rRNA genes in various vertebrate cells [538]. The relatedness of 18 S and 25 S rRNA of plants (several *Vicia* species and *Allium cepa*) has been estimated [540].

Investigation of the 18 S rRNA of Novikoff hepatoma ascites cells has already provided a lot of structural information essential for further and, eventually, complete sequence analysis [492, 525, 537]. A distinctive pattern of 116 spots was obtained by homochromatography fingerprinting of T 1 RNase derived oligonucleotides. The nucleotide composition of most of the nucleotides was estimated. The examination of the oligonucleotides indicates the existence of 176 types of oligonucleotides among which 66 types contain 83 modified nucleotides [492].

Sequences of short segments at the 3′ and 5′ ends of 18 S rRNA of various species have been estimated. At the 3′ end the sequence $GAUCAUUA_{OH}$ was found in 18 S rRNA from HeLa cells [513, 514], rabbit reticulocytes [504], *Drosophila* [507, 508], L 5158 Y mouse cells [520], *Xenopus laevis* [515], *Bombyx mori* [511], yeast [507—510], *Tetrahymena pyriformis* [512], hamster fibroblasts [515] and chicken fibroblasts [506, 514]. The 3′ terminal sequences $AUUA_{OH}$ was also found in the 18 S rRNA of Arthropoda (*Aphis* and *Galleria*) [397]. In plants guanosine was analyzed as the 3′ terminal nucleoside [528, 529]. In Dictyostelium rRNA the 3′ terminal T 1 oligonucleotides were found different from that in rRNA of vertebrates [528]. The sequence of 32 nucleotides in position 5 to 36 of the 3′ end of 18 S rRNA from rat liver has been determined [513]. Interestingly, the sequence CCUCC localized near the 3′ end of *E. coli* 16 S rRNA and involved in mRNA binding by *E. coli* ribosomes [541] was not found at the 3′ end of eukaryotic rRNAs.

At the 5′ end of the rRNA of the small ribosomal subunit of mouse [520, 521], chicken [521] and *Dictyostelium* [521] the sequence UACCUG was found. The similar oligonucleotide AACCUG with adenine instead of uracil at the 5′ terminus was found in *Tetrahymena* [512], whereas UCCUCG was found in *Bombyx mori* [511]. Therefore, the 5′ terminal nucleotide sequences of the RNA of the small ribosomal subunit show some degree of sequence homology but are not identical in all eukaryotes.

Homologous internal sequences characterized be odd bases have been found in 18 S rRNA of Novikoff hepatoma, yeast 17 S rRNA and *E. coli* 16 S rRNA [510, 542]. The sequence (m^6_2A-m^6_2A-C-U)-Gp of a hepatoma 18 S rRNA [492] is homologous to the sequence m^6_2A-m^6_2A-C-C-U-Gp of *E. coli* 16 S rRNA [542] and m^6_2A-m^6_2A-C-U-C-Gp of yeast 17 S rRNA [510], and the sequence A-A-C(m^1cap^3Ψ-C-A-C)A-C-Gp of 18 S rRNA is similar to the sequence A-A-C-m^1cap^3 ΨC-A-C-A-C-Gp of 17 S rRNA of yeast [463]. The structural homologies suggest that the evolution of prokaryotic 16 S rRNA to mammalian 18 S rRNA involves mechanisms by which some prokaryotic and lower eukaryotic sequences may have been conserved.

For the 28 S rRNA, only few data on terminal sequences are available. As 5′ termini have been found pU in *Xenopus laevis* [527], pG in avian myeloblasts [506], and pC in mouse [505] and Novikoff hepatoma cells [681].

The extent of homologous regions within ribosomal RNAs of eukaryotes has been

Table 13. Extent of hybridization between DNA and rRNA of various species.
(From Bendich and Mc Carthy [544] and Birnstiel and Grunstein [545])

Source of DNA	Per cent of rRNA		
	HeLa cells	*Xenopus*	Pea
Human	100	59	
Cattle	75	50	
Xenopus	58	100	
Sea urchin	10	13	18
Pea			100

analyzed also by DNA × rRNA hybridization experiments, including hybridization kinetics and hybridization competition [544—548]. Table 13 show that the rate of hybridization differs, depending on the phylogenetic relationship of the organism from which DNA and rRNA were obtained. It is obvious from the data that many of the rRNA sequences have changed significantly during evolution but that some sequences have been conserved [544—548]. Man and *Xenopus* have about 60% common base sequences, whereas man and sea urchin have only about 10% identical base sequences [545]. Comparative fingerprint analysis has revealed that sequences involved in heterologous hybridization have a higher A + T composition than the bulk of the ribosomal RNA. From competition experiments it appears that two-thirds of the conserved similar regions are present in 18 S rRNA and the remaining one-third in 28 S rRNA.

Even plants and at least lower animals have some common base sequences in their rRNAs [544]. In DNA of various species of eukaryotes (mammals, birds, reptiles, amphibia, fish, echinoderms, arthropods, molluscs, protozoa, yeast, plants), different degrees of homology to rRNA from *Xenopus leavis* were found by hybridization experiments, but no complementarity was found between *X. leavis* rRNA and DNA from prokaryotes [546].

Altogether, the degree of sequence alterations seems to have increased with the evolutionary divergence of the species.

Secondary structure

Because the nucleotide sequences of the high molecular weight eukaryotic rRNA have not yet been determined, models of the secondary structure have been constructed from results obtained so far by means of physical and chemical methods, mainly ultra-violet spectroscopy [549 to 555], optical rotatory dispersion [550, 553, 556], circular dichroism [557, 558], infrared spectroscopy [552, 554], X-ray diffraction [559], acid base titration [553, 560, 561], and electron microscopy [410, 412—416, 437, 555, 562].

The secondary structure of rRNA molecules can be described as an arrangement of base-paired regions, so-called hairpin loops, which are connected by unpaired regions of the RNA chain as symbolized in Fig. 16 [549—555, 557—561, 563]. The bihelical hairpin loops are formed by specific base pairings, in which the single polynucleotide chain doubles back upon itself. In solution of sufficient ionic strength, at physiological pH, and at temperature below 30 °C (depending on the ionic conditions), as well as in the ribosomal particles about 60—80% of the rRNA strand seems to be organized in

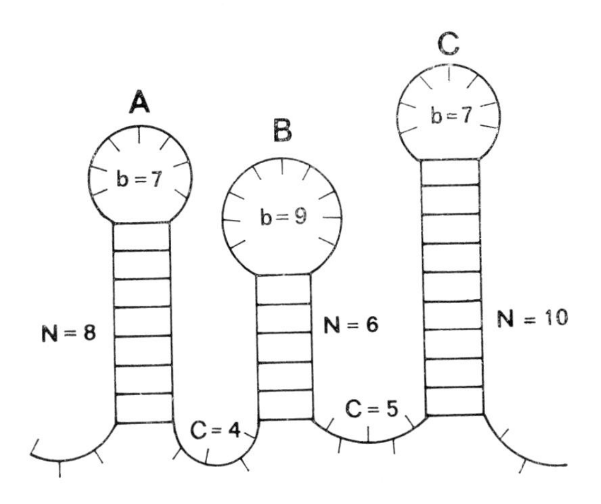

Fig. 16. Simplified model of the secondary structure of rRNA.
N: number of base pairs per loop; b: number of unpaired residues per loop; c: number of residues between the loops.
According to R. A. Cox and E. Goodwin, taken from [5].

hairpin loop structures [550—554, 557]. The number of base pairs per bihelical region in the 28 S type rRNA is higher than in the 18 S rRNA [549, 551, 553, 557]; for RNA from rabbit reticulocyte ribosomes, values of 6—16 and 6—8, respectively, were calculated [549]. The number of unpaired residues within the hairpin loops is about 10—15, and the number of nucleotides connecting the bihelical structures averages 5 [549, 553]. The bihelical regions are characterized by a high content of G-C base pairs, which is in the order of about 40—80% [552, 553, 555, 557, 563]. This points to an uneven distribution of the guanosine and cytosine as well as of the adenosine and cytosine residues along the rRNA chain [554, 555]. Therefore, the total number of bihelical regions in different rRNAs depends mainly on the number of G-C base pairs, independently of the overall nucleotide composition of the RNAs.

Electron microscopic studies of single RNA molecules visualize also bihelical hairpin loop structures [409, 412—416, 437, 555, 562]. Only such hairpins can be detected which are stable enough to be preserved during the experimental procedures. However, also relatively weak hairpin structures can be observed due to hairpin fixation in the rRNA from *Drosophila melanogaster* photoreacted with psoralen. Reproducible patterns of hairpins were shown in both the 26 S and the 18 S rRNA. A 487 nucleotide open loop near one end of the molecule was preserved in the 18 S rRNA; additionally small hairpins were found at specific positions [437].

The conformation of isolated ribosomal RNA depends also on the preparation procedures. RNA of the small subunit of chicken liver ribosomes extracted with acetic acid-urea or with phenol differs in its size as demonstrated by electron microscopy whereas the structure of the 28 S rRNA is rather insensitive against the preparation procedure [416]. The structure of 28 S rRNA is also relatively resistant to formaldehyde denaturation [416] and treatment with urea [408].

Different conformational states were found in 18 S rRNA of HeLa cells [564] and in 40 S precursor rRNA of *Xenopus* [565] by gel electrophoresis. Multiple electrophoretic forms were observed which are probably due to variations in base pairing and stacking that result when the parent form is exposed to temperature differences [564]. A radiochemical method for the investigation of the structure of RNA was suggested by measuring the rate of bisulfite-catalyzed deiodination of pyrimidines in rRNA of rat pituitary tumor cells [566].

Until now only a general and somewhat simplified description of the secondary structure of rRNA can be given on the basis of the available data. More detailed and final conclusions can be expected from the data of sequence analysis and X-ray diffraction of crystallized rRNA molecules.

b) *Low molecular weight ribosomal ribonucleic acids*

The large subunit of cytoplasmic ribosomes contains in addition to the 28 S rRNA two low molecular weight RNA species, 5 S and 5.8 S rRNA, which are present in equimolar ratios to the 28 S rRNA that means one molecule each per 60 S subunit. In the following, some structural properties of both rRNA species are presented, while functional aspects will be described in Chapter IX and their biosynthesis in Chapter VII (for an extended review see [9]). 5 S rRNA was found also in chloroplast ribosomes [567—570] and mitochondrial ribosomes [571, 572], whereas 5.8 S rRNA has not been identified in ribosomes of these organelles. Furthermore, the presence of a so-called 4.5 S rRNA was described for chloroplast ribosomes [570, 573]. Interestingly, eukaryotic 5.8 S rRNA and not 5 S rRNA seems to be closely related to prokaryotic 5 S rRNA with respect to some of their properties.

5 S rRNA

5 S rRNA of cytoplasmic ribosomes from various eukaryotic organisms has been investigated as summarized in Table 14.

Table 14. Survey of various animal and plant species from which 5 S rRNA has been isolated and characterized

Species	References
HeLa cells	[586[a], 587, 589[a], 590, 632]
KB cells	[583, 584[a], 585[a], 588, 592]
Bovine trachea	[595]
Rat liver	[591, 631]
Mouse	[588, 592, 593]
Chinese hamster	[594]
Dolphin kidney	[595]
Chicken	[596[a], 597[a]]
Turtle	[598[a]]
Trout	[599[a], 629[a]]
Iguna	[600[a]]
Drosophila	[589, 601[a], 602[a], 603]
Xenopus laevis (somatic)	[604—606[a]]
Xenopus laevis (ovary)	[605[a], 606[a]]
Xenopus mulleri (somatic)	[607[a]]
Xenopus mulleri (ovary]	[607[a]]
Sea urchin	[608]
Marsupial	[609]
Drosophila melanogaster	[601[a], 602[a], 603]
Rattlesnake	[610]
Yeast	[612, 617, 618]
S. cerevisiae	[574[a], 575[a], 619, 621]
S. carlsbergensis	[575[a], 587, 611[a]]
K. lactis	[575[a]]
P. membranefaciens-1	[575[a]]
P. membranefaciens-2	[575[a]]
T. utilis	[568, 589, 613—615, 616[a], 620]
Rye	[625[a], 626[a]]
Tomato	[625[a], 626[a]]
Sunflower	[625[a], 626[a]]
Dwarf bean	[568, 625[a], 626[a]]
Wheat embryo	[623, 627]
Chlorella	[567, 589, 622, 624[a]]

[a] Estimation of sequences

Most 5 S rRNAs of eukaryotic ribosomes contain 121 nucleotides; a different number of nucleotides has been found in 5 S rRNA from plants (116 to 118 nucleotides).

5 S rRNA is characterized by the lack of odd bases; only in yeast 5 S rRNA one pseudouridine residue has been found [574, 575].

The nucleotide sequence of 5 S rRNA from a considerable number of organisms has been estimated (Table 14). The published sequences have been collected [576, 577] and updated recently [578]. Comparative 5 S rRNA sequence studies were performed mainly to get information about the evolution of ribosomal components [577, 579—582].

Besides complete sequence data also oligonucleotide composition studies revealed that 5 S rRNAs from the following mammalian species are similar to 5 S rRNA from KB cells: Mouse Landschütz tumor cells [592], marsupial [609], Chinese hamster V 79 [594], dolphin kidney [595], embryonic bovine trachea [595], rat pituitary [592] and rabbit reticulocytes [592]. Therefore, it can be concluded that mammalian 5 S rRNA has been highly conserved during evolution. Furthermore, also the sequences of 5 S rRNA from more distant species are fairly similar; e.g., chicken 5 S rRNA differs from mammalian 5 S rRNA in seven positions [596, 597], and from the 5 S rRNA of somatic *Xenopus* cells in eight positions only [605]. An interesting finding was made with the 5 S rRNA from oocytes and somatic cells of *Xenopus* [604, 606, 607]. Kidney cells contain one 5 S rRNA type only, whereas in oocytes at least three different additional 5 S rRNA sequences were found [606, 607]. The major species of the oocyte 5 S rRNA differs in four positions from that of kidney [606, 607], suggesting that the oocyte and somatic genes for 5 S rRNA have been evolved independently.

5 S rRNA from plants contains 116 to 118 nucleotides [622, 624, 625]. 5 S rRNA molecules from different plants are parti-

ally different from one another at both the 3′ and 5′ ends and also in some internal positions [625]. Plant 5 S rRNAs have a GAAC sequence in position 41 to 44; the corresponding sequence in 5 S rRNA of animals and fungi is GAUC. On the basis of the available sequence data of 54 various 5 S rRNAs, a "phylogenic tree" has been constructed from which it was concluded that all eukaryotic 5 S rRNAs including 5 S rRNA from plants (21 known sequences) belong to the same type of molecules, characterized by a principally common secondary structure and a high degree of sequence homology [577].

From the sequence data, physical and optical measurements [589, 608, 617, 618], enzyme digestion studies [589, 614, 616, 621, 623], and chemical modifications [615], possible structural models have been described for 5 S rRNA of KB cells [584], *X. laevis* [605], *Drosophila* [601], yeast [589, 612, 614, 616, 620], wheat embryos [623], and chlorella [624].

For yeast 5 S rRNA the following experimental data have been found: A high ordered phosphodiester back bone structure, indicating a significant content of base-paired helical regions [612]; about 65% base-paired U residues [612]; limited A-stacking [612]; exposed single-stranded regions near positions 40 and 90 [587, 589, 621]; an axial ratio of 5:1, and a radius of gyration of 3.45 nm [618]; 28 ± 3 base pairs [617] (in sea urchin 5 S rRNA, an extensive overall base-pairing of 62 to 64% (about 37 base pairs) was found [608]). From these properties a cloverleaf secondary structure of yeast 5 S rRNA was proposed [612] as shown in Fig. 17. Another secondary structure model has been constructed before mainly on the basis of the primary structure and the nucleolytic degradation behaviour of *Torulopsis utilis* 5 S rRNA [614—616]; this model is shown in Fig. 18. The cloverleaf model [612] and some other suggested models [577, 589, 630] have been claimed to be adaptable to all eukaryotic 5 S rRNA molecules. However, none of these models seem to be compatible with all experimental data; thus alternative interpretations are not excluded. Therefore, further and more precise data are needed before final structural models of the 5 S RNA can be established.

Studies on yeast [617, 618] and sea urchin [608] 5 S rRNA suggest that its tertiary structure depends on the Mg^{++} concentration which especially seems to influence some of the G-C pairs. KB cell [583], HeLa cell [632] and rat liver [591

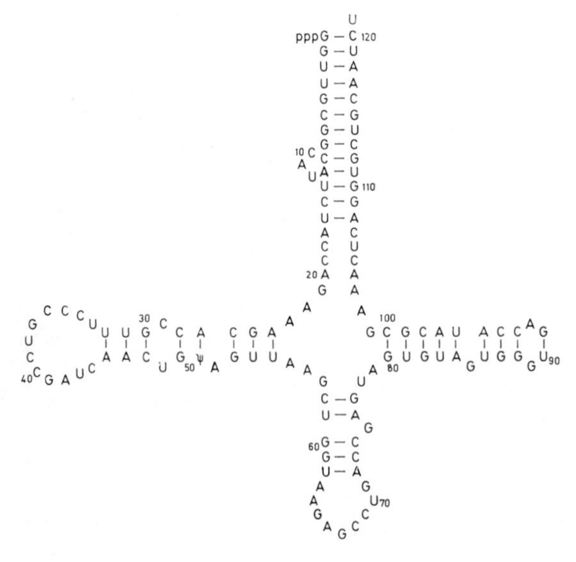

Fig. 17. Proposed cloverleaf secondary structure of *S. cerevisiae* 5 S rRNA.

From G. A. Luoma and A. G. Marshall [612]. With permission of the authors. Copyright by Academic Press Inc. (London) Ltd.

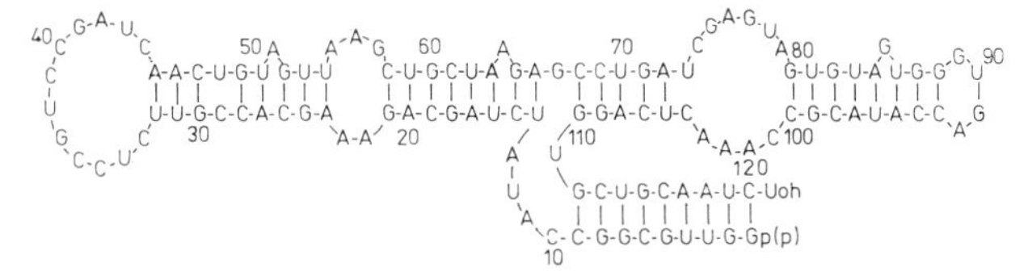

Fig. 18. A possible model of the secondary structure of *Torulopsis utilis* 5 S rRNA. From K. NISHIKAWA and S. TAKEMURA [616]. With permission of the authors and The Japanese Biochemical Society (Tokyo).

5 S RNA where shown to exist in different conformational forms; in gel electrophoresis these 5 S RNA species migrate as two (KB cells, HeLa cells) or three (rat liver) distinct bands representing the native 5 S RNA and thermodynamically stable denaturated forms. The denaturated form can be renatured by incubation of the 5 S rRNA in the presence of sufficient concentrations of Mg^{++} up to 60 °C within 15 min [591]. Renatured and native 5 S RNA have identical electrophoretic mobilities, but not neccesserily identical conformations, which has to be taken into account in investigations on the structure of 5 S RNA.

5.8 S RNA

In eukaryotes the large subunit of cytoplasmic ribosomes contains a 5.8 S rRNA component which is hydrogen bonded to the high molecular-weight rRNA [583, 633]. This RNA species was previously designated as 7 S or 5.5 S RNA, but is now generally named "5.8 S RNA". 5.8 S rRNA contains 158 to 162 nucleotides [576, 578, 634]. In contrast to 5 S rRNA, 5.8. S rRNA contains several modified nucleotides [609, 625, 634—641].

The sequences of the 5.8 S rRNA from the following species have been estimated: HeLa cells [638], Novikoff hepatoma [636], mouse [640], turtle [639], chicken [638], trout [637], *Xenopus laevis* [634, 638], *Xenopus borealis* [634], and yeast [635]. A collection of 5.8 s rRNA sequences has been published [576].

The 5.8 S rRNAs of man, rat, and mouse have practically identical nucleotide sequences [636, 640, 642, 662]. Small differences have been observed on the ends [640]. Turtle 5.8 S rRNA differs from that of mammals in one position close to the 5′ end of the molecule only [639]. Between 5.8 S rRNA of HeLa cells and *Xenopus laevis*, four differences in internal positions and one difference at the 3′ end have been found. Between the 5.8 S rRNA of HeLa cells and chick embryo fibroblasts different nucleotides were found in two positions [638]. Six of these seven interspecific differences are due to base substitutions, while the other difference is one extra nucleotide in the *Xenopus* 5.8 S sequence [638]. The sequence of 5.8 S rRNA of trout differs in eight nucleotides from that of man [637]. 5.8 S rRNA molecules of mammalian and yeast ribosomes show 75% sequence homologies [635, 642]. A small degree of homology only exists between 5.8 S RNA of Novikoff hepatoma ascites cells [643] and 5.8 S RNA of several flowering plants [644]. On the other hand, the nucleotide sequences of 5.8 S rRNA from different plants (broad been, dwarf bean, tomato, sunflower and rye) are very similar [644]. Altogether, from the known sequences it is obvious that the structure of 5.8 S RNA was largely conserved during evolution, from which an important function of 5.8 S RNA within the ribosome can be concluded. There are some suggestions that 5.8 S RNA may be involved in binding of aminoacyl-tRNA and/or in translocation (see Chapter IX).

The genes of prokaryotic 5 S rRNA and eukaryotic 5.8 S rRNA are located in analogous positions of their respective genomes. Both rRNA types are in their final forms part of a common large precursor molecule. The gene of the eukaryotic 5 S rRNA, however, was found in separate specific positions of the genome (see Chapter VII). From these facts it was concluded that prokaryotic 5 S rRNA and eukaryotic 5.8 S rRNA might have a common evolutionary origin [645]. To check this hypothesis, the sequences of 5.8 S rRNA from yeast and hepatoma and of eukaryotic and prokaryotic 5 S rRNAs have been compared using a computer test establishing the best alignment of two sequences allowing a fixed number of insertions or deletions. From the evaluation of possible structural relationships it was concluded that 5.8 S rRNA and prokaryotic (*E. coli*) and eukaryotic 5 S rRNAs (KB cells) have no common evolutionary origin, and they are no more related than random sequences [646]. In this context it is interesting to note that the *E. coli* 5 S rRNA binding proteins L 18 and L 25 interact also with 5.8 S rRNA but not with 5 S rRNA from yeast ribosomes [647]. Interactions have been found also between 5.8 S rRNA but not with 5 S rRNA of rat liver and several proteins of the 50 S ribosomal subunit of *E. coli* [648]. This suggests structural and functional similarities between the 5.8 S rRNA of eukaryotes and the 5 S rRNA of prokaryotes. Pseudouridine and 2′-O-methylated uridine and guanine have been identified. In 5.8 S rRNA from yeast one pseudouridine [635], in that from HeLa cells [638], *Xenopus laevis* [634, 638], and *Xenopus borealis* [634] two pseudouridines and one 2′-O-methylated uridine and one 2′-O-methylated guanine and in the 5.8 S rRNA from turtle [639] one pseudouridine and one 2′-O-methylated uridine and one 2′-O-methylated guanine have been identified. A certain degree of heterogeneity was found in the 5.8 S rRNAs of some species, because not all 5.8 S rRNA molecules contain modified nucleotides. A partially 2′-O-methylated uridine was found in 5.8 S rRNA of chicken [638], turtle [539], *Xenopus* [634, 638], trout [637] and mammals [638, 641, 643, 649]. Significant variations in the levels of 2′-O-methylation of the GGU_m-GGAUp sequence were found in twelve different tissues [649]. The role of this 2′-O-methyluridylic acid residue is unclear. In 5.8 S RNA of *Xenopus* [634] and mammals [649] pseudouridine occurs in about half of the molecules at position 19 and 57, respectively. The pseudouridine content in 5.8 S RNA of different mammalian tissues in relatively constant [649].

Length heterogeneities were observed for 5.8 S RNAs of mammals [642], *Xenopus* [634], trout [637] and yeast [650]. 5.8 S RNA molecules with 160 and 162 nucleotides have been found in *Xenopus laevis* besides the major species with 161 nucleotides [634]. The major species of 5.8 S RNA in yeast contains 158 nucleotides; besides two minor species exist comprising 5% of the total amount of 5.8 S RNA each, which have 6 and 7 additional nucleotides at the 5′ terminus [650].

Preparations of the ribosomal RNA from the large subunits of cytoplasmic eukaryotic ribosomes contain the 5.8 S RNA in a complex with the high molecular weight 28 S type rRNA [583, 619, 632, 633, 637, 651—657]. The 5.8 S rRNA molecule can be released from this complex by gentle heating and by treatment with urea or dimethylsulfoxide [619, 655]. The complex has been restored by appropriate annealing procedures yielding a complex with thermal denaturation properties identical with those of the native 5.8—28 S rRNA complex [627, 654]. The 3′-terminal located 42—43 nucleotides of the 5.8 S RNA molecule are protected against ribonuclease digestion in the complex with 28 S rRNA, but only 20 to 21 3′-terminal nucleotides seem to be associated with the 28 S RNA by complementary base interactions, whereas the other part of the stable 5.8 S RNA fragment forms a G-C rich loop which stabilizes the interaction with the 28 S RNA [654].

5.8 S RNA has been released also from

intact large subunits of rabbit reticulocyte ribosomes [653] and from ribosomes of the rattlesnake by heating and from yeast ribosomes by heating [619, 659] or treatment with 50% formaldehyde [659]. The release of 5.8 S RNA was not accomplished by a loss of proteins from the particles [651, 658, 659] and therefore it was concluded that the interaction between 5.8 S rRNA and 28 S RNA is not influenced decisively by ribosomal proteins. Reassociation of 5.8 S RNA to 5.8 RNA depleted yeast ribosomes was possible by dialysis under appropriate ionic conditions [659].

Several models of the secondary structure of 5.8 S RNA have been proposed [635, 637, 642, 654, 660]. Estimations of the secondary structure of 5.8 S RNA from Novikoff hepatoma ascites cells based on maximal base pairing and the results of partial ribonuclease digestion suggested the presence of five base-paired regions in the molecule [642]. The model contains a very stable G-C rich hairpin loop (residues 116—138), a less stable A-U rich hairpin loop (residues 64—91) and two symmetrical bulges (residues 15—25 and 40—44) as shown in Fig. 19. Yeast 5.8 S RNA seems to represent all essential structural features found also for rat 5.8 rRNA; this was concluded from denaturation spectra, derivative profiles of hyperchromicity vs. temperature and ethidium bromide probing. Despite the much lower G-C content in yeast 5.8 S RNA in comparison to rat 5.8 rRNA, similar models regarding the secondary structure have been proposed [661], differing from a model suggested before for 5.8 S RNA from yeast [635]. The results of S 1 nuclease digestion [662] and of the reaction of specific cytidine residues of 5.8 S rRNA from Hela cells with sodium bisulphite [663] are consistent with the proposed secondary structure model [642] of Novikoff hepatoma 5.8 S rRNA. An estimate of the secondary structure of 5.8 S rRNA from trout retained also essentially all of the features found for mammalian 5.8 S RNA, but the very stable G-C-rich loop is extended by two additional hydrogen bound pairs while the A-U-rich loop is shorter [637].

Laser Raman spectroscopic investigations of 5.8 S RNA from yeast led more recently to the proposal of a new cloverleaf secondary structure for eukaryotic 5.8 S rRNA (see Fig. 20) [660]. The experiments have been performed with renatured 5.8 S rRNA samples obtained by heating for 5 min at 65°C in the presence of Mg^{++}. The RAMAN data suggest that yeast 5.8 S RNA has a highly ordered backbone structure and a high degree of base pairing, an extensive G-C stacking, more than 70% base-paired U residues and significant U stacking, a smaller structural requirement for Mg^{++} than tRNA and only a moderate A stacking. This new model [660] is in better agree-

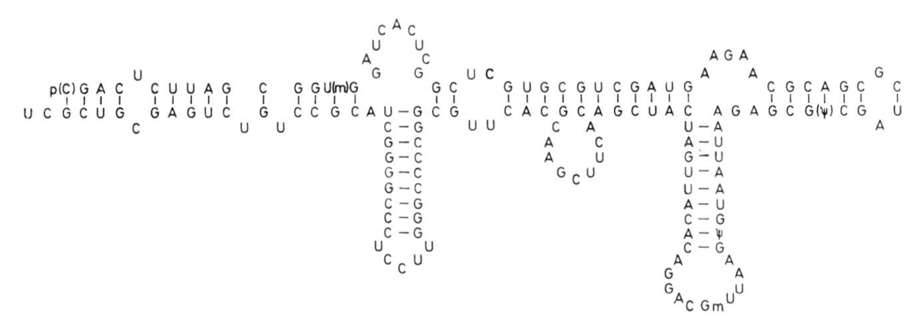

Fig. 19. A possible model of the secondary structure of Novikoff ascites hepatoma 5.8 S rRNA. From R. N. Nazar et al. [642]. With permission of the authors and The American Society of Biological Chemists (Baltimore).

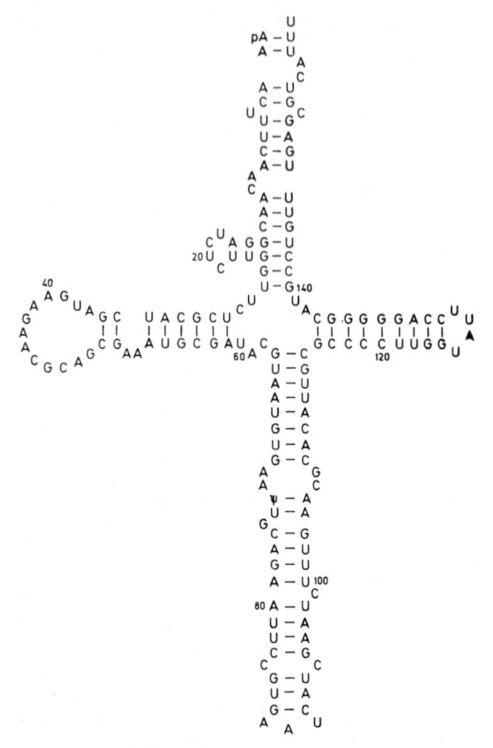

Fig. 20. Proposed cloverleaf secondary structure of *S. cerevisiae* 5.8 S rRNA. From G. A. LUOMA and A. G. MARSHALL [660]. With permission of the authors.

ment with these data than the previously proposed ones [635, 642]; it was also adopted for the 5.8 S rRNA of Novikoff hepatoma ascites cells and is consistent with the results of partial enzymatic hydrolysis of Novikoff hepatoma 5.8 S rRNA [642]. However, it has to be taken into account that alternative interpretations are not excluded.

The structure of 5.8 S RNA seems to be quite different in free form and in complex with 28 S RNA. At the ribosome the stem region of the free form [635, 642] becomes unpaired allowing base pairing of its 3′ end to the 28 S RNA [654] which possibly destabilizes also other regions of the molecule [660].

Multimer forms of 5.8 S RNA from rat and chicken liver and yeast have been observed by gel electrophoresis after heating of the samples in Mg^{++} free buffers [664]. The appearance of such forms can be explained by intermolecular base pairing of terminal sequences in concentrated solutions of 5.8 S RNA after melting of intramolecular hydrogen bonds.

5.8 S RNA contains a particularly stable hairpin helix. This section of the molecule was isolated by partial T 1 ribonuclease digestion of 5.8 S RNA from yeast [665]. The hairpin fragment (containing 26 nucleotides in positions C_{115} to G_{140}), dissociates cooperatively with 18% hyperchromism with a t_m value of 83 °C at 2.7 mM sodium, contains 90% G + C, and 9 base pairs. The fragment seems to be normally folded into a stable helical structure in the form of an unimolecular hairpin helix rather than a bimolecular double helix [643].

5. Proteolytic and nucleolytic activities of ribosomal particles

The presence of an exopeptidase [666] and more recently also of an endopeptidase activity in rat liver [667—672] and HeLa cell ribosomes [673, 674] has been reported. The endopeptidase activity is firmly bound to the ribosomal particles and cannot be washed off with 0.5 M NH_4Cl or 0.5 M KCl [667, 670]. The endopeptidase has an optimum at pH 7, splits preferentially Phe-Tyr-bonds and, more slowly, also other bonds, and can be inhibited by 1-chloro-4-phenyl-3-tosyl-amido-2-butanone and soya bean inhibitor [667, 670]. The protease was found on the 40 S subunit after dissociation of rat liver ribosomes with puromycin [668]. Extracted ribosomal proteins exhibit the same endopeptidase activity [667].

A further proteolytic enzyme attached to rat liver ribosomes has been described, characterized by a very specific effect on two large ribosomal subunit proteins in an autolysis assay. This enzyme could be detached from ribosomes at high ionic strength [674].

Ribonuclease activities were found in preparation of ribosomes from *Euglena* [675], higher plants [676], rat skeletal muscle [677], rat liver [678—680], and rat hepatomas [679]. This enzymatic activity is usually bound very firmly to ribosomes prepared by standard procedures and cannot be detached by repeated washes from the ribosomes. Using modified preparation conditions, however, rat liver ribosomes have been prepared free of any traces of ribonuclease activities [680].

Up to now it has not been possible to correlate any of the ribosome-bound proteolytic or RNAase activities to definite ribosomal proteins, nor is anything known about their possible biological importance.

6. Other components

Monovalent and divalent cations — mainly K^+ and Mg^{++}, respectively — and diamines have been found as integral components in preparations of eukaryotic ribosomes.

Magnesium ions are highly important for structural and functional properties of ribosomes, which has been recognized already in the very first studies on ribosomal preparations (summarized in [682]); e.g., 0.25 to 0.29 μmoles of Mg^{++} and about 0.04 μmoles of Ca^{++} were found in ribosomes from pea seedlings [683]. These values correspond to 1600 to 1800 magnesium ions per 80 S ribosome, which is equivalent to about 1% of the dry mass of the ribosomal particle. Further estimations of the magnesium content of ribosomes gave values of 0.1 to 0.3 μmole per μmole phosphorus in ribosomal preparations from yeast [684], rabbit reticulocytes [685], Guinea pig pancreas [686], and Jensen sarcoma [687]. Animal and plant 80 S ribosomes can bind up to 0.5 μmole of Mg^{++} per μmole of phosphorus in buffers containing magnesium ions in the relatively high concentration of 0.01 M [684, 687]. Consequently, under such conditions the ribosomal RNA seems to be present within the ribosome almost completely in the form of Mg^{++} salt.

Monovalent cations substitute Mg^{++} to some extent as counterions in ribosomal particles. The content of monovalent cations in the media and the presence of polyamines within ribosomal particles influence the amount of Mg^{++} bound to ribosomes and the effects of Mg^{++} on the association-dissociation behaviour of ribosomal subunits and ribosomes, respectively (see Chapter VIII), and on the structural integrity of the particles.

Spermidine and cadaverine and in lower quantities also 1,3-diaminopropane were found in liver ribosomes [688], whereas in ribosomes from the pancreas and liver of guinea pigs only low quantities of spermine and another unidentified polyamine were found [689]. Spermidine, putrescine and traces of spermine have been found in yeast ribosomes [690].

Unwashed rabbit reticulocyte ribosomes contain per one nmol about 240 nmoles of spermine and spermidine. Only 36 nmoles each of spermine and spermidine per one nmol of ribosomes were found in salt-washed reticulocyte ribosomes prepared by centrifugation of the ribosomes from a solution containing 0.5 M KCl [691]. The polyamine content of salt-washed ribosomes from undeveloped cysts of *Artemia salina* was found to be less than half that of 0.5 M KCl-washed reticulocyte ribosomes [691]. No differences in the content of polyamines in the small and in the large ribosomal subunits could be detected [692].

Addition of polyamines to ribosomal preparations increases the stability of the ribosomal particles, prevents their dissociation into subunits at low Mg^{++} and increases their activity in cell-free translation systems [682]. The effect of polyamines on protein synthesis can be substituted partly by Mg^{++}, but, in addition, a specific effect on the initiation of protein synthesis has been described [692]. This follows from the fact that spermine stimulates the binding of globin mRNA to 40 S ribosomal subunits [693]. Experiments with edeine, a peptide antibiotic containing spermidine, suggest that one polyamine binding site on 40 S subunits may be involved in the reaction sequence at the level at which the GTP of the ternary complex, GTP × eIF-2 × Met-$tRNA_f$, is hydrolyzed to GDP [694].

Furthermore it was found in a poly(U) directed wheat germ cell-free system that polyamines increase the fidelity of protein synthesis as measured by the rate of misincorporation of leucine in place of phenylalanine [695, 696], and in experiments with TMV RNA it could be shown that the accumulation of incomplete chains is decreased by spermidine [696].

7. References

[1] Traugh, J. A., and R. R. Traut: Recent advances in the preparation of mammalian ribosomes and analysis of their protein components. Meth. Cell Biol. **7**, 67—103 (1973)

[2] Wool, I. G., and G. Stöffler: Structure and function of eukaryotic ribosomes. In: Ribosomes. (Eds. M. Nomura, A. Tissiéres, and P. Lengyel). Cold Spring Harbor Laboratory, 1974, 417 to 460

[3] Bielka, H., F. Noll, J. Stahl, H. Theise, H. Welfle, and P. Westermann: Proteins of eukaryotic ribosomes. Proc. Nordic Soc. Cell Biol. **9**, 185—196 (1976)

[4] Cox, R. A.: Structure and function of prokaryotic and eukaryotic ribosomes. Progr. Biophys. Mol. Biol. **32**, 193—231 (1977)

[5] Bielka, H., and J. Stahl: Structure and function of eukaryotic ribosomes. Int. Rev. Biochem. **18**, 79—168 (1978)

[6] Leader, D. P.: The control of phosphorylation of ribosomal proteins. In: Recently discovered systems of enzyme regulation by reversible phosphorylation. (Ed.: P. C. Cohen). Elsevier, North-Holland, Amsterdam, 1980, 203—233

[7] Wool, I. G.: The structure and function of eukaryotic ribosomes. Ann. Rev. Biochem. **48**, 719—754 (1979)

[8] Maden, B. E. H., M. Salim, and J. S. Robertson: Progress in the structural analysis of mammalian 45 S and ribosomal RNA. In: Ribosomes (Eds.: M. Nomura, A. Tissiéres, and P. Lengyel). Cold Spring Harbor Laboratory, 1974, 829—839

[9] Erdmann, V. A.: Structure and function of 5 S and 5.8 S RNA. Molecular Biology (Ed.: E. W. Cohn). Academic Press, New York, 1976, **18**, 45—90

[10] Hadjiolov, A. A., and N. Nikolaev: Maturation of ribosomal ribonucleic acids and the biogenesis of ribosomes. Progr. Biophys. Mol. Biol. **31**, 95—144 (1967)

[11] Mutolo, V., G. Giudice, V. Hopps, and G. Donatuti: Species specificity of embryonic ribosomal proteins. Biochim. Biophys. Acta **138**, 214—217 (1967)

[12] Low, R. B., and I. G. Wool: Mammalian ribosomal protein: analysis by electrophoresis on polyacrylamide gel. Science **155**, 330—332 (1967)

[13] Welfle, H., and H. Bielka: Charakterisierung von Proteinen aus Leber- und Hepatom-Ribosomen durch Polyacrylamid-Gelelektrophorese. Z. Naturforsch. **23b**, 690—694 (1968)

[14] Bielka, H., and H. Welfle: Characterization of ribosomal proteins from different tissues and species of animals by electrophoresis on polyacrylamide gel. Molec. Gen. Genetics **102**, 128—131 (1968)

[15] Di Girolamo, M., and P. Cammarano: The protein composition of ribosomes and ribosomal subunits from animal tissues. Electrophoretic studies. Biochim. Biophys. Acta **168**, 181—194 (1968)

[16] Gualerzi, C., and P. Cammarano: Species specificity of ribosomal proteins from chloroplast and cytoplasmic ribosomes of higher plants. Electrophoretic studies. Biochim. Biophsy. Acta **199**, 203—213 (1969)

[17] Gualerzi, C., and P. Cammarano: Comparative electrophoretic studies on the protein of chloroplast and cytoplasmic ribosomes of spinach leaves. Biochim. Biophys. Acta **190**, 170—186 (1969)

[18] ODINTSOVA, M. S., and N. P. YURINA: Proteins of chloroplast and cytoplasmic ribosomes. J. Mol. Biol. **40**, 503–506 (1969)

[19] WELFLE, H., H. BIELKA, and M. BÖTTGER: Studies on proteins of animal ribosomes. II. Separation of ribosomal proteins from rat liver by preparative polyacrylamide gel electrophoresis and some properties of the protein fractions. Molec. Gen. Genetics **104**, 165–171 (1969)

[20] WESTERMANN, P., H. BIELKA, and M. BÖTTGER: Studies on proteins of animal ribosomes. I. Separation of ribosomal proteins from rat liver by chromatography on CM-cellulose and properties of some protein components. Molec. Gen. Genetics **104**, 157–164 (1969)

[21] KANAI, K., J. J. CASTLES, I. G. WOOL, W. S. STIREWALT, and A. KANAI: The proteins of liver and muscle ribosomal subunits: partial separation by carboxymethyl-cellulose column chromatography. FEBS-Lett. **5**, 68–72 (1969)

[22] GUALERZI, C.: Electrophoretic comparison of cytoplasmic and mitochondrial ribosomal proteins from *Neurospora crassa*. Ital. J. Biochem. **18**, 418–425 (1969)

[23] HOOBER, J. K., and G. BLOBEL: Characterization of the chloroplastic and cytoplasmic ribosomes of *Chlamydomonas reinhardii*. J. Mol. Biol. **41**, 121–138 (1969)

[24] LOW, R. B., I. G. WOOL, and T. E. MARTIN: Skeletal muscle ribosomal proteins: general characteristics and effect of diabetes. Biochim. Biophys. Acta **194**, 190 to 202 (1969)

[25] GOULD, H. J.: Proteins of rabbit reticulocyte ribosomal subunits. Nature **227**, 1145–1147 (1970)

[26] BICKLE, T. A., and R. R. TRAUT: Differences on size and number of 80 S and 70 S ribosomal proteins by SDS gel electrophoresis. J. Biol. Chem. **246**, 6828 to 6834 (1971)

[27] BURKA, E. R., and S. I. BULOVA: Heterogeneity of reticulocyte ribosomes. Biochem. Biophys. Res. Commun. **42**, 801 to 805 (1971)

[28] KING, H. W. S., H. J. GOULD, and J. J. SHEARMAN: Molecular weight distribution of proteins in rabbit reticulocyte ribosomal subunits. J. Mol. Biol. **61**, 143–156 (1971)

[29] HOUSSAIS, J.-F.: Nature de l'hétérogénéité interspécifique des protéines des ribosomes de mammifères. Europ. J. Biochem. **24**, 232–241 (1971)

[30] MACINNES, J. W.: Differences between ribosomal subunits from brain and those from other tissues. J. Mol. Biol. **65**, 157 to 161 (1972)

[31] TERAO, K., and K. OGATA: Characterization of the proteins of the small subunits of rat liver ribosomes. Biochim. Biophys. Acta **285**, 473–482 (1972)

[32] OCHIAI, H., F. KANDA, and M. IWABUCHI: The number and size of ribosomal proteins in the cellular slime mold *Dictyostelium discoideum*. J. Biochem. **73**, 163–167 (1973)

[33] KANDA, F., H. OCHIAI, and M. IWABUCHI: Molecular-weight determination and stoichiometric measurements of 40 S and 60 S ribosomal proteins of the cellular slime mold *Dictyostelium discoideum*. Europ. J. Biochem. **44**, 469–479 (1974)

[34] MCCONKEY, E. H., and E. J. HAUBER: Evidence for heterogeneity of ribosomes within the HeLa cell. J. Biol. Chem. **250**, 1311–1318 (1975)

[35] THOMAS, H.: Gel electrophoresis of ribosomal components from seeds of *Pisum sativum*. Exptl. Cell Res. **77**, 298–302 (1973)

[36] LAMBERTSSON, A. G., S. B. RASMUSON, and G. D. BLOOM: The ribosomal proteins of *Drosophila melanogaster*. 1. Characterization in polyacrylamid gel of proteins from larval, adult, and ammonium chloride-treated ribosomes. Molec. Gen. Genetics **108**, 349–357 (1970)

[37] HALLBERG, R. L., and C. A. SUTTON: Nonidentity of ribosomal structural proteins in growing and starved *Tetrahymena*. J. Cell Biol. **75**, 268–276 (1977)

[38] KALTSCHMIDT, E., and H. G. WITTMANN: Ribosomal proteins, XII. Number of proteins of small and large ribosomal subunits of *Escherichia coli* as determined by two-dimensional gel electrophoresis. Proc. Natl. Acad. Sci. U.S. **67**, 1276 to 1282 (1970)

[39] KALTSCHMIDT, E., and H. G. WITTMANN: Ribosomal proteins. VII. Two-dimensional

polyacrylamide gel electrophoresis for fingerprinting of ribosomal proteins. Anal. Biochem. **36**, 401—412 (1970)

[40] Welfle, H.: Studies on proteins of animal ribosomes. XI. A simple method of two-dimensional polyacrylamide gel electrophoresis of ribosomal proteins of rat liver. Acta Biol. Med. Germ. **28**, 547 to 551 (1971)

[41] Hultin, T., and A. Sjöqvist: Analogies in protein pattern and conformation among ribosomes from different classes of vertebrates. Comp. Biochem. Physiol. **40 B**, 1011—1027 (1971)

[42] Martini, O. H. W., and H. J. Gould: Enumeration of rabbit reticulocyte ribosomal proteins. J. Mol. Biol. **62**, 403 to 405 (1971)

[43] Howard, G. A., and R. R. Traut: A modified two-dimensional gel system for the separation and radioautography of microgram amounts of ribosomal proteins. Meth. Enzymol. **30**, 526—539 (1974)

[44] Hamilton, M. G.: Estimation of the number and size of ribosomal proteins by two-dimensional gel electrophoresis. Meth. Enzymol. **30 F**, 540—545 (1974)

[45] Lin, A., E. Collatz, and I. G. Wool: Micro-scale two-dimensional polyacrylamide gel electrophoresis of ribosomal proteins. Molec. Gen. Genetics **144**, 1—9 (1976)

[46] Hoffman, W. L., and J. Ilan: Complete solubilization of ribosomal proteins during the fractionation of mouse liver ribosomal proteins by two-dimensional polyacrylamide gel electrophoresis. Prep. Biochem. **6**, 13—26 (1976)

[47] Hoffman, W. L., and R. M. Dowben: Two-dimensional polyacrylamide gel electrophoresis of ribosomal proteins: improved resolution with Triton X-100. Anal. Biochem. **89**, 540—549 (1978)

[48] Savic, A., and D. Poccia: Separation of histones from contaminating ribosomal proteins by two-dimensional gel electrophoresis. Anal. Biochem. **88**, 573—579 (1978)

[49] Bosselman, R. A., and M. S. Kaulenas: A rapid two-dimensional microelectrophoretic procedure for the analysis of ribosomal proteins. Anal. Biochem. **70**, 281—284 (1976)

[50] Mets, L. J., and L. Bogorad: Two-dimensional polyacrylamide gel electrophoresis: an improved method for ribosomal proteins. Anal. Biochem. **57**, 200 to 210 (1974)

[51] Hultin, T., and A. Sjöqvist: Two-dimensional polyacrylamide gel electrophoresis of animal ribosomal proteins based on charge inversion. Anal. Biochem. **46**, 342—346 (1972)

[52] Bernabeu, C., O. Martinez, D. Vazquez, and J. P. G. Ballesta: A tracking marker for the second dimension of the two dimensional gel electrophoresis of ribosomal proteins. Anal. Biochem. **92**, 203—204 (1979)

[53] Lastick, S. M., and E. H. McConkey: Exchange and stability of HeLa ribosomal proteins in vivo. J. Biol. Chem. **251**, 2867—2875 (1976)

[54] Huynh-Van-Tan, J. Delaunay, and G. Schapira: Eukaryotic ribosomal proteins. Two-dimensional electrophoretic studies. FEBS-Lett. **17**, 163—167 (1971)

[55] Welfle, H., J. Stahl, and H. Bielka: Studies on proteins of animal ribosomes. 13. Enumeration of ribosomal proteins of rat liver. FEBS Lett. **26**, 228—232 (1972)

[56] Welfle, H., J. Stahl, and H. Bielka: Studies on proteins of animal ribosomes. 8. Two-dimensional polyacrylamide gel electrophoresis of ribosomal proteins of rat liver. Biochim. Biophys. Acta **243**, 416—419 (1971)

[57] Sherton, C. C., and I. G. Wool: Determination of the number of proteins in liver ribosomes and ribosomal subunits by two-dimensional polyacrylamide gel electrophoresis. J. Biol. Chem. **247**, 4460 to 4467 (1972)

[58] Chatterjee, S. K., M. Kazemie, and J. H. Matthaei: Separation of the ribosomal proteins by two-dimensional electrophoresis. Hoppe-Seyler's Z. Physiol. Chem. **354**, 481—486 (1973)

[59] Peeters, B., L. Vanduffel, A. Depuydi, and W. Rombauts: The number and size of the proteins in the subunits of human placental ribosomes. FEBS-Lett. **36**, 217—221 (1973)

[60] Pratt, H., and R. A. Cox: Proteins from biologically active ribosomal subparticles of *Xenopus laevis*. Biochim. Biophys. Acta **310**, 188—204 (1973)

[61] HUYNH-VAN-TAN, M. GAVRILOVIC, and G. SCHAPIRA: Eukaryotic ribosomal proteins: The number of proteins in the subunits and their isoelectric points. FEBS-Lett. **45**, 299—303 (1974)

[62] SUNDKVIST, I. C., and G. A. HOWARD: Comparison of activity and protein content of ribosomal subunits prepared by four different methods from rabbit reticulocytes. FEBS-Lett. **41**, 287—291 (1974)

[63] ÜCER, U., and E. BERMEK: Separation of the proteins in human tonsillar cytoplasmic ribosomes by two-dimensional polyacrylamide gel electrophoresis. Europ. J. Biochem **50**, 183—189 (1974)

[64] SHERTON, C. C., and I. G. WOOL: A comparison of the proteins of rat skeletal muscle and liver ribosomes by two-dimensional polyacrylamide gel electrophoresis. J. Biol. Chem. **249**, 2258—2267 (1974)

[65] HOWARD, G. A., J. A. TRAUGH, E. A. CROSER, and R. R. TRAUT: Ribosomal proteins from rabbit reticulocytes: Number and molecular weights of proteins from ribosomal subunits. J. Mol. Biol. **93**, 391—404 (1975)

[66] ANDERSON, W. M., A. GRUNDHOLM, and B. H. SELLS: Modification of ribosomal proteins during liver regeneration. Biochem. Biophys. Res. Commun. **62**, 669 to 676 (1975)

[67] TERAO, K., and K. OGATA: Studies on structural proteins of the rat liver ribosomes. I. Molecular weights of the proteins of large and small subunits. Biochim. Biophys. Acta **402**, 214—229 (1975)

[68] TERAO, K., Y. TAKAHASHI, and K. OGATA: Differences between the protein moieties of active subunits and EDTA-treated subunits of rat liver ribosomes with specific references to a 5 S rRNA · protein complex. Biochim. Biophys. Acta **402**, 230—237 (1975)

[69] HANNA, N., G. BELLEMARE, and C. GODIN: Free and membrane-bound ribosomes. 1. Separation by two-dimensional gel electrophoresis of proteins from rat liver monosomes. Biochim. Biophys. Acta **331**, 141—145 (1973)

[70] HANNA, N., and C. GODIN: Free and membrane-bound ribosomes. 3. Analysis by two-dimensional gel electrophoresis of proteins from liver ribosomal subunits of rats with different dietary intake of phenylalanine. Biochim. Biophys. Acta **374**, 342—349 (1974)

[71] FEHLMANN, M., G. BELLEMARE, and C. GODIN: Free and membrane-bound ribosomes. 2. Two-dimensional gel electrophoresis of proteins from free and membrane-bound rabbit reticulocyte ribosomes. Biochim. Biophys. Acta **378**, 119 to 124 (1975)

[72] FEHLMANN, M., G. BELLEMARE, and C. GODIN: Free and membrane-bound rabbit reticulocyte ribosomes. Proteins from the large and the small subunits. FEBS-Lett. **59**, 8—12 (1975)

[73] DONDI, PH. G.: Analysis, by two-dimensional gel electrophoresis, of ribosome crystal proteins. Biochem. J. **149**, 475—476 (1975)

[74] CREUSOT, F., J. DELAUNAY, and G. SCHAPIRA: Eukaryotic ribosomal proteins. Molecular weights of proteins from rabbit liver subunits. Biochimie **57**, 167 to 173 (1975)

[75] LUGNIER, A. A. J., G. DIRHEIMER, J.-J. MADJAR, J.-P. REBOUD, J. GORDON, and G. A. HOWARD: Action of ricin from *Ricinus communis* L. seeds on eukaryotic ribosomal proteins. FEBS-Lett. **67**, 343 to 347 (1976)

[76] HOFFMAN, W. L., and J. ILAN: Analysis by two-dimensional polyacrylamide gel electrophoresis of liver ribosomal subunit proteins obtained from free and membrane-bound polysomes of unfasted animals. Biochim. Biophys. Acta **474**, 411—424 (1977)

[77] SCHIFFMAN, D., and I. HORAK: Ribosomal proteins of HeLa cells. Europ. J. Biochem. **82**, 91—95 (1978)

[78] WELFLE, H., M. GOERL, and H. BIELKA: Number and molecular weights of the basic proteins of rat liver ribosomes. Molec. Gen. Genetics **163**, 101—112 (1978)

[79] ISSINGER, O.-G., and H. BEIER: Characterisation of ribosomal proteins from HeLa and Kerbs II ascites tumor cells by different two-dimensional polyacrylamide gel electrophoresis techniques. Molec. Gen. Genetics **160**, 297—309 (1978)

[80] MADJAR, J.-J., M. ARPIN, M. BUISSON, and J.-P. REBOUD: Spot position of rat liver ribosomal proteins by 4 different 2-dimensional electrophoreses in polyacryl-

amide gel. Molec. Gen. Genetics **171**, 121—134 (1979)

[81] Cazillis, M., and J.-F. Houssais: The polysomal proteins of L cells. Discrimination between the structural ribosomal proteins, the exchangeable ribosomal proteins and the non-ribosomal proteins by two-dimensional dodecylsulfate electrophoresis and autoradiography. Europ. J. Biochem **93**, 23—30 (1979)

[82] McConkey, E. H., H. Bielka, J. Gordon, S. M. Lastick, A. Lin, K. Ogata, J.-P. Reboud, J. A. Traugh, R. R. Traut, J. R. Warner, H. Welfle, and I. G. Wool: Proposed uniform nomenclature for mammalian ribosomal proteins. Molec. Gen. Genetics **169**, 1—6 (1979)

[83] Lewis, J. A., and D. D. Sabatini: Proteins of rat liver free and membrane-bound ribosomes. Modification of two large subunit proteins by a factor detached from ribosomes at high ionic strength. Biochim. Biophys. Acta **478**, 331 to 349 (1977)

[84] Martini, O. H. W., and H. J. Gould: Characterisation of eukaryotic ribosomal proteins. Molec. Gen. Genetics **142**, 299 to 316 (1975)

[85] Martini, O. H. W., and H. J. Gould: Molecular weight distribution of ribosomal proteins from several vertebrate species. Molec. Gen. Genetics **142**, 317 to 331 (1975)

[86] Ogata, K., and K. Terao: Direct approach to the enumeration of structural proteins of the large ribosomal subunit of rat liver. Acta Biol. Med. Germ. **33**, 539 to 545 (1974)

[87] Kuter, D. J., and A. Rodgers: The synthesis of ribosomal protein and ribosomal RNA in a rat-mouse hybrid cell line. Exptl. Cell Res. **91**, 317—325 (1975)

[88] Kumar, A., and A. R. Subramanian: Ribosome assembly in HeLa cells: Labelling pattern of ribosomal proteins by two-dimensional resolution. J. Mol. Biol. **94**, 409—423 (1975)

[89] Boersma, D., S. M. McGill, J. W. Mollenkamp, and D. J. Roufa: Emetine resistance in chinese hamster cells. Analysis of ribosomal proteins prepared from mutant cells. J. Biol. Chem. **254**, 559 to 567 (1979)

[90] Gumilevskaya, N. A., S. K. Alekhina, L. V. Chumikina, and V. L. Kretovich: Study of basic proteins of 60 S- and 40 S-subunits of pea seed ribosomes by two-dimensional polyacrylamide gel electrophoresis. Biokhimiya (russ.) **42**, 1774 to 1782 (1977)

[91] Bequeret, J., M. Perrot, and M. Crouzet: Ribosomal proteins in the fungus *Podospora anserina*: Evidence for an electrophoretically altered 60 S protein in a cycloheximid resistant mutant. Molec. Gen. Genetics. **156**, 141—144 (1977)

[92] Rodrigues-Pousada, C., and D. H. Hayes: Ribosomal subunits from *Tetrahymena pyriformis*. Isolation and properties of active 40 S and 60 S subunits. Europ. J. Biochem. **89**, 407—415 (1978)

[93] Floyd, G. A., W. C. Merrick, and J. A. Traugh: Identification of initiation factors and ribosome-associated phosphoproteins by two-dimensional polyacrylamide gel electrophoresis. Europ. J. Biochem. **96**, 277—286 (1979)

[94] Gualerzi, C., H. G. Janda, H. Passow, and G. Stöffler: Studies on the protein moiety of plant ribosomes. Enumeration of the proteins of the ribosomal subunits and determination of the degree of evolutionary conservation by electrophoretic and immunochemical methods. J. Biol. Chem. **249**, 3347—3355 (1974)

[95] Kruiswijk, T., and R. J. Planta: Analysis of the protein composition of yeast ribosomal subunits by two-dimensional polyacrylamide gel electrophoresis. Mol. Biol. Rep. **1**, 409—415 (1974)

[96] Kruiswijk, T., and R. J. Planta: Further analysis of the protein composition of yeast ribosomes. FEBS-Lett. **58**, 102 to 105 (1975)

[97] Zinker, S., and J. R. Warner: The ribosomal proteins of *Saccharomyces cerevisiae*. Phosphorylated and exchangeable proteins. J. Biol. Chem. **251**, 1799—1807 (1976)

[98] Grankowski, N., W. Kudlicki, E. Palén, and E. Gașior: Proteins of yeast ribosomal subunits: number and general properties. Acta Biochim. Polon. **23**, 341 to 351 (1976)

[99] Ishiguro, J.: Study on proteins from yeast cytoplasmic ribosomes by two-dimensional gel electrophoresis. Molec. Gen. Genetics **145**, 73—79 (1976)

[100] CODDINGTON, A., and R. FLURI: Characterisation of the ribosomal proteins from *Schizosaccharomyces pombe* by two-dimensional polyacrylamide gel electrophoresis. Molec. Gen. Genetics. **158**, 93 to 100 (1977)

[101] OTAKA, E., and K. KOBATA: Yeast ribosomal proteins. I. Characterization of cytoplasmic ribosomal proteins by two-dimensional gel electrophoresis. Molec. Gen. Genetics. **162**, 259—268 (1978)

[102] KRUISWIJK, T., R. J. PLANTA, and W. H. MAGER: Quantitative analysis of the protein composition of yeast ribosomes. Europ. J. Biochem. **83**, 245—252 (1978)

[103] WALDRON, C., and B. S. COX: Ribosomal proteins of yeast strains carrying mutation which affect the efficiency of nonsense suppression. Molec. Gen. Genetics **159**, 223—225 (1978)

[104] HANSON, M. R., J. N. DAVIDSON, L. J. METS, and L. BOGORAD: Characterization of chloroplast and cytoplasmic ribosomal proteins of *Chlamydomonas reinhardii* by two-dimensional gel electrophoresis. Molec. Gen. Genetics **132**, 105—118 (1974)

[105] BRÜGGER, M., and A. BOSCHETTI: Two-dimensional gel electrophoresis of ribosomal proteins from streptomycin-sensitive and streptomycin-resistant mutants of *Chlamydomonas reinhardii*. Europ. J. Biochem. **58**, 603—610 (1975)

[106] SPIESS, H., and C. G. ARNOLD: Comparative investigation of the cytoplasmic ribosomal proteins of *Chlamydomonas reinhardii* and *Polytoma mirum*. Plant. Sci. Lett. **6**, 267—271 (1976)

[107] SPIESS, H.: Analysis of the chloroplast ribosomal proteins from *Chlamydomonas reinhardii*, streptomycin-resistant and dependent mutants by two-dimensional gel electrophoresis. Plant Sci. Lett. **10**, 103—113 (1977)

[108] FREYSSINET, G.: Characterization of cytoplasmic and chloroplast ribosomal proteins of *Euglena gracilis*. Biochimie **59**, 597—610 (1977)

[109] BRIMACOMBE, R., G. STÖFFLER, and H. G. WITTMANN: Ribosome structure. Ann. Rev. Biochem **47**, 217—249 (1978)

[110] LEISTER, D. E., and I. B. DAWID: Physical properties and protein constituents of cytoplasmic and mitochondrial ribosomes of *Xenopus laevis*. J. Biol. Chem. **249**, 5108—5118 (1974)

[111] VAN DEN BOGERT, C., and H. DE VRIES: The mitochondrial ribosomes of *Neurospora crassa*. II. Comparison of the proteins from *Neurospora crassa* mitochondrial ribosomes with ribosomal proteins from *Neurospora* cytoplasm, from rat liver mitochondria and from bacteria. Biochim. Biophys. Acta **442**, 227—238 (1976)

[112] FAYE, G., and F. SOR: Analysis of mitochondrial ribosomal proteins of *Saccharomyces cerevisiae* by two dimensional polyacrylamide gel electrophoresis. Molec. Gen. Genetics **155**, 27—34 (1977)

[113] MATTHEWS, D. E., R. A. HESSLER, and T. W. O'BRIEN: Rapid evolutionary divergence of proteins in mammalian mitochondrial ribosomes. FEBS-Lett. **86**, 76—80 (1978)

[114] REYES, R., D. VAZQUEZ, and J. P. G. BALLESTA: Activities of nucleoprotein particles derived from rat liver ribosome. Biochim. Biophys. Acta **435**, 317—332 (1976)

[115] SALEEM, M., and B. ATKINSON: Isoelectric points and molecular weights of salt-extractable ribosomal proteins. Can. J. Biochem **54**, 1029—1033 (1976)

[116] TSURUGI, K., E. COLLATZ, K. TODOKORO, and I. G. WOOL: Isolation of eukaryotic ribosomal proteins. Purification and characterization of 60 S ribosomal subunit proteins L 3, L 6, L 7′, L 8, L 10, L 15, L 17, L 18, L 19, L 23′, L 25, L 27′, L 28, L 29, L 31, L 32, L 34, L 35, L 36, L 36′, and L 37′. J. Biol. Chem. **252**, 3961 to 3969 (1977)

[117] TSURUGI, K., E. COLLATZ, K. TODOKORO, N. ULBRICH, H. N. LIGHTFOOT, and I. G. WOOL: Isolation of eukaryotic ribosomal proteins. Purification and characterization of the 60 S ribosomal subunit proteins La, Lb, Lf, P 1, P 2, L 13′, L 14, L 18′, L 20 and L 38. J. Biol. Chem. **253**, 946—955 (1978)

[118] COLLATZ, E., N. ULBRICH, K. TSURUGI, H. N. LIGHTFOOT, W. MACKINLAY, A. LIN, and I. G. WOOL: Isolation of eukaryotic ribosomal proteins. Purification and characterization of the 40 S ribosomal subunit proteins Sa, Sc, S 3a, S 3b, S 5′, S 9, S 10, S 11, S 12, S 14, S 15, S 15′,

S 16, S 17, S 18, S 19, S 20, S 21, S 26, S 27′, and S 29. J. Biol. Chem. **252**, 9071 to 9080 (1977)

[119] Lubsen, N. H., and B. D. Lavis: Use of purified polysomes from rabbit reticulocytes in a specific test for initiation factors. Proc. Natl. Acad. Sci. U.S. **71**, 68–72 (1974)

[120] Freienstein, C., and G. Blobel: Nonribosomal proteins associated with eukaryotic native small ribosomal subunits. Proc. Natl. Acad. Sci. U.S. **72**, 3392 to 3396 (1975)

[121] Olsnes, S.: Removal of structural proteins from ribosomes by treatment with sodium deoxycholate in the presence of EDTA. FEBS-Lett. **7**, 211–213 (1970)

[122] Martin, T. E., and I. G. Wool: Formation of active hybrids from subunits of muscle ribosomes from normal and diabetic rats. Proc. Natl. Acad. Sci. U.S. **60**, 569–574 (1968)

[123] Blobel, G., and D. D. Sabatini: Dissociation of mammalian polyribosomes into subunits by puromycin. Proc. Natl. Acad. Sci. U.S. **68**, 390–394 (1971)

[124] Tsurugi, K., E. Collatz, I. G. Wool, and A. Lin: Isolation of eukaryotic ribosomal proteins. Purification and characterization of the 60 S ribosomal subunit proteins L 4, L 5, L 7, L 9, L 11, L 12, L 13, L 21, L 22, L 23, L 26, L 27, L 30, L 33, L 35′, L 37, and L 39. J. Biol. Chem. **251**, 7940–7946 (1976)

[125] Collatz, E., I. G. Wool, A. Lin, and G. Stöffler: The isolation of eukaryotic ribosomal proteins. The purification and characterization of the 40 S ribosomal subunit proteins S 2, S 3, S 4, S 5, S 6, S 7, S 8, S 9, S 13, S 23/24, S 27, and S 28. J. Biol. Chem. **251**, 4666–4672 (1976)

[126] Rankine, A. D., and D. P. Leader: The variable location of phosphoproteins in eucaryotic ribosomal subunits. Mol. Biol. Rep. **2**, 525–528 (1976)

[127] Leader, D. P., and A. A. Coia: The acidic ribosomal phosphoprotein of eukaryotes and its relationship to ribosomal proteins L 7 and L 12 of *E. coli*. Biochem. J. **176**, 569–572 (1978)

[128] Waller, J.-P., and J. I. Harris: Studies on the composition of the protein from *Escherichia coli* ribosomes. Proc. Natl. Acad. Sci. U.S. **47**, 18–23 (1961)

[129] Hardy, S. J. S., C. G. Kurland, P. Voynow, and G. Mora: The ribosomal proteins of *Escherichia coli*. I. Purification of the 30 S ribosomal proteins. Biochem. **8**, 2897–2905 (1969)

[130] Fujisawa, T., K. Imai, and K. Ogata: Effects of the use of high concentrations of Mg^{++} in the preparation of proteins from rat liver ribosome pretreated with polyvinyl sulfate. Improved methods for preparing ribosomal proteins from ribosomal precursor particles of liver nucleoli. J. Biochem. **85**, 271–276 (1979)

[131] Fogel, S., and P. S. Sypherd: Extraction and isolation of individual ribosomal proteins from *Escherichia coli*. J. Bacteriol. **96**, 358–364 (1968)

[132] Curry, J. B., and R. T. Hersh: Molecular weight of the protein from bovine liver ribosomes. Biochem. Biophys. Res. Commun. **6**, 415–417 (1962)

[133] Leboy, P. S., E. C. Cox, and J. G. Flaks: The chromosomal site specifying a ribosomal protein in *Escherichia coli*. Proc. Natl. Acad. Sci. U.S. **52**, 1367–1374 (1964)

[134] Cox, R. A.: The use of guanidinium chloride in the isolation of nucleic acids. Meth. Enzymol. **12 B**, 120–129 (1968)

[135] Hüvös, P., and R. A. Cox: Solution of ribosomal proteins under mild conditions. Biochim. Biophys. Acta **383**, 421–426 (1975)

[136] Dice, J. F., and R. T. Schimke: Turnover and exchange of ribosomal proteins from rat liver. J. Biol. Chem. **247**, 98 to 111 (1972)

[137] Spahr, P. F.: Amino acid composition of ribosomes from *Escherichia coli*. J. Mol. Biol. **4**, 395–406 (1962)

[138] Miller, R. V., and P. S. Sypherd: Chemical and enzymatic modification of proteins in the 30 S ribosome of *Escherichia coli*. J. Mol. Biol. **78**, 527–538 (1973)

[139] Barritault, D., A. Expert-Bezançon, M.-F. Guérin, and D. Hayes: The use of acetone precipitation in the isolation of ribosomal proteins. Europ. J. Biochem. **63**, 131–135 (1976)

[140] Sherton, C. C., and I. G. Wool: The extraction of proteins from eukaryotic ribosomes and ribosomal subunits. Molec. Gen. Genetics **135**, 97–112 (1974)

[141] Hüvös, P., S. Fey, and P. M. D. Hardwicke: Effect of various methods of pre-

paration on the apparent protein composition of eukaryotic ribosomes. An essential preliminary to stoichiometric measurements. Biochem. J. **176**, 381 to 391 (1978)

[142] Henkel, B., H. Welfle, and H. Bielka: Studies on proteins of animal ribosomes. 21. Tryptic digestion of rat liver ribosomal particles. Acta Biol. Med. Germ. **33**, 691–698 (1974)

[143] Lin, A., T. Tanaka, and I. G. Wool: Isolation of eukaryotic ribosomal proteins: Purification and characterization of S 25 and L 16. Biochem. **18**, 1634–1637 (1979)

[144] Treolar, M. A., M. E. Treolar, and R. Kisilevsky: Ethionine and the phosphorylation of ribosomal protein S 6. J. Biol. Chem. **252**, 6217–6221 (1977)

[145] Hamilton, M. G., and M. E. Ruth: Characterization of some of the proteins of the large subunit of rat liver ribosomes. Biochem. **6**, 2585–2590 (1967)

[146] Westermann, P., H. Bielka, and M. Böttger: Studies on proteins of animal ribosomes. VII. Isolation of proteins from rat liver ribosomes, their molecular weights, amino acid compositions and secondary structure. Molec. Gen. Genetics **111**, 224 to 234 (1971)

[147] Westermann, P., D. Koppitz, and H. Bielka: Studies on proteins of animal ribosomes. V. Splitting of rat liver ribosomes bound to DEAE-Sephadex by LiCl gradient elution. Acta Biol. Med. Germ. **26**, 611–616 (1971)

[148] Peeters, B., L. Vanduffel, and W. Rombauts: Differential extraction of proteins from human placental ribosomes. Arch. Int. Physiol. Biochim. **82**, 192 (1974)

[149] Westermann, P., and H. Bielka: Studies on proteins of animal ribosomes. XV. Proteins of the small subunit of rat liver ribosomes: Isolation, amino acid composition, tryptic peptides and molecular weights. Molec. Gen. Genetics **126**, 349 to 356 (1973)

[150] Westermann, P., and H. Bielka: Preparation and analysis of structural proteins of the small ribosomal subunit of rat liver. Acta Biol. Med. Germ. **33**, 531 to 537 (1974)

[151] Theise, H., F. Noll, and H. Bielka: Studies on proteins of animal ribosomes. XXVII. Preparation and antigenic properties of 40 S subunit protein of rat liver ribosomes. Acta Biol. Med. Germ. **37**, 1353–1362 (1978)

[152] Goerl, M., H. Welfle, and H. Bielka: Preparative two-dimensional polyacrylamide gel electrophoresis of rat liver ribosomal proteins and determination of their amino acid compositions. Biochim. Biophys. Acta **519**, 418–427 (1978)

[153] Collatz, E., A. Lin, G. Stöffler, K. Tsurugi, and I. G. Wool: Group fractionation of eukaryotic ribosomal proteins. J. Biol. Chem. **251**, 1808–1816 (1976)

[154] Du Vernay, V. H., jr., and J. A. Traugh: Two-step purification of the major phosphorylated protein in reticulocyte 40 S ribosomal subunits. Biochem. **17**, 2045 to 2049 (1978)

[155] Welfle, H.: Studies on proteins of animal ribosomes. XII. Determination of molecular weights of ribosomal proteins by SDS gel electrophoresis. Acta Biol. Med. Germ. **27**, 211–215 (1971)

[156] Gaal, Ö., P. Hüvös, and L. Vereczkey: Estimation of molecular weights of rat liver ribosomal proteins. Acta Biochim. Biophys. Acad. Sci. Hung. **8**, 91–96 (1973)

[157] Lin, A., and I. G. Wool: The molecular weights of rat liver ribosomal proteins determined by "three-dimensional" polyacrylamide gel electrophoresis. Molec. Gen. Genetics **134**, 1–6 (1974)

[158] Faye, G., F. Sor, A. Glatigny, F. Lederer, and E. Lesquoy: Comparison of amino acid compositions of mitochondrial and cytoplasmic ribosomal proteins of *Sacharomyces cerevisiae*. Molec. Gen. Genetics **171**, 335–341 (1979)

[159] Behlke, J., H. Theise, F. Noll, and H. Bielka: Size and shape of isolated proteins of the small ribosomal subunit of rat liver. FEBS-Lett. **106**, 223–225 (1979)

[160] Buskirk, J. J. van, and W. M. Kirsch: γ-carboxyglutamic acid in eukaryotic and prokaryotic ribosomes. Biochem. Biophys. Res. Commun. **82**, 1329–1331 (1978)

[161] Buskirk, J. J. van, and W. M. Kirsch: The occurence of γ-carboxyglutamic acid in mammalian ribosomes. Biochem. Bio-

phys. Res. Commun. **80**, 1033—1038 (1978)

[162] Amons, R., A. van Agthoven, W. Pluijms, and W. Möller: A comparison of the alanine-rich sequences of the L 7/L 12-ribosomal proteins from rat liver, *Artemia salina* and *Escherichia coli*, with the amino-terminal region of the alkali light chain A_1 from rabbit myosin. FEBS-Lett. **86**, 282—284 (1978)

[163] Amons, R., A. van Agthoven, W. Pluijms, W. Möller, K. Higo, T. Itoh, and S. Osawa: A comparison of the amino-terminal sequence of the L 7/L 12-type proteins of *Artemia salina* and *Saccharomyces cerevisiae*. FEBS-Lett. **81**, 308 to 310 (1977)

[164] Amons, R., W. Pluijms, and W. Möller: The primary structure of ribosomal protein eL 12/eL 12-P from *Artemia salina* 80 S ribosomes. FEBS-Lett. **104**, 85—89 (1979)

[165] Itoh, T., and B. Wittmann-Liebold: The primary structure of protein 44 from the large subunit of yeast ribosomes. FEBS-Lett. **96**, 399—402 (1978)

[166] Martini, O. H. W., R. Temkin, A. Jones, K. Riley, and H. J. Gould: Quantitation of eukaryotic ribosomal proteins separated by two-dimensional polyacrylamide gel electrophoresis. FEBS-Lett. **56**, 205—211 (1975)

[167] Westermann, P., W. Heumann, and H. Bielka: On the stoichiometry of proteins in the small ribosomal subunit of hepatoma ascites cells. FEBS-Lett. **62**, 132—135 (1976)

[168] Lambowitz, A. M., N. H. Chua, and D. Luck: Mitochondrial ribosome assembly in *Neurospora*. Preparation of mitochondrial ribosomal precursor particles, site of synthesis of mitochondrial ribosomal proteins and studies on the poky mutant. J. Mol. Biol. **107**, 223—253 (1976)

[169] Agthoven, A. van, J. Kriek, R. Amons, and W. Möller: Isolation and characterization of the acidic phosphoproteins of 60 S ribosomes from *Artemia salina* and rat liver. Europ. J. Biochem. **91**, 553—565 (1978)

[170] Reporter, M.: Methylation of basic residues in structural proteins. Mech. Ageing. Develop. **1**, 367—372 (1973)

[171] Reporter, M.: Methylation of structural and total ribosomal proteins of cultured muscle cells. Fed. Proc. **33**, 1584 (1974)

[172] Chang, F. N., I. J. Navickas, C. N. Chang, and B. M. Dancis: Methylation of ribosomal proteins in HeLa cells. Arch. Biochem. Biophys. **172**, 627—633 (1976)

[173] Vandrey, J. P., C. J. Goldenberg, and G. L. Eliceiri: In vivo isotope incorporation patterns into HeLa ribosomal proteins. Biochim. Biophys. Acta **432**, 104—112 (1976)

[174] Goldenberg, C. J., and G. L. Eliceiri: Methylation of ribosomal proteins in HeLa cells. Biochim. Biophys. Acta **479**, 220—234 (1977)

[175] Chang, F. N., I. J. Navickas, C. Au, and C. Budzilowicz: Identification of the methylated ribosomal proteins in HeLa cells and the fluctuation of methylation during the cell cycle. Biochim. Biophys. Acta **518**, 89—94 (1978)

[176] Cannon, M., D. Schindler, and J. Davies: Methylation of proteins in 60 S ribosomal subunits from *Saccharomyces cerevisiae*. FEBS-Lett. **75**, 187—191 (1977)

[177] Hernandez, F., M. Cannon, and J. Davies: Methylation of proteins in 40 S ribosomal subunits from *Saccharomyces cerevisiae*. FEBS-Lett. **89**, 271—275 (1978)

[178] Kruiswijk, T., A. Kunst, R. J. Planta, and W. H. Mager: Modification of yeast ribosomal proteins. Methylation. Biochem. J. **175**, 221—225 (1978)

[179] Comb, D. G., N. Sarkar, and C. J. Pinzino: The methylation of lysine residues in protein. J. Biol. Chem. **241**, 1857—1862 (1966)

[180] Liew, C. C., and A. G. Gornall: Acetylation of ribosomal proteins. I. Characterization and properties of rat liver ribosomal proteins. J. Biol. Chem. **248**, 977—983 (1973)

[181] Pestana, A., and H. C. Pitot: Acetylation of ribosome-associated proteins in vitro by an acetyltransferase bound to rat liver ribosomes. Biochem. **14**, 1397—1403 (1975)

[182] Pestana, A., and H. C. Pitot: Acetylation of nascent polypeptide chains on rat liver polyribosomes in vivo and in vitro. Biochem. **14**, 1404—1412 (1975)

[183] TRACHEWSKY, D.: Effect of aldosterone on acetylation of ribosomal proteins in outer and inner zones of kidney. Proc. Soc. Exp. Biol. Med. **147**, 396–398 (1974)

[184] LIEW, C. C., and C. C. YIP: Acetylation of reticulocyte ribosomal proteins at time of protein biosynthesis. Proc. Natl. Acad. Sci. U.S. **71**, 2988–2991 (1974)

[185] TRAUGH, J. A., and S. B. SHARP: Protein modification enzymes associated with the protein-synthesizing complex from rabbit reticulocytes. Protein kinase, phosphoprotein phosphatase, and acetyltransferase. J. Biol. Chem. **252**, 3738–3744 (1977)

[186] DELAUNAY, J., J. E. LOEB, M. PIERRE, and G. SCHAPIRA: Mammalian ribosomal proteins: studies on the in vitro phosphorylation patterns of ribosomal proteins from rabbit liver and reticulocytes. Biochim. Biophys. Acta **312**, 147–151 (1973)

[187] RODGERS, A.: Ribosomal proteins in rapidly growing and non-proliferating mouse cells. Biochim. Biophys. Acta **294**, 292–296 (1973)

[188] DELAUNAY, J., C. MATHIEU, and G. SCHAPIRA: Eukaryotic ribosomal proteins. Interspecific and intraspecific comparison by two-dimensional polyacrylamide gel electrophoresis. Europ. J. Biochem. **31**, 561–564 (1972)

[189] DELAUNAY, J., and G. SCHAPIRA: Rat liver and hepatoma proteins. Two-dimensional polyacrylamide gel electrophoresis. Biochim. Biophys. Acta **259**, 243–246 (1972)

[190] BIELKA, H., J. STAHL, and H. WELFLE: Studies on proteins of animal ribosomes. IX. Proteins of ribosomal subunits of some tumors chracterized by two-dimensional polyacrylamide gel electrophoresis. Arch. Geschwulstforsch. **38**, 109–112 (1971)

[191] SUBRAMANIAN, A. R., J. M. GILBERT, and A. KUMAR: Comparison of ribosomal proteins from neoplastic and non-neoplastic cells. Resolution by two-dimensional gel electrophoresis. Biochim. Biophys. Acta **383**, 93–96 (1975)

[192] BOSSELMAN, R. A., J. A. PRICE, A. L. BURNS, M. S. KAULENAS, and L. C. NORKIN: Ribosomal proteins in normal simian cells, SV 40-transformed simian cells, and simian cells infected with SV 40, adenovirus 5, and vesicular stomatitis v irus Intervirol. **9**, 8–15 (1978)

[193] SUBRAMANIAN, A. R., and E. C. HENSHAW: Variations in the protein pattern of Ehrlich ascites tumor cell polyribosomes. Biochim. Biophys. Acta **520**, 203 to 209 (1978)

[194] LAMBERTSON, A. G.: The ribosomal proteins of *Drosophila melanogaster*. 2. Comparison of protein patterns of ribosomes from larvae, pupae and adult flies by two-dimensional polyacrylamide gel electrophoresis. Molec. Gen. Genetics **118**, 215–222 (1972)

[195] LAMBERTSON, A. G.: The ribosomal proteins of *Drosophila melanogaster*. 3. Further studies on the ribosomal protein composition during development. Molec. Gen. Genetics **128**, 241–247 (1974)

[196] SCHNEIDER, U., and E. KUBIL: The ribosomal proteins of *Drosophila melanogaster*. Localization of stage specific proteins on the small and large ribosomal subunit. Molec. Gen. Genetics **134**, 277 to 280 (1974)

[197] LAMBERTSON, A. G.: The ribosomal proteins of *Drosophila melanogaster*. 5. Analysis by two-dimensional gel electrophoresis of the ribosomal proteins of the temperature-sensitive lethal allele of suppressor-of-forked,$1(1)su(f)^{ts67g}$. A putative ribosomal protein mutant. Molec. Gen. Genetics **139**, 145–156 (1975)

[198] LAMBERTSON, A. G.: The ribosomal proteins of *Drosophila melanogaster*. 4. Characterization by two-dimensional gel electrophoresis of the ribosomal proteins from nine postembryonic developmental stages. Molec. Gen. Genetics **139**, 133 to 144 (1975)

[199] FEKETE, É., and A. G. LAMBERTSON: Imaginal disc ribosomal proteins of *Drosophila melanogaster*. Molec. Gen. Genetics **159**, 85–87 (1978)

[200] BERGER, E.: The ribosomes of *Drosophila*. 1. Subunit and protein composition. Molec. Gen. Genetics **128**, 1–9 (1974)

[201] GOULD, H. J., O. H. W. MARTINI, and H. S. W. KING: 80 S ribosomal proteins. Biochem. J. **129**, 31p–32p (1972)

[202] DELAUNAY, J., F. CREUSOT, and G. SCHAPIRA: Evolution of ribosomal proteins. Europ. J. Biochem. **39**, 305–312 (1973)

[203] DELAUNAY, J., F. CREUSOT, and G. SCHAPIRA: Phylogenic studies on ribosomal proteins. Biochimie **56**, 1459—1463 (1974)

[204] DELAUNAY, J., F. CREUSOT, and G. SCHAPIRA: Evolution of ribosomal proteins. Acta Biol. Med. Germ. **33**, 559—564 (1974)

[205] THEISE, H., F. NOLL, and H. BIELKA: Studies on proteins of animal ribosomes. 19. Electrophoretic and immunological analysis of ribosomal proteins of different species of vertebrates. Acta Biol. Med. Germ. **33**, 565—569 (1974)

[206] KUTER, D. J., and A. RODGERS: Ribosomal protein differences between animal cells. Exptl. Cell Res. **87**, 186—194 (1974)

[207] FUJISAWA, T., and G. L. ELICEIRI: Ribosomal proteins of hamster, mouse and hybrid cells. Biochim. Biophys. Acta **402**, 238—243 (1975)

[208] RAMJOUE, H.-P. R., and J. GORDON: Evolutionary microdivergence of chick and rat liver ribosomal proteins. J. Biol. Chem. **252**, 9065—9070 (1977)

[209] BERNABEU, C., D. VAZQUEZ, and F. P. CONDE: Comparative study between prokaryotes and eukaryotes by chemical iodination of ribosomal proteins. Biochim. Biophys. Acta **577**, 400—409 (1979)

[210] NAGABHUSHAN, N., A. GULYAS, and S. ZALIK: Comparison of plant cytoplasmic ribosomal proteins by two-dimensional polyacrylamide gel electrophoresis. Plant. Physiol. **53**, 516—518 (1974)

[211] VASLET, C., and E. BERGER: The ribosomes of *Drosophila*. IV. Electrophoretic identity among ribosomal subunit proteins from wild type and mutant *D. melanogaster* and *D. simulans*. Molec. Gen. Genetics **147**, 189—194 (1976)

[212] CHUA, N.-H., and D. J. L. LUCK: Biosynthesis of organelle ribosomes. In: Ribosomes. (Eds.: M. NOMURA, A. TISSIÈRES, and P. LENGYEL). Cold Spring Harbor Laboratory, 1974, 519—539

[213] GRIVELL, L. A., P. NETTER, P. BORST, and P. P. SLONIMSKI: Mitochondrial antibiotic resistance in yeast: ribosomal mutants resistant to chloramphenicol, erythromycin and spiramycin. Biochim. Biophys. Acta **312**, 358—367 (1973)

[214] LEISTER, D. E., and I. B. DAWID: Mitochondrial ribosomal proteins in *Xenopus laevis/X. mulleri* interspecific hybrids. J. Mol. Biol. **96**, 119—123 (1975)

[215] CZEMPIEL, W., J. KLOSE, and R. BASS: Mammalian mitochondrial ribosomes: characterization of ribosomal proteins by two-dimensional gel electrophoresis. FEBS-Lett. **62**, 259—261 (1976)

[216] JONES, B. L., NAGABHUSHAN, N., A. GULYAS, and S. ZALIK: Two dimensional acrylamide gel electrophoresis of wheat leaf cytoplasmic and chloroplast ribosomal proteins. FEBS-Lett. **23**, 167—170 (1972)

[217] FREYSSINET, G., and J. A. SCHIFF: The chloroplast and cytoplasmic ribosomes of *Euglena*. II. Characterization of ribosomal proteins. Plant Physiol. **53**, 543 to 554 (1974)

[218] METS, L. J., and L. BOGORAD: Altered chloroplast ribosomal proteins associated with erythromycin-resistant mutants in two genetic systems of *Chlamydomonas reinhardii*. Proc. Natl. Acad. Sci. U.S. **69**, 3779—3783 (1972)

[219] DAVIDSON, J. N., M. R. HANSON, and L. BOGORAD: An altered chloroplast ribosomal protein in ery-M 1 mutants of *Chlamydomonas reinhardi*. Molec. Gen. Genetics **132**, 119—129 (1974)

[220] OHTA, N., R. SAGER, and M. INOUYE: Identification of a chloroplast ribosomal protein altered by a chloroplast mutation in *Chlamydomonas*. J. Biol. Chem. **250**, 3655—3659 (1975)

[221] FREYSSINET, G.: Changes in chloroplast ribosomal proteins in a streptomycin-resistant mutant of *Euglena gracilis*. Plant Sci. Lett. **5**, 305—311 (1975)

[222] BOERSMA, D., S. M. MCGILL, J. W. MOLLENKAMP, and D. J. ROUFA: Emetine resistance in chinese hamster cells is linked genetically with an altered 40 S ribosomal subunit protein, S 20. Proc. Natl. Acad. Sci. U.S. **76**, 415—419 (1979)

[223] LOEB, J. E., and C. BLAT: Phosphorylation of some rat liver ribosomal proteins and its activation by cyclic AMP. FEBS-Lett. **10**, 105—108 (1970)

[224] KABAT, D.: Phosphorylation of ribosomal proteins in rabbit reticulocytes. Characterization and regulatory aspects. Biochem. **9**, 4160—4174 (1970)

[225] KRYSTOSEK, A., L. F. BITTE, M. F. CAWTHON, and D. KABAT: Phosphorylation of ribosomal proteins in eukaryotes. In: Ribosomes. (Eds.: M. NOMURA, A. TIS-

SIÉRES, and P. LENGYEL). Cold Spring Harbor Laboratory, 1974, 855–870

[226] GORDON, J.: Determination of an upper limit to the phosphorus content of polypeptide chain elongation factor and ribosomal proteins in *Escherichia coli*. Biochem. Biophys. Res. Commun. **44**, 579 to 586 (1971)

[227] RAHMSDORF, H. J., S. H. PAI, H. PONTA, P. HERRLICH, R. ROSKOSKI, jr., M. SCHWEIGER, and F. W. STUDIER: Protein kinase induction in *Escherichia coli* by bacteriophage T 7. Proc. Natl. Acad. Sci. U.S. **71**, 586–589 (1974)

[228] RAHMSDORF, H. J., P. HERRLICH, S. H. PAI, M. SCHWEIGER, and H. G. WITTMANN. Ribosomes after infection with bacteriophage T 4 and T 7. Molec. Gen. Genetics **127**, 259–271 (1973)

[229] TRAUGH, J. A., and R. R. TRAUT: Phosphorylation of ribosomal proteins of *Escherichia coli* by protein kinase from rabbit skeletal muscle. Biochem. **11**, 2503–2509 (1972)

[230] ISSINGER, O.-G., and R. R. TRAUT: Selective phosphorylation from GTP of proteins L 7 and L 12 of E. coli 50 S ribosomes by a protein kinase from rabbit reticulocytes. Biochem. Biophys. Res. Commun. **59**, 829–836 (1974)

[231] BLAT, C., and J. E. LOEB: Effect of glucagon on phosphorylation of some rat liver ribosomal proteins in vivo. FEBS-Lett. **18**, 124–126 (1971)

[232] EIL, C., and I. G. WOOL: Phosphorylation of rat liver ribosomal subunits: partial purification of two cyclic AMP activated protein kinases. Biochem. Biophys. Res. Commun. **45**, 1001–1009 (1971)

[233] STAHL, J., H. WELFLE, and H. BIELKA: Studies on proteins of animal ribosomes. XIV. Analysis of phosphorylated rat liver ribosomal proteins by two-dimensional polyacrylamide gel electrophoresis. FEBS-Lett. **26**, 233–236 (1972)

[234] YAMAMURA, H., Y. INOUE, R. SHINOMURA, and Y. NISHIZUKA: Similarity and pleiotropic actions of adenosine 3′, 5′-monophosphate-dependent protein kinases from mammalian tissues. Biochem. Biophys. Res. Commun. **46**, 589–596 (1972)

[235] CORREZE, C., P. PINELL, and J. NUNEZ: Effects of thyroid hormones an phosphorylation of liver ribosomal proteins and on protein phosphokinase activity. FEBS-Lett. **23**, 87–91 (1972)

[236] EIL, C., and I. G. WOOL: Phosphorylation of liver ribosomal proteins. Characteristics of the protein kinase reaction and studies of the structure of phosphorylated ribosomes. J. Biol. Chem. **248**, 5122 to 5129 (1973)

[237] EIL, C., and I. G. WOOL: Function of phosphorylated ribosomes. The activity of ribosomal subunits phosphorylated in vitro by protein kinase. J. Biol. Chem. **248**, 5130–5136 (1973)

[238] BARELA, T. D., and D. E. KIZER: In vivo phosphorylation of free and membrane-bound ribosomal protein. Biochim. Biophys. Acta **335**, 218–225 (1974)

[239] BÖHM, H., and J. STAHL: Studies on proteins of animal ribosomes. XVII. Quantitative aspects of enzymatic phosphorylation of ribosomal proteins and particles from rat liver. Acta Biol. Med. Germ. **32**, 449–461 (1974)

[240] STAHL, J., H. BÖHM, and H. BIELKA: Enzymatic phosphorylation of eukaryotic ribosomal proteins and factors of protein biosynthesis. Acta Biol. Med. Germ. **33**, 667–676 (1974)

[241] GRESSNER, A. M., and I. G. WOOL: The phosphorylation of liver ribosomal proteins in vivo. J. Biol. Chem. **249**, 6917 to 6925 (1974)

[242] VENTIMIGLIA, F. A., and I. G. WOOL: A kinase that transfers the γ-phosphoryl group of GTP to proteins of eukaryotic 40 S ribosomal subunits. Proc. Natl. Acad. Sci. U.S. **71**, 350–354 (1974)

[243] GRESSNER, A. M., and I. G. WOOL: The stimulation of the phosphorylation of ribosomal protein S 6 by cycloheximide and puromycin. Biochem. Biophys. Res. Commun. **60**, 1482–1490 (1974)

[244] GRESSNER, A. M., and I. G. WOOL: Influence of glucagon and cyclic adenosine 3′:5′-monophosphate on the phosphorylation of rat liver ribosomal protein S 6. J. Biol. Chem. **251**, 1500–1504 (1976)

[245] GRESSNER, A. M., and I. G. WOOL: Effect of experimental diabetes and insulin on phosphorylation of rat liver ribosomal protein S 6. Nature **259**, 148–150 (1976)

[246] Hoffman, W. L., and J. Ilan: Analysis by two-dimensional polyacrylamide gel electrophoresis of the in vivo phosphorylation of ribosomal proteins derived from free and membrane-bound polysomes. Mol. Biol. Rep. **2**, 219—224 (1975)

[247] Gressner, A. M., and H. Greiling: The phosphorylation of liver ribosomal protein S 6 during the development of acute hepatic cell injury induced by D-galactosamine. FEBS-Lett. **74**, 77—81 (1977)

[248] Gressner, A. M., and H. Greiling: The state of ribosomal protein phosphorylation during thioacetamide-induced liver injury. Exptl. Mol. Pathol. **28**, 39—47 (1978)

[249] Cenatiempo, Y., A. J. Cozzone, A. Genot, and J.-P. Reboud: In vitro phosphorylation of proteins from free and membrane-bound rat liver polysomes. FEBS-Lett. **79**, 165—169 (1977)

[250] Genot, A., J.-P. Reboud, Y. Cenatiempo, and A. J. Cozzone: Endogenous phosphorylation of ribosomal proteins from membrane-free rat liver polysomes. FEBS-Lett. **86**, 103—107 (1978)

[251] Arpin, M., J.-J. Madjar, and J.-P. Reboud: Occurrence of phosphorylated forms of an acidic protein of the large ribosomal subunit of rat liver. Biochim. Biophys. Acta **519**, 537—541 (1978)

[252] Horak, I., and D. Schiffmann: Acidic phosphoproteins of HeLa and rat 60 S ribosomal subunits. FEBS-Lett. **82**, 82—84 (1977)

[253] Tas, P. W. L., and B. H. Sells: Accessibility of ribosomal proteins to lactoperoxidase-catalyzed iodination following phosphorylation and during subunit interaction. Europ. J. Biochem. **92**, 271—278 (1978)

[254] Pierre, M., C. Creuzet, and J. E. Loeb: Localization of phosphoproteins in ribosomal subunits and in free and bound polysomes of rat liver. FEBS-Lett. **45**, 88 to 91 (1974)

[255] Perlis, B., and J. E. Loeb: Dephosphorylation of rat liver ribosomal proteins by a phosphoprotein phosphatase. Biochimie **56**, 1007—1010 (1975)

[256] Treloar, M. A., and R. Kisilevsky: Protein kinase activity and ribosome phosphorylation in ethionine-treated rats. Can. J. Biochem. **57**, 209—215 (1979)

[257] Genot, A., J.-P. Reboud, Y. Cenatiempo, and A. J. Cozzone: Differential phosphorylation of basic and acidic ribosomal proteins by protein kinase bound to membrane free rat liver polysomes. FEBS-Lett. **99**, 261—264 (1979)

[258] Stöffler, G., I. G. Wool, A. Lin, and K. H. Rak: The identification of the eukaryotic ribosomal proteins homologous with *Escherichia coli* proteins L 7 and L 12. Proc. Natl. Acad. Sci. U.S. **71**, 4723—4726 (1974)

[259] Ashby, C. D., and S. Roberts: Phosphorylation of ribosomal proteins in rat cerebral cortex in vitro. J. Biol. Chem. **250**, 2546—2555 (1975)

[260] Leader, D. P., A. A. Coia, and L. H. Fahmy: The extent of phosphorylation of the acidic 60 S ribosomal phosphoprotein, Lγ, in Krebs 2 ascites cells and in the skeletal muscle of normal and diabetic rats. Biochem. Biophys. Res. Commun. **83**, 50—58 (1978)

[261] Roberts, S., and C. D. Ashby: Ribosomal protein phosphorylation in rat cerebral cortex in vitro. Influence of cyclic adenosine 3′:5′-monophosphate. J. Biol. Chem. **253**, 288—296 (1978)

[262] Hill, A. M., and D. Trachewsky: Effect of aldosterone on renal ribosomal protein phosphorylation. J. Steroid Biochem. **5**, 561—568 (1974)

[263] Kabat, D.: Phosphorylation of ribosomal proteins in rabbit reticulocytes. A cell-free system with ribosomal protein kinase activity. Biochem. **10**, 197—203 (1971)

[264] Kabat, D.: Turnover of phosphoryl groups in reticulocyte ribosomal phosphoproteins. J. Biol. Chem. **247**, 5338 to 5344 (1972)

[265] Martini, O. H. W., and H. J. Gould: Phosphorylation of rabbit reticulocyte ribosomal protein in vitro. Biochim. Biophys. Acta **295**, 621—629 (1973)

[266] Traugh, J. A., M. Mumby, and R. R. Traut: Phosphorylation of ribosomal proteins by substrate-specific protein kinase from rabbit reticulocytes. Proc. Natl. Acad. Sci. U.S. **70**, 373—376 (1973)

[267] Cawthon, M. L., L. F. Bitte, A. Krystosek, and D. Kabat: Effect of cyclic adenosine 3′:5′-monophosphate on ribo-

somal protein phosphorylation in reticulocytes. J. Biol. Chem. **249**, 275–278 (1974)

[268] Traugh, J. A., and G. G. Porter: A comparison of ribosomal proteins from rabbit reticulocytes phosphorylated in situ and in vitro. Biochem. **15**, 610–615 (1976)

[269] Issinger, O.-G.: Phosphorylation of acidic ribosomal proteins from rabbit reticulocytes by a ribosome-associated casein kinase. Biochim. Biophys. Acta **477**, 185 to 189 (1977)

[270] Monier, D., K. Santhanam, and S. R. Wagle: Studies on the inhibition of amino acid incorporation into protein by isolated rat liver ribosomes by protein kinases. Biochem. Biophys. Res. Commun. **46**, 1881–1886 (1972)

[271] Lightfoot, H. N., M. Mumby, and J. A. Traugh: Dephosphorylation of 40 S ribosomal subunits by phosphoprotein phosphatase activity from rabbit reticulocytes. Biochem. Biophys. Res. Commun. **66**, 1141–1146 (1975)

[272] Grancharova, T. V., T. A. Getova, and T. K. Nikolov: Phosphorylation of ribosomal proteins from eukaryotes in homologous and heterologous cell-free systems. Biochim. Biophys. Acta **418**, 397–403 (1976)

[273] Sanecka-Obacz, M., and T. Borkowski: Phosphorylation of protein from rabbit brain ribosomes. Acta Biochim. Polon. **21**, 397–401 (1974)

[274] Walton, G. M., G. N. Gill, I. B. Abrass, and L. D. Garren: Phosphorylation of ribosome-associated protein by an adenosine 3′:5′-cyclic monophosphate-dependent protein kinase: Location of the microsomal receptor and protein kinase. Proc. Natl. Acad. Sci. U.S. **68**, 880–884 (1971)

[275] Walton, G. M., and G. N. Gill: Adenosine 3′,5′-monophosphate and protein kinase dependent phosphorylation of ribosomal protein. Biochem. **12**, 2604 to 2611 (1973)

[276] Barden, N., and F. Labrie: Cyclic adenosine 3′,5′-monophosphate dependent phosphorylation of ribosomal proteins from bovine anterior pituitary gland. Biochem. **12**, 3096–3102 (1973)

[277] Azhar, S., and K. M. J. Menon: Adenosine 3′,5′-monophosphate dependent phosphorylation of ribosomes and ribosomal subunits from bovine corpus luteum. Biochim. Biophys. Acta **392**, 64 to 74 (1975)

[278] Majumder, G. C., and R. W. Turkington: Hormone-dependent phosphorylation of ribosomal and plasma membrane proteins in mouse mammary gland in vitro. J. Biol. Chem. **247**, 7207–7217 (1972)

[279] Rupp, R. G., R. M. Humphrey, and J. R. Shaeffer: Phosphorylation of ribosome-associated proteins during the mammalian cell cycle. Unique phosphorylation of a specific protein during mitosis. Biochim. Biophys. Acta **418**, 81–92 (1976)

[280] Leader, D. P., A. D. Rankine, and A. A. Coia: The phosphorylation of ribosomal protein S 6 in baby hamster kidney fibroblasts. Biochem. Biophys. Res. Commun. **71**, 966–974 (1976)

[281] Leader, D. P., and A. A. Coia: Characterization of the acidic phosphoprotein of eukaryotic ribosomes using a new system of two-dimensional gel electrophoresis. Biochim. Biophys. Acta **519**, 213–223 (1978)

[282] Leader, D. P., and A. A. Coia: The phosphorylation of ribosomal protein S 6 on the monoribosomes and polyribosomes of baby hamster kidney fibroblasts. FEBS-Lett. **90**, 270–274 (1978)

[283] Schubart, U. K., S. Shapiro, N. Fleischer, and O. M. Rosen: Cyclic adenosine 3′:5′-monophosphate-mediated insulin secretion and ribosomal protein phosphorylation in a hamster islet cell tumor. J. Biol. Chem. **252**, 92–101 (1977)

[284] Bitte, L., and D. Kabat: Phosphorylation of ribosomal proteins in sarcoma 180 tumor cells. J. Biol. Chem. **247**, 5345 to 5350 (1972)

[285] Olson, M. O. J., A. W. Prestayko, C. E. Jones, and H. Busch: Phosphorylation of proteins of ribosomes and nucleolar preribosomal particles from Novikoff hepatoma ascites cells. J. Mol. Biol. **90**, 161 to 168 (1974)

[286] Prestayko, A. W., M. O. J. Olson, and H. Busch: Phosphorylation of proteins of ribosomes and nucleolar preribosomal particles in vivo in Novikoff hepatoma

ascites cells. FEBS-Lett. **44**, 131—135 (1974)

[287] LEADER, D. P., and A. A. COIA: The phosphorylation of an acidic protein of the large ribosomal subunit of Krebs II ascites cells. Biochem. J. **162**, 199—200 (1977)

[288] RANKINE, A. D., D. P. LEADER, and A. A. COIA: The phosphorylation of the ribosomal proteins of Krebs II ascites cells. Biochim. Biophys. Acta **474**, 293—307 (1977)

[289] LEADER, D. P., and A. A. COIA: The phosphorylation of ribosomal protein L 14 and S 3 in Krebs 2 ascites cells. Biochim. Biophys. Acta **519**, 224—232 (1978)

[290] ROSNITSCHEK, I., U. TRAUB, and P. TRAUB: Enhanced phosphorylation of ribosomal protein S 6 and other cytoplasmic proteins after mengovirus infection of Ehrlich ascites tumor cells. Hoppe-Seyler's Z. Physiol. Chem. **359**, 593—600 (1978)

[291] LASTICK, S. M., P. J. NIELSEN, and E. H. McCONKEY: Phosphorylation of ribosomal protein S 6 in suspension cultured HeLa cells. Molec. Gen. Genetics **152**, 223—230 (1977)

[292] KAERLEIN, M., and I. HORAK: Phosphorylation of ribosomal proteins in HeLa cells infected with vaccinia virus. Nature **259**, 150—151 (1976)

[293] BLAIR, G. E., and I. HORAK: Phosphorylation of ribosomes in adenovirus infection. Biochem. Soc. Trans. **5**, 660—661 (1977)

[294] HORAK, I., and D. SCHIFFMANN: Acidic phosphoproteins of the 60 S ribosomal subunits from HeLa cells. Europ. J. Biochem. **79**, 375—380 (1977)

[295] KAERLEIN, M., and I. HORAK: Identification and characterization of ribosomal proteins phosphorylated in vaccinia-virus infected HeLa cells. Europ. J. Biochem. **90**, 463—469 (1978)

[296] HORAK, I., and K. KOSCHEL: Does cAMP control protein synthesis? FEBS-Lett. **83**, 68—70 (1977)

[297] MARVALDI, J., and J. LUCAS-LENARD: Differences in the ribosomal protein gel profile after infection of L cells with wild type or temperature-sensitive mutants of vesicular stomatitis virus. Biochem. **16**, 4320—4327 (1977)

[298] HOUSTON, L. L.: Effect of the toxic Castor bean protein, ricin, on the phosphorylation pattern of [^{32}P]-labelled ribosomes from mouse L cells. Biochem. Biophys. Res. Commun. **85**, 131—139 (1978)

[299] SAMUEL, C. E., Mechanism of interferon action. Kinetics of interferon action in mouse L_{929} cells: Phosphorylation of protein synthesis initiation factor eIF-2 and ribosome-associated protein P 1. Virology **93**, 281—285 (1979)

[300] KRUPPA, J., and O. H. W. MARTINI: Dephosphorylation of one 40 S ribosomal protein in MPC 11 cells induced by hypertonic medium. Biochem. Biophys. Res. Commun. **85**, 428—435 (1978)

[301] MARTINI, O. H. W., and J. KRUPPA: Ribosomal phosphoproteins of mouse myeloma cells. Changes in the degree of phosphorylation induced by hypertonic initiation block. Europ. J. Biochem. **95**, 349—358 (1979)

[302] LI, C.-C., and H. AMOS: Alteration of phosphorylation or ribosomal proteins as a function of variation of growth conditions of primary cells. Biochem. Biophys. Res. Commun. **45**, 1398—1407 (1971)

[303] HASELBACHER, G. K., R. E. HUMBLE, and G. THOMAS: Insuline-like growth-factor: insulin or serum increase phosphorylation of ribosomal protein S 6 during transition of stationary chick embryo fibroblasts into early G 1 phase of the cell cycle. FEBS-Lett. **100**, 185—190 (1979)

[304] JERGIL, B.: Protein kinase from rainbow-trout-testis ribosomes. Partial purification and characterization. Europ. J. Biochem. **28**, 546—554 (1979)

[305] TREWAVAS, A.: The phosphorylation of ribosomal protein in *Lemna minor*. Plant Physiol. **51**, 760—767 (1973)

[306] TEPFER, D. A., and D. E. FOSKET: Phosphorylation of ribosomal protein in soybean. Phytochem. **14**, 1161—1165 (1975)

[307] BECKER-URSIC, D., and J. DAVIES: In vivo and in vitro phosphorylation of ribosomal proteins by protein kinases from *Saccharomyces cerevisiae*. Biochem. **15**, 2289—2296 (1976)

[308] KUDLICKI, W., N. GRANKOWSKI, and E. GAȘIOR: Ribosomal protein as substrate

for a GTP-dependent protein kinase from yeast. Mol. Biol. Rep. **3**, 121–129 (1976)

[309] HEBERT, J., M. PIERRE, and J. E. LOEB: Phosphorylation in vitro and in vivo of ribosomal proteins from *Saccharomyces cerevisiae*. Europ. J. Biochem. **72**, 167 to 174 (1977)

[310] KRUISWIJK, T., J. T. DE HEY, and R. J. PLANTA: Modification of yeast ribosomal proteins. Phosphorylation. Biochem. J. **175**, 213–219 (1978)

[311] GRANKOWSKI, N., and E. GAȘIOR: An in vivo and in vitro phosphorylation of yeast ribosomal proteins. Acta Biochim. Polon. **32**, 45–56 (1975)

[312] KUDLICKI, W., N. GRANKOWSKI, and E. GAȘIOR: Isolation and properties of two protein kinases from yeast which phosphorylate casein and some ribosomal proteins. Europ. J. Biochem. **84**, 493–498 (1978)

[313] SANCHEZ-MADRID, F., P. CONDE, D. VAZQUEZ, and J. P. G. BALLESTA: Acidic proteins from *Saccharomyces cerevisiae* ribosomes. Biochem. Biophys. Res. Commun. **87**, 281–291 (1979)

[314] RANKINE, A. D., and D. P. LEADER: The phosphorylation of ascites cell ribosomes in vivo: identification of a phosphorylated protein of the small ribosomal subunit by two-dimensional gel electrophoresis. FEBS-Lett. **52**, 284–287 (1975)

[315] AGTHOVEN, A. J. VAN, J. A. MAASSEN, and W. MÖLLER: Structure and phosphorylation of an acidic protein from 60 S ribosomes and its involvement in elongation factor 2 dependent GTP hydrolysis. Biochem. Biophys. Res. Commun. **77**, 989–998 (1977)

[316] MÖLLER, W., L. J. SLOBIN, R. AMONS, and D. RICHTER: Isolation and structure of eucaryotic L 7 and L 12. Proc. Natl. Acad. Sci. U.S. **72**, 4744–4748 (1975)

[317] KALTHOFF, H., and D. RICHTER: In vivo studies on the incorporation of microinjected acidic proteins of the large ribosomal subunit from *Escherichia coli* and *Artemia salina* into oocyte ribosomes from *Xenopus laevis*. Biochem. **18**, 4144 to 4147 (1979)

[318] KRISTIANSEN, K., P. PLESNER, and A. KRÜGER: Phosphorylation in vivo of ribosomes in *Tetrahymena pyriformis*. Europ. J. Biochem. **83**, 395–403 (1978)

[319] KRISTIANSEN, K., and A. KRÜGER: Ribosomal proteins in growing and starved *Tetrahymena pyriformis*. Starvation-induced phosphorylation of ribosomal proteins. Biochim. Biophys. Acta **521**, 435 to 451 (1978)

[320] KRISTIANSEN, K., and A. KRÜGER: Phosphorylation and degradation of ribosomes in starved *Tetrahymena pyriformis*. Exptl. Cell Res. **118**, 159–169 (1979)

[321] BELANGER, G., G. BELLEMARE, and G. LEMIEUX: Ribosomal phosphoproteins in *Physarum polycephalum*. Biochem. Biophys. Res. Commun. **86**, 862–868 (1979)

[322] RUBIN, C. S., and O. M. ROSEN: Protein phosphorylation. Ann. Rev. Biochem. **44**, 831–887 (1975)

[323] USCHEWA, A.: Speziesspezifität ribosomaler Proteine. Ph. D. thesis. Academy of Sciences, GDR, 1980

[324] ODA, G., A. R. STRØM, P. L. VISENTIN, and M. YAGUCHI: An acidic, alanine-rich 50 S ribosomal protein from *Halobacterium cutirubrum*: amino acid sequence homology with *Escherichia coli* proteins L 7 and L 12. FEBS-Lett. **43**, 127–130 (1974)

[325] TISCHENDORF, G. W., H. ZEICHARDT, and G. STÖFFLER: Architecture of the *Escherichia coli* ribosome as determined by immune electron microscopy. Proc. Natl. Acad. Sci. U.S. **72**, 4820–4824 (1975)

[326] TING SHIH, C.-Y., J. E. TOIVONEN, and G. R. CRAVEN: Partial purification and characterization of the proteins from the 40 S-ribosomes of *Artemia salina* and wheat germ. Europ. J. Biochem. **97**, 189 to 196 (1979)

[327] MADRZAK, C. J., U. SZYBIAK, and A. B. LEGOCKI: Acidic proteins from wheat germ ribosomes. FEBS-Lett. **103**, 304 to 307 (1979)

[328] LAMBOWITZ, A. M., R. J. LAPOLLA, and A. R. COLLINS: Mitochondrial ribosome assembly in *Neurospora*. Two-dimensional gel electrophoretic analysis of mitochondrial ribosomal proteins. J. Cell. Biol. **82**, 17–31 (1979)

[329] LEMIEUX, G., G. BÉLANGER, L. NICOLE, and G. BELLEMARE: Ribosomes of *Physarum polycephalum*. Subunits and protein composition. Biochim. Biophys. Acta. **578**, 357–364 (1979)

[330] HATAWAY, G. M., T. S. LUNDAK, S. M. TAHARA, and J. A. TRAUGH: Isolation of protein kinases from reticulocytes and phosphorylation of initiation factors. Meth. Enzymol. **60**, 511–521 (1979)

[331] VISENTIN, L. P., M. YAGUCHI, and A. T. MATHESON: Structural homologies in alanine-rich acidic ribosomal proteins from procaryotes and eucaryotes. Can. J. Biochem. **57**, 719–726 (1979)

[332] WEBER, L., E. BERGER, C. VASLET, and B. YEDVOBNICK: The ribosomes of *Drosophila*. III. RNA and protein homology between *D. melanogaster* and *D. virilis*. Genetics **84**, 573–585 (1976)

[333] WETTENHALL, R. E. H., G. J. HOWLETT, and D. GREIG: Phosphorylation of a specific ribosomal protein S 6 in thymocytes and the regulation of protein synthesis by concanavalin-A and prostaglandin-E 1. Proc. Austral. Biochem. Soc. **12**, 41 (1979)

[334] THOMAS, G., M. SIEGMANN, and J. GORDON: Multiple phosphorylation of ribosomal protein-S 6 during transition of quiescent 3T 3 cells into early G 1, and cellular compartmentalization of the phosphate donor. Proc. Natl. Acad. Sci. U.S. **76**, 3952–3956 (1979)

[335] SANCHEZ-MADRID, F., R. REYES, P. CONDE, and J. P. G. BALLESTA: Acidic proteins from eukaryotic cells. Effect on ribosomal functions. Europ. J. Biochem. **98**, 409–416 (1979)

[336] PANIJEL, J., and M. C. DELAUNAY: An immunological approach to the structure of the ribosomal particles. Biochem. Biophys. Res. Commun. **29**, 158–165 (1967)

[337] STÖFFLER, G.: Structure and function of the *Escherichia coli* ribosome: immunochemical analysis. In: Ribosomes. (Eds.: M. NOMURA, A. TISSIÈRES, and P. LENGYEL), Cold Spring Harbor Laboratory, 1974, 615–667

[338] WITTMANN, H.-G.: Structure, function and evolution of ribosomes. Europ. J. Biochem. **61**, 1–13 (1976)

[339] STÖFFLER, G., and H. G. WITTMANN: Primary structure and three-dimensional arrangement of proteins within the *Escherichia coli* ribosome. In: Protein biosynthesis (Eds.: H. WEISSBACH and S. PESTKA), Acad. Press, New York–London, 1977, 117–202

[340] NOLL, F., H. THEISE, and H. BIELKA: Studies on proteins of animal ribosomes. XVIII. Reaction of ribosomes and ribosomal proteins with antibodies against ribosomal proteins. Acta Biol. Med. Germ. **33**, 547–553 (1974)

[341] LUTSCH, G., F. NOLL, H. THEISE, and H. BIELKA: Localization of ribosomal protein S 2 in rat liver ribosomes by immune electron microscopy. Acta Biol. Med. Germ. **36**, 287–290 (1977)

[342] NOLL, F., U.-A. BOMMER, G. LUTSCH, H. THEISE, and H. BIELKA: Localization of rat liver ribosomal protein S 2 and its involvement in initiation factor eIF-2 binding to the 40 S ribosomal subunit. FEBS-Lett. **87**, 129–131 (1978)

[343] BIELKA, H., F. NOLL, H. WELFLE, P. WESTERMANN, G. LUTSCH, J. STAHL, H. THEISE, U.-A. BOMMER, B. GROSS, M. GOERL, and B. HENKEL: Structural and functional organization of the eukaryotic ribosome. Studies on proteins of rat liver ribosomes. FEBS Symp. **51**, 387–399 (1979)

[344] LUTSCH, G., F. NOLL, H. THEISE, G. ENZMANN, and H. BIELKA: Localization of proteins S 1, S 2, S 16 and S 23 on the surface of small subunits of rat liver ribosomes by immune electron microscopy. Molec. Gen. Genetics **176**, 281–291 (1979)

[345] LACEUR, F., J. HAREL, L. HAREL, and J. HERMET: Étude antigénique des ribosomes des cellules de mammiféres. Compt. Rend. Acad. Sci. **255**, 1161–1163 (1962)

[346] BIGLEY, N. J., M. C. DODD, and V. B. GEYER: The immunological specificity of antibodies to liver ribosomes and nuclei. J. Immunol. **90**, 416–423 (1963)

[347] ALBERGHINA, F. A. M., and S. R. SUSKIND: Ribosomes and ribosomal protein from *Neurospora crassa*. I. Physical, chemical, and immunochemical properties. J. Bacteriol. **94**, 630–649 (1967)

[348] ROTHSCHILD, H., H. ITAKAWA, and S. R. SUSKIND: Ribosomes and ribosomal proteins from *Neurospora crassa*. II. Ribosomal proteins in different wild-type strains and during various stages of development. J. Bacteriol. **94**, 1800–1801 (1967)

[349] MEYER-BERTENRATH, J. G., and H. WÜRZ: Aggregationsstufen von Rattenleber-Ribosomen unter dem Einfluß Ribosomen-spezifischer Antikörper. Z. Naturforsch. **22b**, 1153–1156 (1967)

[350] MEYER-BERTENRATH, J. G.: Alterationen der Ribosomenstruktur während der durch Nitrosomorpholin induzierten Carcinogenese. Hoppe-Seyler's Z. Physiol. Chem. **348**, 645–650 (1967)

[351] CURRY, J. B.: Physical, chemical and immunochemical properties of ribosomal proteins. Diss. Abstr.-B **28**, 3156 (1968)

[352] BIELKA, H., and F. NOLL: Antigenic properties of ribosomes from animal tissues. Acta Biol. Med. Germ. **21**, K 1–K 3 (1968)

[353] NOLL, F., and H. BIELKA: Studies on proteins of animal ribosomes. III. Immunochemical analysis of ribosomes from different tissues and species of animals. Molec. Gen. Genetics **106**, 106–113 (1970)

[354] NOLL, F., and H. BIELKA: Untersuchungen über Proteine tierischer Ribosomen IV. Antigeneigenschaften von Ribosomen und ribosomalem Protein aus Leber und Hepatom. Arch. Geschwulstforsch. **35**, 338–346 (1970)

[355] BIELKA, H., H. WELFLE, P. WESTERMANN, F. NOLL, F. GRUMMT, and J. STAHL: Isolation and properties of proteins from animal ribosomes. FEBS Symp. **23**, 19 to 40 (1972)

[356] JUNGHAHN, I.: Untersuchungen über Protein tierischer Ribosomen. X. Antigeneigenschaften von Split-Proteinen und Core-Partikeln aus Rattenleberribosomen. Acta Biol. Med. Germ. **28**, 445–450 (1972)

[357] NOLL, F., W. HEUMANN, and H. BIELKA: On the nature of anti-ribosome antibodies and some properties of their reaction with ribosomes. Immunochem. **10**, 9–14 (1973)

[358] HALLERMAYER, G., and W. NEUPERT: Immunological difference of mitochondrial and cytoplasmic ribosomes of *Neurospora crassa*. FEBS-Lett. **41**, 264–268 (1974)

[359] STÖFFLER, G., N. FISCHER, K.-H. RAK, R. EHRLICH, and I. G. WOOL: Immunochemical analysis of the structure of eukaryotic ribosomes. Antigenic properties of rat liver ribosomes and ribosomal proteins and characterization of the antisera. Molec. Gen. Genetics **166**, 167 to 179 (1978)

[360] FISCHER, N., G. STÖFFLER, and I. G. WOOL: Immunological comparison of the proteins of chicken and rat liver ribosomes. J. Biol. Chem. **253**, 7355–7360 (1978)

[361] BUSCH, H., R. K. BUSCH, W. H. SPOHN, J. WIKMAN, and Y. DASKAL: Antibodies produced by immunization of goats with 60 S ribosomal subunits from Novikoff hepatoma ascites cells. Proc. Soc. Exptl. Biol. Med. **137**, 1470–1478 (1972)

[362] BUSCH, R. K., W. H. SPOHN, Y. DASKAL, and H. BUSCH: Antibodies produced by immunization of goats with 40 S ribosomal subunits from Novikoff hepatoma ascites cells. Proc. Soc. Exptl. Biol. Med. **140**, 1030–1033 (1972)

[363] WIKMAN-COFFELT, J., G. A. HOWARD, and R. R. TRAUT: Comparison of antigenic properties of ribosomal proteins from Novikoff hepatoma and normal liver. Biochim. Biophys. Acta **277**, 671 to 676 (1972)

[364] WOOL, I. G., and G. STÖFFLER: Determination of the size of the pool of free ribosomal proteins in rat liver cytoplasm. J. Mol. Biol. **108**, 201–218 (1976)

[365] ALBERGHINA, F. A. M., and L. MAZZETTI: Immunochemical studies of ribosomal proteins using passive hemagglutination techniques. Experientia **25**, 1256–1257 (1969)

[366] DELAUNAY, J., and G. SCHAPIRA: Immunogenetic properties of eukaryotic ribosomal proteins. Biochim. Biophys. Acta **386**, 138–143 (1974)

[367] NOLL, F., and H. BIELKA: Antigeneigenschaften ribosomaler Proteine aus tierischen Geweben. Acta Biol. Med. Germ. **23**, K 15–K 18 (1969)

[368] SARKAR, P. K., an S. SOM: Species specificity of ribosomal proteins in eukaryotic and prokaryotic organisms. Biochem. Biophys. Res. Commun. **37**, 767–770 (1969)

[369] FREY, J. R., A. VON DER DECKEN, and T. HULTIN: Antigenic properties of acid-soluble ribosomal proteins of *Artemia salina*. Exptl. Cell Res. **53**, 615–625 (1968)

[370] VON DER DECKEN, A., and T. HULTIN: Immunological properties of isolated

protein fractions from *Artemia salina* ribosomes. Exptl. Cell Res. **64**, 179–189 (1971)

[371] Bielka, H., F. Noll, J. Stahl, H. Theise, H. Welfle, and P. Westermann: Localization and function of proteins in rat liver ribosomes. In: Translation of natural and synthetic polynucleotides (Ed.: A. B. Legocki). University of Agriculture, Poznan 1977, 312–316

[372] Franek, F., and F. Noll: unpublished results

[373] Schur, P. H., L. A. Moroz, and H. G. Kunkel: Precipitating antibodies to ribosomes in the serum of patients with systemic *Lupus erythematosus*. Immunochem. **4**, 447–452 (1967)

[374] Lamon, E. W., and J. C. Bennett: Antibodies to homologous RNA in the rabbit following stimulation by exogeneous RNA. Proc. Soc. Exptl. Biol. Med. **134**, 968–970 (1970)

[375] Panijel, J.: Use of anti-RNA antibodies in the study of the structure of polynucleotides and ribosomal particles. In: Nucleic acids in immunology (Eds.: O. J. Plescia, and W. Braun) Springer Verlag, Berlin–Heidelberg–New York, 1968, 174 to 200

[376] Cayeux, P., J. Panijel, and C. Souleil: Action des anticorps anti-RNA sur les ribosomes et les phages. Biochim. Biophys. Acta **108**, 605–618 (1965)

[377] Howard, G. A., R. L. Smith, and J. Gordon: Chicken liver ribosomes: Characterization of cross-reaction and inhibition of some functions by antibodies prepared against *Escherichia coli* ribosomal proteins L 7 and L 12. J. Mol. Biol. **106**, 623–637 (1976)

[378] Noll, F.: unpublished results

[379] Wool, I. G.: personal communication

[380] Cavanagh, D.: A solid-phase radioimmunoassay for the detection of antibodies to ribosomes. Anal. Biochem. **79**, 217 to 225 (1977)

[381] Koffler, D., I. Feiferman, and M. A. Gerber: Radioimmunoassay for antibodies to cytoplasmic ribosomes in human serum. Science **198**, 741–743 (1977)

[382] Wikman-Coffelt, J.: Radioimmunoassay for ribosomes. Anal. Biochem. **48**, 339–345 (1972)

[383] Sturgill, B. C., and R. R. Carpenter: Antibody to ribosomes in systemic *Lupus erythematosus*. Arthr. Rheum. **8**, 213–218 (1965)

[384] Sturgill, B. C., and M. R. Preble: Antibody to ribosomes in systemic *Lupus erythematosus*: Demonstration by immunofluorescence and precipitation in agar. Arthr. Rheum. **10**, 538–543 (1967)

[385] Lamon, E. W., and J. C. Bennett: Antibodies to ribosomal ribonucleic acid (rRNA) in patients with systemic *Lupus erythematosus* (SLE). Immunology **19**, 439–442 (1970)

[386] Homberg, J.-C., M. Rizzetto, and D. Doniach: Ribosomal antibodies detected by immunofluorescence in systemic *Lupus erythematosus* and other collagenoses. Clin. Exptl. Immunol. **17**, 617–628 (1974)

[387] Cammarano, P., S. Pons, and P. Londei: Discontinuity of the large ribosomal subunit RNA and rRNA molecular weights in eukaryote evolution. Acta Biol. Med. Germ. **34**, 1123–1135 (1975)

[388] Verma, J. M., M. Edelman, M. Herzberg, and U. Z. Littauer: Size determination of mitochondrial ribosomal RNA from *Aspergillus nidulans* by electron microscopy. J. Mol. Biol. **52**, 137–140 (1970)

[389] Loening, U. E.: Molecular weights of ribosomal RNA in relation to evolution. J. Mol. Biol. **38**, 355–365 (1968)

[390] Miller, M. J., and D. McMahon: Synthesis and maturation of chloroplast and cytoplasmic ribosomal RNA in *Chlamydomonas reinhardi*. Biochim. Biophys. Acta **366**, 35–44 (1974)

[391] Grierson, D.: Characterisation of ribonucleic acid components from leaves of *Phaseolus aureus*. Europ. J. Biochem. **44**, 509–515 (1974)

[392] Pring, D. R., and R. W. Thornburg: Molecular weights of maize mitochondrial and cytoplasmic ribosomal RNAs under denaturing conditions. Biochim. Biophys. Acta **383**, 140–146 (1975)

[393] Loening, U. E.: The determination of the molecular weight of ribonucleic acid by polyacrylamide-gel electrophoresis. The effects of changes in conformation. Biochem. J. **113**, 131–118 (1969)

[394] Perry, R. P., T.-Y. Cheng, J. J. Freed, J. R. Greenberg, D. E. Kelley, and

K. D. TARTOF: Evolution of the transcription unit of ribosomal RNA. Proc. Natl. Acad. Sci. U.S. **65**, 609–616 (1970)

[395] CUNNINGHAM, R. S., and M. W. GRAY: Isolation and characterization of ^{32}P-labeled mitochondrial and cytosol ribosomal RNA from germinating wheat embryos. Biochim. Biophys. Acta **475**, 476 to 491 (1977)

[396] REIJNDERS, L., P. SLOOF, J. SIVAL, and P. BORST: Gel electrophoresis of RNA under denaturing conditions. Biochim. Biophys. Acta **324**, 320–333 (1973)

[397] ISHIKAWA, H.: Integrity of ribosomal ribonucleic acids from aphids and water fleas. Biochim. Biophys. Acta **435**, 258 to 268 (1976)

[398] LEVIS, R., and S. PENMAN: Processing steps and methylation in the formation of the rRNA of cultured *Drosophila* cells. J. Mol. Biol. **121**, 219–238 (1978)

[399] BELMONTE, E., A. CUPELLO, G. LAZZARINI, and A. GIUDITTA: Electrophoretic characterization of rRNA in the squid. Comp. Biochem. Physiol. **63**, 373 (1979)

[400] KLOOTWIJK, J., J. KLEIN, and L. A. GRIVELL: Minimal post-transcriptional modification of yeast mitochondrial ribosomal RNA. J. Mol. Biol. **97**, 337–350 (1975)

[401] REIJNDERS, L., P. SLOOF, and P. BORST: The molecular weights of the mitochondrial ribosomal RNAs of *Saccharomyces carlsbergensis*. Europ. J. Biochem. **35**, 266–269 (1973)

[402] SCHUTGENS, R. B. H., L. REIJNDERS, S. P. HOEKSTRA, and P. BORST: Transcription of *Tetrahymena* mitochondrial DNA in vivo. Biochim. Biophys. Acta **308**, 372–380 (1973)

[403] BORST, P., and L. A. GRIVELL: Mitochondrial ribosomes. FEBS-Lett. **13**, 73–88 (1971)

[404] JÄCKLE, H., and O. SCHMIDT: Characterization of rRNA from insect eggs (*Enscelis plebejus, Cicadina, Smittia spec., Chironomidae, Diptera*). Experientia **34**, 1260–1261 (1978)

[405] ATTARDI, B., and G. ATTARDI: Expression of the mitochondrial genome in HeLa cells. I. Properties of the discrete RNA components from the mitochondrial fraction. J. Mol. Biol. **55**, 231–243 (1971)

[406] LEAVER, C. J., and J. INGLE: The molecular integrity of chloroplast ribosomal ribonucleic acid. Biochem. J. **123**, 235 to 243 (1971)

[407] MACHE, R., M. JALLIFIER-VERNE, C. ROZIER, and S. LOISEAUX: Molecular weight determination of precursor, mature and post-mature plastid ribosomal RNA from spinach using fully denaturing conditions. Biochim. Biophys. Acta **517**, 390–399 (1978)

[408] INABA, K., H. MISUMI, and T. ODA: Molecular forms and size determination of 18 S RNA from rat liver microsomes by electron microscopy. J. Electron Microscopy **23**, 45–48 (1974)

[409] SCHIBLER, U., T. WYLER, and O. HAGENBÜCHLE: Changes in size and secondary structure of the ribosomal transcription unit during vertebrate evolution. J. Mol. Biol. **94**, 503–517 (1975)

[410] GRANBOULAN, N., and K. SCHERRER: Visualization in the electron microscope and size of RNA from animal cells. Europ. J. Biochem. **9**, 1–20 (1969)

[411] MURPHY, W. J., and G. ATTARDI: Use of a DNA probe for mapping by electron microscopy the ribosomal sequences in ribosomal RNA precursors from duck cells. J. Mol. Biol. **90**, 65–76 (1974)

[412] WINICOV, I.: Alternate temporal order in ribosomal RNA maturation. J. Mol. Biol. **100**, 141–155 (1976)

[413] WELLAUER, P. K., and I. B. DAWID: Secondary structure maps of RNA: processing of HeLa ribosomal RNA. Proc. Natl. Acad. Sci. U.S. **70**, 2827–2831 (1973)

[414] WELLAUER, P. K., I. B. DAWID, D. E. KELLEY, and R. P. PERRY: Secondary structure maps of ribosomal RNA. II. Processing of mouse-L-cell ribosomal RNA and variations in the processing pathway. J. Mol. Biol. **89**, 397–407 (1974)

[415] WELLAUER, P. K., and I. B. DAWID: Secondary structure maps of ribosomal RNA and DNA. I. Processing of *Xenopus laevis* ribosomal RNA and structure of single-stranded ribosomal DNA. J. Mol. Biol. **89**, 379–395 (1974)

[416] HOCHKEPPEL, H.-K., and J. GORDON: Preparation-sensitive conformations in ribosomal RNAs. Nature **273**, 560–561 (1978)

[417] SOERGEL, M. E., and F. L. SCHAFFER: Ri-

bosomal ribonucleic acids of cultured cells: A preliminary survey of differences among mammalian species detectable by polyacrylamide gel electrophoresis. In vitro **8**, 13–18 (1972)

[418] ATTARDI, G., and F. AMALDI: Structure and synthesis of ribosomal RNA. Ann. Rev. Biochem. **39**, 183–351 (1970)

[419] STAYNOV, D. Z., J. C. PINDER, and W. B. GRATZER: Molecular weight determination of nucleic acids by gel electrophoresis in non-aqueous solution. Nature New Biol. **235**, 108–110 (1972)

[420] SPOHR, G., M.-E. MIRAULT, T. IMAIZUMI, and K. SCHERRER: Molecular-weight determination of animal-cell RNA by electrophoresis in formamide under fully denaturing conditions on exponential polyacrylamide gels. Europ. J. Biochem. **62**, 313–322 (1976)

[421] BOEDTKER, H.: Dependence of the sedimentation coefficient on molecular weight of RNA after reaction with formaldehyde. J. Mol. Biol. **35**, 61–70 (1968)

[422] STRAUSS, J. H., jr. and R. L. SINSHEIMER: Ribonuclease digestion of R 17 viral RNA. J. Mol. Biol. **11**, 84–89 (1965)

[423] HAMILTON, M. G.: The molecular weight of the 30 S RNA of Jensen sarcoma ribosomes as determined by equilibrium centrifugation. Biochim. Biophys. Acta **134**, 473–475 (1967)

[424] MCCONKEY, E. H., and J. W. HOPKINS: Molecular weights of some HeLa ribosomal RNAs. J. Mol. Biol. **39**, 545–550 (1969)

[425] STANLEY, W. M. jr., and R. M. BOCK: Isolation and physical properties of the ribosomal ribonucleic acid of *Escherichia coli*. Biochem. **4**, 1302–1311 (1965)

[426] BRUENING, G., and R. M. BOCK: Covalent integrity and molecular weights of yeast ribosomal ribonucleic acid components. Biochim. Biophys. Acta **149**, 377–386 (1967)

[427] HUNT, J. A.: Molecular weight and chain length of rabbit reticulocyte rRNA. Nature **226**, 950–952 (1970)

[428] HAEGEMAN, C., W. MIN-JOU, and W. FIERS: Studies on the bacteriophage MS 2. IX. The heptanucleotide sequences present in the pancreatic ribonuclease digest of the viral RNA. J. Mol. Biol. **57**, 597–613 (1971)

[429] VAN DEN BOS, R. C., J. RETÈL, and R. J. PLANTA: The size and the location of the ribosomal RNA segments in ribosomal precursor RNA of yeast. Biochim. Biophys. Acta **232**, 494–508 (1971)

[430] PLOUMHANS, M.-T., and E. FREDERICQ: Visometric behavior of yeast ribosomal RNA in mixed water-organic solvent solutions and its application to a simple determination of the molecular weight. Biochim. Biophys. Acta **477**, 379–393 (1977)

[431] KOSER, R. B., and J. R. COLLIER: The molecular weight and thermolability of *Ilyanassa* ribosomal RNA. Biochim. Biophys. Acta **254**, 272–277 (1971)

[432] SHINE, J., and L. DALGARNO: Occurrence of heat-dissociable ribosomal RNA in insects: The presence of three polynucleotide chains in 26 S RNA from cultured *Aedes aepypti* cells. J. Mol. Biol. **75**, 57 to 72 (1973)

[433] APPLEBAUM, S. W., R. P. EBSTEIN, and G. R. WYATT: Dissociation of ribosomal ribonucleic acid from silkmoth pupae by heat and dimethylsulfoxide: Evidence for specific cleavage points. J. Mol. Biol. **21**, 29–34 (1966)

[434] GREENBERG, J. R.: Synthesis and properties of ribosomal RNA in *Drosophila*. J. Mol. Biol. **46**, 85–98 (1969)

[435] ISHIKAWA, H., and R. W. NEWBURGH: A rapidly-labeled RNA species from the silkgland of the wax moth *Galleria mellonella* (L.). Biochem. Biophys. Res. Commun. **40**, 654–660 (1970)

[436] ISHIKAWA, H., and R. W. NEWBURGH: A DNA-like RNA fraction in the posterior silkgland of the wax moth, *Galleria mellonella* (L.). Biochim. Biophys. Acta **232**, 661–670 (1971)

[437] WOLLENZIEN, P. L., D. C. YOUVAN, and J. E. HEARST: Structure of psoralen-crosslinked ribosomal RNA from *Drosophila melanogaster*. Proc. Natl. Acad. Sci. U.S. **75**, 1642–1646 (1978)

[438] JORDAN, B. R., R. JOURDAN, and B. JACQ: Late steps in the maturation of *Drosophila* 26 S ribosomal RNA. Generation of 5.8 S and 2 S RNAs by cleavages occurring in the cytoplasm. J. Mol. Biol. **101**, 85–105 (1976)

[439] Bostock, C. J., D. M. Prescott, and M. Lauth: Lability of 26 S ribosomal RNA in *Tetrahymena pyriformis*. Exptl. Cell Res. **66**, 260—262 (1971)

[440] Stevens, A. R., and P. F. Pachler: Discontinuity of 26 S rRNA in *Acanthamoeba castellani*. J. Mol. Biol. **66**, 225—237 (1972)

[441] Garvin, R. T., R. C. Hill, and M. M. Weber: The atypical RNA components of cytoplasmic ribosomes from *Crithidia fasciculata*. Arch. Biochem. Biophys. **191**, 774—781 (1978)

[442] Gray, M. W.: The rRNA of the trypanosomatid protozoan *Crithidia fasciculata*: physical characteristics and methylated sequences. Can. J. Biochem. **57**, 914—926 (1979)

[443] Spencer, R., and G. A. M. Cross: J. Gen. Microbiol. **93**, 82—88 (1976)

[444] Simpson, L., and A. M. Simpson: Kinetoplast RNA of Leishmania tarentolae. Cell **14**, 169—178 (1978)

[445] Rawson, J. R., E. J. Crouse, and E. Stutz: The integrity of the 25 S ribosomal RNA from *Euglena gracilis* 87 S ribosomes. Biochim. Biophys. Acta **246**, 507—516 (1971)

[446] Higo, S., M. Higo, and S. Tanifuji: Specific dissociation of pea 25 S rRNA by hot-phenol treatment. Biochim. Biophys. Acta **246**, 499—506 (1971)

[447] Leaver, C. J.: Molecular integrity of chloroplast ribosomal ribonucleic acid. Biochem. J. **135**, 237—240 (1973)

[448] Munsche, D., and R. Wollgiehn: Altersabhängige Labilität der ribosomalen RNA aus Chloroplasten von *Nicotiana rustica*. Biochim. Biophys. Acta **340**, 437—445 (1974)

[449] Kokileva, L., I. Mladenova, and R. Tsanev: Differential thermal stability of old and new ribosomal RNA of rat liver. FEBS-Lett. **12**, 313—316 (1971)

[450] Awata, S., and Y. Natori: Turnover of rat liver 28 S ribosomal RNA. Nicking as the initial step of degradation. Biochim. Biophys. Acta **478**, 486—494 (1977)

[451] Pellegrini, M., J. Manning, and N. Davison: Sequence arrangement of the rDNA of *Drosophila melanogaster*. Cell **10**, 213—224 (1977)

[452] Lava-Sanchez, P. A., F. Amaldi, and A. Posta: Base composition of ribosomal RNA and evolution. J. Mol. Evol. **2**, 44—55 (1972)

[453] Amaldi, F.: Non-random variability in evolution of base compositions of ribosomal RNA. Nature **221**, 95—96 (1969)

[454] Cox, B. J., and G. Turnock: Synthesis and processing of ribosomal RNA in cultured plant cells. Europ. J. Biochem. **37**, 367—376 (1973)

[455] Gray, M. W.: The presence of $O^{2'}$-methylpseudouridine in the 18 S + 26 S ribosomal ribonucleates of wheat embryo. Biochem. **13**, 5453—5463 (1974)

[456] Miassod, R., and J.-P. Cecchini: Partial base-methylation and other structural differences in the 17 S rRNA of sycamore cells during growth in cell culture. Biochim. Biophys. Acta **562**, 292—301 (1979)

[457] Cecchini, J.-P., and R. Miassod: Studies on the methylation of cytoplasmic rRNA from cultured higher plant cells. Europ. J. Biochem. **98**, 203—214 (1979)

[458] Lau, R. Y., T. D. Kennedy, and B. G. Lane: Wheat-embryo ribonucleates. III. Modified nucleotide constituents in each of the 5.8 S, 18 S and 26 S ribonucleates. Can. J. Biochem. **52**, 1110—1123 (1974)

[459] Retèl, J., R. C. van den Bos, and R. J. Planta: Characteristics of the methylation in vivo of ribosomal RNA in yeast. Biochim. Biophys. Acta **195**, 370—380 (1969)

[460] Klootwijk, J., R. C. van den Bos, and R. J. Planta: Secondary methylation of yeast ribosomal RNA. FEBS-Lett. **27**, 102—106 (1972)

[461] Klootwijk, J., and R. J. Planta: Analysis of the methylation sites in yeast ribosomal RNA. Europ. J. Biochem. **39**, 325 to 333 (1973)

[462] Klootwijk, J., and R. J. Planta: Mol. Biol. Rep. **1**, 187—191 (1973)

[463] Maden, B. E. H., J. Forbes, P. de Jonge, and J. Klootwijk: Presence of a hypermodified nucleotide in HeLa cell 18 S and *Saccharomyces* 17 S rRNA. FEBS-Lett. **59**, 60—63 (1975)

[464] Leick, V.: Formation of subribosomal particles on the macronuclei of *Tetrahymena pyriformis* Europ. J. Biochem. **8**, 221 to 228 (1969)

[465] Gray, M. W.: Dinucleotide sequences containing both base and sugar modifications

in the ribosomal RNA of *Crithidia fasciculata.* Biochim. Biophys. Acta **374**, 253 to 257 (1974)

[466] Slack, J. M. W., and U. E. Loening: 28-S RNA from *Xenopus laevis* contains a sequence of three adjacent 2′-O-methylations. Europ. J. Biochem. **43**, 69—72 (1974)

[467] Weinberg, R. A., and S. Penman: Processing of 45 S nucleolar RNA. J. Mol. Biol. **47**, 169—178 (1970)

[468] Salim, M., and B. E. H. Maden: Early and late methylations in HeLa cell ribosome maturation. Nature **244**, 334—336 (1973)

[469] Klagsbrun, M.: An evolutionary study of the methylation of transfer and ribosomal ribonucleic acid in prokaryote and eukaryote organisms. J. Biol. Chem. **248**, 2612—2620 (1973)

[470] Maden, B. E. H., and M. Salim: The methylated nucleotide sequences in HeLa cell ribosomal RNA and its precursors. J. Mol. Biol. **88**, 133—164 (1974)

[471] Amaldi, F., and G. Attardi: Partial sequence analysis of ribosomal RNA from HeLa cells. I. Oligonucleotide pattern of 28 S and 18 S RNA after pancreatic ribonuclease digestion. J. Mol. Biol. **33**, 737—755 (1968)

[472] Jeanteur, P., F. Amaldi, and G. Attardi: Partial sequence analysis of ribosomal RNA from HeLa cells. II. Evidence for sequences of non-ribosomal type in 45 and 32 S ribosomal RNA precursors. J. Mol. Biol. **33**, 757—775 (1968)

[473] Maden, B. E. H., and J. Forbes: Standard and non standard products in combined T_1 plus pancreatic RNase fingerprints of HeLa cell rRNA and its precursors. FEBS-Lett. **28**, 289—292 (1972)

[474] Maden, B. E. H., C. D. Lees, and M. Salim: Some methylated sequences and the numbers of methylgroups in HeLa cell rRNA. FEBS-Lett. **28**, 293—296 (1972)

[475] Lane, B. G., and T. Tamaoki: Methylated bases and sugars in 16-S and 28-S RNA from L cells. Biochim. Biophys. Acta **179**, 332—340 (1969)

[476] Saponara, A. G., and M. D. Enger: The isolation from ribonucleic acid of substituted uridines containg α-aminobutyrate moieties derived from methionine. Biochim. Biophys. Acta **349**, 61—77 (1974)

[477] Egawa, K., Y. C. Choi, and H. Busch: Studies on the role of 23 S nucleolar RNA as an intermediate in the synthesis of 18 S ribosomal RNA. J. Mol. Biol. **56**, 565 to 577 (1971)

[478] Grummt, I., U. E. Loening, and J. M. W. Slack: Methylation of ribosomal precursor RNA, synthesized in vitro, by isolated rat-liver nucleoli. Europ. J. Biochem. **59**, 313—318 (1975)

[479] Liau, M. C., and R. B. Hurlbert: Interrelationships between synthesis and methylation of ribosomal RNA in isolated Novikoff tumor nucleoli. Biochem. **14**, 127—134 (1975)

[480] Tiollais, P., M.-A. Auger, and F. Galibert: Methylation process of mammalian ribosomal precursor RNA. Biochimie **57**, 1315—1321 (1975)

[481] Khan, M. S. N., M. Salim, and B. E. H. Maden: Extensive homologies between the methylated nucleotide sequences in several vertebrate rRNAs. Biochem. J. **169**, 531—542 (1978)

[482] Judes, C., and M. Jacob: Compartmentation of the δ-adenosylmethionine pool in developing chick embryo cerebral hemispheres, as demonstrated by a fingerprint study of 18 S ribosomal RNA. FEBS-Lett. **27**, 289—292 (1972)

[483] Hashimoto, S., M. Sakai, and M. Muramatsu: 2′-O-methylated oligonucleotides in ribosomal 18 S and 28 S RNA of a mouse hepatoma, MH 134. Biochem. **14**, 1956—1964 (1975)

[484] Khan, M. S. N., and B. E. H. Maden: Conformation of methylated sequences in HeLa cell 18 S ribosomal RNA: Nuclease S_1 as a probe. Europ. J. Biochem. **84**, 241 to 250 (1978)

[485] Goddard, J. P., and B. E. H. Maden: Reaction of HeLa cell methyl-labelled 28 S rRNA with sodium bisulphite: a conformational probe for methylated sequences. Nucl. Acids Res. **3**, 341—440 (1976)

[486] Fuke, M., and H. Busch: Partial methylation of 18 S ribosomal RNA detected by T_1 ribonuclease digestion and homochromatography fingerprinting. FEBS-Lett. **77**, 287—290 (1977)

[487] MADEN, B. E. H., and M. S. N. KHAN: Methylated nucleotide sequence in HeLa-cell ribosomal ribonucleic acid. Biochem. J. **167**, 211–221 (1977)

[488] NAZAR, R. N., and H. BUSCH: Structural analyses of mammalian ribosomal ribonucleic acid and its precursors. J. Biol. Chem. **249**, 919–929 (1974)

[489] HUGHES, D. G., and B. E. H. MADEN: The pseudouridine contents of the ribosomal ribonucleic acids of three vertebrate species. Numerical correspondence between pseudouridine residues and 2′-O-methyl groups is not always conserved. Biochem. **171**, 781–786 (1978)

[490] HUGHES, D. G., S. HUGHES, and B. E. H. MADEN: The pseudouridine content of HeLa cell ribosomal RNA. FEBS-Lett. **72**, 304–308 (1976)

[491] CHOI, Y. C., and H. BUSCH: Comparison of large fragments obtained by T_1 RNase digestion of ribosomal and nucleolar preribosomal RNA of Novikoff hepatoma ascites cells: The 5′-terminal eicosanucleotide. Biochem. Biophys. Res. Commun. **58**, 674–682 (1974)

[492] CHOI, Y. C., and H. BUSCH: Modified nucleotides in T_1 RNase oligonucleotides of 18 S ribosomal RNA of the Novikoff hepatoma. Biochem. **17**, 2551–2560 (1978)

[493] ELADARI, M. E., A. HAMPE, and F. GALIBERT: Nucleotide sequence neighbouring a late modified guanylic residue within 28 S rRNA of several eukaryotic cells. Nucl. Acids Res. **4**, 1759–1767 (1977)

[494] CHOI, Y. C., and H. BUSCH: Structural analysis of nucleolar precursors of ribosomal ribonucleic acid. J. Biol. Chem. **245**, 1954–1961 (1970)

[495] THOMAS, H., J. GORDON, and H. ROGG: N^4-acetylcytidine. A previously unidentified labile component of the small subunit of eukaryotic ribosomes. J. Biol. Chem. **253**, 1101–1105 (1978)

[496] BRAND, R. C., J. KLOOTWIJK, T. J. M. V. STEENBERGEN, A. J. D. KOK, and R. J. PLANTA: Secondary methylation of yeast ribosomal precursor RNA. Europ. J. Biochem. **75**, 311–318 (1977)

[497] RIJVEN, A. H. G. C., and J. A. ZWAR: Methylation patterns of ribonucleic acids from chloroplasts and cytoplasm of femigreek (*Trigonella foenumgraecum L.*) cotyledons. Biochim. Biophys. Acta **299**, 564–567 (1973)

[498] KURIYAMA, Y., and D. J. L. LUCK: Methylation and processing of mitochondrial ribosomal RNAs in poky and wild-type *Neurospora crassa*. J. Mol. Biol. **83**, 253–266 (1974)

[499] DUBIN, D. T., and R. H. TAYLOR: Modification of mitochondrial ribosomal RNA from hamster cells: The presence of GmG and late-methylated UmGmU in the large subunit (17 S) RNA. J. Mol. Biol. **121**, 523–540 (1978)

[500] DUBIN, D. T.: Methylated nucleotide content of mitochondrial ribosomal RNA from hamster cells. J. Mol. Biol. **84**, 257 to 273 (1974)

[501] KLEIN, J., and J. KLOOTWIJK: Determination of 2′-O-methyl groups in ^{32}P-labeled RNAs. Anal. Biochem. **74**, 263 to 272 (1976)

[502] VAUGHAN, M. H., R. SOEIRO, J. R. WARNER, and J. E. DARNELL: The effects of methionine deprivation on ribosome synthesis in HeLa cells. Proc. Natl. Acad. Sci. U.S. **58**, 1527–1534 (1967)

[503] GRISWOLD, M. D., R. D. BROWN, and G. P. TOCCHINI-VALENTINI: An analysis of the degree of homology between 28 S rRNA from *Xenopus laevis* and *Xenopus mulleri*. Biochem. Biophys. Res. Commun. **58**, 1093–1103 (1974)

[504] HUNT, J. A.: Terminal-sequence studies of high-molecular-weight ribonucleic acid. The 3′-termini of rabbit reticulocyte ribosomal RNA. Biochem. J. **120**, 353–363 (1970)

[505] KOMINAMI, R., H. HAMADA, Y. FUJII-KURIYAMA, and M. MURAMATSU: 5′-terminal processing of ribosomal 28 S RNA. Biochem. **17**, 3965–3970 (1978)

[506] AHMAD, M. S., P. D. MARKHAM, and D. G. GLITZ: Terminal nucleotides of avian myeloblastosis virus RNA and of ribosomal RNA from chicken leukemic myeloblasts. Biochim. Biophys. Acta **281**, 554–563 (1972)

[507] SHINE, J., and L. DALGARNO: Identical 3′-terminal octanucleotide sequence in 18 S ribosomal ribonucleic acid from different eukaryotes. Biochem. J. **141**, 609 to 615 (1974)

[508] DALGARNO, L., and J. SHINE: Conserved terminal sequence in 18 S rRNA may re-

present terminator anticodons. Nature New Biol. **245**, 261—262 (1973)

[509] Jonge, P. de, J. Klootwijk, and R. J. Planta: Terminal nucleotide sequences of 17-S ribosomal RNA and its immediate precursor 18-S RNA in yeast. Europ. J. Biochem. **72**, 361—369 (1977)

[510] Jonge, P. de, J. Klootwijk, and R. J. Planta: Sequence of 3′ terminal 21 nucleotides of yeast 17 S rRNA. Nucl. Acids Res. **4**, 3655—3663 (1977)

[511] Sprague, K. U., R. A. Kramer, and M. B. Jackson: The terminal sequences of *Bombyx mori* 18 S ribosomal RNA. Nucl. Acids Res. **2**, 2111—2118 (1975)

[512] Niles, E. G.: 5′- and 3′-terminal nucleotide sequences of *Tetrahymena pyriformis* 17 S rRNA. Biochem. **16**, 2380—2383 (1977)

[513] Alberty, H., M. Raba, and H. J. Gross: Isolation from rat liver and sequence of a RNA fragment containing 32 nucleotides from position-5 to position-36 from 3′-end of ribosomal 18 S RNA. Nucl. Acids. Res. **5**, 425—434 (1978)

[514] Eladari, M. E., and F. Galibert: Sequence determination of 3′-terminal T_1 oligonucleotide of 18 S ribosomal RNA. Nucl. Acids Res. **3**, 2749—2756 (1976)

[515] Vass, J. K., and B. E. H. Maden: Studies on the conformation of the 3′ terminus of 18 S rRNA. Europ. J. Biochem. **85**, 241—247 (1978)

[516] Maden, B. E. H., and J. S. Robertson: Demonstration of the "5.8 S" ribosomal sequence in HeLa cell ribosomal precursor RNA. J. Mol. Biol. **87**, 227—235 (1974)

[517] Choi, Y. C., N. R. Ballal, R. K. Busch, and H. Busch: Homochromatographic and immunological analysis of controls of nucleolar gene function. Cancer Res. **36**, 4301—4306 (1976)

[518] Inagaki, A., and H. Busch: Marker nucleotides for non-ribosomal spacer segments of preribosomal ribonucleic acid. Biochem. Biophys. Res. Commun. **49**, 1398—1406 (1972)

[519] Inagaki, A., and H. Busch: Structural analysis of nucleolar precursors of ribosomal ribonucleic acids. J. Biol. Chem. **247**, 3327—3335 (1972)

[520] Eladari, M.-E., and F. Galibert: Sequence determination of 5′-terminal and 3'-terminal T_1 oligonucleotides of 18 S ribosomal RNA of a mouse cell line (L 5178 Y). Europ. J. Biochem. **55**, 247—255 (1975)

[521] Sakuma, K., R. Kominami, M. Muramatsu, and M. Sugiura: Conservation of the 5′-terminal nucleotide sequences of ribosomal 18 S RNA in eukaryotes. Differential evolution of large and small ribosomal RNA. Europ. J. Biochem. **63**, 339—350 (1976)

[522] Nazar, R. N., and H. Busch: A comparison of polypyrimidine fragments in 28 S ribosomal RNA from rat liver and the Novikoff ascites hepatoma. Cancer Res. **32**, 2322—2331 (1972)

[523] Nazar, R. N., and H. Busch: Structural analyses of mammalian ribosomal ribonucleic acid and its precursors. The distribution of polypyrimidine sequences in ribosomal 18-S RNA. Biochim. Biophys. Acta **299**, 428—443 (1973)

[524] Kanamaru, R., Y. C. Choi, and H. Busch: Structural analysis of ribosomal 28 S ribonucleic acid of Novikoff hepatoma cells. J. Biol. Chem. **249**, 2453—2465 (1974)

[525] Fuke, M., and H. Busch: A T_1 ribonuclease fragment present in 18 S ribosomal RNA of Novikoff rat ascites hepatoma cells and absent from 18 S ribosomal RNA of HeLa cells. J. Mol. Biol. **99**, 277—282 (1975)

[526] Khan, M. S. N., and B. E. H. Maden: Nucleotide sequences within the ribosomal ribonucleic acids of HeLa cells, *Xenopus laevis* and chick embryo fibroblasts. J. Mol. Biol. **101**, 235—254 (1976)

[527] Slack, J. M. W., and U. E. Loening: 5′-ends of ribosomal and ribosomal precursor RNAs from *Xenopus laevis*. Europ. J. Biochem. **43**, 59—67 (1974)

[528] Hagenbüchle, O., M. Santer, J. A. Steitz, and R. J. Mans: Conservation of the primary structure at the 3′ end of 18 S rRNA from eukaryotic cells. Cell **13**, 551—563 (1978)

[529] Oakden, K. M., and B. G. Lane: Wheat embryo ribonucleates. VI. Comparison of the 3′-hydroxyl termini in rapidly labelled RNA from metabolizing wheat embryos with the corresponding termini in ribosomal RNA from differentiating embryos of wheat, barley, corn and pea. Can. J. Biochem. **54**, 261—271 (1976)

[530] FUKE, M., and H. BUSCH: Sequence analysis of T 1 ribonuclease fragments of 18 S ribosomal RNA by 5'-terminal labeling, partial digestion and homochromatography fingerprinting. Nucl. Acids Res. **4**, 339–352 (1977)

[531] MAXAM, A. M., and W. GILBERT: A new method for sequencing DNA. Proc. Natl. Acad. Sci. U.S. **74**, 560–564 (1977)

[532] SANGER, F., and A. R. COULSON: A rapid method for determining sequences in DNA by primed synthesis with DNA polymerase. J. Mol. Biol. **94**, 441–448 (1975)

[533] DAUBERT, S., and M. E. DAHMUS: Synthesis and characterization of a DNA probe complementary to rat liver 28 S ribosomal DNA. Biochem. Biophys. Res. Commun. **68**, 1037–1044 (1976)

[534] MORROW, J. F., S. N. COHEN, A. C. Y. CHANG, M. W. BOYER, H. M. GOODMAN, and R. B. HELLING: Replication and transcription of eukaryotic DNA in *Escherichia coli*. Proc. Natl. Acad. Sci. U.S. **71**, 1743 to 1747 (1974)

[535] MCCLEMENTS, W., and A. M. SKALKA: Analysis of chicken ribosomal RNA genes and construction of lambda hybrids containing gene fragments. Science **196**, 195–197 (1977)

[536] KHAN, M. S. N., and B. E. H. MADEN: Comparison between the ribosomal ribonucleic acids from free and membrane-bound ribosomal fractions of HeLa cells. Biochem. J. **155**, 197–200 (1976)

[537] FUKE, M., H. BUSCH, and P. N. RAO: Evolutionary trends in 18 S ribosomal RNA nucleotide sequences of rat, mouse, hamster and man. Nucl. Acids Res. **3**, 2939–2957 (1976)

[538] HASHIMOTO, S., and M. MURAMATSU: Differences in nucleotide sequences of ribosomal RNA between the liver and a hepatoma of C3H/He mice. Europ. J. Biochem. **33**, 446–458 (1973)

[539] HIGASHI, K., S. GOTOH, K. NISHINAGA, and Y. SAKAMOTO: Further evidence of heterogeneity of ribosomal RNA in various rat tissues. Biochim. Biophys. Acta **262**, 320–327 (1972)

[540] KNÄLMANN, M., and E. C. BURGER: Degree of relatedness of ribosomal 18/25 S RNA of *Vicia species* and *Allium cepa*. Cytobiol. **18**, 422–430 (1979)

[541] SHINE, J., and L. DALGARNO: The 3'-terminal sequence of *Escherichia coli* 16 S ribosomal RNA: Complementarity to nonsense triplets and ribosome binding sites. Proc. Natl. Acad. Sci. U.S. **71**, 1342 to 1346 (1974)

[542] FELLNER, P.: Nucleotide sequence from specific areas of the 16 S and 23 S ribosomal RNAs of *E. coli*. Europ. J. Biochem. **11**, 12–27 (1969)

[543] GALIBERT, F., P. TIOLLAIS, and M. E. ELADARI: Fingerprinting studies of the maturation of ribosomal RNA in mammalian cells. Europ. J. Biochem. **55**, 239 to 245 (1975)

[544] BENDICH, A. J., and B. J. MCCARTHY: Ribosomal RNA homologies among distantly related organisms. Proc. Natl. Acad. Sci. U.S. **65**, 349–356 (1970)

[545] BIRNSTIEL, M. L., and M. GRUNSTEIN: The ribosomal cistrons of eukaryotes — a model system for the study of evolution of serially repeated genes. FEBS Symp. **23**, 349–365 (1972)

[546] SINCLAIR, J. H., and D. D. BROWN: Retention of common nucleotide sequences in the ribosomal deoxyribonucleic acid of eukaryotes and some of their physical characteristics. Biochem. **10**, 2761–2769 (1971)

[547] MARTIN, T. E., J. N. BICKNELL, and A. KUMAR: Hybrid 80 S monomers formed from subunits of ribosomes from protozoa, fungi, plants and mammals. Biochem. Genetics **4**, 603–615 (1970)

[548] GARBI, S. A.: Fine structure of ribosomal RNA. I. Conservation of homologous regions within ribosomal RNA of eukaryotes. J. Mol. Biol. **106**, 791–816 (1976)

[549] COX, R. A., H. J. GOULD, and K. KANAGALINGAM: A study of the alkaline hydrolysis of fractionated reticulocyte ribosomal ribonucleic acid and its relevance to secondary structure. Biochem. J. **106**, 733–741 (1968)

[550] COTTER, R. J., P. MCPHIE, and W. B. GRATZER: Internal organization of the ribosome. Nature **216**, 864–868 (1967)

[551] COX, R. A.: A spectrophotometric study of the secondary structure of ribonucleic acid isolated from the smaller and larger ribosomal subparticles of rabbit reticulocytes. Biochem. J. **117**, 101–118 (1970)

[552] Thomas, G. J., jr., and M. Spencer: Studies on ribosomal RNA structure. II. Secondary structures in solution of rRNA and crystallizable fragments. Biochim. Biophys. Acta **179**, 360—368 (1969)

[553] Gould, H. J., and H. Simpkins: Studies on the secondary structure of ribosomal ribonucleic acid components of rabbit reticulocytes. Biopolymers. **7**, 223—239 (1969)

[554] Cox, R. A., E. Godwin, and J. R. B. Hastings: Spectroscopic evidence for the uneven distribution of adenine and uracil residues in ribosomal ribonucleic acid of *Drosophila melanogaster* and of *Plasmodium knowlesi* and its possible evolutionary significance. Biochem. J. **155**, 465 to 475 (1976)

[555] Godwin, E., R. A. Cox, and P. Hüvös: Studies of the RNA and protein moieties of the larger subribosomal particle of rabbit reticulocytes. Acta Biol. Med. Germ. **33**, 733—752 (1974)

[556] Watson, J. D., and C. Kidson: Effects of ionic conditions on the structure of ribosomal RNA components. J. Mol. Biol. **43**, 331—335 (1969)

[557] Cox, R. A., W. Hirst, E. Godwin, and I. Kaiser: The circular dichroism of ribosomal ribonucleic acids. Biochem. J. **155**, 279—295 (1976)

[558] Cox, R. A., and W. Hirst: Calculation of the circular dichroism of bihelical ribonucleic acids as a function of nucleotide composition. Biochem. J. **155**, 292—295 (1976)

[559] Spencer, M., W. J. Pigram, and J. Littlechild: Studies on ribosomal RNA structure. I. The isolation of crystallizable fragments. Biochim. Biophys. Acta **179**, 348—359 (1969)

[560] Cox, R. A., and H. R. V. Arnstein: The isolation, characterization and acid-base properties of ribonucleic acid from rabbit reticulocyte ribosomes. Biochem. J. **89** 574—585 (1963)

[561] Bielka, H., and G. Lutsch: Studies on the structure of ribosomal ribonucleic acid from normal and tumour tissues. VI. Dissociation properties. Z. Naturforschg. **22b**, 1035—1043 (1967)

[562] Ladhoff, A.-M., and S. Rosenthal: Electron microscopic studies on the substructure of eukaryotic ribosomal RNA. Acta Biol. Med. Germ. **33**, 761—769 (1974)

[563] Cox, R. A.: Hydrolysis of polynucleotides and the characterization of their secondary structure. A theoretical study. Biochem. J. **106**, 725—731 (1968)

[564] Peacock, A. C., S. L. Bunting, and K. Nishinaga: Conformational states of 18 S rRNA from HeLa cells detected by gel electrophoresis. Biochim. Biophys. Acta **475**, 352—365 (1977)

[565] Slack, J. M. W., M.-L. Sartirana, and U. E. Loening: Multiple conformations of ribosomal precursor RNA. Nature **253**, 282—284 (1975)

[566] Scherberg, N., and S. Refetoff: Iodination — deiodination. A radiochemical method for detection of structure and changes in structure in RNA. Biochim. Biophys. Acta **475**, 337—351 (1977)

[567] Galling, G., and B. R. Jordan: Isolation and characterization of cytoplasmic and chloroplastic 5 S RNAs in the unicellular green alga *Chlorella*. Biochimie **54**, 1257—1265 (1972)

[568] Payne, P. J., and T. A. Dyer: Characterization of cytoplasmic and chloroplast 5 S ribosomal ribonucleic acid from broadbean leaves. Biochem. J. **124**, 83—89 (1971)

[569] Gray, P. W., and R. B. Hallick: Isolation of *Euglena gracilis* chloroplast 5 S rRNA and mapping the 5 S rRNA gene on chloroplast DNA. Biochem. **18**, 1820—1825 (1979)

[570] Whitfeld, P. R., C. J. Leaver, W. Bottomley, and B. A. Atchinson: Low-molecular-weight (4.5 S) RNA in higher-plant chloroplast ribosomes. Biochem. J. **175**, 1103—1112 (1978)

[571] Leaver, C. J., and M. A. Harmey: Higherplant mitochondrial ribosomes contain a 5 S ribosomal ribonucleic acid component. Biochem. J. **157**, 275—277 (1976)

[572] Cunningham, R. S., L. Bonen, W. F. Doolittle, and W. M. Gray: Unique species of 5 S, 18 S and 26 S ribosomal RNA in wheat mitochondria. FEBS-Lett. **69**, 116—122 (1976)

[573] Hartley, M. R.: The synthesis and origin of chloroplast low-molecular weight

ribosomal ribonucleic acid in spinach. Europ. J. Biochem. **96**, 311—320 (1979)

[574] MIYAZAKI, M.: Studies on the nucleotide sequence of pseudouridine-containing 5 S RNA from *Saccharomyces cerevisiae*. J. Biochem **75**, 1407—1410 (1974)

[575] MIYAZAKI, M.: Nucl. Acids. Res., Suppl. **3**, 153—156 (1977)

[576] ERDMANN, V. A.: Collection of published 5 S and 5.8 S rRNA sequences. Nucl. Acids Res. **5**, r 1—r 13 (1978)

[577] HORI, H., and S. OSAWA: Evolutionary change in 5 S RNA secondary structure and a phylogenic tree of 54 5 S RNA species. Proc. Natl. Acad. Sci. U.S. **76**, 381 to 385 (1979)

[578] ERDMANN, V. A.: Collection of published 5 S and 5.8 S RNA sequences and their pIecursors. Nucl. Acids. Res. **6**, r 29 to r 44 (1979)

[579] SANKOFF, D., R. J. CEDERGREN, and G. LAPALME: Frequency of insertion, deletion, transversion, and transition in the evolution of 5 S rRNA. J. Molec. Evol. **7**, 133—149 (1976)

[580] HORI, H.: Molecular evolution of 5 S RNA. Molec. Gen. Genetics **145**, 119 to 123 (1976)

[581] HORI, H.: Evolution of 5 S RNA. J. Molec. Evol. **7**, 75—86 (1975)

[582] HORI, H., K. HIGO, and S. OSAWA: The rates of evolution in some ribosomal components. J. Molec. Evol. **9**, 191—201 (1977)

[583] FORGET, B. G., and S. M. WEISSMAN: Low molecular weight RNA components from KB cells. Nature **213**, 878—882 (1967)

[584] FORGET, B. G., and S. M. WEISSMAN: The nucleotide sequence of ribosomal 5 S ribonucleic acid from KB cells. J. Biol. Chem. **244**, 3148—3165 (1969)

[585] FORGET, B. G., and S. M. WEISSMAN: Nucleotide sequence of KB cell 5 S RNA. Science **158**, 1695—1699 (1967)

[586] HATLEN, L. E., F. AMALDI, and G. ATTARDI: Oligonucleotide pattern after pancreatic ribonuclease digestion and the 3′ and 5′ termini of 5 S ribonucleic acid from HeLa cells. Biochem. **8**, 4989—5005 (1969)

[587] VIGNE, R., B. R. JORDAN, and R. MONIER: A common conformational feature in several prokaryotic and eukaryotic 5 S RNAs: A highly exposed, single-stranded loop around position 40. J. Mol. Biol. **76**, 303—311 (1973)

[588] WILLIAMSON, R., and G. G. BROWNLEE: 5 S ribosomal ribonucleic acid from two mouse cell lines gives 'fingerprints' identical with human 5 S ribosomal ribonucleic acid. Biochem. J. **114**, 29 P (1969)

[589] VIGNE, R., and B. R. JORDAN: Partial enzyme digestion studies on *E. coli*, *Pseudomonas*, *Chlorella*, *Drosophila*, HeLa and yeast 5 S RNA supports a general class of 5 S RNA models. J. Molec. Evol. **10**, 77—86 (1977)

[590] YAMAMOTO, M., and K. H. SEIFART: Heterogeneity in the 3′-terminal sequence of ribosomal 5 S RNA synthesized by isolated HeLa cell nuclei in vitro. Biochem. **17**, 457—461 (1978)

[591] KEL'VE, M., A. METSPALU, A. LIND, M. SAARMA, and R. VILLEMS: Conformational isomers of the 5 S RNA of rat liver ribosomes. Molek. Biol. (russ.) **12**, 695 to 699 (1978)

[592] WILLIAMSON, R., and G. G. BROWNLEE: The sequence of 5 S ribosomal RNA from two mouse cell lines. FEBS-Lett. **3**, 306 to 308 (1969)

[593] TAKAI, K., S. HASHIMOTO, and M. MURAMATSU: Oligonucleotide sequences of pancreatic and T_1 ribonuclease digests of 5 S ribosomal RNA from mouse. Biochem. **14**, 536—542 (1975)

[594] STAMBROOK, P., and R. WILLIAMSON: Error frequency in 5 S RNA from cells grown in 5-bromodeoxyuridine. Europ. J. Biochem. **48**, 297—302 (1974)

[595] WALKER, T. A., J. L. BETZ, J. OLAH, and N. R. PACE: The nucleotide sequence of dolphin and bovine 5 S ribosomal ribonucleic acid. FEBS-Lett. **54**, 241—244 (1975)

[596] PACE, N. R., T. A. WALKER, B. PACE, and R. L. ERICKSON: J. Molec. Evol. **3**, 151—159 (1974)

[597] BROWNLEE, G. G., and E. M. CARTWRIGHT: The nucleotide sequence of the 5 S RNA of chicken embryo fibroblasts. Nucl. Acids Res. **2**, 2279—2288 (1975)

[598] ROY, K. L.: The nucleotide sequence of turtle (*Terrapene carolina*) 5 S rRNA. FEBS-Lett. **80**, 266—270 (1977)

[599] KOMIYA, H., and S. TAKEMURA: Nucleotide sequence of 5 S ribosomal RNA from

rainbow trout (*Salmo gairdneri*) liver. J. Biochem. **86**, 1067–1080 (1979)

[600] Roy, K. L., and L. Enns: Nucleotide sequence of 5 S ribosomal ribonucleic acid of *Iguana iguana*. J. Biol. Chem. **251**, 6352–6354 (1976)

[601] Benhamou, J., and B. R. Jordan: Nucleotide sequence of *Drosophila melanogaster* 5 S RNA: Evidence for a general 5 S RNA model. FEBS-Lett. **62**, 146 to 149 (1976)

[602] Benhamou, J., R. Jourdan, and B. R. Jordan: Sequence of *Drosophila* 5 S RNA synthesized by cultured cells and by the insect at different developmental stages. J. Molec. Evol. **9**, 279–298 (1977)

[603] Tartof, K. D., and R. P. Perry: The 5 S RNA genes of *Drosophila melanogaster*. J. Mol. Biol. **51**, 171–183 (1970)

[604] Wegnez, M., R. Monier, and H. Denis: Sequence heterogeneity of 5 S RNA in *Xenopus laevis*. FEBS-Lett. **25**, 13–20 (1972)

[605] Brownlee, G. G., E. Cartwright, T. McShane, and R. Williamson: The nucleotide sequence of somatic 5 S RNA from *Xenopus laevis*. FEBS-Lett. **25**, 8–12 (1972)

[606] Ford, P. J., and E. M. Southern: Different sequences for 5 S RNA in kidney cells and ovaries of *Xenopus laevis*. Nature **241**, 7–12 (1973)

[607] Ford, P. J., and R. D. Brown: Sequences of 5 S ribosomal RNA from *Xenopus mulleri* and the evolution of 5 S gene-coding sequences. Cell **8**, 485–493 (1976)

[608] Bellemare, G., R. J. Cedergren, and G. H. Cousineau: Comparison of the physical and optical properties of *Escherichia coli* and sea urchin 5 S ribosomal RNAs. J. Mol. Biol. **68**, 445–454 (1972)

[609] Averner, M. J., and N. R. Pace: The nucleotide sequence of marsupial 5 S ribosomal ribonucleic acid. J. Biol. Chem. **247**, 4491–4493 (1972)

[610] Rodrigues, E. M. B., and F. L. de Lucca: Distribution of 5 S RNA in the liver of the South American rattlesnake, *Crotalus durissus terrificus*. Experientia **29**, 37–39 (1973)

[611] Hindley, J., and S. M. Page: Nucleotide sequence of yeast 5 S ribosomal RNA. FEBS-Lett. **26**, 157–160 (1972)

[612] Luoma, G. A., and A. G. Marshall: Laser raman evidence for a new cloverleaf secondary structure for eucaryotic 5 S RNA. J. Mol. Biol. **125**, 95–105 (1978)

[613] Nishikawa, K., and S. Takemura: Structure and function of 5 S ribosomal ribonucleic acid from *Torulopsis utilis*. I. Purification and complete digestion with pancreatic ribonuclease A and ribonuclease T 1. J. Biochem. **76**, 925–934 (1974)

[614] Nishikawa, K., and S. Takemura: Structure and function of 5 S ribosomal RNA from *Torulopsis utilis*. III. Detection of single-tranded regions by digestion with nuclease S 1. J. Biochem. **81**, 995–1003 (1977)

[615] Nishikawa, K., and S. Takemura: Structure and function of 5 S ribosomal ribonucleic acid from *Torulopsis utilis*. IV. Detection of exposed guanine residues by chemical modification with kethoxal. J. Biochem. **84**, 259–266 (1978)

[616] Nishikawa, K., and S. Takemura: Structure and function of 5 S ribosomal ribonucleic acid from *Torulopsis utilis*. II. Partial digestion with ribonucleases and derivation of the complete sequence. J. Biochem. **76**, 935–947 (1974)

[617] Wong, Y. P., D. R. Kearns, B. R. Reid, R. G. Shulman: The extent of base pairing in 5 S RNA. Yeast 5 S RNA. J. Mol. Biol. **72**, 741–749 (1972)

[618] Connors, P. G., and W. W. Beeman: Size and shape of 5 S ribosomal RNA. J. Mol. Biol. **71**, 31–37 (1972)

[619] Udem, S. A., K. Kaufman, and J. R. Warner: Small ribosomal ribonucleic acid species of *Saccharomyces cerevisiae*. J. Bacteriol. **105**, 101–106 (1971)

[620] Nishikawa, K., and S. Takemura: Nucleotide sequence of 5 S RNA from *Torulopsis utilis*. FEBS-Lett. **40**, 106–109 (1974)

[621] Nichols, J. L., and L. Welder: S 1 nuclease as a probe of yeast ribosomal 5 S RNA conformation. Biochim. Biophys. Acta **561**, 445–451 (1979)

[622] Jordan, B. R., and G. Galling: Nucleotide sequence of *Chlorella* cytoplasmic 5 S RNA. FEBS-Lett. **37**, 333–334 (1973)

[623] Barber, C., and J. L. Nichols: Conformational studies on wheat embryo 5 S RNA using nuclease S 1 as a probe. Can. J. Biol. **56**, 357–364 (1978)

[624] JORDAN, B. R., G. GALLING, and R. JOURDAN: Sequence and conformation of 5 S RNA from *Chlorella* cytoplasmic ribosomes: Comparison with other 5 S RNA molecules. J. Mol. Biol. **87**, 205–225 (1974)

[625] PAYNE, P. I., M. J. CORRY, and T. A. DYER: Nucleotide sequence analysis of cytoplasmic 5 S ribosomal ribonucleic acid from five species of flowering plants. Biochem. J. **135**, 845–851 (1973)

[626] PAYNE, P. I., and T. A. DYER: Evidence for the nucleotide sequence of 5 S rRNA from the flowering plant *Secale cereale* (Rye). Europ. J. Biochem. **71**, 33–38 (1976)

[627] AZAD, A. A., and B. G. LANE: A possible role of 5 S rRNA in the reversible association of ribosomal subunits. Proc. Austral. Biochem. Soc. **9**, 41 (1976)

[628] BARRELL, B. G., and B. F. C. CLARK: Handbook of nucleic acid sequences. Joynson-Bruvvers, Ltd., Oxford, England, 1974

[629] ROY, K. L.: A proposed nucleotide sequence for the 5 S rRNA of rainbow trout (*Salmo gairdneri*). Can. J. Biochem. **56**, 60–65 (1978)

[630] FOX, G. E., and C. R. WOESE: 5 S RNA secondary structure. Nature **256**, 505 to 507 (1975)

[631] ACKERMAN, S., A. A. KESHGEGIAN, D. HENNER, and J. J. FURTH: Enzymic polyadenylation of 5 S rRNA and synthesis of a complementary DNA. Biochem. **18**, 3232–3242 (1979)

[632] WEINBERG, R. A., and S. PENMAN: Small molecular weight monodisperse nuclear RNA. J. Mol. Biol. **38**, 289–304 (1968)

[633] PENE, J. J., E. KNIGHT, and J. E. DARNELL: Characterization of a low molecular weight RNA in HeLa cell ribosomes. J. Mol. Biol. **33**, 609–623 (1968)

[634] FORD, P. J., and T. MATHIESON: The nucleotide sequences of 5.8 S ribosomal RNA from *Xenopus laevis* and *Xenopus borealis*. Europ. J. Biochem. **87**, 199–214 (1978)

[635] RUBIN, G. M.: The nucleotide sequence of *Saccharomyces cerevisiae* 5.8 S ribosomal ribonucleic acid. J. Biol. Chem. **248**, 3860–3875 (1973)

[636] NAZAR, R. N., T. O. SITZ, and H. BUSCH: Sequence homologies in mammalian 5.8 S ribosomal RNA. Biochem. **15**, 505–508 (1976)

[637] NAZAR, R. N., ad K. L. ROY: Nucleotide sequence of rainbow trout (*Salmo gairdneri*) ribosomal 5.8 S ribonucleic acid. J. Biol. Chem. **253**, 395–399 (1978)

[638] KHAN, M. S. N., and B. E. H. MADEN: Nucleotide sequence relationships between vertebrate 5.8 S rRNAs. Nucl. Acids Res. **4**, 2495–2505 (1977)

[639] NAZAR, R. N., and K. L. ROY: The nucleotide sequence of turtle 5.8 S rRNA. FEBS-Lett. **72**, 111–116 (1976)

[640] HAMPE, A., M.-E. ELADARI, and F. GILBERT: Nucleotide sequence study of mouse 5.8 S RNA. Biochimie **8**, 943 to 951 (1976)

[641] NAZAR, R. N., T. O. SITZ, and H. BUSCH: Heterogeneity in the methylation and 5′ termini of Novikoff ascites hepatoma 5.8 S ribosomal RNA. FEBS-Lett. **45**, 206 to 212 (1974)

[642] NAZAR, R. N., T. O. SITZ, and H. BUSCH: Structural analyses of mammalian ribosomal ribonucleic acid and its precursors. Nucleotide sequence of ribosomal 5.8 S ribonucleic acid. J. Biol. Chem. **250**, 8591 to 8597 (1975)

[643] NAZAR, R. N., T. O. SITZ, and H. BUSCH: Homologies in eukaryotic 5.8 S ribosomal RNA. Biochem. Biophys. Res. Commun. **62**, 736–743 (1975)

[644] WOLEDGE, J., M. J. CORRY, and P. I. PAYNE: Ribosomal RNA homologies in flowering plants: comparison of the nucleotide sequences in 5.8 S rRNA from broad been, dwarf bean, tomato, sunflower and rye. Biochim. Biophys. Acta **349**, 339–350 (1974)

[645] DOOLITTLE, W. F., and N. R. PACE: Transcriptional organization of the ribosomal RNA cistrons in *Escherichia coli*. Proc. Natl. Acad. Sci. U.S. **68**, 1786 to 1790 (1971)

[646] CEDERGREN, R. J., and D. SANKOFF: Evolutionary origin of 5.8 S ribosomal RNA. Nature **259**, 74 (1976)

[647] WREDE, P., and V. A. ERDMANN: *Escherichia coli* 5 S RNA binding protein L 18 and L 25 interact with 5.8 S RNA but not with 5 S RNA from yeast ribosomes. Proc.

Natl. Acad. Sci. U.S. **74**, 2706–2709 (1977)

[648] LIND, A., A. METSPALU, M. SAARMA, I. TOOTS, M. USTAV, and R. VILLEMS: Universal structure of peptidyl transferase center in ribosomes and the function of eukaryotic 5.8 S ribosomal ribonucleic acid. Biorgan. Khim. (russ.) **3**, 1138–1140 (1977)

[649] NAZAR, R. N., T. O. SITZ, and H. BUSCH: Tissue specific differences in the 2′-O-methylation of eukaryotic 5.8 S ribosomal RNA. FEBS-Lett. **59**, 83–87 (1975)

[650] RUBIN, G. M.: Three forms of the 5.8 S ribosomal RNA species in *Saccharomyces cerevisiae*. Europ. J. Biochem. **41**, 197 to 202 (1974)

[651] KING, H. W. S., and H. GOULD: Low molecular weight ribonucleic acid in rabbit reticulocyte ribosomes. J. Mol. Biol. **51**, 687–702 (1970)

[652] PAYNE, P. I., and T. A. DYER: Plant 5.8 S RNA is a component of 80 S but not 70 S ribosomes. Nature New Biol. **235**, 145–147 (1972)

[653] PRESTAYKO, A. W., M. TONATO, and H. BUSCH: Low molecular weight RNA associated with 28 S nucleolar RNA. J. Mol. Biol. **47**, 505–515 (1970)

[654] PACE, N. R., TH. A. WALKER, and E. SCHROEDER: Structure of the 5.8 S–28 S ribosomal RNA junction complex. Biochem. **16**, 5321–5328 (1977)

[655] SY, J., and K. S. MCCARTY: Characterization of 5.8 S RNA from a complex with 26 S ribosomal RNA from *Arbacia punctulata*. Biochim. Biophys. Acta **199**, 86–94 (1970)

[656] ISHIKAWA, H.: Re-joining of the 18 S fragments dissociated from the 28 S ribosomal RNA of insect: A structural role of 5.8 S RNA. Biochem. Biophys. Res. Commun. **90**, 417–424 (1979)

[657] AZAD, A. A., and B. G. LANE: Wheat-embryo ribonucleates. I. Subcellular localization of a satellite polyribonucleotide. Can. J. Biochem. **51**, 606–612 (1973)

[658] GIORGINI, J. F., and F. L. DE LUCCA: Studies on low molecular weight RNA associated with 28 S-ribosomal RNA from *Crotalus durissus terrificus* liver. Nucl. Acids Res. **3**, 165–175 (1976)

[659] NAZAR, R. N.: The release and reassociation of 5.8 S rRNA with yeast ribosomes. J. Biol. Chem. **253**, 4505–4507 (1978)

[660] LUOMA, G. A., and A. G. MARSHALL: Laser raman evidence for new cloverleaf secondary structures for eukaryotic 5.8 S RNA and prokaryotic 5 S RNA. Proc. Natl. Acad. Sci. U.S. **75**, 4901–4905 (1978)

[661] VAN, N. T., R. N. NAZAR, and TH. O. SITZ: Comparative studies on the secondary structure of eukaryotic 5.8 S ribosomal RNA. Biochem. **16**, 3754–3759 (1977)

[662] KHAN, M. S. N., and B. E. H. MADEN: Conformation of mammalian 5.8 S ribosomal RNA: S_1 nuclease as a probe. FEBS-Lett. **72**, 105–110 (1976)

[663] KELLY, J. M., J. P. GODDARD, and B. E. H. MADEN: Evidence on the conformation of HeLa cell 5.8 S ribosomal ribonucleic acid from the reaction of specific cytidine residues with sodium bisulphite. Biochem. J. **173**, 521–532 (1978)

[664] SITZ, T. O., S.-C. KUO, and R. N. NAZAR: Multimer forms of eukaryotic 5.8 S ribosomal RNA. Biochem. **17**, 5811–5815 (1978)

[665] LIGHTFOOT, D.: Thermodynamics of a stable yeast 5.8 S rRNA hairpin helix. Nucl. Acids Res. **5**, 3565–3577 (1978)

[666] KERWAR, S. S., H. WEISSBACH, and G. G. GLENNER: An aminopeptidase activity associated with brain ribosomes. Arch. Biochem. Biophys. **143**, 336–337 (1971)

[667] LANGNER, J., S. ANSORGE, P. BOHLEY, H. WELFLE, and H. BIELKA: Presence of an endopeptidase activity in rat liver ribosomes. Acta Biol. Med. Germ. **36**, 1729 to 1733 (1977)

[668] BYLINKINA, V. S., M. I. LEVYANT, G. G. GORACH, and V. N. OREKHOVICH: On the localization of ribosomal proteinase. Biokhimiya (russ.) **43**, 100–102 (1978)

[669] LEVYANT, M. I., V. S. BYLINKINA, M. G. TRUDOLYUBOVA, and V. N. OREKHOVICH: On the proteolytic activity of ribosomes. Biokhimiya (russ.) **40**, 1322–1324 (1975)

[670] LEVYANT, M. I., V. S. BYLINKINA, M. G. TRUDOLYUBOVA, and V. N. OREKHOVICH: On the presence of proteinase in polyso-

mes from rat liver. Molek. Biol. (russ.) **10**, 770—776 (1976)

[671] Levyant, M. I., V. S. Bylinkina, V. A. Spivak, and V. N. Orekhovich: On the specificity of neutral proteinase from ribosomes. Biokhimiya (russ.) **43**, 1423—1428 (1978)

[672] Langner, J., B. Wiederanders, S. Ansorge, P. Bohley, and H. Kirschke: The ribosomal serine proteinase: Cathepsin R. Acta Biol. Med. Germ. **38**, 1527 to 1538 (1979)

[673] Korant, B. D.: Protein cleavage in virus-infected cells. Acta Biol. Med. Germ. **36**, 1565—1573 (1977)

[674] Korant, B. D.: Protease activity associated with HeLa cell ribosomes. Biochem. Biophys. Res. Commun. **74**, 926—933 (1977)

[675] Heizmann, P.: Proprietes des ribosomes et des RNA ribosomiques d'*Euglena gracilis*. Biochim. Biophys. Acta **224**, 144 to 154 (1970)

[676] Acton, G. J.: Phytochrome controlled acid RNase: an "attached" protein of ribosomes. Phytochem. **13**, 1303—1310 (1974)

[677] Shishkin, S. S., B. A. Kitavtsev, and S. S. Debov: Distribution of latent ribonuclease activity between ribosome subunits and protein fractions separated during dissociation of polysomes. Biokhimiya (russ.) **43**, 399—406 (1978)

[678] Reboud, A. M., M. Buisson, and J.-P. Reboud: Specific ribonucleoprotein fragments from 40-S ribosomal subunits. Biochim. Biophys. Acta **432**, 176—184 (1976)

[679] Sukhova, T. I., V. S. Shapot, and G. O. Krechetova: Nuclease activity of cytoplasmic ribosomes from hepatocytes and some experimental hepatomas. Biokhimiya (russ.) **43**, 1838—1844 (1978)

[680] Dessev, G., and K. Grancharov: Degradation of ribonucleic acid in rat liver ribosomes. J. Mol. Biol. **76**, 425—444 (1973)

[681] Nazar, R. N.: Studies on the 5′ termini of Novikoff ascites hepatoma ribosomal precursor RNA. Biochem. **16**, 3215—3219 (1977)

[682] Spirin, A. S., and L. P. Gavrilova: The ribosome. Springer-Verlag, Berlin—Heidelberg—New York, 1969

[683] Ts'o, P. O. P., J. Bonner, and J. Vinograd: Structure and properties of microsomal nucleoprotein particles from pea seedlings. Biochim. Biophys. Acta **30**, 570—582 (1958)

[684] Lausink, A. G. W. J.: Yeast ribosomes and magnesium ions. Nijmegen, Netherlands: Doctor Thesis, 1964

[685] Edelman, I. S., P. O. P. Ts'o, and J. Vinograd: The binding of magnesium to microsomal nucleoprotein and ribonucleic acid. Biochim. Biophys. Acta **43**, 393—403 (1960)

[686] Siekevitz, P., and G. E. Palade: A cytochemical study on the pancreas of the guinea pig. VI. Release of enzymes and ribonucleic acid from ribonucleoprotein particles. J. Biophys. Biochem. Cytol. **7**, 631—644 (1960)

[687] Petermann, M. L.: Ribonucleoprotein from a rat tumor, the Jensen sarcoma. I. The effect of magnesium binding on ultracentrifugal and electrophoretic properties. J. Biol. Chem. **235**, 1998—2003 (1960)

[688] Zillig, W., W. Krone, and M. Albers: Untersuchungen zur Biosynthese der Proteine, III. Beitrag zur Kenntnis der Zusammensetzung und Struktur der Ribosomen. Hoppe-Seyler's Z. Physiol. Chem. **317**, 131—143 (1959)

[689] Siekevitz, P., and G. E. Palade: Cytochemical study on the pancreas of the guinea pig. VII. Effects of spermine on ribosomes. J. Cell. Biol. **13**, 217—232 (1962)

[690] Ohtaka, Y., and K. Uchida: The chemical structure and stability of yeast ribosomes. Biochim. Biophys. Acta **76**, 94 to 104 (1963)

[691] Konecki, D., G. Kramer, P. Pinphanichakarn, and B. Hardesty: Polyamines are necessary for maximum in vitro synthesis of globin peptides and play a role in chain initiation. Arch. Biochem. Biophys. **169**, 192—198 (1975)

[692] Kramer, G., O. W. Odom, and B. Hardesty: Polyamines in eukaryotic peptide initiation. Meth. Enzymol. **60**, 555—565 (1979)

[693] Henderson, A. B., G. Kramer, and B. Hardesty: Binding of iodine-labeled mRNA to 40 S ribosomal subunits. Meth. Enzymol. **60**, 401—410 (1979)

[694] Odom, O. W., G. Kramer, A. B. Henderson, P. Pinphanichakarn, and B. Hardesty: GTP hydrolysis during methionyl-tRNA$_f$ binding to 40 S ribosomal subunits and the site of edeine inhibition. J. Biol. Chem. **253**, 1807—1816 (1978)

[695] Igarashi, K., K. Kashiwagi, R. Aoki, M. Kojima, and S. Hirose: Comparative studies on the increase by polyamines of fidelity of protein synthesis in *Echerichia coli* and wheat germ cell-free systems. Biochem. Biophys. Res. Commun. **91**, 440—448 (1979)

[696] Abraham, A. K., S. Olsnes, and A. Pihl: Fidelity of protein synthesis in vitro is increased in the presence of spermidine. FEBS-Lett. **101**, 93—96 (1979)

VII. Biosynthesis of ribosomal components and biogenesis of ribosomal particles

The process of ribosome biogenesis involves the synthesis of the constituent molecules and their assembly to biologically active ribosomal subunits. The proteins are synthesized on cytoplasmic polysomes and thereafter transported into the nucleus. The rRNA molecules are initially transcribed on special genes as higher molecular weight precursor molecules and then processed to the mature forms. The process of pre-rRNA maturation involves furthermore the binding of ribosomal proteins to pre-rRNA resulting in the formation of pre-rRNP particles. The final step of ribosome biogenesis is the transport of ribosomal subparticles to the cytoplasm. The main results obtained in the last few years will be described in the following; for more comprehensive reviews see [1—11].

1. Ribosomal RNA

a) *Primary pre-rRNA*

The ribosomal RNAs of the 28 S-, 18 S-, and 5.8 S-type are synthesized as parts of a common high molecular weight primary transcript (pre-rRNA), which is processed posttranscriptionally by various steps and mechanisms into the mature rRNAs. Such large pre-rRNA molecules containing the sequences for 28 S-, 18 S-, and 5.8 S-type rRNA were found in the nucleoli of all eukaryotes studied so far. The size of the

Table 15. Molecular weights ($\cdot 10^{-6}$) of pre-rRNA molecules in various eukaryotes

Species		References
Algae	2.1—3.5	[12, 14, 15]
Yeast	2.5—2.8	[16—18]
Higher plants	2.3—2.9	[19—29]
Echinodermata	2.5	[30]
Insects	2.5—3.8	[28, 31—35]
Fishes	2.6—2.7	[13, 28]
Amphibia	2.5—2.8	[13, 28, 34—38]
Reptiles	2.7—2.8	[13, 28]
Birds	3.7—3.9	[13, 28]
Marsupials	4.1	[28]
Mammals	4.1—4.7	[13, 28, 39]

Contents

pre-rRNA molecules analyzed ranges from $2.1 \cdot 10^6$ in *Acetabularia* [12] to about $4.7 \cdot 10^6$ in mammalian cells [13] (see Table 15). Whether these pre-rRNA molecules represent primary transcripts or stable intermediate processing products has not yet been established.

Triphosphate residues at the 5′-ends of primary pre-rRNA molecules have been found in 40 S pre-rRNA of *Xenopus laevis* [40], in 35 S pre-rRNA of *Tetrahymena pyriformis* [41], in 34 S pre-rRNA of *Drosophila* [42], and in 37 S pre-rRNA of *Saccharomyces* [43, 44]. pppAp was detected at the 5′-end of the immediate precursor to 17 S rRNA from *Dictyostelium* [45, 46] indicating that the 37 S pre-rRNA, from which the precursor of the 17 S rRNA originates, has preserved its 5′-initiation sequence. These data suggest that the largest and first labeled pre-rRNAs found in these species are likely to be primary transcription products.

On the other hand, the 5′-end nucleotides in pre-rRNAs of mouse hepatoma [47] and Novikoff hepatoma cells [48] were found heterogeneous supporting the possibility that the 5′-end of the primary transcript is already processed before transcription is terminated. From electron microscopic studies of transcriptionally active nucleolar chromatin from various plant and animal cells it has also been concluded that pre-rRNA molecules do not represent the proper primary products of rDNA transcription but are rather stable intermediates that have already been processed during transcription [49]. Therefore, further studies are necessary to clarify whether processing in eukaryotes starts when transcription is terminated or, as in prokaryotes [50, 51], already during transcription.

b) 5′→3′ polarity of the pre-rRNA molecule

Pre-rRNA of eukaryotes has the general sequence:

Spacer — 18 S rRNA — spacer — 5.8 S RNA — 28 S rRNA [4, 5]. The analysis of the location of the 18 S and 28 S rRNA regions relative to the 5′- or 3′- ends of the precursor molecule yielded conflicting results. Most of the experiments, however, have demonstrated that the 18 S rRNA sequence is located proximal to the 5′-end of the pre-rRNA molecule and the 28 S rRNA sequence proximal to the 3′-end. This polarity could be proven in studies with *Euglena gracilis* [14], *Chlamydomonas reinhardi* [52] and *Saccharomyces cerevisiae* [17], by different sensitivities of the synthesis of 18 S and 28 S rRNA to cordycepin [53] or UV [54, 55] and by studies of the sequence of methylation of the growing rRNA transcript [56]. A similar topology of the 18 S and 28 S rRNA within the transcriptional unit of *Xenopus laevis* was proposed from studies on the synthesis of rRNA in vitro [57, 58].

The 5′→3′ polarity of the 40 S pre-rRNA molecule of *Xenopus laevis* was also estimated by hybridization of 18 S and 28 S rRNA molecules with different rDNA fragments partially digested enzymatically at the 5′- or at the 3′-end, respectively, [59]. The same has been concluded from the visualization of growing transcripts on rRNA genes in *Xenopus laevis* cleaved with a restriction enzyme [60].

c) Methylation and pseudouridylation of pre-rRNA

Ribosomal RNA is characterized by the presence of methylated 2′-OH groups of the ribose moieties and, to a smaller extent, also by methylated bases (see Chapter VI, Section 4). Methylation of pre-rRNA and rRNA has been studied mainly in HeLa cells and yeast cells [3]. Most of the methylations take place at the level of the primary pre-rRNA [61—64] along the sequences corresponding to the mature 18 S, 5.8 S and 28 S rRNA molecules whereas the spacer sequences are not methylated. All early methylations are found also in the mature rRNAs. In HeLa cell 18 S rRNA 6 additional bases are methylated at later steps of the rRNA maturation process [61], and a further methyl group is introduced finally into the 18 S rRNA within the 40 S ribosomal subunit in the cytoplasm [65].

Methylation of pre-rRNA in yeast takes place in the same general way as found in HeLa cells and is also characterized by the introduction of the majority of methyl groups into the 37 S pre-rRNA followed by some later methylations of the 17 S rRNA [64].

Methylated sequences have been obviously conserved during evolution. The methylated sequence $m_2^6Am_2^6ACUG$ of the rRNA of the small ribosomal subunit, e.g., has been identified in bacteria [66], yeast cells [67], and Novikoff hepatoma cells [68].

Modifications of the pre-rRNA by methylation seem to be essentially required both for their accurate processing and for the formation of pre-rRNP particles [69 to 71] (see also Chapter VI, Section 4).

Pseudouridylation is another posttranscriptional modification event of pre-rRNA molecules, also restricted to sequences found in the mature rRNAs as demonstrated for pre-rRNA of HeLa cells [72] and yeast [73].

d) Processing of pre-rRNA

Synthesis and processing of pre-rRNA to mature rRNA molecules has been investigated in various species, e.g., HeLa cells [72, 74—77], BHK cells [78, 79], rat liver [80—83], rooster liver [84], rat uterus [85], Novikoff ascites hepatoma cells [86], *Xenopus* [87], *Drosophila* [42, 88—90], mouse L cells [91—93], yeast [16—18, 94 to 96], *Acetabularia* [12], plants [22, 29, 97 to 101], *Chlamydomonas* [52], lymphocytes [102], fibroblasts [103], sea urchin [30, 104, 105], *Neurospora crassa* [106, 107], and *Tetrahymena* [108].

Processing of the pre-rRNA molecule precedes preferentially within the nucleolus and the basic events of these processes are very similar in all eukaryotes [4, 5]. However, some differences were found in various species as will be demonstrated in the following for HeLa cells, L cells and yeast.

In **HeLa cells** maturation of the pre-rRNA molecule to the mature rRNAs can be described schematically as shown in Fig. 21. The pre-rRNA is cleaved in four steps starting at the 5′ region of the molecule. By this first cleavage the long spacer region at the 5′-end of the 45 S precursor molecule is removed; the cleavage products are a spacer segment of $1.33 \cdot 10^6$ and the intermediate 41 S pre-rRNA molecule of $3.3 \cdot 10^6$ molecular weight. The following cleavage at site 2 results in a 32 S pre-rRNA molecule of $2.3 \cdot 10^6$ and a 20 S pre-rRNA of $1.0 \cdot 10^6$, containing the sequences of the 28 S and 5.8 S rRNA, and of the 18 S rRNA molecule, respectively.

Cleavage at site 3 (see Fig. 21) leads to

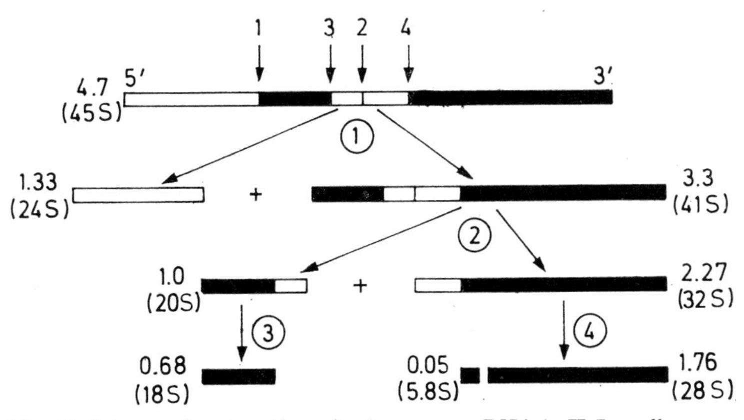

Fig. 21. Scheme of maturation of primary pre-rRNA in HeLa cells.

mature 18 S rRNA. From isolated 32 S rRNA of HeLa cells several products of lower molecular weight are released after brief thermal treatment. One of the products liberated contains nucleotide sequences derived from the 5.8 S rRNA and the transcribed spacer region of the 32 S rRNA. This putative precursor of 5.8 S rRNA is characterized by a sedimentation coefficient of about 10 S [75]. In *Drosophila* cells the generation of 5.8 S and 26 S rRNAs results from cleavages occurring in the cytoplasm [88].

The positions of cleavage sites 2 and 3 (Fig. 21) are basically the same in all eukaryotes studied so far, but their temporal order varies among different organisms [87, 93].

In mouse **L cells** [91] (see Fig. 22) pre-rRNA molecules of 45 S, 41 S, and 32 S have been found similar to those in HeLa cells. However, in L cells the 20 S pre-rRNA, the precursor molecule to 18 S rRNA in HeLa cells, is missing, while an additional 36 S pre-rRNA of $2.7 \cdot 10^6$ molecular weight was found. The 36S pre-rRNA contains the sequences present in the 32 S pre-rRNA (28 S, 5.8 S, and the spacer) and, additionally, the spacer sequence found in the 20 S pre-rRNA of HeLa cells.

The processing pathway of pre-rRNA in *Xenopus laevis* [87] and cultured *Drosophila* cells [42] was found similar to that in mouse L cells. In some types of cells, e.g., rat liver [81, 83], lymphocytes [102], BHK cells [78], and mouse L cells [93] both temporal orders of cleavages at sites 2 and 3 of the primary pre-rRNA molecule may occur depending on the physiological conditions. Alterations in processing of rat liver pre-rRNA are caused, e.g., by inhibition of protein synthesis by cycloheximide [81]. This suggests that the channelling of the nucleolar pre-rRNA along alternative processing pathways is under stringent control by a continuous supply with special proteins.

The processing scheme for **yeast cells** is demonstrated in Fig. 23. The 35 S- or 37 S pre-rRNA from yeast ($2.5 \cdot 10^6$ molecular weight) has no spacer segments at the 5′-end and the processing starts with a central endonucleolytic cleavage at site 2 corresponding to mammalian pre-rRNA. This cleavage generates the intermediate 29 S pre-rRNA ($1.6 \cdot 10^6$) and 20 S pre-rRNA ($0.9 \cdot 10^6$ molecular weight) of the mature 26 S and 17 S rRNA, respectively [16, 17]. The 20 S pre-rRNA becomes then converted to the mature 17 S rRNA. The

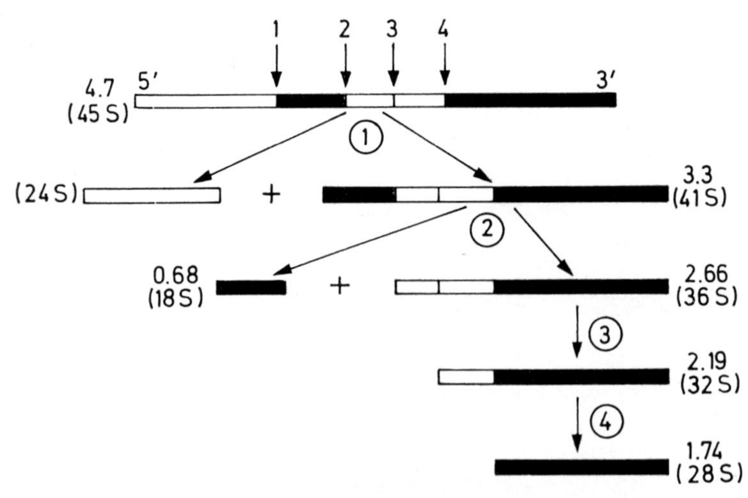

Fig. 22. Scheme of maturation of primary pre-rRNA in mouse L cells. For further explanations see the text.

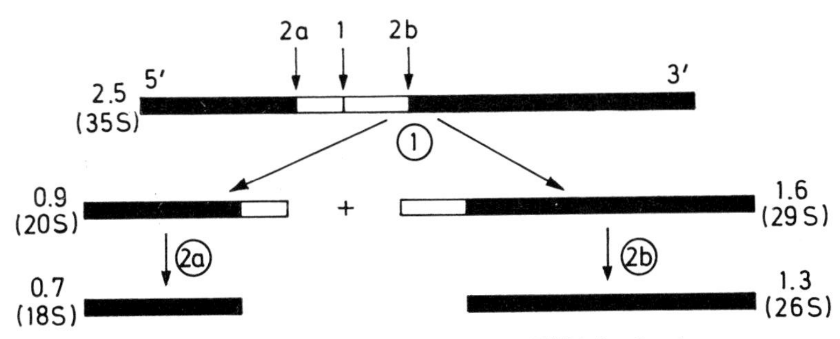

Fig. 23. Scheme of maturation of primary pre-rRNA in *Saccharomyces*. For further explanations see the text.

pool size of the 27 S pre-rRNA is about twice that of the 20 S and 37 S pre-rRNA due to the slower conversion rate of 27 S pre-rRNA to the mature 26 S rRNA and 5.8 S rRNA species [17]. The 5.8 S rRNA molecule is generated from the 27 S pre-rRNA via an immediate precursor molecule characterized by a sedimentation coefficient of 7 S and non-ribosomal nucleotide sequences [94, 95].

The final processing events converting pre-rRNA into mature rRNA seem to be variable with respect to the exact location of the cleavage sites. This is suggested by the finding that 5.8 S rRNA preparations from yeast contain, besides the majority of 5.8 S rRNA, about 10% of molecules which have 6 to 8 additional nucleotides at the 5'-terminus [109]. Heterogeneity was also found at the 5'-end of the 28 S rRNA of mouse hepatoma cells [110]. The presence of pCp, typical for mature 28 S rRNA, increases with the prolongation of the labeling period, and a chase with cold phosphate makes the 5'-terminus almost homogeneously pCp. Obviously apart from the major cleavages of pre-rRNA, unique 5'-terminal processing takes place on newly made 28 S rRNA in the cytoplasm [110].

The rate of pre-rRNA processing seems to be significantly correlated with the generation time of cells. In yeast cells with a 3 h generation time the processing rate is about five to ten times faster than in the mammalian cells with generation times of 12 to 24 h [111]. This suggests that the processing of pre-rRNA is obviously used for regulating the rate of ribosome production. The rate-limiting step in the processing of pre-rRNA is very probably the cleavage of the 32 S pre-rRNA, because a substantial accumulation of this precursor molecule has been found under various conditions.

e) *Biosynthesis of 5 S rRNA*

The eukaryotic 5 S rRNA is transcribed on special genes independently of DNA sequences coding for the high molecular pre-rRNA described before [9, 112—121]. Special precursors to 5 S rRNA have not been observed in most of the investigated systems; this is in accordance with the finding that nucleoside di- and triphosphates have been identified at the 5'-end of the 5 S rRNA [46, 122, 123]. Thus, 5 S rRNA, as present in the large ribosomal subunit, is considered as the proper primary transcription product. 5 S rRNA genes can be transcribed with fidelity also in isolated nuclei and even in purified chromatin preparations of HeLa cells [124—130], mouse plasmacytoma cells [129] and *Xenopus laevis* ovary cells [131, 132].

The detailed analysis of 5 S rRNA, synthesized in isolated nuclei of rat liver and HeLa cells has shown, however, that the initial synthesis product is at least slightly larger than the mature 5 S rRNA, characterized by an extra sequence of 8 nucleotides at the 3′-terminus [129]. Also short-time pulse-labeling of cultured mouse hepatoma cells point to the presence of a slightly larger 5 S rRNA precursor [129]. 5 S rRNA precursor molecules have been found furthermore in *Xenopus* oocytes [133] and in cultured *Drosophila* cells at supra optimal temperatures [134]. In the latter case the accumulation of a RNA molecule was found containing 5 S rRNA sequences and 15 additional nucleotides at the 3′-end [134]. The sequence of this segment has been determined; it contains a stretch of 4 uridine residues at the 3′-end which could be a polymerase termination signal [135]. The half-life of this RNA is less than 1 min and labeling kinetics indicate that it is an obligatory precursor to all mature 5 S rRNA molecules [136]. An (A + T)-rich region was found directly adjacent to the 3′-end of the gene in *Xenopus laevis* 5 S rDNA [137] which resembles the sequence of the 15 additional nucleotides in the 5 S pre-rRNA in *Drosophila* cells [134].

The synthesis of 5 S rRNA is normally coordinated with the synthesis of 18 S and 28 S rRNA [9]. This coordination is abolished under certain conditions; e.g., when protein synthesis is inhibited by cycloheximide the synthesis of the high molecular weight pre-rRNA is suppressed significantly stronger than the synthesis of 5 S rRNA [138].

Different types of 5 S rRNA were found in somatic and oocyte cells of *Xenopus laevis* (see Chapter VI, Section 4). Normal *Xenopus* liver cells ar 'ed in organ culture do not detectably synthesize ovary-type 5 S rRNA sequences, but transformed somatic kidney cells synthesize ovary-type 5 S rRNA in amounts of 10 to 20% of the total 5 S rRNA [201]. The control of oocyte 5 S rRNA gene repression in somatic cells is possibly connected with the location of the 5 S rRNA genes in different chromosome regions [201].

2. Ribosomal proteins

Ribosomal proteins are synthesized on cytoplasmic ribosomes [139—143], preferentially on free and on so-called loosely bound polysomes [144] (see Chapter X, Section 4c). Also mitochondrial ribosomal proteins, as e.g. found for *Neurospora crassa* [145, 146], and chloroplast ribosomal proteins (*Euglena gracilis* [147, 148]) are synthesized on cytoplasmic ribosomes.

After their synthesis ribosomal proteins are transported to the nucleolus very rapidly, at least compared with their rate of synthesis [143, 149]. Due to the different kinetics of both processes, the cytoplasmic pool of free ribosomal proteins is very small as found, e.g., in rat liver [150—152] and in Drosophila cell cultures [90]. The majority of newly synthesized ribosomal proteins are bound after their transport to the nucleus within the nucleolus to the pre-rRNA molecules, thus forming ribonucleoprotein particles (see Section 3 of this chapter).

Ribosomal proteins can be synthesized in vitro by corresponding poly(A)$^+$ mRNA [153—156] and also by mRNAs lacking poly(A) [155]. Wheat germ cell free systems were found to be able to translate completely almost all rat liver mRNAs for individual ribosomal proteins in equimolar amounts in the presence of saturating concentrations of mRNAs [153]. mRNAs which code for ribosomal proteins in Ehrlich ascites tumor cells [154] and in rat liver [153] are of small size indicating that they are monocistronic. This is in agreement with the finding that yeast polysomes engaged in the synthesis of ribosomal proteins are of relatively small molecular weight [157]. Furthermore the synthesis of 35 ribosomal proteins as studied in *Saccharomyces cerevisiae* was nearly equally affected by the inhibition of initiation of protein synthesis in a temperature sensitive mutant indicating also the monocistronic nature of ribosomal protein mRNA [158].

The rate of synthesis of ribosomal proteins can be regulated depending on the needs of the cell; e.g., the rate of synthesis is significantly higher in regenerating liver than in normal liver [139]. On the other hand, in mutant strains of *Saccharomyces cerevisiae* which are unable to synthesize ribosomes the synthesis of each ribosomal protein slows down suggesting that their synthesis is coordinately controlled in vivo [159]. The stop of production of new mRNA for each of the ribosomal proteins seems to be responsible for the decline of ribosomal protein synthesis under nonpermissive conditions [159].

Like their biosynthesis, also the catabolism of ribosomal proteins takes place coordinately [160–162]. A half-live of 4.5 days has been determined for rat liver ribosomal proteins [162]. In regenerating rat liver newly synthesized ribosomal proteins are degraded with an apparent half-live of about 20 to 40 min when rRNA synthesis is selectively inhibited by the administration of low doses of actinomycin D to partially hepatectomized rats [163]; degradation of free ribosomal proteins was observed also in HeLa cells [164, 165] and in *Saccharomyces cerevisiae* in the absence of 60 S subunits assembly [166].

3. Formation of preribosomal particles

Newly synthesized ribosomal proteins are transported to the nucleus and associated within the nucleolus with newly transcribed pre-rRNA thus forming specific ribonucleoprotein particles [6]. Pre-rRNA is synthesized independently of protein synthesis, but in the absence of protein synthesis the high molecular weight pre-rRNA is rapidly cleaved to 32 S pre-rRNA which accumulates [7]. Likewise, ribosomal protein synthesis proceeds independently of rRNA synthesis [167], but ribosomal proteins synthesized under conditions of inhibited rRNA synthesis are not found in the nucleolus and do not become associated with pre-rRNA after RNA synthesis is reinitiated. Thus, the formation of preribosomal particles in the nucleolus seems to be controlled by the synthesis of pre-rRNA and ribosomal proteins.

Also in vitro ribosomal proteins are taken up by isolated rat liver [168] and HeLa cell nuclei [169] and were found mainly in the nucleolus, associated with nucleolar 80 S particles containing newly synthesized 45 S pre-rRNA [168]. The pool of free ribosomal proteins within the nucleolus of eukaryotic cells is low [164] and obviously different for the individual proteins [164].

The processing of pre-rRNA as described in Section 1d of this chapter takes place mainly in the nucleolus within ribonucleoprotein particles formed by binding of ribosomal and nucleolar proteins to the pre-rRNA molecule. Therefore, different RNA-particles can be expected as the result of the maturation process starting from very early precursor particles containing newly synthesized 45 S pre-rRNA to immediate precursors of mature ribosomal subunits. The isolation and characterization of such ribosomal precursor particles is rather difficult and therefore their number and properties have not been finally estimated.

Two main species of preribosomal particles have been prepared from the nucleoli of animal cells: Particles characterized by sedimentation coefficients of 80 S and 55 S were extracted, e.g., from isolated HeLa cell nucleoli with dithiothreitol in the presence of EDTA [170–172]. At $5 \cdot 10^{-4}$ M Mg^{++} these particles have $s^0_{20,w}$ values of 110 S [172] and 65–75 S [173], respectively. 80 S nucleolar particles contain 45 S pre-rRNA, 5 S rRNA, as well as ribosomal and nucleolar proteins [172]. 55 S particles contain 32 S pre-rRNA, 5 S rRNA, large ribosomal subunit proteins, and nucleolar proteins [172]. Therefore, the 80 S particles are assumed to be precursors of both ribosomal subunits, whereas the 55 S particles are obviously precursors of the large ribosomal subunit.

The protein composition of nucleolar preribosomal particles has been analyzed first by one dimensional polyacrylamide gel electrophoresis [139, 174—176] and by fingerprints of tryptic digestion products [177]. Two classes of proteins have been found; the bulk of proteins consists of newly formed ribosomal proteins, while another set of proteins seems to be of nucleolar origin which are not transported to the cytoplasm [6].

Proteins of nucleolar preribosomal particles from Novikoff hepatoma ascites cells [178], HeLa cells [164], mouse leukemia cells [179, 180] and from rat liver [160, 181] have been analyzed later on by two-dimensional polyacrylamide gel electrophoresis. The results confirm the presence of the bulk of ribosomal proteins in isolated nucleolar preribosomal particles. In nucleolar 80 S preribosomal particles from mouse leukemia cells were found about 55 to 60 proteins; most of them were identical to cytoplasmic ribosomal subunit proteins [179, 180]. A set of 11 proteins, different from ribosomal proteins, has been analyzed in nucleolar preribosomal particles [179]. Nucleolar 60 S preribosomal particles, the precursors to large ribosomal subunits, from mouse leukemia cells contain all large ribosomal subunit proteins found in 80 S particles and one additional protein of the large subunit [179]. Kinetic studies on ribosomal protein assembly in preribosomal particles and ribosomal subunits of mouse leukemia cells, in which the labeling patterns of ribosomal proteins after short and long periods of incubation of the cells with $^3H^-$ labeled amino acids has been compared, revealed 9 ribosomal proteins and one non-ribosomal protein as candidates to become early associated with newly transcribed pre-rRNA [180].

110 S preribosomal particles have been prepared also from rat liver nucleoli [181]. These particles correspond to 80 S precursor particles but were prepared at higher Mg^{++} concentration than 80 S preribosomal particles. They contain about 30 and 13 proteins of the large and small subunit, respectively, while 5 and 12 proteins of the large and small subunit, respectively were missing completely [181]. Additionally, 11 non-ribosomal proteins were found in the 110 S particle preparation. Total preribosomal particle preparations from rat liver nucleoli contain one additional protein of the large ribosomal subunit (protein L 5) and four additional proteins of the small one [181]. Obviously these proteins are attached to the 110 S particle during the processing of this particle type in the nucleolus.

In yeast three distinct preribosomal particles were detected sedimenting at 90 S, 66 S and 43 S, respectively [182]. The 90 S particles contain 37 S pre-rRNA and are believed to be precursors of the 66 S and 43 S particles. The 66 S and 43 S particles contain 29 S and 18 S pre-rRNA, respectively. The 66 S particles are processed within the nucleus to 60 S ribosomal precursor particles; this process is accompanied by a loss of protein. The final maturation of the 43 S precursor particle takes place in the cytoplasm [182]. Labeling kinetics of ribosomal proteins isolated from cytoplasmic 40 S and 60 S ribosomal subunits have revealed that a limited number of ribosomal proteins of the 60 subunit (L 7, L 9, L 24, L 30) and also some proteins of the 40 S subunit are incorporated into newly formed ribosomes only after their entry into the cytoplasm. These proteins were not found in nuclear precursor particles [183]. Most of the ribosomal proteins, however, appear to be assembled already with the preribosomal particles within the nucleus [183]. Phosphorylated ribosomal proteins become incorporated late into preribosomal particles [184].

Studies on the assembly of ribosomes in rat liver [160, 176], HeLa cells [185—188], mouse leukemia cells [179, 180], and Novikoff hepatoma ascites cells [178] also revealed the existence of a number of ribosomal protein species which are associated with the immature ribosome at a very late stage during ribosome biogenesis. This is indicated by the finding that a few ribosomal proteins are predominantly labeled in cytoplasmic ribosomal subunits after short periods of labeling [180, 185—188].

In isolated 60 S and 40 S ribosomal subunits from mouse leukemia cells, five proteins of the large and four proteins of the small subunit were found incorporating labeled amino acids more quickly than the majority of ribosomal proteins [180]. These proteins were shown to be absent or only very faintly labeled in 80 S preribosomal particles. Furthermore, these proteins are incorporated into ribosomal subunits in the absence of 28 S and 18 S rRNA formation in mouse leukemia cells treated with the adenosine analogue toyocamycin [189]. This drug allows the synthesis of 45 pre-rRNA and the formation of nucleolar 80 S preribosomal particles [190] of normal protein composition [189], but inhibits the processing into mature ribosomal subunits [190, 191]. Therefore, at least in this case, labeling of several structural ribosomal proteins seems to take place by cytoplasmic exchange of ribosomal proteins and by maturation of incomplete preribosomal particles [189].

Little is known about the mechanism and control of transport of ribosomal precursor particles from the nucleus to the cytoplasm. During the transport a few nucleolar and ribosomal proteins seem to be exchanged either in the nucleoplasm or in the cytosol [6]. Ribosomal subunits are released also from isolated nuclei (regenerating rat liver [192, 193], HeLa cells [196]); this process can be stimulated by poly (U) [194, 195], 18 S and 28 S rRNA, ribosomes and large ribosomal subunits [195]. It has been suggested that the transport of ribosomal subunits from the nucleus [195] may be controlled by the amount of free polysomes in the cytoplasm. In agreement with this suggestion is the finding that ribosomal RNAs and proteins associated with free polysomes are more highly labeled than those of membrane-bound polysomes during early periods of labeling [196]. Newly transported ribosomal subunits have been found to enter polysomes close to the nuclear envelope in *Chironomus salivary* gland cells [197]. Specific inhibition of the nuclear export of ribosomal subunits by puromycin and cycloheximide can be explained by the possibility that immediate involvement in polysome formation is a prerequisite for the export of ribosomal subunits [198]. The entry times for 40 S and 60 S subunits into polysomes were found to be 3 min and 8 min, respectively, in yeast cells [199, 200]. Small and large subunits show different kinetics of their entry into large polysomes. This cannot be explained by simple entering of mature subunits from a subunit pool which are withdrawn randomly to enter polysomes [199], but possibly reflects the final maturation of new subunits in the cytoplasm [6].

Although progress has been achieved in the last years in the analysis of biogenesis of ribosomal particles, many problems are still unsolved and further studies are absolutely necessary.

4. References

[1] Attardi, G., and F. Amaldi: Structure and synthesis of ribosomal RNA. Ann. Rev. Biochem. **39**, 183—351 (1970)

[2] Chua, N.-H., and D. J. L. Luck: Biosynthesis of organelle ribosomes. In: Ribosomes (Eds.: M. Nomura, A. Tissiéres, and P. Lengyel). Cold Spring Harbor Laboratory, 1974, 519—539

[3] Maden, B. E. H., M. Salim, and J. S. Robertson: Progress in the structural analysis of mammalian 45 S and ribosomal RNA. In: Ribosomes (Eds.: M. Nomura, A. Tissiéres, and P. Lengyel). Cold Spring Harbor Laboratory, 1974, 829—839

[4] Hadjiolov, A. A., and N. Nikolaev: Maturation of ribosomal ribonucleic acids and the biogenesis of ribosomes. Progr. Biophys. Mol. Biol. **31**, 95–144 (1976)

[5] Perry, R. P.: Processing of RNA. Ann. Rev. Biochem. **45**, 605–629 (1976)

[6] Warner, J. R.: The assembly of ribosomes in eukaryotes. In: Ribosomes (Eds.: M. Nomura, A. Tissiéres, and P. Lengyel). Cold Spring Harbor Laboratory, 1974, 461–488

[7] Maden, B. E. H.: The structure and formation of ribosomes in animal cells. Progr. Biophys. Mol. Biol. **22**, 127–177 (1971)

[8] Rungger, D. W., and M. Crippa: The primary ribosomal DNA transcript in eukaryotes. Progr. Biophys. Mol. Biol. **31**, 247–269 (1977)

[9] Reeder, R. H.: Ribosomes from eukaryotes: Genetics. In: Ribosomes (Eds.: M. Nomura, A. Tissiéres, and P. Lengyel). Cold Spring Harbor Laboratory, 1974, 489–518

[10] Maden, B. E. H.: Ribosomal precursor RNA and ribosome formation in eukaryotes. Trends Biochem. Sci. **1**, 196–199 (1976)

[11] Mc Laughlin, C. S.: Yeast ribosomes: Genetics. In: Ribosomes (Eds.: M. Nomura, M. Tissiéres, and P. Lengyel). Cold Spring Harbor Laboratory, 1974, 815–827

[12] Spring, H., M. F. Trendelenburg, U. Scheer, W. W. Franke, and W. Herth: Structural and biochemical studies of the primary nucleus of two green alga species, *Acetabularia mediterranea* and *Acetabularia major*. Cytobiol. **10**, 1–65 (1974)

[13] Schibler, U., T. Wyler, and O. Hagenbüchle: Changes in size and secondary structure of the ribosomal transcription unit during vertebrate evolution. J. Mol. Biol. **94**, 503–517 (1975)

[14] Brown, R. D., and R. Haselkorn: Synthesis and maturation of cytoplasmic ribosomal RNA in *Euglena gracilis*. J. Mol. Biol. **59**, 491–503 (1971)

[15] Woodcock, C. L. F., J. E. Stanchfield, and R. R. Gould: Morphology and size of ribosomal cistrons in two plant species: *Acetabularia mediterranea* and *Chlamydomonas reinhardi*. Plant Sci. Lett. **4**, 17–23 (1975)

[16] Udem, S. A., and J. R. Warner: Ribosomal RNA synthesis in *Saccharomyces cerevisiae*. J. Mol. Biol. **65**, 227–242 (1972)

[17] Trapman, J., and R. J. Planta: Detailed analysis of the ribosomal RNA synthesis in yeast. Biochim. Biophys. Acta **414**, 115–125 (1975)

[18] Dudov, K. P., M. D. Dabeva, and A. A. Hadjiolov: Simple agar-urea gel electrophoretic fractionation of high molecular weight ribonucleic acids. Anal. Biochem. **76**, 250–258 (1976)

[19] Grierson, D., M. E. Rogers, M. L. Sartirana, and U. E. Loening: The synthesis of ribosomal RNA in different organisms: Structure and evolution of the rRNA precursor. Cold Spring Harbor Sym. Quant. Biol. **35**, 589–598 (1970)

[20] Rogers, M. E., U. E. Loening, and R. S. S. Fraser: Ribosomal RNA precursors in plants. J. Mol. Biol. **49**, 681–692 (1970)

[21] Grierson, D., and U. E. Loening: Distinct transcription products of ribosomal genes in two different tissues. Nature New Biol. **235**, 80–82 (1972)

[22] Miassod, R., and J. P. Cecchini: Hormone effect on the half-life of the 42 S pre-rRNA of cultured sycamore cells. FEBS-Lett. **96**, 277–282 (1978)

[23] Melanson, D. C., and J. Ingle: Regulation of rRNA accumulation by auxin in artichoke tissue. Plant Physiol. **62**, 761–765 (1978)

[24] Seitz, U., and U. Seitz: The molecular weight of rRNA precursor molecules and their processing in higher plant cells. Z. Naturforsch. **34c**, 253–258 (1979)

[25] Leaver, C. J., and J. L. Key: Ribosomal RNA synthesis in plants. J. Mol. Biol. **49**, 671–680 (1970)

[26] Richter, G.: Precursors of rRNA in freely suspended callus cells of parsley (*Petroselinum sativum*). Planta **113**, 79 to 96 (1973)

[27] Seitz, U., and U. Seitz: Kern-Plasma-Transport neusynthetisierter rRNA in Zellen einer Suspensionskultur von *Petroselinum sativum*. Planta **106**, 141–148 (1972)

[28] Perry, R. P., T.-Y. Cheng, J. J. Freed, J. R. Greenberg, D. E. Kelley, and K. D. Tartof: Evolution of the tran-

scription unit of ribosomal RNA. Proc. Natl. Acad. Sci. U.S. **65**, 609–616 (1970)

[29] Cecchini, J. P., and R. Miassod: Ribosomal cistrons in higher plant cells. I. A definitive scheme for the maturation pathway of the primary transcriptional product of ribosomal cistrons in *Acer pseudoplantanus* L. cells. Biochim. Biophys. Acta **418**, 104–116 (1976)

[30] Griffith, J. K., and T. D. Humphreys: Ribosomal RNA synthesis and processing in embryos of the hawaiian sea urchin *Tripneustes gratilla*. Biochem. **18**, 2178 to 2185 (1979)

[31] Trendelenburg, M. F., U. Scheer, and W. W. Franke: Structural organization of the transcription of ribosomal DNA in oocytes of the house cricket. Nature New Biol. **245**, 167–170 (1973)

[32] Rubinstein, L., and U. Clever: Nonconservative processing of ribosomal RNA in an insect, *Chironomus tentans*. Biochim. Biophys. Acta **246**, 517–529 (1971)

[33] Ringborg, U., B. Daneholt, J. E. Edström, E. Egyhazi, and B. Lambert: Electrophoretic characterization of nucleolar RNA from *Chironomus tentans* salivary gland cells. J. Mol. Biol. **51**, 327 to 340 (1970)

[34] Anderson, D., and L. D. Smith: Synthesis of heterogeneous nuclear RNA in full-grown oocytes of *Xenopus laevis* (Daudin). Cell **11**, 663–671 (1977)

[35] Griffith, J. K.: Dev. Biol. **65**, 353–371 (1978)

[36] Rogers, M. E., and G. Klein: Amphibian ribosomal ribonucleic acids. Biochem. J. **130**, 281–288 (1972)

[37] Scheer, U., M. F. Trendelenburg, and W. W. Franke: Transcription of ribosomal RNA cistrons. Correlation of morphological and biochemical data. Exptl. Cell. Res. **80**, 175–190 (1973)

[38] Caston, J. D., and P. H. Jones: Synthesis and processing of high molecular weight RNA by nuclei isolated from embryos of *Rana pipiens*. J. Mol. Biol. **69**, 19–38 (1972)

[39] McConkey, E. H., and J. W. Hopkins: Molecular weights of some HeLa ribosomal RNAs. J. Mol. Biol. **39**, 545–550 (1969)

[40] Reeder, R. H., B. Sollner-Webb, and H. L. Wahn: Sites of transcription initiation in vivo on *Xenopus laevis* ribosomal DNA. Proc. Natl. Acad. Sci. U.S. **74**, 5402 to 5406 (1977)

[41] Niles, E. G.: Isolation of a high specific activity 35 S rRNA precursor from *Tetrahymena pyriformis* and identification of its 5′ terminus, pppAp. Biochem. **17**, 4839–4844 (1978)

[42] Levis, R., and S. Penman: Processing steps and methylation in the formation of the ribosomal RNA of cultured *Drosophila* cells. J. Mol. Biol. **121**, 219–238 (1978)

[43] Nikolaev, N., O. I. Georgiev, P. V. Venkov, and A. A. Hadjiolov: The 37 S precursor to rRNA is the primary transcript of rRNA genes in *Saccharomyces cerevisiae*. J. Mol. Biol. **127**, 297–308 (1979)

[44] Klootwijk, J., P. de Jonge, and R. J. Planta: The primary transcript of the ribosomal repeating unit in yeast. Nucl. Acids Res. **6**, 27–39 (1979)

[45] Batts-Young, B., N. Maizels, and H. F. Lodish: Precursors of ribosomal RNA in the cellular slime mold *Dictyostelium discoideum*. Isolation and characterization. J. Biol. Chem. **252**, 3952–3960 (1977)

[46] Batts-Young, B., and H. F. Lodish: Triphosphate residues at the 5′ ends of rRNA precursor and 5 S RNA from *Dictyostelium discoideum*. Proc. Natl. Acad. Sci. U.S. **75**, 740–744 (1978)

[47] Kominami, R., and M. Muramatsu: Heterogeneity of 5′ termini of nucleolar 45 S, 32 S and 28 S RNA in mouse hepatoma. Nucl. Acids Res. **4**, 229–240 (1977)

[48] Nazar, R. N.: Studies on the 5′ termini of Novikoff ascites hepatoma ribosomal precursor RNA. Biochem. **16**, 3215 to 3219 (1977)

[49] Franke, W. W., U. Scheer, H. Spring, M. F. Trendelenburg, and G. Krohne: Morphology of transcriptional units of rDNA. Exptl. Cell Res. **100**, 233–244 (1976)

[50] Nikolaev, N., L. Silengo, and D. Schlessinger: Synthesis of a large precursor to ribosomal RNA in a mutant of *Escherichia coli*. Proc. Natl. Acad. Sci. U.S. **70**, 3361–3365 (1973)

[51] Dunn, J. J., and F. W. Studier: T 7

early RNAs and *Escherichia coli* ribosomal RNAs are cut from large precursor RNAs in vivo by ribonuclease III. Proc. Natl. Acad. Sci. U.S. **70**, 3296–3300 (1973)

[52] Miller, M. J., and D. Mc Mahon: Synthesis and maturation of chloroplast and cytoplasmic ribosomal RNA in *Chlamydomonas reinhardi*. Biochim. Biophys. Acta **366**, 35–44 (1974)

[53] Siev, M., R. Weinberg, and S. Penman: The selective interruption of nucleolar RNA synthesis in HeLa cells by cordycepin. J. Cell. Biol. **41**, 510–520 (1969)

[54] Van den Bos, R. C., J. Retel, and R. J. Planta: The size and the location of the ribosomal RNA segments in ribosomal precursor RNA of yeast. Biochim. Biophys. Acta **232**, 494–508 (1971)

[55] Hackett, P. B., and W. Sauerbier: The transcriptional organization of the ribosomal RNA genes in mouse L cells. J. Mol. Biol. **91**, 235–256 (1975)

[56] Liau, M. C., and R. B. Hurlbert: The topographical order of 18 S and 20 S ribosomal ribonucleic acids within the 45 S precursor molecule. J. Mol. Biol. **98**, 312–332 (1974)

[57] Reeder, R. H., and D. D. Brown: Transcription of the ribosomal RNA genes of an amphibian by the RNA polymerase of a bacterium. J. Mol. Biol. **51**, 361–377 (1970)

[58] Hecht, R. M., and M. L. Birnstiel: Integrity of the DNA template, a prerequisite for the faithful transcription of *Xenopus* rDNA in vitro. Europ. J. Biochem. **29**, 489–499 (1979)

[59] Dawid, I. B., and P. K. Wellauer: A reinvestigation of 5′→3′ polarity in 40 S ribosomal RNA precursor of *Xenopus laevis*. Cell **8**, 443–448 (1976)

[60] Reeder, R. H., T. Higashinakagawa, and O. Miller, Jr.: The 5′→3′ polarity of the *Xenopus* ribosomal RNA precursor molecule. Cell **8**, 449–454 (1976)

[61] Maden, B. E. H., and M. Salim: The methylated nucleotide sequences in HeLa cell ribosomal RNA and its precursors. J. Mol. Biol. **88**, 133–164 (1974)

[62] Tiollais, P., M.-A. Auger, and F. Galibert: Methylation process of mammalian ribosomal precursor RNA. Biochimie **57**, 1315–1321 (1975)

[63] Galibert, F., P. Tiollais, and M. E. Eladari: Fingerprinting studies of the maturat on of ribosomal RNA in mammalian cells. Europ. J. Biochem. **55**, 239 to 245 (1975)

[64] Klootwijk, J., R. C. Van den Bos, and R. J. Planta: Secondary methylation of yeast ribosomal RNA. FEBS-Lett. **27**, 102–106 (1972)

[65] Maden, B. E. H., M. Salim, and D. Summers: Matuıation pathway for ribosomal RNA in the HeLa cell nucleolus. Nature New Biol. **237**, 5–9 (1972)

[66] Fellner, P.: Nucleotide sequence from specific areas of the 16 S and 23 S RNAs of *E. coli*. Europ. J. Biochem. **11**, 12–27 (1969)

[67] De Jonge, P., J. Klootwijk, and R. J. Planta: Sequence of 3′ terminal 21 nucleotides of yeast 17 S rRNA. Nucl. Acids Res. **4**, 3655–3663 (1977)

[68] Choi, Y. C., and H. Busch: Modified nucleotides in T_1 RNase oligonucleotides of 18 S ribosomal RNA of the Novikoff hepatoma. Biochem. **17**, 2551–2560 (1978)

[69] Ouellette, A. J., E. Bandman, and A. Kumar: Regulation of ribosomal RNA methylation in a temperature-sensitive mutant of BHK cells. Nature **262**, 619 to 621 (1976)

[70] Caboche, M., and J. P. Bachellerie: RNA methylation and control of eukaryotic RNA biosynthesis. Effects of cycloleucine, a specific inhibitor of methylation, on ribosomal RNA maturation. Europ. J. Biochem. **74**, 19–29 (1977)

[71] Liau, M. C., M. E. Hunt, and R. B. Hurlbert: Role of ribosomal RNA methylases in the regulation of ribosome production in mammalian cells. Biochem. **15**, 3158–3164 (1976)

[72] Robertson, J. S., and B. E. H. Maden: Nucleotide sequences from the non-conserved region of HeLa cell 32 S ribosomal precursor RNA. Biochim. Biophys. Acta **331**, 61–70 (1973)

[73] Brand, R. C., J. Klootwijk, C. P. Sibum, and R. J. Planta: Pseudouridylation of yeast ribosomal precursor RNA. Nucl. Acids Res. **7**, 121–134 (1979)

[74] Weinberg, R. A., and S. Penman: Processing of 45 S nucleolar RNA. J. Mol. Biol. **47**, 169–178 (1970)

[75] KHAN, M. S. N., and B. E. H. MADEN: Maturation of 5.8 S RNA in HeLa cells. FEBS-Lett. **61**, 10—13 (1976)

[76] WILLEMS, M., M. PENMAN, and S. PENMAN: The regulation of RNA synthesis and processing in the nucleolus during inhibition of protein synthesis. **41**, 177 to 187 (1969)

[77] PENMAN, M., R. HUFFMAN, and A. KUMAR: Regulation of ribosomal RNA synthesis and processing during inhibition of protein synthesis by 1.3-bis(2-chloroethyl)-1-nitrosourea. Biochem. **15**, 2661 to 2668 (1976)

[78] TONIOLO, D., and C. BASILICO: Processing of ribosomal RNA in a temperature sensitive mutant of BHK cells. Biochim. Biophys. Acta **425**, 409—418 (1976)

[79] KUMAR, A., E. BANDMAN, and W. T. MELVIN: Ribosome metabolism in temperature-sensitive mutant of BHK cells. Nature **259**, 692—694 (1976)

[80] DUDOV, K. P., M. D. DABEVA, A. A. HADJIOLOV, and B. N. TODOROV: Processing and migration of ribosomal ribonucleic acids in the nucleolus and nucleoplasm of rat liver nuclei. Biochem. J. **171**, 375—383 (1978)

[81] STOYANOVA, B. B., and A. A. HADJIOLOV: Alterations in the processing of rat-liver ribosomal RNA caused by cycloheximide inhibition of protein synthesis. Europ. J. Biochem. **96**, 349—362 (1979)

[82] DABEVA, M. D., K. P. DUDOV, A. A. HADJIOLOV, and A. S. STOYKOVA: Quantitative analysis of rat liver nucleolar and nucleoplasmic ribosomal ribonucleic acids. Biochem. J. **171**, 367—374 (1978)

[83] DABEVA, M. D., K. P. DUDOV, A. A. HADJIOLOV, J. EMANUILOV, and B. N. TODOROV: Intranuclear maturation pathways of rat liver ribosomal ribonucleic acids. Biochem. J. **160**, 495—503 (1976)

[84] VAN DEN BERG, J. A., M. GRUBER, and G. AB: Estradiol-induced enhancement of the processing of the 32 S ribosomal precursor in rooster liver. FEBS-Lett. **63**, 65—70 (1976)

[85] KNECHT, D., and D. N. LUCK: Synthesis and processing of ribosomal RNA by the uterus of the ovariectomised adult rat during early oestrogen action. Nature **266**, 563—564 (1977)

[86] NAZAR, R. N., T. W. OWENS, T. O. SITZ, and H. BUSCH: Maturation pathway for Novikoff ascites hepatoma 5.8 S ribosomal ribonucleic acid. Evidence for its presence in 32 S nuclear ribonucleic acid. J. Biol. Chem. **250**, 2475—2481 (1975)

[87] WELLAUER, P. K., and I. B. DAWID: Secondary structure maps of ribosomal RNA and DNA. I. Processing of *Xenopus laevis* ribosomal RNA and structure of single-stranded ribosomal DNA. J. Mol. Biol. **89**, 379—395 (1974)

[88] JORDAN, B. R., R. JOURDAN, and B. JACQ.: Late steps in the maturation of *Drosophila* 26 S ribosomal RNA: Generation of 5.8 S and 26 S RNAs by cleavages occurring in the cytoplasm. J. Mol. Biol. **101**, 85—105 (1976)

[89] GARZIANI, F., and S. GARGANO: Ribosomal DNA transcription products during the first steps of magnification in *Drosophila melanogaster*. J. Mol. Biol. **100**, 59—71 (1976)

[90] BERGER, E.: The ribosomes of Drosophila. 5. Normal and defective ribosome biosynthesis in *Drosophila* cell cultures. Molec. Gen. Genetics **155**, 35—40 (1977)

[91] WELLAUER, P. K., I. B. DAWID, D. E. KELLEY, and R. P. PERRY: Secondary structure maps of ribosomal RNA. II. Processing of mouse L-cell ribosomal RNA and variations in the processing pathway. J. Mol. Biol. **89**, 397—407 (1974)

[92] TIOLLAIS, P., F. GALIBERT, and M. BOIRON: Evidence for the existence of several molecular species in the "45 S fraction" of mammalian ribosomal precursor RNA. Proc. Natl. Acad. Sci. U.S. **68**, 1117 to 1120 (1971)

[93] WINICOV, I.: Alternate temporal order in ribosomal RNA maturation. J. Mol. Biol. **100**, 141—155 (1976)

[94] DE JONGE, P., R. A. KASTELEIN, and R. J. PLANTA: Non-ribosomal nucleotide sequences in 7 S RNA, the immediate precursor of 5.8 S ribosomal RNA in yeast. Europ. J. Biochem. **83**, 537—546 (1978)

[95] TRAPMAN, J., P. DE LONG, and R. J. PLANTA: On the biosynthesis of 5.8 S ribosomal RNA in yeast. FEBS-Lett. **57**, 26—30 (1975)

[96] SHULMAN, R. W., and J. R. WARNER: rRNA transcription in a mutant of *Saccharomyces cerevisiae* defective in ribo-

somal protein synthesis. Mol. Gen. Genetics **161**, 221–223 (1978)

[97] Hartley, M. R.: The synthesis and origin of chloroplast low-molecular-weight ribosomal ribonucleic acid in spinach. Europ. J. Biochem. **96**, 311–320 (1979)

[98] Hartley, M. R., and C. Head: The synthesis and origin of chloroplast high-molecular-weight ribosomal ribonucleic acid in spinach. Europ. J. Biochem. **96**, 301–310 (1979)

[99] Gebauer, H. U., U. Seitz, and U. Seitz: Transport and processing of ribosomal RNA in plant cells after treatment with cycloheximide. Z. Naturforsch. **30c**, 213 to 218 (1975)

[100] Zaitseva, G., I. L. Mett, and A. L. Mett: rRNA synthesis in kinetoplasts of *Crithidia oncopelty* and *C. fasciculata*. Zitologiya (russ.) **21**, 310–317 (1979)

[101] Ramagopal, S., and A. Marcus: RNA synthesis in growing and stationary cells of a culture of scarlet rose. Disproportionate synthesis of ribosomal subunits in the stationary state. J. Cell. Physiol. **98**, 603–611 (1979)

[102] Purtell, M. J., and D. D. Anthony: Changes in ribosomal RNA processing paths in resting and phytohemagglutinin-stimulated guinea pig lymphocytes. Proc. Natl. Acad. Sci. U.S. **72**, 3315–3319 (1975)

[103] Johnson, L. F., R. Levis, H. T. Abelson, H. Green, and S. Penman: Changes in RNA in relation to growth of the fibroblast. IV. Alteration in the production and processing of mRNA and rRNA in resting and growing cells. J. Cell Biol. **71**, 933–938 (1976)

[104] Surrey, S., J. Ginzburg, and M. Nemer: Ribosomal RNA synthesis in pre- and postgastrula-stage sea urchin embryos. Develop. Biol. **71**, 83–89 (1979)

[105] Aronson, A. J.: Alterations of RNA metabolism in sea urchin embryos by an inhibitor of protein synthesis initiation. Biochim. Biophys. Acta **477**, 334–342 (1977)

[106] Russel, P. J., J. P. Hammett, and E. U. Selker: J. Bacteriol. **127**, 785–793 (1976)

[107] Free, S. J., P. W. Rice, and R. L. Metzenberg: J. Bacteriol. **137**, 1219–1226 (1979)

[108] Din, N., J. Engberg, W. Kaffenberger and W. A. Eckert: The intervening sequence in the 26 S rRNA coding region of *T. thermophila* is transcribed within the largest stable precursor for rRNA. Cell **18**, 525–532 (1979)

[109] Rubin, G. M., Three forms of the 5.8 S ribosomal RNA species in *Saccharomyces cerevisiae*. Europ. J. Biochem. **41**, 197 to 202 (1974)

[110] Kominami, R., H. Hamada, Y. Fujii-Kuriyama, and M. Maramatsu: 5′-terminal processing of ribosomal 28 S RNA. Biochem. **17**, 3965–3970 (1978)

[111] Taber, R. L. jr., and W. S. Vincent: The synthesis and processing of ribosomal RNA precursor molecules in yeast. Biochim. Biophys. Acta **186**, 317–325 (1969)

[112] Brown, D. D., and C. S. Weber: Gene linkage by RNA-DNA hybridization. I. Unique DNA sequences homologous to 4 S RNA, 5 S RNA and ribosomal RNA. J. Mol. Biol. **34**, 661–680 (1968)

[113] Wimber, D. E., and D. M. Steffensen: Localization of 5 S RNA genes on *Drosophila* chromosomes by RNA-DNA hybridization. Science **170**, 639–641 (1970)

[114] Retel, J., and R. J. Planta: Nuclear satellite DNAs of yeast. Biochim. Biophys. Acta **281**, 299–309 (1972)

[115] Rubin, G., and J. E. Sulston: Physical linkage of the 5 S cistrons to the 18 S and 28 S ribosomal RNA cistrons in *Saccharomyces cerevisiae*. J. Mol. Biol. **79**, 521 to 530 (1973)

[116] Pardue, M. L., D. D. Brown, and M. L. Birnstiel: Location of the genes for 5 S ribosomal RNA in *Xenopus laevis*. Chromosoma **42**, 191–203 (1973)

[117] Tønnesen, T., J. Engberg, and V. Leick: Studies on the amount and location of the tRNA and 5-S rRNA genes in *Tetrahymena pyriformis* GL. Europ. J. Biochem. **63**, 399–407 (1976)

[118] Henderson, A. S., K. C. Atwood, M. T. Yu, and D. Warburton: The site of the 5 S RNA genes in Primates. The great apes. Chromosoma **56**, 29–32 (1976)

[119] Knälmann, M., and E.-Ch. Burger: Cytologische Lokalisation von 5 S and 18/25 S RNA Genorten in Mitose-Chromosomen von *Vicia faba*. Chromosoma **61**, 172–192 (1977)

[120] Szabo, P., R. Elder, D. M. Steffensen, and O. C. Uhlenbeck: Quantitative

in situ hybridization of ribosomal RNA species to polytene chromosomes in *Drosophila melanogaster*. J. Mol. Biol. **115**, 539–563 (1977)

[121] Szabo, P., M. R. Lee, F. B. Elder, and W. Prensky: Localization of 5 S RNA and rRNA genes in the norway rat. Chromosoma **65**, 161 – 172 (1978)

[122] Hindley, J., and S. M. Page: Nucleotide sequence of yeast 5 S ribosomal RNA. FEBS-Lett. **26**, 157–160 (1972)

[123] Takai, K., S. Hashimoto, and M. Muramatsu: Oligonucleotide sequences of pancreatic and T_1 ribonuclease digests of 5 S ribosomal RNA from mouse. Biochem. **14**, 536–542 (1975)

[124] Yamamoto, M., and K. H. Seifart: Synthesis of ribosomal 5 S RNA by isolated nuclei from HeLa cells in vitro. Biochem. **16**, 3201–3208 (1977)

[125] McReynolds, L., and S. Penman: A polymerase activity forming 5 S and pre-4 S RNA in isolated HeLa cell nuclei. Cell **1**, 139–145 (1974)

[126] Udvardy, A., and K. H. Seifart: Transcription of specific genes in isolated nuclei from HeLa cells in vitro. Europ. J. Biochem. **62**, 353–363 (1976)

[127] Yamamoto, M., and K. H. Seifart: Heterogeneity in the 3′-terminal sequence of ribosomal 5 S RNA synthesized by isolated HeLa cell nuclei in vitro. Biochem. **17**, 457–461 (1978)

[128] Yamamoto, M., D. Jones, and K. H. Seifart: Transcription of ribosomal 5 S-RNA by RNA polymerase C in isolated chromatin from HeLa cells. Europ. J. Biochem. **80**, 243–253 (1977)

[129] Hamada, H., M. Muramatsu, Y. Urano, T. Ouishi, and R. Kominami: In vitro synthesis of a 5 S RNA presursor by isolated nuclei of rat liver and HeLa cells. Cell **17**, 163–173 (1979)

[130] Sklar, V. E. F., and R. G. Roeder: Transcription of specific genes in isolated nuclei by exogeneous RNA polymerase. Cell **10**, 405–414 (1977)

[131] Parker, C. S., and R. G. Roeder: Selective and accurate transcription of the *Xenopus laevis* 5 S RNA genes in isolated chromatin by purified RNA polymerase III. Proc. Natl. Acad. Sci. U.S. **74**, 44 to 48 (1977)

[132] Sy, N. G., C. S. Parker, and R. G. Roeder: Transcription of cloned Xenopus 5 S RNA genes by *Xenopus laevis* RNA polymerase III in reconstituted systems. Proc. Natl. Acad. Sci. U.S. **76**, 136–140 (1979)

[133] Denis, H., and M. Wegnez: Recherches biochimiques sur l'oogenese. 7. Synthèse et maturation du RNA 5 S dans les petits oocytes de *Xenopus laevis*. Biochimie **55**, 1137–1151 (1975)

[134] Rubin, G. M., and D. S. Hogness: Effect of heat shock on the synthesis of low molecular weight RNAs in *Drosophila*: Accumulation of a novel form of 5 S RNA. Cell **6**, 207–213 (1975)

[135] Jacq, B., R. Jourdan, and B. R. Jordan: Structure and processing of precursor 5 S RNA in *Drosophila melanogaster*. J. Mol. Biol. **117**, 785–795 (1977)

[136] Levis, R.: Kinetic analysis of 5 S RNA formation in *Drosophila* cells. J. Mol. Biol. **122**, 279–283 (1978)

[137] Brown, R. D., and D. D. Brown: The nucleotide sequence adjoining the 3′ end of the genes coding for oocyte type 5 S ribosomal RNA in *Xenopus*. J. Mol. Biol. **102**, 1–14 (1976)

[138] Hayashi, Y., R. Kominami, and M. Muramatsu: Effect of cycloheximide on the synthesis and processing of 5 S RNA in HeLa cells. J. Biochem. **81**, 451–459 (1977)

[139] Tsurugi, K., T. Morita, and K. Ogata: Studies on the metabolism of ribosomal structural proteins of regenerating rat liver.
Sites of biosynthesis of structural proteins of large subunit and of their assembly with RNA moiety. Europ. J. Biochem. **25**, 117–128 (1972)

[140] Heady, J. E., and E. H. Mc Conkey: Completion of nascent HeLa ribosomal proteins in a cell-free system. Biochem. Biophys. Res. Commun. **40**, 30–36 (1970)

[141] Craig, N., and R. P. Perry: Persistent cytoplasmic synthesis of ribosomal proteins during the selective inhibition of ribosomal RNA synthesis. Nature New Biol. **229**, 75–80 (1971)

[142] Mager, W. H., and R. J. Planta: Cellular site of synthesis of ribosomal proteins in yeast. Biochim. Biophys. Acta **402**, 105–112 (1975)

[143] Wu, R. S., and J. R. Warner: Cytoplasmic synthesis of nuclear proteins. Kinetics of accumulation of radioactive proteins in various cell fractions after brief pulses. J. Cell Biol. **51**, 643–652 (1971)

[144] Nabeshima, Y., K. Tsurugi, and K. Ogata: Preferential biosynthesis of ribosomal structural proteins by free and loosely bound polysomes from regenerating rat liver. Biochim. Biophys. Acta **414**, 30–43 (1975)

[145] Lambowitz, A. M., N.-H. Chua, and D. J. L. Luck: Mitochondrial ribosome assembly in *Neurospora crassa*; preparation of ribosomal precursor particles, site of synthesis of mitochondrial ribosomal proteins and studies on the poky mutant. J. Mol. Biol. **107**, 223–253 (1976)

[146] Lizardi, P. M., and D. J. L. Luck: The intracellular site of synthesis of mitochondrial ribosomal proteins in *Neurospora crassa*. J. Cell Biol. **54**, 56–79 (1972)

[147] Freyssinet, G.: Determination of the sites of synthesis of some *Euglena* cytoplasmic and chloroplast ribosomal ribosomal proteins. Exptl. Cell Res. **115**, 207–219 (1978)

[148] Ledoigt, G., and C. Louvel: Study of the regulation of plastid development in *Euglena gracilis*. II. Functional localization and synthesis of ribosomal chloroplast particles. Biochim. Biophys. Acta **563**, 432–444 (1979)

[149] Warner, J. R.: Distribution of newly formed ribosomal proteins in HeLa cell fractions. J. Cell Biol. **80**, 767–772 (1979)

[150] Delaunay, J., and G. Schapira: Questioning the in vitro protein exchange between ribosomes and post-microsomal supernatant. Biochim. Biophys. Acta **349**, 262–268 (1974)

[151] Terao, K., K. Tsurugi, and K. Ogata: Non-exchangeability of ribosomal structural proteins with cell sap proteins of rat liver in vitro. J. Biochem. **76**, 1113–1122 (1974)

[152] Wool, I. G., and G. Stöffler: Determination of the size of the pool of free ribosomal proteins in rat liver cytoplasm. J. Mol. Biol. **108**, 201–218 (1976)

[153] Nabeshima, Y.-J., K. Imai, and K. Ogata: Biosynthesis of ribosomal proteins by poly(A)-containing mRNAs from rat liver in a wheat germ cell-free system and sizes of mRNAs coding ribosomal proteins. Biochim. Biophys. Acta **564**, 105 to 121 (1979)

[154] Hackett, P. B., E. Egberts, and P. Traub: Characterization of Ehrlich ascites tumor cell messenger RNA specifying ribosomal proteins by translation in vitro. J. Mol. Biol. **119**, 253–267 (1978)

[155] Mager, W. H., R. Hoving, and R. J. Planta: In vitro synthesis of yeast ribosomal proteins. FEBS-Lett. **58**, 219–221 (1975)

[156] Warner, J. R., and C. Gorenstein: The synthesis of eukaryotic ribosomal proteins in vitro. Cell **11**, 201–212 (1977)

[157] Mager, W. H., and R. J. Planta: Yeast ribosomal proteins are synthesized on small polysomes. Europ. J. Biochem. **62**, 193–196 (1976)

[158] Gorenstein, C., and J. R. Warner: The monocistronic nature of ribosomal protein genesis in yeast. Current Genetics **1**, 9–12 (1979)

[159] Gorenstein, C., and J. R. Warner: Coordinate regulation of the synthesis of eukaryotic ribosomal proteins. Proc. Natl. Acad. Sci. U.S. **73**, 1547–1551 (1976)

[160] Tsurugi, K., T. Morita, and K. Ogata: Identification and metabolic relationship between proteins of nucleolar 60-S particles and of ribosomal large subunits of rat liver by means of two-dimensional disc electrophoresis. Europ. J. Biochem. **32**, 555–562 (1973)

[161] Tsurugi, K., T. Morita, and K. Ogata: Effects of the inhibition of ribosomal RNA synthesis on the synthesis of ribosomal structural proteins in regenerating rat liver. Europ. J. Biochem. **29**, 585–592 (1972)

[162] Tsurugi, K., T. Morita, and K. Ogata: Mode of degradation of ribosomes in regenerating rat liver in vivo. Europ. J. Biochem. **45**, 119–126 (1974)

[163] Tsurugi, K., and K. Ogata: Preferential degradation of newly synthesized ribosomal proteins in rat liver treated with a low dose of Actinomycin D. Biochem. Biophys. Res. Commun. **75**, 525–531 (1977)

[164] Phillips, W. P., and E. H. McConkey: Relative stoichiometry of ribosomal pro-

teins in HeLa cells nucleoli. J. Biol. Chem. **251**, 2876–2881 (1976)

[165] WARNER, J. R.: In the absence of rRNA synthesis, the ribosomal proteins of HeLa cells are synthesized normally and degraded rapidly. J. Mol. Biol. **115**, 315–333 (1977)

[166] GORENSTEIN, C., and J. R. WARNER: Synthesis and turnover of ribosomal proteins in the absence of 60 S subunit assembly in *Saccharomyces cerevisiae*. Molec. Gen. Genetics **157**, 327–332 (1977)

[167] CRAIG, N., D. E. KELLEY, and R. P. PERRY: Lifetime of the messenger RNA which codes for ribosomal protein in L cells. Biochim. Biophys. Acta **246**, 493–498 (1971)

[168] BOLLA, R., H. E. ROTH, H. WEISSBACH, and N. BROT: Effect of ribosomal proteins on synthesis and assembly of preribosomal particles in isolated rat liver nuclei. J. Biol. Chem. **252**, 721–725 (1977)

[169] ROTH, H. E., R. BOLLA, G. S. COX, B. REDFIELD, H. WEISSBACH, and N. BROT: Uptake of ribosomal proteins by isolated HeLa nuclei. Biochem. Biophys. Res. Commun. **69**, 608–612 (1976)

[170] VAUGHAN, M. H., J. R. WARNER, and J. E. DARNELL: Ribosomal precursor particles in the HeLa cell nucleus. J. Mol. Biol. **25**, 235–251 (1967)

[171] WARNER, J. R., and R. SOEIRO: Nascent ribosomes from HeLa cells. Proc. Natl. Acad. Sci. U.S. **58**, 1984–1990 (1967)

[172] KUMAR, A., and J. R. WARNER: Characterization of ribosomal precursor particles from HeLa cells. J. Mol. Biol. **80**, 265–276(1972)

[173] LIAU, M. C., and R. P. PERRY: Ribosome precursor particles in nucleoli. J. Cell. Biol. **42**, 272–283 (1969)

[174] KUMAR, A., and J. R. WARNER: Characterization of ribosomal precursor particles from HeLa cell nucleoli. J. Mol. Biol. **63**, 233–246 (1972)

[175] SOEIRO, R., and C. BASILE: Non-ribosomal nucleolar proteins in HeLa cells. J. Mol. Biol. **79**, 507–519 (1973)

[176] HIGASHINAKAGAWA, T., and M. MURAMATSU: Ribosome precursor particles in the nucleolus of rat liver. Easily extractable nucleolar 60-S ribonucleoprotein particles and their relation to cytoplasmic large ribosomal subunits. Europ. J. Biochem. **42**, 245–258 (1974)

[177] SEPHERD, J., and B. E. H. MADEN: Ribosome assembly in HeLa cells. Nature **236**, 211–214 (1972)

[178] PRESTAYKO, A. W., G. R. KLOMP, D. J. SCHMOLL, and H. BUSCH: Comparison of proteins of ribosomal subunits and nucleolar preribosomal particles from Novikoff hepatoma ascites cells by two-dimensional polyacrylamide gel electrophoresis. Biochem. **13**, 1945–1951 (1974)

[179] AUGER-BUENDIA, M.-A., and M. LONGUET: Characterization of proteins from nucleolar preribosomes of mouse leukemia cells by two-dimensional polyacrylamide gel electrophoresis. Europ. J. Biochem. **85**, 105–114 (1978)

[180] AUGER-BUENDIA, M.-A., M. LONGUET, and A. TAVITIAN: Kinetic studies on ribosomal proteins assembly in preribosomal particles and ribosomal subunits of mammalian cells. Biochim. Biophys. Acta **563**, 113–128 (1979)

[181] FUJISAWA, T., K. IMAI, Y. TANAKA, and K. OGATA: Studies on the protein components of 110 S and total ribonucleoprotein particles of rat liver nucleoli. J. Biochem. **85**, 277–286 (1979)

[182] TRAPMAN, J., J. RETEL, and R. J. PLANTA: Ribosomal precursor particles from yeast. Exptl. Cell Res. **90**, 95–104 (1975)

[183] KRUISWIJK, T., R. J. PLANTA, and J. M. KROP: The course of the assembly of ribosomal subunits in yeast. Biochim. Biophys. Acta **517**, 378–389 (1978)

[184] KRUISWIJK, T., J. T. DE HEY, and R. J. PLANTA: Modification of yeast ribosomal proteins. Phosphorylation. Biochem. J. **175**, 213–219 (1978)

[185] KUMAR, A., and A. SUBRAMANIAN: Ribosome assembly in HeLa cells: Labeling pattern of ribosomal proteins by two-dimensional resolution. J. Mol. Biol. **94**, 409–423 (1975)

[186] VANDREY, J. P., C. J. GOLDENBERG, and G. L. ELICEIRI: In vivo isotope incorporation patterns into HeLa ribosomal proteins. Biochim. Biophys. Acta **432**, 104–112 (1976)

[187] AIELLO, L. O., C. J. GOLDENBERG, and G. L. ELICEIRI: In vivo incorporation of ribosomal proteins into HeLa cell

ribosomal particles. Biochim. Biophys. Acta **475**, 652—658 (1977)

[188] Lastick, S. M., and E. H. McConkey: Exchange and stability of HeLa ribosomal proteins in vivo. J. Biol. Chem. **251**, 2867—2875 (1976)

[189] Auger-Buendia, M.-A., and A. Tavitian: Ribosomal protein synthesis and exchange in the absence of 28-S and 18-S ribosomal RNA synthesis in L 5178 Y cells. Biochim. Biophys. Acta **563**, 129 to 142 (1979)

[190] Auger-Buendia, M. A., R. Hamelin, and A. Tavitian: Influence of toyocamycin on the assembly and processing of preribosomal ribonucleoproteins in the nucleolus of mammalian cells. Biochim. Biophys. Acta **521**, 241—250 (1978)

[191] Venkov, P. V., L. I. Stateva, and A. A. Hadjiolov: Toyocamycin inhibition of ribosomal ribonucleic acid processing in an osmotic-sensitive adenosine-utilizing *Saccharomyces cerevisiae* mutant. Biochim. Biophys. Acta **474**, 245—253 (1977)

[192] Racevskis, J., and T. E. Webb: Processing and release of ribosomal RNA from isolated nuclei: Analysis of the ATP-dependence and cytosol-dependence. Europ. J. Biochem. **49**, 93—100 (1974)

[193] Sato, T., K. Ishikawa, and K. Ogata: Characterization of ribonucleoprotein particles released from isolated nuclei of regenerating rat liver in two different in vitro systems. Biochim. Biophys. Acta **474** 536—548 (1977)

[194] Goidl, J. A., D. Canaani, M. Boublik, H. Weissbach, and H. Dickerman: Polyanion-induced release of polyribosomes from HeLa cell nuclei. J. Biol. Chem. **250**, 9198—9205 (1975)

[195] Sato, T., K. Ishikawa, and K. Ogata: Factors causing release of ribosomal subunits from isolated nuclei of regenerating rat liver in vitro. Biochim. Biophys. Acta **474**, 549—561 (1977)

[196] Murty, C. N. and H. Sidransky: Studies on the metabolism of ribosomal RNA and protein of free and membrane-bound polyribosomes of rat liver. Biochim. Biophys. Acta **335**, 226—235 (1974)

[197] Edström, J.-E., and U. Lönn: Cytoplasmic zone analysis. RNA flow studied by micromanipulation. J. Cell. Biol. **70**, 562—572 (1976)

[198] Lönn, U., and J.-E. Edström: Protein synthesis inhibitors and export of ribosomal subunits. Biochim. Biophys. Acta **475**, 677—679 (1977)

[199] Petersen, N. S., C. S. McLaughlin, and D. P. Nierlich: Ribosomal subunit entry into polysomes in yeast. Biochim. Biophys. Acta **447**, 294—303 (1976)

[200] Petersen, N. S., and D. P. Nierlich: Yeast mutant, rna 1, affects the entry into polysomes of ribosomal RNA as well as messenger RNA. Molec. Gen. Genetics **162**, 319—322 (1978)

[201] Ford, P. J., and T. Mathieson: Control of 5 S RNA synthesis in *Xenopus laevis*. Nature **261**, 433—435 (1976)

VIII. Dissociation — reassociation processes of ribosomal particles

Contents

Dissociation of cytoplasmic ribosomes of eukaryotes into 60 S and 40 S subunits is a physiological event occurring during the termination step of the biosynthesis of each protein molecule (see Chapter XI). The subunits formed in this way are generally named "native subunits".

Before the present scheme of eukaryotic protein synthesis became increasingly substantiated, in vitro dissociation and reassociation of polysomes from eukaryotes posed major problems. The subunits obtained by various rather drastic procedures were unsatisfactory in several respects. In 1968 a successful attempt to prepare reassociable ribosomal subunits from rat muscle has been described [1]. This discovery was an important basis for many in vitro studies of eukaryotic protein synthesis.

1. Dissociation by withdrawal of magnesium

Although bacterial ribosomes [2] as well as most ribosomes from mitochondria [3 to 7] and chloroplasts [8—12] dissociate reversibly to about 50 S and 30 S subunits upon lowering the magnesium or increasing the potassium concentration of the suspension medium, similar treatments leave the majority of 80 S type eukaryotic polysomes undissociated. Instead, these ribosomes unfold via various intermediates of decreasing sedimentation coefficients, ultimately resulting in ribosomal particles of about 55 S to 59 S after dialysis versus Mg^{2+}-free media [13—16]. Even ultracentrifugation at moderate speed brings about dissociation of only part of the magnesium-deprived polysomal population. The few subunits separable in this way were generally found unable to reassociate and synthesize polyphenylalanine upon recombination.

Complete removal of magnesium by chelating agents such as EDTA [17—31] or pyrophosphate [20] affords dissociation of eukaryotic ribosomes. However, the application of chelating agents, even in slightly higher amounts than ribosome-bound magnesium, is accompanied by considerable unfolding of both subunits, and during the subsequent separation procedure (ultracentrifugation or gel filtration) partial disassembly of protein and RNA components seems unavoidable. The S values of such subunits are generally in the order of 50 S and 30 S for the large and the small subunit, respectively. The large subunit is deficient in 5 S RNA and at least one ribosomal protein, and the small subunit also seems to undergo irreversible changes very easily, such as unfolding and RNA degradation. Therefore, 50 S and 30 S subunits prepared in this way are generally unable to reassociate to 80 S particles active in polyphenylalanine synthesis. However, under certain carefully controlled conditions, EDTA treatment seems to damage the small ribosomal subunit less than the large one [26], because it is able to reassociate with 60 S subunits obtained by KCl-puromycin treatment to ribosomes active in polyphenylalanine synthesis. Furthermore, the mRNA is released from polysomes upon EDTA treatment as an mRNA-protein (mRNP) complex of heterogeneous size between 20 S and 120 S depending on the origin and the preparation procedures applied. The density of such mRNP complexes of about 1.4 g/cm^3 suggests a protein to RNA ratio of about 4:1 (for review see [32]).

Table 16. Dissociation of ribosomes of various eukaryotic tissues resulting in reassociable subunits

Tissue	Conditions applied					References
	High K^+	Gradient temperature $> 19\,°C$	Antioxydant	Puromycin	Runoff	
Rat skeletal muscle	+	+	+	+	—	[1, 62—65, 95]
	(+)	—	—	+	+	[73]
Rat liver	+	+	+	+	—	[63—65, 71, 75, 76, 79, 85]
	(+)	—	cell sap	+	(+)	[69, 73, 92, 93]
	+	+	+	—	±	[68, 72]
Mouse liver	+	—	+	—	+	[87, 96]
Rabbit reticulocytes	+	+	+	±	—	[67, 80]
Human tissues	+	+	+	+	—	[90, 91]
Mouse plasmocytoma	+	+	+	+	—	[81, 88]
Sea urchin eggs and embryos	+	—	—	—	—	[70]
Xenopus laevis oocytes	+	+	+	—	—	[82]
Pea seedlings, rice	+	—	+	—	±	[84, 87]
Neurospora crassa	+	+	+	+	±	[78]
Yeast	±	—	+	—	±	[66, 89]
Reticulocytes, liver, muscle	(+)	—	cell sap	+	(+)	[93]
Tetrahymena pyriformis	+	—	+	—	—	[63]
	—	—	+	—	+	[94]
Dog pancreas	+	—	+	—	(+)	[97—99]

2. Other agents for dissociating eukaryotic ribosomes

The electrostatic binding forces of magnesium responsible for subunit association can be repressed by increasing the concentrations of certain monovalent cations. Thus, at low Mg^{2+} concentrations (1 to 10 mM) about 100-fold excess of K^+ or NH_4^+ ions causes dissociation of 80 S ribosomes, and polysomes carrying nascent proteins are partially unfolded [18, 30, 31, 33—43].

The dissociating effects of either p-chloromercuribenzoate [44] or Hg^{2+} [45] suggest the participation of sulfhydryl groups in the subunit association into eukaryotic ribosomes. In addition, hydrogen bonds seem to play a considerable role in subunit interactions, because urea also has dissociating activity [46, 47]. Further papers report on dissociating influences of silver [48], proteases [49, 50], diethylpyrocarbonate [51], oxidizing agents and sodium deoxycholate [52], and alkaline pH values [53, 54]. Furthermore, pressure-induced dissociation has been demonstrated, e.g., when eukaryotic ribosomes are centrifuged under specific conditions at high rotor speeds [55—57]. In addition, sonication leads to dissociation of polysomes [58].

Most treatments mentioned before result in more or less deficient subunits. An important step toward the preparation of complete, active ribosome subunits was the discovery of the effect of puromycin, which terminates the nascent peptide chain by interaction with peptidyl-tRNA at the ribosomal peptidyltransferase center. Absence of nascent peptide in eukaryote ribosomes considerably facilitates dissociation as can be judged also from the observation that ribosomes of fasted [59] or ethionine-treated animals [60] dissociate much easier than those of untreated controls. Disappearance of polysomes in soybean roots [61] upon anaerobiosis or DNP also simplifies dissociation.

3. Preparation of ribosomal subunits reassociable to active ribosomes

Most of the irreversible changes during subunit preparation can be avoided by the following methods [1, 62—65]:

1. Increasing the potassium concentration to about 1 M at a relatively high Mg^{2+} concentration of 12.5—15 mM.
2. Increasing the incubation temperature to 28—37 °C.
3. Including β-mercaptoethanol as an antioxidant in all steps.
4. Application of puromycin, an antibiotic which terminates the nascent protein chain and releases it from the ribosome.
5. Separation of the subunits on sucrose gradients at elevated temperatures (28 °C) containing high potassium, magnesium, and β-mercaptoethanol.

The basic principles of this preparation scheme have been applied with more or less extensive modifications by various groups to a great variety of different eukaryotic ribosomes and were in most cases found suitable for the preparation of ribosomal subunits able to reform active ribosomes or polysomes (see [62—101], Table 17). In many cases, temperatures around 20 °C and potassium concentrations of 300—500 mM with 3—5 mM Mg^{2+} simultaneously were sufficient and resulted in even more active ribosomal subunits.

Regarding puromycin it has been shown [79] that chain termination by puromycin, i.e., formation and release of peptidyl-puromycin, takes place at 0° C, and so-called pseudo-polysomes still carrying mRNA and uncharged tRNA can be isolated. Final dissociation of this complex may be achieved at temperatures above 20 °C. Furthermore a procedure which enables preparation of very active ribosomal subunits on a larger scale by combination of puromycin treatment at low ionic strength with subsequent separation of the subunits from all smaller compounds in-

Table 17. Proteosynthetic activity of hybrids of ribosomal subunits of different origin

Source of ribosomes		Per cent of incorporation activity				References
Tissue A	Tissue B	A_{60S}/A_{40S}	B_{60S}/A_{40S}	A_{60S}/B_{40S}	B_{60S}/A_{40S}	
Skeletal muscle from normal rats	Skeletal muscle from diabetic rats					
	9 mM Mg^{2+}	100	50	107	39	[1]
	15 mM Mg^{2+}	100	96	106	86	[1]
Skeletal muscle from normal rats	Skeletal muscle from diabetic rats	100	28	122	41	[95]
Rat muscle	Rat liver	100	78	92	67	[63]
Rabbit muscle	Rat muscle	100	63	59	50	[63]
Rat liver (normal)	Rat liver (hypophysectomized)	100	79	92	85	[68]
Rat liver (normal)	Rat liver (hypophysectomized)	100	77	77	99	[85]
Rat liver	Rat muscle	100	95	81	89	[73]
Sea urchin eggs	Sea urchin embryos					
	8 mM Mg^{2+}	100	90	103	88	[70]
	10 mM Mg^{+2}	100	105	96	91	[70]
Rat liver	Pea seedlings	100	75	50	121	[87]
Mouse liver	Pea seedlings	100	76	58	106	[87]
Rat muscle	Tetrahymena pyriformis	100	45	53	18	[63]

cluding puromycin by Sephadex gel filtration has been developed [69, 93].

The use of puromycin can be avoided if the ribosome population consists of particles mainly or completely free of nascent proteins [70, 81, 93] or if precautions have been taken to terminate the nascent chains, e.g., by prior incubation of polysomes with all components necessary for in vitro protein synthesis. In this way probably the most active ribosomal subunits have been prepared [96], which up to 80% were able to take part in protein synthesis.

Besides the most common procedures mentioned above, rat liver ribosomes can also be dissociated reversibly in the presence of urea, but the subunits obtained seem to have lost part of their activity [46, 47].

Major problems in the preparation of active ribosomal subunits were caused by the high content of ribonuclease in many eukaryotic tissues. Apart from reticulocytes which have a low nuclease activity, dog pancreas deserves special emphasis because of its extremely high content of nuclease inhibitors. A rather simple method for preparation of active ribosome subunits from this tissue includes the application of DEAE cellulose chromatography [97–99].

Altogether, dissociation of eukaryotic ribosomes into reassociable subunits is more difficult than for prokaryotic ribosomes, and precautions must be taken so that the dissociation procedure does not damage the binding forces within the ribosomal subunits to a greater extent than the forces interacting between the small and the large subunit. Noninitiated 80 S ribosomes dissociate more easily than polysomes carrying nascent protein chains. Thus, the dissociation behavior of eukaryotic polysomes is different from that of prokaryotic polysomes. Because the interactions between the nonribosomal components promoting protein synthesis (mRNA, aminoacyl-tRNA, and peptidyl-tRNA) are basically the same in both eukaryotes and prokaryotes, the observed different response to dissociating agents may be attributable to the more complex structure of the eukaryotic ribosome. Besides possibly stronger interactions between the nonribosomal components mentioned and ribosomal proteins and/or ribosomal RNA during eukaryotic protein synthesis, stronger interactions with elongation factors may account for some of the observed dissociation differences. However, because eukaryotic ribosomes and polysomes pretreated with KCl, deficient in factors, also reveal different dissociation behavior, the more complex structure of eukaryotic ribosomes seems to be the main reason for the higher stability of eukaryotic polysomes in comparison to prokaryotic polysomes.

4. Reassociation of ribosomal subunits

Several criteria have been applied to test the ability of the various ribosomal subunit preparations to reassociate. Appearance of 80 S particles at appropriate ionic conditions in the ultracentrifuge distribution pattern is one piece of evidence that reassociation of ribosomal subunits has occurred. 50 S and 30 S subunits obtained from eukaryotes by EDTA treatment form only a few 80 S particles after being mixed at elevated magnesium concentrations. However, ultracentrifugal analysis is only a rough estimate for reassociation because it does not reflect minor changes in the particles. Therefore, tests for reassociability generally include determination of proteosynthetic activity. Most conveniently, poly(U)-dependent polyphenylalanine synthesis in the presence of cell sap or purified elongation factors at 6—12 mM Mg^{2+} has been used as a standard procedure to check the quality of subunits. Because poly(U) is only an artificial messenger, not reflecting the initiation events, testing the AUG-dependent binding of initiator-tRNA and/or translation of added intact mRNA at low magnesium concentrations is a better check for subunit integrity.

For prokaryotes, formation of active hybrids between ribosomal subunits of distinct species was repeatedly reported. Even 50 S subunits from *E. coli* were found to associate with 30 S subunits from chloroplasts of *Euglena* and to synthesize polyphenylalanine [100].

From a variety of different eukaryotic tissues, active cytoplasmic ribosomal subunits have been prepared, and the ability to form hybrids between the 40 S of one and the 60 S subunit of another tissue has been tested in various cases. The results summarized in Table 17 indicate that eukaryote cytoplasmic ribosomal subunits are widely exchangeable between different tissues and, thus, that their basic construction seems to be quite similar. However, 60 S subunits of *Tetrahymena pyriformis* exhibit an absence of hybrid formation between very distant species because they are unable to reassociate with 40 S subunits of rat muscle [63] or a variety of other animal, plant, and yeast cells [101]. Although the 40 S subunits of *Tetrahymena* reassociate with all kinds of eukaryotic 60 S subunits tested so far and although the protozoan 60 S subunits are able to reassociate with homologous 40 S subunits, no formation of 80 S hybrids between protozoan 60 S and heterologous 40 S particles was observed by sucrose gradient or electron microscopic and functional analyses. These results are in agreement with the relatively strong evolutionary changes of the large ribosomal subunit in comparison to the relatively slight alterations of the small one (see Chapter IV).

5. Native ribosomal subunits

In the course of the so-called ribosome cycle within the cell, ribosomes dissociate after releasing the synthesized peptide chain into their subunits (see Chapter XI). Analyses of polysome profiles of various cell types, therefore, often reveal two rather small peaks in the areas corresponding to 60 S and 40 S subunits. When prepared at low ionic strength, these native subunits [102—123] contain a variety of extra proteins when compared to the "derived" ribosomal subunits. Analysis of these additional proteins of native 40 S subunits by various investigators showed the presence of at least 10 different proteins [104—111, 114—119, 121, 122] among which a considerable number if not all initiation factors were identified. Thus, the native small ribosomal subunits can be considered as a complex of 40 S subunits which, depending on the preparation procedure, contains a smaller or larger number of proteins attached among which initiation factors seem to play an important role.

6. References

[1] Martin, T. E., and I. G. Wool: Formation of active hybrids from subunits of muscle ribosomes from normal and diabetic rats. Proc. Natl. Acad. Sci. U.S. **60**, 569—574 (1968)

[2] Tissiéres, A., J. D. Watson, D. Schlessinger, and B. R. Hollingworth: Ribonucleoprotein particles from *Escherichia coli*. J. Mol. Biol. **1**, 221—233 (1959)

[3] Vignais, P. V., B. J. Stevens, J. Huet, and J. André: Mitoribosomes from *Candida utilis*. Morphological, physical, and chemical characterization of the monomer form and its subunits. J. Cell Biol. **54**, 468—492 (1972)

[4] Kleinow, W., and W. Neupert: The mitochondrial ribosome from *Locusta migratoria*: Dissociation into subunits. FEBS-Lett. **15**, 359—364 (1971)

[5] Kleinow, W., W. Neupert, and F. Miller: Electron microscope study of mitochondrial 60 S and cytoplasmic 80 S ribosomes from *Locusta migratoria*. J. Cell Biol. **62**, 860—875 (1974)

[6] Grivell, L. A., L. Reijnders, and P. Borst: Isolation of yeast mitochondrial ribosomes highly active in protein synthesis. Biochim. Biophys. Acta **247**, 91 to 103 (1971)

[7] Curgy, J.-J., G. Ledoigt, B. J. Stevens, and J. André: Mitochondrial and cytoplasmic ribosomes from *Tetrahymena pyriformis*. Correlative analysis by gel electrophoresis and electron microscopy. J. Cell Biol. **60**, 628—640 (1974)

[8] Hoober, J. K., and G. Blobel: Characterization of the chloroplastic and cytoplasmic ribosomes of *Chlamydomonas reinhardi*. J. Mol. Biol. **41**, 121—138 (1969)

[9] Mets, L., and L. Bogorad: Altered chloroplast ribosomal proteins associated with erythromycin-resistant mutants in two genetic systems of *Chlamydomonas reinhardi*. Proc. Natl. Acad. Sci. U.S. **69**, 3779—3783 (1972)

[10] Chua, N. H., G. Blobel, and P. Siekevitz: Isolation of cytoplasmic and chloroplast ribosomes and their dissociation into active subunits from *Chlamydomonas reinhardi*. J. Cell Biol. **57**, 798 to 814 (1973)

[11] Hanson, M. R., J. N. Davidson, L. Mets, and L. Bogorad: Characterization of chloroplast and cytoplasmic ribosomal proteins of *Chlamydomonas reinhardi* by two-dimensional gel electrophoresis. Molec. Gen. Genet. **132**, 105—118 (1974)

[12] Grivell, L. A., and G. S. P. Groot: Spinach chloroplast ribosomes active in protein synthesis. FEBS Lett. **25**, 21—24 (1972)

[13] Bielka, H., H. Welfle, M. Böttger, and W. Förster: Strukturveränderungen und Dissoziation von Leberribosomen in Abhängigkeit von der Mg^{++}-Konzentration. Europ. J. Biochem. **5**, 183—190 (1968)

[14] Reisner, A. H., J. Rowe, and H. M. MacIndoe: Structural studies on the ribosomes of *Paramecium*: Evidence for a primitive animal ribosome. J. Mol. Biol. **32**, 587—610 (1968)

[15] Bont, W. S., M. de Vries, J. Geels, and A. Huizinga: The effect of the Mg^{2+} con-

centration on the sedimentation coefficient of ribosomes from rat liver. Europ. J. Biochem. **19**, 211–217 (1971)

[16] Wallach, Z., P. E. Zeelon, and D. Gershon: Biochemical studies of the free nematode *Turbatrix aceti*. II. Isolation and partial characterization of polysomes, ribosomes and ribosomal subunits. Comp. Biochem. Physiol. **46 B**, 337–347 (1973)

[17] Ts'o, P. O. P., and J. Vinograd: Studies on ribosomes from reticulocytes. Biochim. Biophys. Acta **49**, 113–129 (1961)

[18] Hamilton, M. G., and M. L. Petermann: Ultracentrifugal studies on ribonucleoprotein from rat liver ribosomes. J. Biol. Chem. **234**, 1441–1446 (1959)

[19] Kuff, E. L., and R. F. Zeigel: Cytoplasmic ribonucleoprotein components of the Novikoff hepatoma. J. Biophys. Biochem. Cytol. **7**, 465–478 (1960)

[20] Lamfrom, H., and E. R. Glowacki: Controlled dissociation of rabbit reticulocyte ribosomes and its effect on hemoglobin synthesis. J. Mol. Biol. **5**, 97–108 (1962)

[21] Tashiro, Y., and P. Siekevitz: Ultracentrifugal studies on the dissociation of hepatic ribosomes. J. Mol. Biol. **11**, 149 to 165 (1965)

[22] Tashiro, Y., and T. Morimoto: Sedimentation studies on the interaction of ribosomal subunits from liver. Biochim. Biophys. Acta **123**, 523–533 (1966)

[23] Hamilton, M. G., and M. E. Ruth: The dissociation of rat liver ribosomes by ethylenediaminetetraacetic acid; molecular weights, chemical composition, and buoyant densities of the subunit. Biochem. **8**, 851–856 (1969)

[24] Petermann, M. L., A. Pavlovec, and M. G. Hamilton: Effects of agents that influence hydrogen bonding on the structure of rat liver ribosomes. Biochem. **11**, 3925–3933 (1972)

[25] Stahl, J., G. R. Lawford, B. Williams, and P. N. Campbell: A requirement for the presence of cell sap in the reversible dissociation of rat liver polysomes to ribosomal subunits. Biochem. J. **109**, 155 to 157 (1968)

[26] Stahl, J., H. Welfle, K. Dressler, and H. Bielka: Some properties of rat liver ribosomal subunits prepared by EDTA- and by puromycin-KCl-treatment. Acta Biol. Med. Germ. **28**, 451–457 (1972)

[27] Nolan, R. D., and H. R. V. Arnstein: The dissociation of rabbit reticulocyte ribosomes with EDTA and the location of mRNA. Europ. J. Biochem. **9**, 445–450 (1969)

[28] Blobel, G.: Isolation of a 5 S RNA-protein complex from mammalian ribosomes. Proc. Natl. Acad. Sci. U.S. **68**, 1881–1885 (1971)

[29] Terao, K., Y. Takahashi, and K. Ogata: Differences between the protein moieties of active subunits and EDTA-treated subunits of rat liver ribosomes with specific references to a 5 S rRNA · protein complex. Biochim. Biophys. Acta **402**, 230–237 (1975)

[30] Gravela, E.: The dissociation of Yoshida hepatoma ribosomes into active subparticles. Biochem. J. **121**, 145–150 (1971)

[31] Gander, E. S., B. Luppis, A. Stewart, and K. Scherrer: Dissociation and reassociation of globin-synthesizing polyribosomes from immature avian red cells. Europ. J. Biochem. **29**, 369–376 (1972)

[32] Brawerman, G.: Eukaryotic messenger RNA. Ann. Rev. Biochem. **43**, 621–642 (1974)

[33] Hamada, K., P. Yang, R. Heintz, and R. Schweet: Some properties of reticulocyte ribosomal subunits. Arch. Biochem. Biophys. **125**, 598–603 (1968)

[34] Bonanou, S., R. A. Cox, B. Higginson, and K. Kanagalingam: The production of biologically active subparticles from rabbit reticulocyte ribosomes. Biochem. J. **110**, 87–98 (1968)

[35] Schoentjes, M., and E. Fredericq: Propriétés physicochimiques des ribosomes de thymus. Arch. Int. Physiol. Biochim. **76**, 947–948 (1968)

[36] Godin, C., J. Kruh, and J. C. Dreyfus: Reconstituted reticulocyte ribosomes active in hemoglobin synthesis. Biochim. Biophys. Acta **182**, 175–179 (1969)

[37] Fais, D., R. S. Shakulov, and E. V. Klyachko: Dissociation of the monoribosomes from the animal cells. Molek. Biol. (russ.) **4**, 232–239 (1970)

[38] Clegg, J. C. S., and H. R. V. Arnstein: The controlled dissociation of proteins from rat liver ribosomes by potassium chloride. Europ. J. Biochem. **13**, 149–157 (1970)

[39] Reboud, A. M., J. P. Reboud, C. Witt-

mann, and M. Arpin: Action des sels monocationiques sur la dissociation des ribosomes de foie de rat. Propriétés des sous-unités. Biochim. Biophys. Acta **213**, 437–450 (1970)

[40] Martin, T. E., and L. H. Hartwell: Resistance of active yeast ribosomes to dissociation by KCl. J. Biol. Chem. **245**, 1504–1508 (1970)

[41] Cammarano, P., S. Pons, A. Romeo, M. Galdieri, and C. Gualerzi: Characterization of unfolded and compact ribosomal subunits from plants and their relationship to those of lower and higher animals: evidence for physicochemical heterogeneity among eukaryotic ribosomes. Biochim. Biophys. Acta **281**, 571–596 (1972)

[42] Reboud, A.-M., M. Arpin, and J.-P. Reboud: Ribosomal subunits from rat liver. 1. Isolation and properties of active 40 S and 60 S subunits. Europ. J. Biochem. **26**, 347–353 (1972)

[43] Reboud, A.-M., M. Buisson, and J.-P. Reboud: Ribosomal subunits from rat liver. 2. Effect of cations on isolated subunits. Europ. J. Biochem. **26**, 354–359 (1972)

[44] Beeley, J. A.: Dissociation of pancreatic ribosomes by p-hydroxymercuri [^{14}C] benzoate. Biochim. Biophys. Acta **259**, 112 to 116 (1972)

[45] Roskoski, R., jr.: Role of divalent cations on the association of rat liver ribosomal subunits. Arch. Biochem. Biophys. **130**, 561–566 (1969)

[46] Petermann, M. L., and A. Pavlovec: Dissociation of rat liver ribosomes to active subunits by urea. Biochem. **10**, 2770 to 2776 (1971)

[47] Petermann, M. L.: The dissociation of rat liver ribosomes to active subunits by urea. Meth. Enzymol. 20, 429–433 (1971)

[48] Hardy, S. J. S., and G. Turnock: Silver plated ribosomes. Nature New Biol. **232**, 152–153 (1971)

[49] Dickman, S. R., and E. Bruenger: Structure of canine pancreas polysomes. Effects of proteases on sedimentation behavior and incorporation of amino acids into polypeptides. Biochem. **4**, 2335 to 2339 (1965)

[50] Valez, R., N. L. Farrell, and M. L. Freedman: Selective proteolytic dissociation of rabbit reticulocyte single ribosomes not attached to messenger RNA. Biochim. Biophys. Acta **228**, 719 to 727 (1971)

[51] Hüvös, P., Ö. Gaal, L. Vereczkey, and F. Solymosy: Dissociation of rat liver polyribosomes upon treatment with diethylpyrocarbonate. Acta Biochim. Biophys. **6**, 317–325 (1971)

[52] Golub, A. L., and J. S. Clegg: Dissociation of oxidized 80 S ribosomes by sodium deoxycholate. Biochim. Biophys. Acta **182**, 121–134 (1969)

[53] Gould, H. J., H. R. V. Arnstein, and R. A. Cox: The dissociation of reticulocyte polysomes into subunits and the location of messenger RNA. J. Mol.Biol. **15**, 600–618 (1966)

[54] Zehavi-Willner, T.: The release of 5 S RNA from reticulocyte ribosomes. Biochem. Biophys. Res. Commun. **39**, 161–169 (1970)

[55] Hauge, J. G.: Pressure-induced dissociation of ribosomes during ultracentrifugation. FEBS Lett. **17**, 168–172 (1971)

[56] Baierlein, B., and A. A. Infante: Pressure induced dissociation of ribosomes. Meth. Enzymol. **30**, 328–349 (1974)

[57] Nieuwenhuysen, P., J. Clauwaert, and K. Heremans: Pressure-induced dissociation of ribosomes isolated from *Artemia salina* observed by light scattering. Arch. Int. Physiol. Biochim. **83**, 983 to 984 (1975)

[58] Zylber, E. A., and S. Penman: The disruption of polyribosomes by scnication. Biochim. Biophys. Acta **204**, 230–236 (1970)

[59] Norman, M. R., and S. Gamulin: The distinction between various functional states of hepatic ribosomes and different distributions of ribosomes between these states in fed and starved mice. Biochem. J. **129**, 38p (1972)

[60] Abakumova, O. J., D. A. Ivanov, T. J. Ugarova, V. G. Schelechova, and M. I. Lerman: Characteristics of liver ribosome fraction obtained from ethionine poisened rats. Isolation and properties of active ribosomal subunits from single ribosomes accumulated under the effect of ethionine. Biokhim. (russ.) **38**, 35–51 (1973)

[61] Lin, C. Y., ad J. L. Key: Dissociation

and reassembly of polyribosomes in relation to protein synthesis in the soybean root. J. Mol. Biol. **26**, 237—247 (1967)
[62] MARTIN, T. E., F. S. ROLLESTON, R. B. LOW, and I. G. WOOL: Dissociation and reassociation of skeletal muscle ribosomes. J. Mol. Biol. **43**, 135—149 (1969)
[63] MARTIN, T. E., and I. G. WOOL: Active hybrid 80 S particles formed from subunits of rat, rabbit and protozoan (*Tetrahymena pyriformis*) ribosomes. J. Mol. Biol. **143**, 151—161 (1969)
[64] MARTIN, T. E., I. G. WOOL, and J. J. CASTLES: Dissociation and reassociation of ribosomes from eukaryotic cells. Meth. Enzymol. **20**, 417—429 (1971)
[65] SHERTON, C. C., R. F. DICAMELLI, and I. G. WOOL: Separation of large quantities of eukaryotic ribosomal subunits by zonal ultracentrifugation. Meth. Enzymol. **30**, 354—367 (1974)
[66] BATTANER, E., and D. VAZQUEZ: Preparation of active 60 S and 40 S subunits from yeast ribosomes. Meth. Enzymol. **20**, 446—449 (1971)
[67] BONANOU, S. A., and H. R. V. ARNSTEIN: The reassociation of fractionated rabbit reticulocyte ribosomal subunits into particles active in polyphenylalanine synthesis. FEBS Lett. **3**, 348—350 (1969)
[68] FOSTER, L. B., and B. H. SELLS: Functional capacity of intact and hybrid ribosomes from livers of normal and hypophysectomized rats. Arch. Biochem. Biophys. **132**, 561—564 (1969)
[69] LAWFORD, G. R.: The effect of incubation with puromycin on the dissociation of rat liver ribosomes into active subunits. Biochem. Biophys. Res. Commun. **37**, 143—150 (1969)
[70] KEDES, L. H., and L. STAVY: Structural and functional identity of ribosomes from eggs and embryos of sea urchins. J. Mol. Biol. **43**, 337—340 (1969)
[71] RAO, P., and K. MOLDAVE: Interaction of polypeptide chain elongation factor with rat liver ribosomal subunits. J. Mol. Biol. **46**, 447—457 (1969)
[72] TERAO, K., and K. OGATA: Preparation and some properties of active subunits from rat liver ribosomes. Biochem. Biophys. Res. Commun. 38, 80—85 (1970)
[73] VON DER DECKEN, A., P. ASHBY, D. MCILREAVY, and P. N. CAMPBELL: The reversible dissociation and activity of ribosomal subunits of liver and skeletal musle from rats. Biochem. J. **120**, 815—918 (1970)
[74] JUDEWICZ, N. D., and I. D. ALGRANATI: Association and dissociation of *Neurospora crassa* cytoplasmatic ribosomes. J. Bacteriol. **133**, 418—421 (1978)
[75] BORGHETTI, A. F., P. COMI, R. FRANCHI, A. M. GIANNI, B. GIGLIONI, S. OTTOLENGHI, and G. G. GUIDOTTI: Ribosomal subunits from rat liver. Isolation and recombination into active ribosomal particles. Ital. J. Biochem. **19**, 397—416 (1971)
[76] WELFLE, H., J. STAHL, and H. BIELKA: Studies on proteins of animal ribosomes. XIII. Enumeration of ribosomal proteins of rat liver. FEBS Lett. **26**, 228—232 (1972)
[77] BUSIELLO, E., M. DIGIROLAMO, and L. FELICETTI: Role od mammalian ribosomal subunits and elongation factors in poly-U-directed protein synthesis. Biochim. Biophys. Acta **228**, 289—298 (1971)
[78] STURANI, E., F. A. M. ALBERGHINA, and F. CASACCI: *Neurospora crassa* ribosomes: Effects of bound peptidyl-tRNA on dissociation into subunits and on phenylalanine polymerization at low Mg^{2+} concentration. Biochim. Biophys. Acta **254**, 296—303 (1971)
[79] BLOBEL, G., and D. SABATINI: Dissociation of mammalian polyribosomes into subunits by puromycin. Proc. Natl. Acad. Sci. U.S. **68**, 390—394 (1971)
[80] BLOBEL, G.: Release, identification, and isolation of messenger RNA from mammalian ribosomes. Proc. Natl. Acad. Sci. U.S. **68**, 832—835 (1971)
[81] MECHLER, B., and B. MACH: Preparation and properties of ribosomal subunits from mouse plasmacytoma tumors. Europ. J. Biochem. **21**, 552—564 (1971)
[82] PRATT, H., and R. A. COX: Dissociation of ribosomes from oocytes of *Xenopus laevis* into active subparticles. Biochem. J. **124**, 897—903 (1971)
[83] VAN ETTEN, J. L.: Preparation of biologically active ribosomal subunits from fungal spores. J. Bacteriol. **106**, 704—706 (1971)
[84] APP, A. A., M. G. BULIS, and W. J. MCCARTHY: Dissociation of ribosomes and seed germination. Plant Physiol. **47**, 81 to 86 (1971)

[85] Barden, N., and A. Korner: A decreased aminoacyl-transfer-ribonucleic acid-binding capacity of 40 S ribosomal subunits resulting from hypophysectomy of the rat. Biochem. J. **127**, 411–417 (1972)

[86] Cammarano, P., A. Romeo, M. Gentile, A. Felsani, and C. Gualerzi: Size heterogeneity of the large ribosomal subunits and conservation of the small subunits in eucaryote evolution. Biochim. Biophys. Acta **281**, 597–624 (1972)

[87] Cammarano, P., A. Felsani, M. Gentile, C. Gualerzi, A. Romeo, and G. Wolf: Formation of active hybrid 80 S particles from subunits of pea seedlings and mammalian liver ribosomes. Biochim. Biophys. Acta **281**, 625–642 (1972)

[88] Faust, C. H., jr., and H. Matthaei: A systematic study of the isolation of murine plasma cell ribosomal subunits, their sedimentation properties and activity in polyphenylalanine synthesis. Biochem. **11**, 2682–2691 (1972)

[89] Van der Zeijst, B. A. M., A. J. Kool, and H. P. J. Bloemers: Isolation of active ribosomal subunits from yeast. Europ. J. Biochem. **30**, 15–25 (1972)

[90] Mönkemeyer, H., and E. Bermek: Preparation and properties of ribosomal subunits from human tissue (tonsils). Hoppe-Seyler's Z. Physiol. Chem. **354**, 949–956 (1973)

[91] Vanduffel, L., B. Peeters, and W. Rombauts: Isolation and characterization of human ribosomal subunits. Arch. Int. Physiol. Biochem. **81**, 392 (1973)

[92] Von der Decken, A.: Conditions affecting the dissociation reaction in vitro of rat liver ribosomes. Europ. J. Biochem. **33**, 475–480 (1973)

[93] Brown, G. E., A. J. Kolb, and W. M. Stanley, jr.: A general procedure for the preparation of highly active eukaryotic ribosomes and ribosomal subunits. Meth. Enzymol. **30**, 368–387 (1974)

[94] Rodrigues-Pousada, C., and D. H. Hayes: Ribosomal subunits from *Tetrahymena pyriformis*. Isolation and properties of active 40 S and 60 S subunits. Europ. J. Biochem. **89**, 407–415 (1978)

[95] Wettenhall, R. E. H., K. Nakaja, and I. G. Wool: The reassociation of ribosomal subunits from the muscle of normal and diabetic animals. Biochem. Biophys. Res. Commun. **59**, 230–236 (1974)

[96] Falvey, A. K., and T. Staehelin: Structure and function of mammalian ribosomes. I. Isolation and characterization of active liver ribosomal subunits. J. Mol. Biol. **53**, 1–19 (1970)

[97] Dickman, S. R., and E. Bruenger: Purification and properties of dog pancreas ribosomes. Biochem. **8**, 3295–3302 (1969)

[98] Dickman, S. R.: Purification and isolation of active mammalian ribosomal subunits. J. Cell. Physiol. **74**, 253–255 (1969)

[99] Dickman, S. R., and E. Bruenger: Purification and isolation of active mammalian ribosomal subunits. Fed. Proc. **28**, 726 (1969)

[100] Lee, S. G., and W. R. Evans: Hybrid ribosome formation from *E. coli* and chloroplast ribosome subunits. Science **173**, 241–242 (1971)

[101] Martin, T. E., J. N. Bicknell, and A. Kumar: Hybrid 80 S monomers formed from subunits of ribosomes from protozoa, fungi, plants and mammals. Biochem. Genet. **4**, 603–615 (1970)

[102] Perry, R. P., and D. E. Kelley: Messenger RNA-protein complexes and newly synthesized ribosomal subunits: Analysis of free particles and components of polyribosomes. J. Mol. Biol. **35**, 37–59 (1968)

[103] Henshaw, E. C., D. G. Guiney, and C. A. Hirsch: The ribosome cycle in mammalian protein synthesis. I. The place of monomeric ribosomes and ribosomal subunits in the cycle. J. Biol. Chem. **248**, 4367 to 4376 (1973)

[104] Hirsch, C. A., M. A. Cox, W. J. W. van Venrooij, and E. C. Henshaw: The ribosome cycle in mammalian protein synthesis. II. Association of the native smaller subunits with protein factors. J. Biol. Chem. **248**, 4377–4385 (1973)

[105] Ayuso-Parilla, M., E. C. Henshaw, and C. A. Hirsch: The ribosome cycle in mammalian protein synthesis. III. Evidence that the nonribosomal proteins bound to the native smaller subunit are initiation factors. J. Biol. Chem. **248**, 4386–4393 (1973)

[106] Ayuso-Parilla, M., C. A. Hirsch, and E. C. Henshaw: Release of the nonribo-

somal proteins from the mammalian native 40 S ribosomal subunit by aurintricarboxylic acid. J. Biol. Chem. **248**, 4394—4399 (1973)

[107] LUBSEN, N. H., and B. DAVIS: Use of purified polysomes from rabbit reticulocytes in a specific test for initiation factors. Proc. Natl. Acad. Sci. U.S. **71**, 68—72 (1974)

[108] SUNDKVIST, I. C., W. L. MCKEEHAN, M. H. SCHREIER, and T. STAEHELIN: Initiation factor activity associated with free 40 S subunits from rat liver and rabbit reticulocytes. J. Biol. Chem. **249**, 6512—6516 (1974)

[109] SUNDKVIST, I. C., and G. A. HOWARD: Comparison of activity and protein content of ribosomal subunits prepared by four different methods from rabbit reticulocytes. FEBS-Lett. **41**, 287—291 (1974)

[110] SUNDKVIST, I. C., and T. STAEHELIN: Structure and function of free 40 S ribosome subunits: Characterization of initiation factors. J. Mol. Biol. **99**, 401 to 418 (1975)

[111] PAIN, V. M., and E. C. HENSHAW: Initiation of protein synthesis in Ehrlich ascites tumour cells. Evidence for physiological variation in the association of methionyl-$tRNA_f$ with native 40 S ribosomal subunits in vivo. Europ. J. Biochem. **57**, 335—342 (1975)

[112] SAMESHIMA, M., and M. IZAWA: Properties and synthesis of multiple components in native small ribosomal subunits of mouse ascites tumor cells. Biochim. Biophys. Acta **378**, 405—414 (1975)

[113] SMITH, K. E., and E. C. HENSHAW: Binding of Met-$tRNA_f$ to native 40 S ribosomal subunits in Ehrlich ascites tumor cells. J. Biol. Chem. **256**, 6880—6884 (1975)

[114] FREIENSTEIN, C., and G. BLOBEL: Nonribosomal proteins associated with eukaryotic native small ribosomal subunits. Proc. Natl. Acad. Sci. U.S. **72**, 3392—3396 (1975)

[115] EMANUILOV, I., D. D. SABATINI, J. A. LAILE, and C. FREIENSTEIN: Localization of eukaryotic initiation factor 3 on native small ribosomal subunits. Proc. Natl. Acad. Sci. U.S. **75**, 1389—1393 (1978)

[116] SADNIK, I., F. HERRERA, J. MCCUISTON, H. A. THOMPSON, and K. MOLDAVE: Studies on native ribosomal subunits from rat liver. Evidence for activities associated with native 40 S subunits that affect interactions with acetyl-phe-tRNA, met-tRNA and 60 S subunits. Biochem. **14**, 5328—5335 (1975)

[117] HERRERA, F., I. SADNIK, G. GOUGH, and K. MOLDAVE: Studies on native small ribosomal subunits from rat liver. Purification and characterization of three eukaryote binding factors specific for initiator tRNA. Biochem. **16**, 4664—4671 (1977)

[118] THOMPSON, H. A., I. SADNIK, J. SCHEINBUKS, and K. MOLDAVE: Studies on native ribosomal subunits from rat liver. Purification and characterization of a ribosome dissociation factor. Biochem. **16**, 2221—2230 (1977)

[119] MOLDAVE, K., H. A. THOMPSON, and I. SADNIK: Preparation of a ribosome dissociation factor from native ribosomal subunits of rat liver. Meth. Enzymol. **60**, 290 to 297 (1979)

[120] MOLDAVE, K., and I. SADNIK: Preparation of derived and native ribosomal subunits from rat liver. Meth. Enzymol. **59**, 402—410 (1979)

[121] VAN VENROOIJ, W. J., A. P. M. JANSSEN, J. H. HOEYMAKERS, and B. M. DEMAN: On the heterogeneity of native ribosomal subunits in Ehrlich-ascites-tumor cells cultured in vitro. Europ. J. Biochem. **64**, 429—435 (1976)

[122] AMESZ, H., T. HAUBRICH, and H. O. VOORMA: Postribosomal complexes containing eukaryotic initiation factor eIF-2. Mol. Biol. Rep. **5**, 121—125 (1979)

[123] FRESNO, M., and D. VAZQUEZ: Initiation of protein synthesis in eukaryotic systems with native 40 S ribosomal subunits: Effects of translation inhibitors. Meth. Enzymol. **60**, 566—577 (1979)

IX. Interaction, topography and function of ribosomal components

The analysis of the ribosomal structure, especially of the interactions between the individual components and their spatial arrangement within the particles are important prerequisites for the understanding of the function of the whole ribosome during translation. Various methods, such as splitting off proteins, chemical substitution, enzymatic degradation, affinity labeling, chemical cross-linking, and immune electron microscopy have mainly been used to investigate these supramolecularly organized organelles. Furthermore, studies of complexes between ribosomes or their subunits with mRNA, tRNA, elongation and initiation factors allow conclusions as to the function of individual proteins and the RNA moieties in the translation process and their participation in the organization of functional sites in the ribosome.

Contents

1. Interactions of proteins and RNA

From experiments about the release of proteins from ribosomal particles by increasing concentrations of monovalent cations [1—18], by urea [5, 7, 8] or by formamide [24] it is obvious that the intact ribosome structure is maintained mainly by ionic forces and by hydrogen bonds.

Successive splitting off of individual proteins can be achieved by stepwise increasing concentrations of LiCl or KCl [4, 16]. Several groups of proteins were obtained by extraction of small and large ribosomal subunits by such procedures. The proteins identified by 2 D-gel electrophoresis are listed in Table 18 [16]; simidlar results were published also by other authors [14, 17] The amount of proteins released increases with decreasing Mg^{2+} concentration and is higher when Na^+ and Li^+ are used in comparison to Cs^+ and K^+ at the same concentrations [10, 11, 13, 16]. 2-D electrophoretic analysis of the split protein fraction from rat liver ribo-

Table 18. Classification of ribosomal proteins according to their splitting behaviour [16]

LiCl concentration	Proteins of 40 S subunits extracted	Proteins of 60 S subunits extracted
0.4 M	S 7, S 10, S 17, S 20, S 21	L 5, L 9, L 22, L 23/23a, L 38, L 39
0.6 M	S 3, S 3a, S 3b, S 4, S 30	L 11, L 30, L 31, L 36, L 36a
0.8 M	S 2, S 9, S 14, S 15, S 23, S 24, S 26, S 29	L 7, L 8, L 12, L 13/13a, L 14, L 15, L 17, L 19, L 21, L 24, L 26, L 28, L 29, L 32, L 33, L 35, L 37
1.0 M		L 16, L 18
Remaining core particle	S 5, S 6, S 8, S 11, S 13/16, S 15a, S 18, S 19, S 25, S 27, S 28	L 3, L 4, L 6, L 7a, L 18a, L 27/27a, L 34

somes obtained with increasing concenrrations of LiCl [14, 16] or KCl [16, 17] did not reveal any general significant relationships between the order of their removal from the particles and their electrophoretic behaviour, although the ratio of arginine, histidine and lysine to aspartic acid and glutamic acid in the extracted protein fractions increases with increasing LiCl concentrations demonstrating a higher basicity of the split proteins [7]. Thus the binding forces of the proteins in the particles are obviously different for the various proteins. Although some proteins are split off at well defined ion concentrations, others are observed in the split fractions over a wide range of LiCl [16] or KCl concentrations [17]. One explanation for the different splitting behaviour may be that the proteins of the first group are bound mainly by ionic interactions, which are loosened within a narrow range of LiCl or KCl concentration. Besides ionic interactions, other bonds (e.g. hydrophobic or hydrogen bonds) may contribute to some extent to the binding of proteins of the second group. If the ionic interactions are loosened by cations, the remaining nonionic interactions are obviously not strong enough to prevent partial splitting but are influenced with increasing ion concentrations indirectly by changes of the particle structure, thus causing a delayed release of these proteins in the following steps of the splitting procedure. The observations that urea supports the splitting of proteins [5] points to hydrogen bonds, responsible, along with ionic interactions, for the binding of the proteins within the ribosomal particles. Moreover, the different sensitivity of ribosomal proteins to removal by salts could result from a more or less internal localization of the proteins in the particle structure, thus reflecting in a first approximation their interaction with RNA and the order of their assembly during biogenesis of the ribosomal subunits. Core proteins resistant to extraction with 3 M LiCl were found specifically bound to the rRNA [15].

Differences in the strength of binding of individual proteins to 18 S RNA were also demonstrated by binding of proteins S 3b, S 4, S 6, S 11, S 17, S 18, and S 23/24 to 18 S RNA in the presence of 0.5 M KCl, whereas proteins S 2, S 3, S 7, S 13/16, S 14, S 15, S 15a, S 19, S 20, S 21, and S 26 did not bind under the same conditions [18]. Although the binding forces for the second group of proteins are not strong enough to maintain binary complexes with ribosomal RNA, it might be that some of them are in contact with RNA within the ribosomal subunit, because all proteins can be attached to 18 S RNA by UV irradiation of the 40 S subunit [19].

Treatment of 60 S ribosomal subunits of rat liver [21—23], rabbit reticulocytes [21, 25], *Xenopus laevis* oocytes [26], and human placenta [27] with EDTA releases a 7 S complex, which consists of 5 S RNA and protein L 5. This bimolecular complex is absolutely necessary for the biological function of the large subunit. Treatment of 60 S ribosomal subunits of rat liver with formamide also results in the removal of the 7 S ribonucleoprotein complex [24], identical with that obtained by EDTA treatment. The 7 S complex is active in ATP and GTP hydrolysis [22], and therefore it is suggested that this complex should be located at or near the P site of the large ribosomal subunit.

A second low molecular weight ribosomal RNA, 5.8 S RNA, is released from 60 S subunits of yeast ribosomes together with 5 S RNA by brief heat treatment or by 50% formamide [28]. This RNA, which interacts with the 26 S rRNA, can be reassociated to the large ribosomal subunit in a buffer containing 0.4 M K^+ and 6 mM Mg^{2+} [28].

Interactions of 5 S RNA as well as of 5.8 S RNA with ribosomal proteins were furthermore studied using immobilized RNAs. 5.8 S RNA bound via its oxidized 3′-end to Sepharose 6 B retains ribosomal proteins L 5, L 6, L 7, and L 18 out of the total mixture [29, 30] (see Table 19). Protein L 5, which was identified as a component of the 7 S complex, binds also to 5.8 S RNA, whereas it was not retained by immobilized 5 S RNA [29, 31]. An explanation for this discrepancy might be that the 3′-end of the Sepharose-bound 5 S RNA is substituted and thus not accessible for the reaction with proteins.

Using this technique is was found that also tRNA reacts with proteins of both ribosomal subunits [27] (Table 19). The binding of proteins S 14, L 35, and L 36 to tRNA was confirmed by an experiment in which it was shown that charged tRNAs are retained on millipore filters by a split protein fraction consisting of S 10, S 14, S 15, S 19, L 35, and L 36 [32].

The partial or total in vitro reconstitution of ribosomal subunits from core particles and split proteins or from rRNA and ribosomal proteins, which needs special ionic conditions, was one of the most efficient methods for the structural analysis of ribosomal particles and for the analysis of the functional role of proteins in ribosomal particles of *E. coli* [33—35]. Similar studies performed with subunits of rat liver ribosomes [1—3, 9, 12] have so far been less successful. Although it was possible in a few cases to obtain particles with chemical and physical properties comparable to those of ribosomal subunits, further experiments are needed to prove the functional activity of the reconstituted particles.

Interactions between ribosomal RNAs have been demonstrated for 5.8 S and 28 S rRNA as well as for 5 S and 18 S rRNA of wheat germ [36, 40], barley embryo [38,

Table 19. Binding of ribosomal proteins to immobilized rRNAs and tRNA

RNA type	Proteins bound to RNA	References
5 S rRNA	L 6, L 18/18a (L 7, L 8, L 35)	[29]
	L 6, L 19 (L 7, L 23′, L 27/27′, L 35′, L 39)	[31]
5.8 S rRNA	L 5, L 6, L 7, L 18	[29]
tRNA	S 6, S 14, S 23/24 (S 3b, S 8, S 9, S 11, S 18) L 6, L 13, L 19, L 21, L 26, L 32, L 35, L 36 (L 5, L 7, L 8, L 14, L 18, L 25, L 27, L 29, L 31)	[29]

39], Crotalus [37], and mouse sarcoma [38, 39]. 5.8 S RNA can be released from 28 S RNA by brief heat treatment or by treatment with dimethylsulfoxide or formamide [37]. The meaning of these interactions with regard to the functions of ribosomal RNAs (compare Section 4 of this chapter) is not yet clear and needs further investigations.

2. Topography of proteins and RNA within the ribosome

The topography of ribosomal components was investigated preferentially by four different methods: (1) Chemical substitution of amino acid side chains as well as (2) enzymatic degradation of proteins and RNA allow some conclusions concerning proteins and RNA on the surface of the particles. (3) Investigations using antibodies specific against individual ribosomal proteins, especially in combination with electron microscopy, the so-called immune electron microscopy, provides information about the localization of antigenic sites on the surface of the particles with respect to the three-dimensional arrangement of the particles. (4) Cross-linking can be applied to estimate protein neighbourhoods within the ribosome and to identify proteins in the vicinity of ribosomal RNAs.

a) Chemical substitution of proteins in ribosomal particles

Chemical substitution of amino acid side chains is strongly dependent on the three-dimensional arrangement of the macromolecules and the quaternary structure of the supramolecularly organized complexes. Studies on the chemical substitution of ribosomes and their subunits were performed with two basic aims, namely: estimation of the localization of proteins with respect to the particle surface and the contact area between the ribosomal subunits and, secondly, in order to identify protein(s) responsible for biochemical functions which are inhibited by the substitution reaction (see also Section 3 of this chapter).

Substitutions by procion brilliant blue [41], p-hydroxymercuribenzoate [42], iodoacetamide [44, 45], dansylchloride [43], 2-methoxy-5-nitrotropone [46, 47], glutaraldehyde [14], methyl acetimidate [48, 49], p-nitrophenylacetate [48] and by reductive methylation using formaldehyde and sodium borohydride [50—52] have shown that all proteins in the ribosomal subunits of rat liver are accessible to substitution, although to a different extent. These findings are in accordance with data on the iodination of proteins in ribosomal particles which also demonstrate that all [53] or most of the proteins [54, 55] are located at least partially on the surface of the subunits. According to the accessibilities of individual proteins within the ribosomal subunits for substitution by iodoacetamide, methyl acetimidate, p-nitrophenyl acetate and for reductive methylation proteins S 2, S 3, S 3b, S 7, S 10, L 5, L 11, L 19, L 26, L 28, L 29, L 30, L 31, L 36, L 36a, and L 38 were found to be exposed and proteins S 6, S 11, S 18, S 19, S 23, S 24, S 28, L 3, L 13, L 13a, L 15, L 18a, L 27, L 27a, and L 32 more or less buried.

When the substitution of proteins in the ribosomal subunits by iodoacetamide [44, 45], methyl acetimidate [48, 49], and 2-methoxy-5-nitrotropone [46, 47] are compared with the rate of substitution in 80 S ribosomes, it becomes obvious that a group of proteins of both subunits is less accessible in the 80 S ribosome. To this group belong proteins which are protected from substitution by their localization at the subunit contact area as well as other proteins being less reactive as a consequence of partial shielding by mRNA or peptidyl-tRNA. An additional reduction of substitution may be caused also by structural changes of the subunits during their reassociation to 80 S particles [45, 46, 58]. By comparing the effects of attachment of

60 S ribosomal subunits to 40 S subunits or to 40 S poly(U)-complexes, proteins with reduced accessibilities for several reagents have been estimated. These proteins may be located at the subunit contact area [45], a conclusion that is partly supported by iodination studies [58, 59]. However, the results are not yet conclusive and, therefore, the final determination of ribosomal components located at the subunit contact area needs further studies.

b) Enzymatic degradation of proteins and RNA in ribosomal particles

Enzymatic degradation of proteins and RNA, respectively, within ribosomal particles has been performed by using different proteases and ribonucleases and subsequent comparison of the digestion products with those obtained when the isolated components were degraded. It was shown by degradation with trypsin and chymotrypsin [60, 68] and analysis of the products by one- [60, 61] and two-dimensional polyacrylamide gel electrophoresis [62, 63] that proteins organized in subunit particles differ in their accessibility to proteolytic degradation, largely independently of their content of arginine and lysine residues. The different rates of protection of proteins in the subunit structures against proteolytic digestion may predominantly depend on their structure, the conformation of the whole ribosomal particles, the localization of the proteins in the ribosomal particles, and their interaction with rRNA and other proteins. When the structure of ribosomal subunits is loosened by EDTA, pyrophosphate or other reagents, more proteins are unmasked and thus become more easily cleavable by proteolytic enzymes (61, 63—67].

Analysis of ribosomal particles after partial tryptic digestion of their proteins revealed that the structural integrity of the small subunit of rat liver ribosomes is largely destroyed, whereas the sedimentation coefficient of the large subunit is nearly identical with that of control subunits even under conditions under which all proteins were more or less degraded [68].

In experiments with pancreatic ribonuclease [69—75], T 1 ribonuclease [73 to 77], and endogeneous nucleases [73, 78], it was shown that the RNAs within the ribosomal subunits are accessible to nuclease attack, even under ionic conditions that hold the ribosomal particles in a tight conformation. Therefore, it has been concluded that the polynucleotide chain of the RNA is located at least partially on the surface of the ribosomal particles. About 20—30% of the rRNA can be removed from ribosomes without altering the sedimentation coefficient of the particles significantly [69, 73, 76]. Loosening of ribosomal particles, e.g. by potassium deficiency [70, 71, 76] or EDTA treatment [73, 78], results in an increased sensitivity of the RNAs to nucleases. Analysis of the digestion products has revealed that the cleavable sites in the RNA within the ribosome are distributed irregularly along the RNA chain [73—75] and that the unprotected oligonucleotides located on the surface of the ribosomes are rich in pyrimidines [74, 75]. In addition, the finding that pancreatic ribonuclease, T 1 ribonuclease, and endogeneous ribonuclease split RNA in the ribosomes at similar sites [73] suggests that the specificity of the cleavage is determined mainly by the conformation of the polynucleotide chains within the particles and their interactions with proteins.

When 40 S ribosomal subunits from rat liver treated with nucleases are incubated subsequently with EDTA or urea [77], or high concentrations of LiCl [78], specific nucleoprotein particles consisting of rRNA fragments and a few proteins can be separated.

c) In vitro phosphorylation of proteins in ribosomal particles

Ribosomal proteins can be phosphorylated in vivo and in vitro [79—111] (see also Chapter VI, Section 2). In vitro phosphorylation of proteins has been described for ribosomes or ribosomal subunits of rat liver [80, 82, 83, 88, 92, 93, 104, 106, 109, 110], rabbit reticulocytes [79, 85, 87, 89, 90, 95, 98, 100, 102, 107], rabbit heart, liver, and muscle [79, 89], rat tumors [86, 93], bovine adrenal glands [81, 91], bovine corpus luteum [97], mouse mammary glands [84], mouse sarcoma [86], HeLa cells [104, 105], yeast [96, 99, 101, 103, 108], wheat [100], pea [94], and *Lemna minor* [94].

When the proteins extracted from ribosomes are used as substrates for the phosphate transfer from ATP to proteins catalyzed by ATP-dependent protein kinases, a larger number of proteins becomes phosphorylated, whereas on the intact ribosomal subunits fewer proteins are accessible to phosphorylation [83, 89, 90, 92, 93]. 6 to 12 ribosomal proteins of the large subunit and 1 to 5 of the small one are accessible to in vitro phosphorylation when organized in the particle structure. The most strongly labeled protein of the small ribosomal subunit is protein S 6 [83, 93, 98, 106]. Besides also proteins S 3a and S 4 [93], S 7, S 9, S 10, S 11, S 13, S 14, and S. 15 [83, 93, 106] were found phosphorylated, but to a lesser extent. When large ribosomal subunits are used, proteins L 3 [106], L 6, L 18, L 24, L 28, L 29, L 34 (to a smaller amount also L 27, L 35, L 36) and acidic proteins are phosphorylated [83, 93, 104—106].

When 80 S monomeres or polysomes were subjected to phosphorylation, only proteins S 6, L 6 and L 18 or L 18a [83, 87, 106, 110] and acidic proteins of the large ribosomal subunit [87, 102, 110] are mainly [83] or exclusively labeled [87, 102, 106, 110]. Corresponding results were obtained with yeast ribosomes [101, 103, 108]. A comparison of the phosphorylation patterns of proteins in polysomes and 80 S ribosomes with those of the 60 S and 40 S subunits shows that the phosphorylatable sites of proteins S 3a, S 4, S 7, S 9, S 10, S 11, S 13, S 14, and S 15 of the small and of proteins L 3, L 24, L 28, L 29, and L 34 of the large ribosomal subunit are influenced in their reactivities depending, for example, on their localization at the interphase, on their interaction with messenger-RNA and/or peptidyl-tRNA, and on conformational changes during dissociation and reassociation.

d) Cross-linking of proteins within ribosomal subunits

Direct cross-linking of protein sulfhydryl groups can be observed by incubation of rat liver ribosomes in thiol-free media at increased ionic strength, which results in the formation of a disulfide bridge between proteins L 6 and L 29 of the large ribosomal subunit [65, 170].

By using the cleavable bifunctional reagent dimethyl 3.8-diaza-4.7-dioxo-5.6-dihydroxydecanbisimidate [112], a large number of protein complexes is formed in the small ribosomal subunit of rat liver [113]. Up to now 4 protein complexes have been isolated electrophoretically, and their components were identified as S 2 to S 3, S 3—S 3a, S 15—S 15a, and S 5—S 25 after cleavage of the complexes by periodate [113].

Subjecting the quaternary initiation complex consisting of the small ribosomal subunit, initiation factor eIF-2, Met-tRNA$_f$ and GTP to the same reaction, proteins S 3, S 3a, S 15 (S 6, S 13/16) were found cross-linked to eIF-2 [114], thus confirming furthermore the neighbourhood of proteins S 3 and S 3a.

Protein S 3a could be bound, on the other hand, also to the oxidized 3′-end of 18 S RNA in small ribosomal subunits of polysomes [115]. In the same experiment the 28 S RNA could be attached to protein L 3, and 5.8 S RNA to two other proteins of the large ribosomal subunit indicating the close vicinity of these proteins and the 3′-ends of the corresponding ribosomal RNAs. By using UV-induced cross-linking between proteins and RNA it was demonstrated that all proteins of the small ribosomal subunit are in contact with the 18 S RNA [19] and that the proteins of the large ribosomal subunit, which were divided into a low-dose and a high-dose reactive group, can be bound to 28 S RNA [116]. Cross-linking between proteins of yeast ribosomes and ^{14}C-spermidine was also suggested as an indication for the neighbourhood between individual ribosomal proteins and ribosomal RNA [117].

e) *Localization of ribosomal proteins by immune electron microscopy*

Ribosomal subunits of rat liver exhibit readily recognizable asymmetric structures in the electron microscope (see

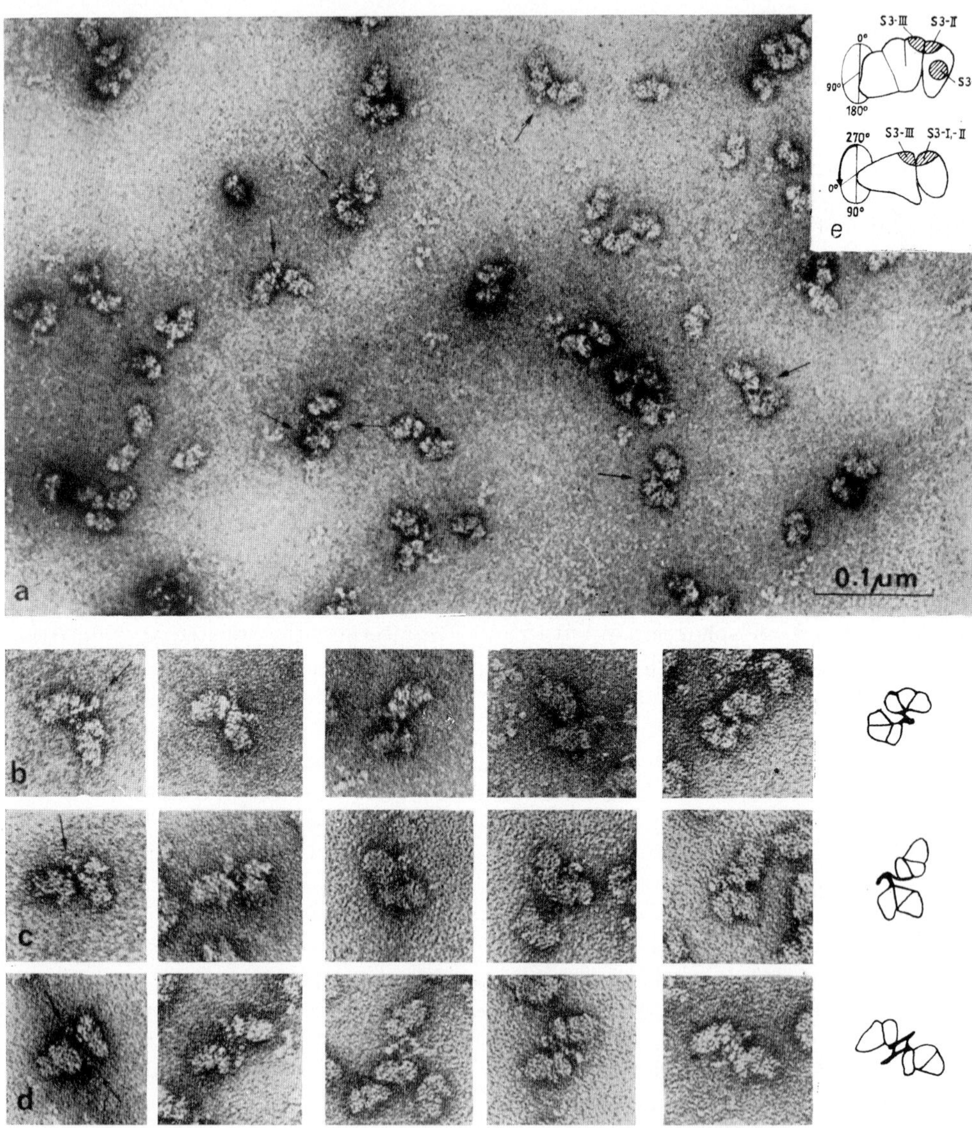

Fig. 24. Electron micrographs of negatively stained 40 S ribosomal subunits reacted with anti-S 3 antibodies from chicken.
a) General view (×200000); b)—d) selected images (×300000) and interpretative drawings; e) schematic representation of the observed binding sites of anti-S 3 antibodies. From G. Lutsch et al. [122].

Chapter V) that can serve as orientation marks for the determination of binding sites of monospecific antibodies against individual ribosomal proteins. The reaction of antibodies with ribosomal subunits results in complexes of two subunits joined by the bivalent IgG molecule (subunit-IgG-subunit) [118—122]. The Fab arms of the bivalent antibody molecules are bound to an antigenic determinant of

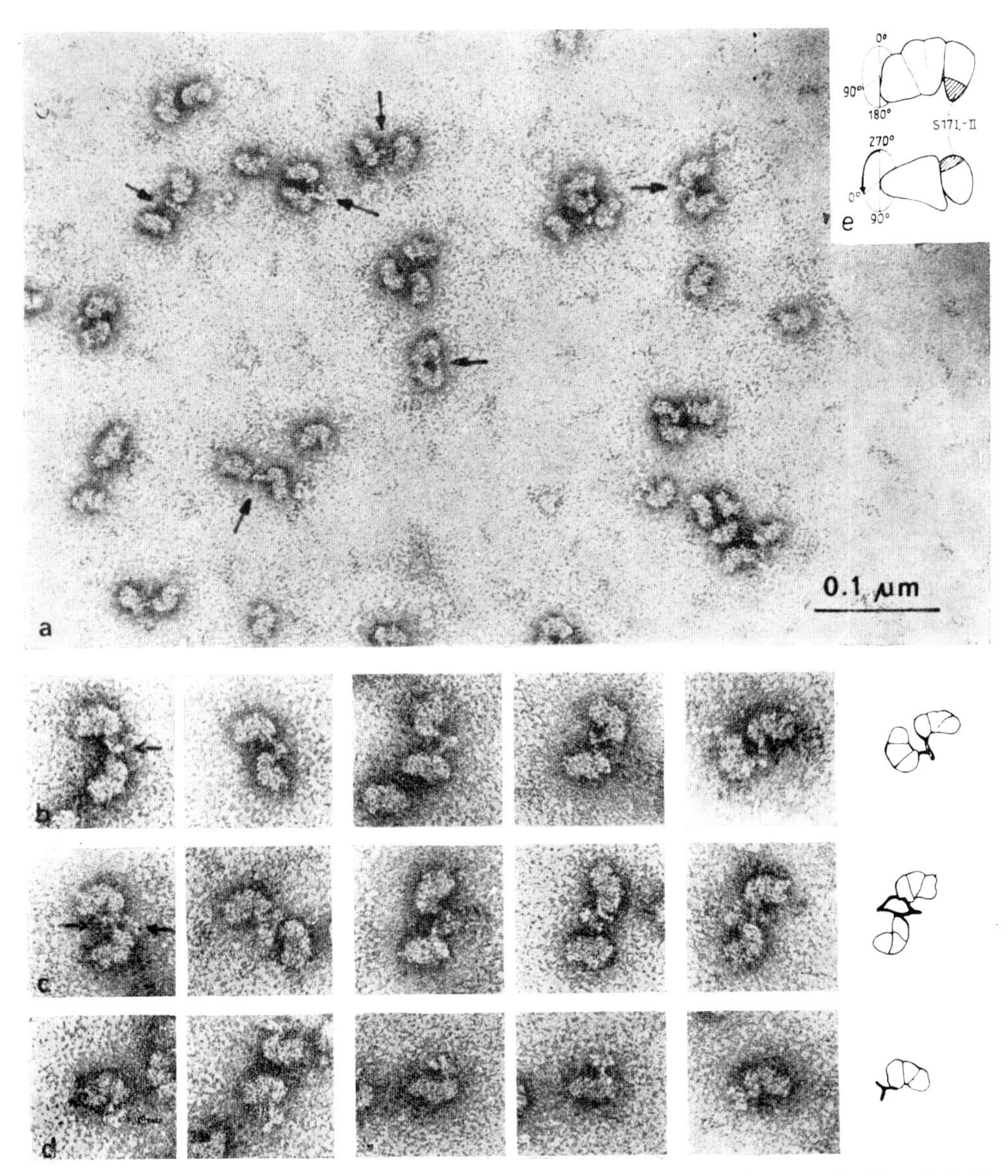

Fig. 25. Electron micrographs of negatively stained 40 S ribosomal subunits reacted with anti-S 17 antibodies from rabbits.

a) General view (× 200 000); b)—d) selected images (× 300 000) and the corresponding interpretative drawings; e) schematic representation of the observed binding sites of anti-S 17 antibodies.
From G. Lutsch et al. [122].

a given ribosomal protein exposed on the ribosomal surface. As far as studied, all proteins in the small subunit of rat liver ribosomes are accessible to reaction with their corresponding antibodies (unpublished results).

Up to now the location of 9 proteins, that means their antigenic determinants, has been determined by immune electron microscopy on the surface of the small ribosomal subunit of rat liver [121, 122]. Fig. 24 shows negatively stained 40 S—anti-S 3 complexes formed by antibodies against protein S 3 raised in chicken. Besides monomeric ribosomes numerous dimeric complexes linked by a "Y"-shaped antibody molecule can be observed at different binding sites. A gallery of selected images (Fig. 24) demonstrates the distribution of the three detected attachment sites of the Fab arms of the antibody molecules on the surface of 40 S ribosomal subunit. The antigenic determinants of protein S 3 are located mainly in the head

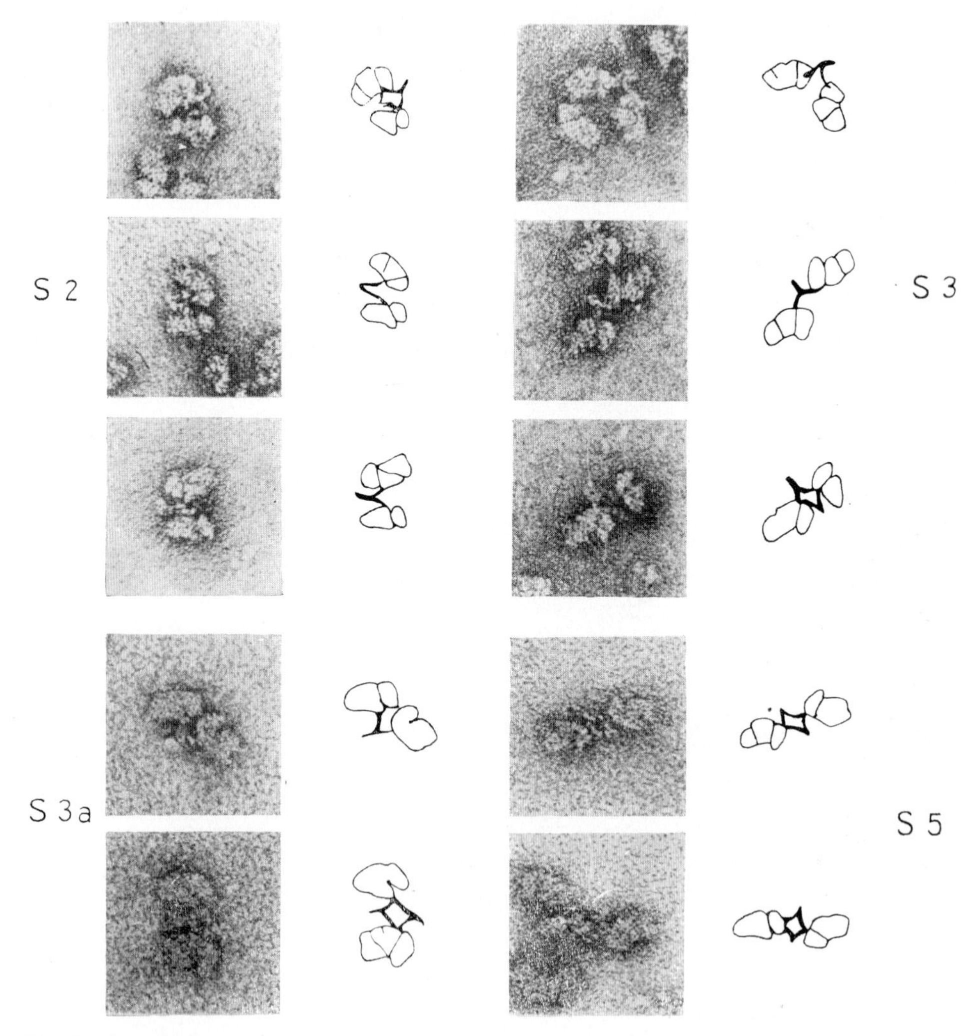

Fig. 26. Selected images and interpretative drawings of 40 S—IgG—40 S subunit complexes obtained with antibodies against proteins S 2, S 3, S 3a, S 5, S 6, S 7, S 9, S 17, and S 21 demonstrating the different antibody attachment sites.
From G. LUTSCH et al. [122] and unpublished results.

region of the small subunit. Two of them are very closely neighboured as seen from images of subunit dimers joined by two antibody molecules. Another example (Fig. 25) illustrates the location of antigenic determinants of protein S 17 using antibodies against this protein produced in rabbits. This figure shows that 40 S subunits are linked by one (Fig. 25b) or two antibody molecules (Fig. 25c) at the lower part of the head.

Sel ected original images and schematic drawings of 40 S subunits in complex with antibodies against ribosomal proteins S 2, S 3, S 3a, S 5, S 6, S 7, S 9, S 17, and S 21 are shown in Fig. 26. The location of the antigenic determinants of these 9 proteins on the surface of the small ribosomal subunit is demonstrated schematically by means of a two-dimensional model of the 40 S subunit in Fig. 27.

The topographical studies demonstrate that some proteins (S 5, S 17) are located in the head region of the small ribosomal

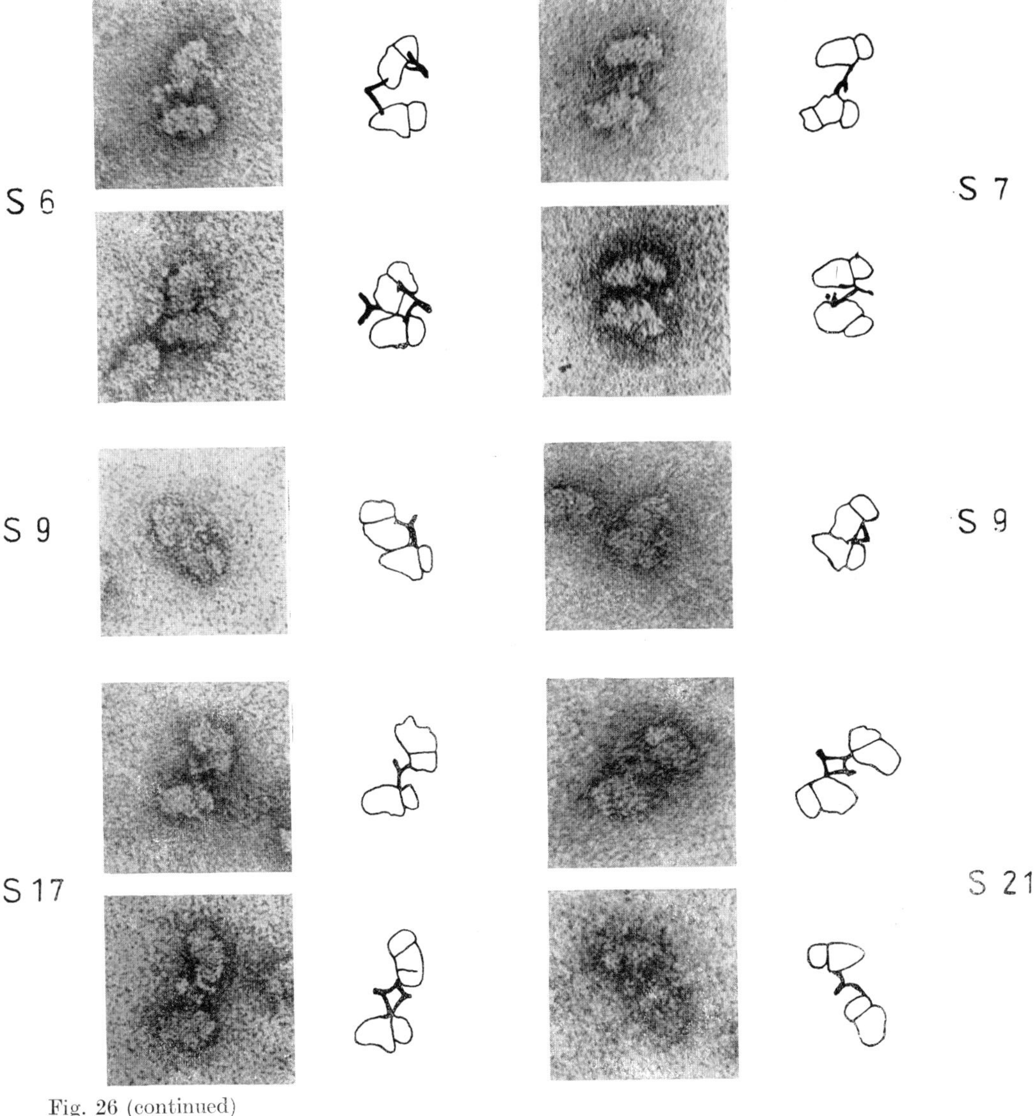

Fig. 26 (continued)

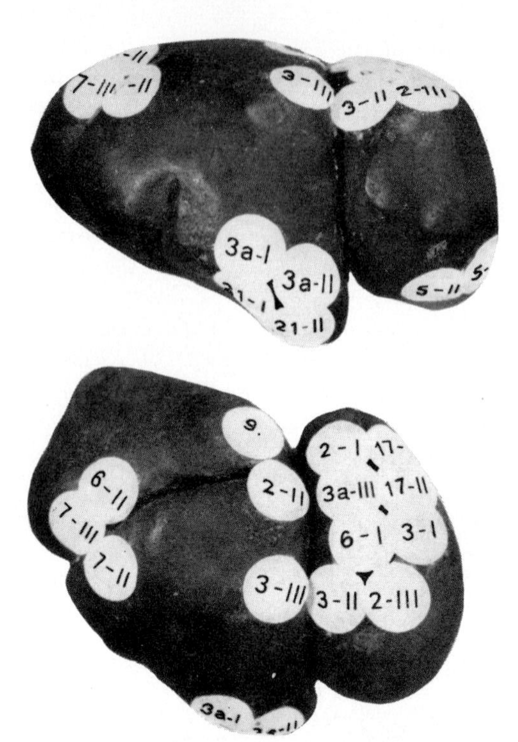

Fig. 27. Three-dimensional model of the small subunit of rat liver ribosomes with the locations of antibody binding sites for individual ribosomal proteins in two different views (see Figs. 6 and 9A). H. BIELKA, G. LUTSCH and F. NOLL, unpublished

subunit only, whereas others (S 7, S 9 and S 21) were found in the body part only. Proteins S 2, S 3, S 3a, and S 6 have antigenic determinants in both regions of the small ribosomal subunit.

All proteins studied so far have two (S 5, S 6, S 17, S 21) or three (S 2, S 3, S 3a, and S 7) antibody binding sites located on the subunit surface, except S 9, for which one antibody binding site could be found only (Fig. 27, Table 20). The fact that each protein is present in the small ribosomal subunit in one copy only [123], and that non-cross reacting antibodies were used [124, 125] allows the conclusion that the multiple antibody binding sites for a given protein really represent different antigenic determinants of the same protein molecule on the surface of the ribosomal particle.

Immune electron microscopy allows furthermore also conclusions as to the in situ shape of ribosomal proteins. Antigenic determinants of some proteins (S 2, S 3a, S 6, and S 7) are widely spaced apart, whereas those of others (S 5, S 17, and S 21) are in very close vicinity to each other within the ribosomal particle (Fig. 27, Table 20). Therefore, it can be assumed that some proteins have a preferentially elongated shape within the ribosomal particle, while others are of a more or less globular conformation. This conclusion is in good agreement with findings about the shape of isolated proteins in solution as estimated by hydrodynamic methods (sedimentation and diffusion measurements) [126]. By these methods it has been found that proteins S 2, S 3, S 6, and S 9 are rather extended molecules, whereas proteins S 17, S 19, and S 21 have a more ellipsoidal or globular shape.

From the data obtained by the hydrodynamic methods and by immune electron microscopy the conclusion can be drawn that no major differences seem to exist in the shape of ribosomal proteins in solution and when assembled in the ribosomal particle structure.

3. Functions of ribosomal proteins

Ribosomal functions can be inhibited by splitting off proteins (see Section 1 of this chapter), by chemical substitution (see Section 2a) and by specific antibodies against ribosomal proteins (see Section 2e). Therefore, these methods are suitable tools to analyze the functional significance of ribosomal constituents. Furthermore the accessibilities of proteins in ribosomes of different functional states to chemical substitutions as well as affinity labeling and

Table 20. Number of antigenic determinants and size of proteins of the small subunit of rat liver ribosomes as estimated by immune electron microscopy and ultracentrifuge measurements

Protein	Immune electron microscopy		Hydrodynamic measurements
	Number of antigenic determinants	Maximal distances [nm] (center to center) of antigenic determinants	Maximal lenght [nm] of proteins in solution
S 2	3	~18.0	19.4
S 3	3	~10.0	13.6
S 3a	3	~20.0	n.d.
S 5	2	directly neighboured	n.d.
S 6	2	~18.0	15.4
S 7	3	~20.0	11.7
S 9	1		12.8
S 17	2	directly neighboured	7.7
S 21	2	directly neighboured	8.0

cross-linking techniques can be successfully used to analyze proteins in active sites of ribosomal particles.

a) Functions of ribosomal split proteins

Treatment of rat liver ribosomes with 1.0 M ammonium chloride and 50% ethanol [127] or increasing concentrations of KCl [17] results in the removal of specific proteins, which causes loss of protein synthesis activity, an inhibition of Ac-Phe-tRNA and EF-T-dependent Phe-tRNA binding, as well as a reduced peptidyltransferase activity. A partial restoration of the so-called fragment reaction was obtained with a split protein fraction consisting of L 21, L 24, L 27, L 28, and L 29 [127]. From yeast ribosomes acidic proteins were extracted, which could be substituted by the bacterial ribosomal proteins L 7/L 12, demonstrating a similar function of bacterial and yeast acidic ribosomal proteins during protein synthesis [128]. The extraction of acidic proteins from ribosomes of *Artemia salina* [129, 130] or yeast [128] results in an inhibition of EF-G dependent GTP hydrolysis [129], polyphenylalanine synthesis [128—130] or EF-T dependent AA-tRNA binding [129—130] suggesting that acidic ribosomal proteins are involved in the binding of both elongation factors. The extraction of acidic proteins from rat liver ribosomes on the other hand, does not influence the peptidyltransferase activity of core particles [127]. These findings are in good agreement with the observation that the binding of both elongation factors to ribosomes from chicken liver is blocked by antibodies against proteins L 7/L 12 of *E. coli*, while the peptidyltransferase activity was not impaired [131].

b) Analysis of functions of ribosomal proteins by chemical substitution

Reaction of p-hydroxymercuribenzoate [42, 132, 134], N-ethylmaleimide (NEM) [134], showdomycin [135] or heavy metal ions [134] with 40 S ribosomal subunits of rat liver [42] or human tonsils [132] and with ribosomes from rabbit reticulocytes [134] or of human lymphatic tissue [135], respectively, inhibits poly(U)-dependent Phe-tRNA binding. The inhibition of this reaction at the lowest inhibiting reagent concentration in 40 S rat liver ribosomal subunits is connected with the substitution of proteins S 4 and S 10 [42]; one of them, S 10, was independently identified as an A site protein [144]. Pretreatment of ribosomes or 60 S subunits with NEM [133, 136—139], or iodoacetamide [139] reduces the ability of the particles to bind aminoacyl-tRNA [133—138] and GTP [137 to 139] and inhibits polyphenylalanine synthesis [133, 135—137, 139—140). Because NEM does not inactivate EF-T nor reacts with rRNA or impairs poly(U) binding [134], the inhibitions seem to be due to modifications of proteins located at or near the A site of the large ribosomal subunit [134, 137]. On the other hand, it was found that mammalian ribosomes pretreated with NEM or other reagents, specifically reacting with sulfhydryl groups, show an increase in the fragment reaction assay for peptide bond formation, i.e. a stimulation of the peptidyltransferase activity [140, 141].

The number of sulfhydryl groups available to reaction with NEM depends on the structural and functional states of the ribosomal particles [142]. Ribosomes carrying the peptidyl-tRNA in the A site bind less NEM than so-called P site ribosomes, reflecting conformational changes in the ribosome during translocation of the peptidyl-tRNA into the P site. Binding of poly(U) and AA-tRNA, on the other hand, results in an impaired reactivity to

Table 21. Tentative data on proteins and RNA involved in the organization of functional sites of rat liver ribosomes

Function	Small ribosomal subunit	Large ribosomal subunit	
	Protein	Protein	RNA
Poly(U) binding	S 9, S 15, S 18, S 23 [44, 144]; S 3/3a, S 26 (see Addendum)		
Elongation factor eEF-T binding	S 5, S 9, S 10 [44, 144]	L 18, L 24, L 38 [44, 144]	
Elongation factor eEF-G binding	S 10, S 14 [44, 144]	L 3, L 32 [44, 144]	
A site organization	S 10, S 23, S 24 [44, 144]	L 4, L 28, L 29, L 30 [44, 144]	5.8 S RNA (?) [22, 30]
P site organization	S 15, S 15a, S 19 [44, 144]	L 5 [21—23] L 8 [44, 144]	5 S RNA (?) [22, 164]
Peptidyltransferase center		L 10 [146] L 21, L 26 [47] L 21, L 24, L 27, L 28, L 36 [127] L 21, L 26, L 31 [17] L 28, L 29 [145] L 21, L 23, L 32/33 L 26 [147]	28 S RNA [146, 169]
Initiation factor eIF-2 binding	S 3, S 3a, S 6, S 13/16, S 15 [114, 127, 128]		
Initiation factor eIF-3 binding (studied with rabbit reticulocyte ribosomes)	S 2, S 3, S 3a, S 4, S 5, S 6, S 7, S 13, S 14 [149]		

NEM, whereas the GTP-dependent binding of EF-G increases the number of reactive sulfhydryl groups [142].

Differences in the accessibility of individual proteins in ribosomal particles (polysomes, 80 S monosomes, ribosomal subunits) in the presence or absence, respectively, of various components of the translation complex (mRNA, AA-tRNA, elongation factors) were detected in experiments with iodoacetamide [44, 45, 143, 144], methylacetimidate [44, 48, 49, 144], p-nitrophenylacetate [48], and 2-methoxy-5-nitrotropone [47]. The conclusions obtained by these experiments, regarding the involvement of proteins in the organization of functional sites or ribosomal subunits, are summarized in Table 21.

c) *Affinity labeling*

Affinity labeling has turned out to be a very suitable tool for the characterization of functional sites in biomacromolecules and higher ordered structures. In contrast to bacterial ribosomes there are up to now only relatively few affinity labeling studies with ribosomes from eukaryotes.

N-bromoacetylpuromycin, an analogue to the product of the puromycin reaction, N-peptidylpuromycin, reacts with protein(s) L 28/29 of the large subunit of rat liver ribosomes [145]. By using the photolabel N-(2-nitro-4-azidobenzoyl)puromycin, in addition to L 28/29, also L 4, L 6, L 10 and L 24 are labeled [171]. After poly(U) dependent binding of N-(2-nitro-4-azidobenzoyl)-Phe-tRNA to rat liver ribosomes exclusively protein L 10 was found labeled [146]. Thus, all these proteins seem to be located near the peptidyltransferase center of rat liver ribosomes. When applying p-nitrophenoxy-carbonyl (PNPC)-Phe-tRNA to rat liver ribosomes, proteins L 21, L 23, L 32/33 and L 36 were described to be specifically labeled [147]. With PNPC-Phe-tRNA and also with iodoacetyl-Phe-tRNA ribosomal proteins at the peptidyltransferase center of yeast ribosomes were detected [148].

d) *Interactions between the small ribosomal subunit and initiation factors*

The assembly of the translation initiation complex is characterized by stepwise interactions of the ribosomal subunits with initiation factors and other non-ribosomal components (see Chapter XI).

First attempts to analyze proteins and structures of the small ribosomal subunit involved in interactions with initiation factors eIF-2 and eIF-3 have been carried out by different methods.

From cross-linking experiments it has been suggested that proteins S 3, S 3a, S 6, S 13/16, and S 15 of the small subunit of rat liver ribosomes are located at or near the binding site for eIF-2 [114].

Furthermore, interactions of the ternary initiation complex (Met-$tRNA_f \times$ eIF-2 $\times$ GTP) with 40 S ribosomal subunits were studied by using monospecific antibodies against individual ribosomal proteins [119 151]. Because antibodies against protein S 3, S 6, and S 13 block very strongly the eIF-2 dependent binding of Met-$tRNA_f$ to 40 S subunits it has been concluded that at least these proteins, as also estimated by cross-linking, are involved in the interaction with components of the ternary initiation complex and thus should contribute to the formation of the P site on the small subunit of eukaryotic ribosomes [119, 121, 151].

Initiation factor eIF-3 can be cross-linked to proteins S 2, S 3, S 3a, S 4, S 5, S 6, S 7, S 13, and S 14 of the small subunit of rabbit reticulocyte ribosomes ([149] and personal communication by these authors). Thus, proteins S 3, S 3a and S 13 seem to be involved both in the organization of the eIF-2 as well as of the eIF-3 binding site.

The interaction of the small ribosomal subunit of *Artemia salina* with highly purified eukaryotic initiation factor eIF-3 has furthermore been studied by electron microscopy [152]. Electron micrographs of (40 S subunit $\times$ eIF-3) complexes show eIF-3 as a round particle (diameter of approx. 8 nm) attached to the 40 S subunit between its head and body region (see Chapter V). A visible fine structure of eIF-3 confirms the existence of several subunits. Similar findings were obtained by electron microscopic analysis of so-called native small ribosomal subunits (see Chapters V and VIII) from rabbit reticulocytes [153].

e) Ribosomal proteins involved in antibiotic resistance and lectin binding

Information about the functional role of proteins in ribosomes are furthermore available by cultivation of drug-resistant mutant cells and analysis of their ribosomal proteins (see also Chapter VI, Section 1k).

Chinese hamster cells, e.g. resistant against emetine, a potent inhibitor of eEF-G dependent translocation [154], possess a protein of the small ribosomal subunit with an altered electrophoretic mobility. This protein is therefore regarded as being part of the emetine binding site and thus being involved in the translocation reaction [155, 156].

In ribosomes of a cycloheximide-resistant fungus (*Podospora anserina*) [157] and of yeast mutant cells [*Schizosaccharomyces pompe*) [158] a protein of the large ribosomal subunit has been found to be modified.

The binding of ricin to 80 S ribosomes [159—162] or to a single site on the large ribosomal subunit of rat liver [161] inhibiting the interaction of ribosomes with elongation factors [159, 160] indicate the presence of a glycoprotein in the vicinity of the ribosomal A site [160]. The occurence of a glycoprotein in large ribosomal subunits of chicken liver and rabbit reticulocytes was confirmed by binding of a ribosomal protein with a molecular weight of 31000 to immobilized concanavalin A [163].

4. Functions of ribosomal ribonucleic acids

In comparison to bacterial ribosomes very little is known about the biological role of the RNA moieties in the function of eukaryotic ribosomes.

The eukaryote-specific 5 S ribosomal RNA (see Chapter VI, Section 3b) possesses a Py-G-A-U sequence which is highly conserved. This structure is complementary to the G-C-U-A sequence in loop IV of the eukaryotic initiator-tRNA [164]. Therefore, it has been concluded that in eukaryotes the 5 S RNA is involved in initiation of protein synthesis by interacting with Met-$tRNA_f$ and thus should be located at or near the P site of the large ribosomal subunit. In this context it should be recalled that the 5 S RNA is associated with protein L 5 of the large ribosomal subunit, that this complex exhibits ATP-ase activity [22] (see also Section 1 of this chapter), and that the initiation of protein synthesis in eukaryotes needs ATP, which is hydrolyzed during this process (see Chapter XI, Section 2). The proposed specific role of 5 S RNA in eukaryotes is supported by the fact that this RNA type is the transcript of a special gene of eukaryotic cells, not present in prokaryotes. Furthermore, the existence and special function of this RNA type underlines the conclusion that the initiation process in eukaryotes is more complex than in prokaryotes. Furthermore, from the finding that the 5 S RNA can interact with the 18 S RNA of the small ribosomal subunit (see also Section 1 of this chapter) is has been discussed that the 5 S RNA molecule might be involved in the association of the ribosomal subunits during the initiation process of translation [165].

There are no clear experimental data yet available on the function of the ribosomal 5.8 S RNA in eukaryotes. Like 5 S RNA of prokaryotes, 5.8 S RNA is a processed transcription product and has also some other properties in common with the prokaryotic 5 S RNA (see Chapter VI, Section 4b). Therefore, the 5.8 S RNA of eukaryotes seems to be a candidate for the interaction with AA-tRNA at the ribosomal A site in the elongation process, similar to prokaryotic 5 S rRNA [22, 30]. More experiments, however, are needed to verify this assumption.

As already mentioned (Chapter VI, Section 4a), the 18 S RNA of most eukaryotes has a GpApUpCpApUpUpA_{OH} sequence at its 3'-end. Because this sequence possesses trinucleotides (AUC, AUU) which are complementary to the terminator codons UAG and UAA in the mRNA, the possibility that the 3'-end of 18 S RNA may have a specific function in the termination of protein synthesis has been discussed [166, 167]. On the other hand, in various eukaryotic mRNAs nucleotide sequences within the leader parts have been analyzed that can theoretically base-pair with parts of the 3'terminal region of the 18 S rRNA. Therefore, as it is most likely the case with the 16 S rRNA in prokaryotes (see Chapter XI, Section 1), the 18 S rRNA might be involved also in the initiation process of translation [168]. Also from the attachment of the 3'-end of the 18 S RNA molecule to protein S 3a [115], which also can be cross-linked together with ribosomal protein S 3 to initiation factor eIF-2 [114] and eIF-3 [149], and from the finding that initiation factor eIF-2 binds to 18 S rRNA (see p. 280) it can be assumed that 18 S RNA contributes to initiation of protein synthesis.

Data on possible functions of the 28 S RNA molecule of the large ribosomal subunit are not yet available, except for the observation that remarkable parts of the affinity reagents, chlorambucilyl- [169], iodoacetyl- [169], and 2-nitro-4-azidobenzoyl-phenylanalyl-tRNA [146], bind to specific sites at the 28 S RNA molecule indicating that these oligonucleotide sequences are located at or near the peptidyl transferase center of the large ribosomal subunit.

5. References

[1] Lerman, M. I.: Dissociation of stru - tural proteins from rat liver ribosomes and in vitro reconstitution of biologically active ribosomes from derived protein-depleted particles and stripped proteins. Molek. Biol. (russ.) **2**, 209—221 (1968)

[2] Reboud, A.-M., M. G. Hamilton, and M. L. Petermann: The partial extraction of proteins from rat liver ribosomes, and the physical properties of the residual ribonucleoprotein particles. Biochem. **8**, 843—850 (1969)

[3] Grummt, F., and H. Bielka: Studies on the reconstitution of rat liver ribosomes from different types of core particles and split proteins. Acta Biol. Med. Germ. **23**, 937—941 (1969)

[4] Clegg, J. C. S., and H. R. V. Arnstein: The controlled dissociation of protein from rat liver ribosomes by potassium chloride. Europ. J. Biochem. **13**, 149 to 157 (1970)

[5] Grummt, F., and H. Bielka: Stepwise dissociation of rat liver ribosomes into core particles and split proteins. Biochim. Biophys. Acta **199**, 540—542 (1970)

[6] Bielka, H., F. Grummt, and I. Schneiders: Stepwise degradation of rat liver ribosomes by LiCl gradient centrifugation. Acta Biol. Med. Germ. **24**, 705—709 (1970)

[7] Westermann, P., and D. Koppitz: Über Wechselwirkungen von Proteinen mit RNS in Leberribosomen. Studia biophys. **24/25**, 419—428 (1970)

[8] Westermann, P., D. Koppitz, and H. Bielka: Studies on proteins of animal ribosomes. V. Splitting of rat liver ribosomes bound to DEAE-Sephadex by LiCl gradient elution. Acta Biol. Med. Germ. **26**, 611—616 (1971)

[9] Terao, K., and K. Ogata: Studies on the small subunit of rat liver ribosomes: Some biochemical properties with specific reference to the reconstruction of the small subunit. Biochim. Biophys. Acta **254**, 278—295 (1971)

[10] Reboud, A.-M., J.-P. Reboud, M. Arpin, and C. Wittmann: Extraction partielle des proteines de ribosomes de foie de rat par different cations monovalents. Proprietes des particules obtenues. Bio-

chim. Biophys. Acta **232**, 171–183 (1971)

[11] Arpin, M., A.-M. Reboud, and J.-P. Reboud: Conformational changes of large ribosomal subunit of rat liver induced by some monovalent cations. Biochim. Biophys. Acta **277**, 134–139 (1972)

[12] Reboud, A.-M., M. Buisson, M. J. Amoros, and J.-P. Reboud: Partial in vitro reconstruction of active 40 S ribosomal subunits from rat liver. Biochem. Biophys. Res. Commun. **46**, 2012–2018 (1972)

[13] Reboud, A.-M., M. Buisson, and J.-P. Reboud: Ribosomal subunits from rat liver. 2. Effects of cations on isolated subunits. Europ. J. Biochem. **26**, 354–359 (1972)

[14] Reboud, A.-M., M. Buisson, J.-J. Madjar, and J.-P. Reboud: Study of mammalian ribosomal protein reactivity in situ. II. Effect of glutaraldehyde and salts. Biochimie **57**, 295–302 (1975)

[15] Reboud, J.-P., A.-M. Reboud, J.-J. Madjar, and M. Buisson: Study of protein reactivity in rat liver ribosomes. Effect of RNA addition to autodigested ribosomes. Acta Biol. Med. Germ. **33**, 661 to 666 (1974)

[16] Welfle, H., B. Henkel, and H. Bielka: Ionic interactions in eukaryotic ribosomes: Splitting of the subunits of rat liver ribosomes by treatment with monovalent cations. Acta Biol. Med. Germ. **35**, 401 to 411 (1976)

[17] Reyes, R., D. Vazquez, and J. P. G. Ballesta: Acitivities of nucleoprotein particles derived from rat liver ribosomes. Biochim. Biophys. Acta **435**, 317–332 (1976)

[18] Westermann, P., and H. Bielka: Preparation and analysis of structural proteins of the small ribosomal subunit of rat liver. Acta Biol. Med. Germ. **33**, 531 to 537 (1974)

[19] Reboud, A.-M., M. Buisson, M.-J. Marion, and J.-P. Reboud: Photoinduced protein-RNA cross-linking in mammalian 40 S ribosomal subunits. Europ. J. Biochem. **90**, 421–426 (1978)

[20] Peeters, B., L. Vanduffel, and W. Rombauts: Isolation and characterization of a 5 S RNA-protein complex from human placental ribosomes. Arch. Int. Physiol. Biochem **82**, 193 (1974)

[21] Blobel, G.: Isolation of a 5 S RNA-protein complex from mammalian ribosomes. Proc. Natl. Acad. Sci. U.S. **68**, 1881–1885 (1971)

[22] Grummt, F., I. Grummt, and V. Erdmann: ATPase and GTPase activities isolated from rat liver ribosomes. Europ. J. Biochem. **43**, 343–348 (1974)

[23] Terao, K., Y. Takahashi, and K. Ogata: Differences between the protein moieties of active subunits and EDTA-treated subunits of rat liver ribosomes with specific references to a 5 S rRNA protein complex. Biochim. Biophys. Acta **402**, 230–237 (1975)

[24] Petermann, M. L., M. G. Hamilton, and A. Pavlovec: A 5 S RNA-protein complex extracted from rat liver ribosomes by formamide. Biochem. **11**, 2323–2326 (1972)

[25] Zehavi-Willner, T.: The release of 5 S RNA from reticulocyte ribosomes. Biochem. Biophys. Res. Commun. **39**, 161 to 169 (1970)

[26] Picard, B., and M. Wegnez: Isolation of 7 S particle from Xenopus laevis oocytes – 5 S RNA-protein complex. Proc. Natl. Acad. Sci. U. S. **76**, 241–245 (1969)

[27] Peeters, B., L. Vanduffel, and W. Rombauts: Isolation and characterization of a 5 S RNA-protein complex from human placental ribosomes. Mol. Biol. Rep. **1**, 349–354 (1974)

[28] Nazar, R. N.: The release and reassociation of 5.8 S rRNA with yeast ribosomes. J. Biol. Chem. **253**, 4505–4507 (1978)

[29] Metspalu, A., M. Saarma, R. Villems, M. Ustav, and A. Lind: Interaction of 5 S RNA, 5.8 S RNA and tRNA with rat liver ribosomal proteins. Europ. J. Biochem. **91**, 73–81 (1978)

[30] Lind, A., A. Metspalu, M. Saarma, I. Toots, M. Ustav, and R. Villems: Universal structure of peptidyl transferase center in ribosomes and the function of eukaryotic 5.8 S ribosomal ribonucleic acid. Bioorgan. Khim. **3**, 1138–1140 (1977)

[31] Ulbrich, N., and I. G. Wool: Identification by affinity chromatography of the eukaryotic ribosomal proteins that bind

to 5 S ribosomal ribonucleic acid. J. Biol. Chem. **253**, 9049—9052 (1978)

[32] Reyes, R., L. Carrasco, and D. Vazquez: Binding of aminoacyl tRNA to rat liver ribosomal proteins. Mol. Biol. Rep. **2**, 471—477 (1976)

[33] Nomura, M.: Assembly of bacterial ribosomes. In vitro reconstitution systems facilitate study of ribosome structure, function, and assembly. Science **179**, 864 to 873 (1973)

[34] Nomura, M., and W. A. Held: Reconstitution of ribosomes: Studies of ribosome structure, function and assembly. Cited in M. Nomura, A. Tissieres, and P. Lengyel (eds.) Ribosomes. Cold Spring Harbor Laboratory, p. 193—233 (1974)

[35] Nierhaus, K. H., and F. Dohme: Total reconstruction of functionally active 50 S ribosomal subunits from *Escherichia coli*. Proc. Natl. Acad. Sci. U.S. **71**, 4713—4717 (1974)

[36] Oakden, K. M., A. A. Azad, and B. G. Lane: Wheat embryo ribonucleates. VII. Rapid, efficient and selective formation of 5 S—18 S and 5.8 S—26 S hybrids in an aequeous solution of the four ribosomal polynucleotides, and the result of a search for the corresponding hybrids in wheat embryo ribosomes. Canad. J. Biochem. **55**, 99—109 (1977)

[37] Giorgini, J. F., and F. L. DeLucca: Studies on the low molecular weight RNA associated with 28 S RNA from *Crotalus durissus terrificus* liver. Nucl. Acids Res. **3**, 165—175 (1976)

[38] Azad, A. A.: Specific intermolecular complexing between high and low molecular weight ribosomal RNAs from mouse ascites sarcoma 180 cells and barley embryos. Proc. Austral. Biochem. Soc. **10**, 58 (1977)

[39] Azad, A. A.: Hybridization between 5 S RNA and 18 S RNA from barley embryos and mouse sarcoma 180 ascites cells. Biochem. Biophys. Res. Commun. **83**, 259—265 (1978)

[40] Nichols, J. L., and W. Wijesinghe: Identification of the 5 S RNA binding site in intermolecular complexes of wheat embryo ribosomal 5 S and 18 S RNA. Canad. J. Biochem. **56**, 760—764 (1978)

[41] Hultin, T.: The use of procion blue as a molecular probe in the study of ribosomal structure. Europ. J. Biochem. **9**, 579 to 584 (1969)

[42] Bielka, H., P. Westermann, H. Welfle, and J. Stahl: Isolation and properties of proteins of the small subunit of rat liver ribosomes. In "Ribosomes and RNA metabolism". Proc. Symp. Smolenice Castle, Slovac Acad. Sci., Bratislava 1973, p. 69 to 79

[43] Gross, B., P. Westermann, and H. Bielka: Studies on proteins of animal ribosomes. XVI. Reaction of proteins of ribosomal subunits and polysomes from rat liver with dansyl chloride. Acta Biol. Med. Germ. **31**, 803—806 (1973)

[44] Westermann, P., B. Gross and W. Heumann: Studies on proteins of animal ribosomes. XXII. Localization and function of ribosomal proteins studied by chemical substitution with iodoacetamide and methyl acetimidate. Acta Biol. Med. Germ. **33**, 699—707 (1974)

[45] Westermann, P., W. Heumann, and H. Bielka: Changes in the reactivity of proteins of rat liver ribosomes against [^{14}C]iodoacetamide depending on their organization in ribosomal subparticles, 80 S ribosomes and on the attachment of poly(U). Chem.-Biol. Interactions **10**, 429—439 (1975)

[46] Reboud, A.-M., J.-J. Madjar, M. Buisson, and J.-P. Reboud: Study of mammalian ribosomal protein reactivity in situ. I. Effects of 2-methoxy-5-nitrotropone on 40 S and 60 S subunits. Biochimie **57**, 285—293 (1975)

[47] Reyes, R., D. Vazquez, and J. P. G. Ballesta: Structure and function of rat liver ribosomes. Modification by 2-methoxy-5-nitrotropone treatment. Europ. J. Biochem. **67**, 267—274 (1976)

[48] Gross, B., and P. Westermann: Studies on proteins of animal ribosomes. XXIV. Localization of proteins in ribosomal subunits of rat liver studied by chemical substitution with p-nitrophenyl acetate and methyl acetimidate. Chem.-Biol. Interactions **15**, 309—317 (1976)

[49] Gross, B., and P. Westermann: Proteins of animal ribosomes. XXV. Proteins of rat liver ribosomes protected by polyuridylic acid from methyl acetimi-

date substitution. Acta Biol. Med. Germ. **35**, 1437–1442 (1976)

[50] REBOUD, A.-M., M. BUISSON, M. ARPIN, and J.-P. REBOUD: Reductive alkylation of mammalian ribosomes. Biochim. Biophys. Acta **474**, 578–587 (1977)

[51] REBOUD, A.-M., M. BUISSON, M.-J. MARION, and J.-P. REBOUD: Change of protein reactivity in mammalian ribosomal subunits as a function of temperature. Europ. J. Biochem. **81**, 141–149 (1977)

[52] KISILEVSKY, R., L. WEILER, and M. TRELOAR: An analysis of alteration in ribosomal conformation using reductive methylation. J. Biol. Chem. **253**, 7101–7108 (1978)

[53] WELFLE, J.: Studies on proteins of animal ribosomes. XX. Iodination of rat liver ribosomal proteins. Acta Biol. Med. Germ. **33**, 677–683 (1974)

[54] LEADER, D. P.: Enzymic iodination of ribosomal subunits from rat liver. Acta Biol. Med. Germ. **33**, 685–689 (1974)

[55] LEADER, D. P.: Enzymic iodination of eukaryotic ribosomal subunits. Characterization and analysis by two-dimensional gel electrophoresis. Biochem. J. **152**, 373–378 (1975)

[56] LEWIS, J. A., and D. D. SABATINI: Accessibility of proteins in rat liver-free and membrane-bound ribosomes to lactoperoxidase-catalyzed iodination. J. Biol. Chem. **252**, 5547–5555 (1977)

[57] BIELKA, H., F. NOLL, H. WELFLE, P. WESTERMANN, G. LUTSCH, J. STAHL, H. THEISE, U.-A. BOMMER, B. GROSS, M. GOERL, and B. HENKEL: Structural and functional organization of the eukaryotic ribosome. Studies on proteins of rat liver ribosomes. FEBS Symp. Vol. **51**, 387–399 (1979)

[58] TAS, P. W. L., and B. H. SELLS: Accessibility of ribosomal proteins to lactoperoxidase-catalyzed iodination following phosphorylation and during subunit interaction. Europ. J. Biochem. **92**, 271–278 (1978)

[59] LEWIS, J. A., and D. D. SABATINI: Protein of rat liver free and membrane bound ribosomes. Modification of two large subunit proteins by a factor detached from ribosomes at high ionic strength. Biochim. Biophys. Acta **478**, 331–349 (1977)

[60] ÖSTNER, U., and T. HULTIN: The use of proteolytic enzymes in the study of ribosomal structure. Biochim. Biophys. Acta **154**, 376–387 (1968)

[61] HULTIN, T., and U. ÖSTNER: Specific unmasking of ribosomal proteins under the influence of chelating agents and increased ionic strenght. Biochim. Biophys. Acta **160**, 229–238 (1968)

[62] HENKEL, B., H. WELFLE, and H. BIELKA: Untersuchungen über Proteine tierischer Ribosomen. XXI. Tryptischer Abbau ribosomaler Partikel aus Rattenleber. Acta Biol. Med. Germ. **33**, 691–698 (1974)

[63] ARPIN, M., J.-P. REBOUD, and A.-M. REBOUD: Study on mammalian ribosomal protein reactivity in situ. III. Effect of trypsin on 40 S and 60 S subunits. Biochimie **57**, 1177–1184 (1975)

[64] HULTIN, T.: Effects of aminoacridines and related compounds on the conformation of rat liver ribosomes. Chem.-Biol. Interactions **2**, 61–77 (1970)

[65] HULTIN, T.: Evidence for disulfide interaction in situ between two adjacent proteins in mammalian 60 S ribosomal subunits. Biochim. Biophys. Acta **269**, 118 to 129 (1972)

[66] SJÖQUIST, A., and T. HULTIN: Conformational effects of mercurials on rat liver ribosomes: A comparison between the unmasking of a shielded protein in the 60 S subunit by phenylmercurials and EDTA. Chem.-Biol. Interactions **6**, 131 to 148 (1973)

[67] SJÖQUIST, A., and T. HULTIN: Evolutionary aspects of a conformationally flexible region in the ribosomes of higher vertebrates. Comp. Biochem. Physiol. **52**, 277 to 292 (1975)

[68] HENKEL, B.: Dissertation, Akademie d. Wiss. d. DDR, Berlin, 1975

[69] COX, R. A.: The effect of pancreatic ribonuclease on rabbit reticulocyte ribosomes and its interpretation in terms of ribosome structure. Biochem. J. **114**, 753–767 (1969)

[70] NÄSLUND, P. H., and T. HULTIN: Effects of potassium deficiency on mammalian ribosomes. Biochim. Biophys. Acta **204**, 237–247 (1970)

[71] NÄSLUND, P. H., and T. HULTIN: Structural and functional defects in mammalian ribosomes after potassium deficiency.

Biochim. Biophys. Acta **254**, 104–116 (1971)

[72] Hüvös, P., Ö. Gaal, and L. Vereczkey: Effect of degradation of rat liver ribosomal RNA on the structure of the ribosomal subunits. Europ. J. Biochem. **26**, 518–527 (1972)

[73] Dessev, G. N., and K. Grancharov: Degradation of ribonucleic acid in rat liver ribosomes. J. Mol. Biol. **76**, 425–444 (1973)

[74] Lind, M., R. Villems, and A. Saarma: Pyrimidine rich surface of ribosomes. Acta Biol. Med. Germ. **33**, 715–724 (1974)

[75] Lind, A., R. Villems, and M. Saarma: Pyrimidine-rich oligonucleotides from rat-liver ribosome surface. Europ. J. Biochem. **51**, 529–536 (1975)

[76] Marion, M.-J., C. Marion, M. Arpin, A.-M. Reboud, and J.-P. Reboud: Effect of ribonuclease T 1 on ribosomal subunits of rat liver. Nucl. Acids Res. **3**, 2563 to 2574 (1976)

[77] Mendel, R. R.: Über die Isolierung und Charakterisierung von Ribonucleoprotein-Fragmenten aus Ribonuclease T 1-behandelten Untereinheiten von Rattenleberribosomen. Diplomarbeit, Humboldt-Universität Berlin, 1974

[78] Reboud, A.-M., M. Buisson, M.-J. Marion, and J.-P. Reboud: Specific ribonucleoprotein fragments from 40 S ribosomal subunits. Biochim. Biophys. Acta **432**, 176–184 (1976)

[79] Kabat, D.: Phosphorylation of ribosomal proteins in rabbit reticulocytes. Characterization and regulatory aspects. Biochem. **9**, 4160–4175 (1970)

[80] Loeb, J. E., and C. Blat: Phosphorylation of some rat liver ribosomal proteins and its activation by cyclic AMP. FEBS. Lett. **10**, 105–108 (1970)

[81] Walton, G. M., G. N. Gill, J. B. Abrass, and L.-D. Garren: Phosphorylation of ribosome-associated protein by an adenosine 3′:5′-cyclic monophosphate-dependent protein kinase: location of the ribosomal receptor and protein kinase. Proc. Natl. Acad. Sci. U.S. **68**, 880–884 (1971)

[82] Eil, C., and I. G. Wool: Phosphorylation of rat liver ribosomal subunits: Partial purification of two cyclic AMP activated protein kinases. Biochem. Biophys. Res. Commun. **43**, 1001–1009 (1971)

[83] Stahl, J., H. Welfle, and H. Bielka: Studies on proteins of animal ribosomes. XIV. Analysis of phosphorylated rat liver ribosomal proteins by two-dimensional polyacrylamide gel electrophoresis. FEBS Lett. **26**, 233–236 (1972)

[84] Majumder, G. C., and R. W. Turkington: Hormone-dependent phosphorylation of ribosomal and plasma membrane proteins in mouse mammary gland in vitro. J. Biol. Chem. **247**, 7207–7217 (1972)

[85] Kabat, D.: Turnover of phosphoryl groups in reticulocyte ribosomal phosphoproteins. J. Biol. Chem. **247**, 5338 to 5344 (1972)

[86] Bitte, L., and D. Kabat: Phosphorylation of ribosomal proteins in sarcoma 180 tumor cells. J. Biol. Chem. **247**, 5345 to 5350 (1972)

[87] Martini, O. H. W., and H. J. Gould: Phosphorylation of rabbit reticulocyte ribosomal proteins in vitro. Biochim. Biophys. Acta **295**, 621–629 (1973)

[88] Eil, C., and I. G. Wool: Phosphorylation of liver ribosomal proteins. Characteristics of the protein kinase reaction and studies of the structure of phosphorylated ribosomes. J. Biol. Chem. **248**, 5122 to 5129 (1973)

[89] Delauney, J., J. E. Loeb, M. Pierre, and G. Schapira: Mammalian ribosomal proteins: Studies on the in vitro phosphorylation patterns of ribosomal proteins from rabbit liver and reticulocytes. Biochim. Biophys. Acta **312**, 147–151 (1973)

[90] Traugh, J. A., M. Mumby, and R. R. Traut: Phosphorylation of ribosomal proteins by substracte specific protein kinase from rabbit reticulocytes. Proc. Natl. Acad. Sci. U.S. **70**, 373–376 (1973)

[91] Walton, G. M., and G. N. Gill: Adenosine 3′,5′-monophosphate and protein kinase dependent phosphorylation of ribosomal proteins. Biochem. **12**, 2604–2611 (1973)

[92] Böhm, H., and J. Stahl: Studies on proteins of animal ribosomes. XVII. Quantitative aspects of enzymatic phosphorylation of ribosomal proteins and particles

from rat liver. Acta Biol. Med. Germ. **32**, 449–461 (1974)

[93] Stahl, J., H. Böhm, and H. Bielka: Enzymatic phosphorylation of eukaryotic ribosomal proteins and factors of protein biosynthesis. Acta Biol. Med. Germ. **33**, 667–676 (1974)

[94] Keates, R. A. B., and A. J. Trevavas: Protein kinase activity associated with isolated ribosomes from pease and *Lemna*. Plant. Physiol. **54**, 95–99 (1974)

[95] Lightfoot, H. N., M. Mumby, and J. A. Traugh: Dephosphorylation of 40 S ribosomal subunits by phosphoprotein phosphatase. Biochem. Biophys. Res. Commun. **66**, 1141–1146 (1975)

[96] Grankowski, N., and E. Gasior: In vivo and in vitro phosphorylation of yeast ribosomal proteins. Acta Biochim. Polon. **22**, 45–56 (1975)

[97] Azhar, S., and K. J. M. Menon: Adenosine 3′,5′-monophosphate dependent phosphorylation of ribosomes and ribosomal subunits from bovine corpus luteum. Biochim. Biophys. Acta **392**, 64 to 74 (1975)

[98] Traugh, J. A., and G. G. Porter: A comparison of ribosomal proteins from rabbit reticulocytes phosphorylated in situ and in vitro. Biochem. **15**, 610–616 (1976)

[99] Becker-Ursic, D., and J. Davies: In vivo and in vitro phosphorylation of ribosomal proteins by protein kinase from *Saccharomyces cerevisiae*. Biochem. **15**, 2289–2296 (1976)

[100] Grancharova, T. V., T. A. Getova, and T. K. Nikolov: Phosphorylation of ribosomal proteins from eukaryotes in homologous and heterologous cell-free systems. Biochim. Biophys. Acta **418**, 397–403 (1976)

[101] Kudlicki, W., N. Grankowski, and E. Gasior: Ribosomal proteins as substrate for a GTP-dependent protein kinase from yeast. Mol. Biol. Rep. **3**, 121–129 (1976)

[102] Issinger, O. G.: Phosphorylation of acidic ribosomal proteins from rabbit reticulocytes by a ribosome-associated casein kinase. Biochim. Biophys. Acta **477**, 185–189 (1977)

[103] Herbert, J., M. Pierre, and J. E. Loeb: Phosphorylation in vivo and in vitro of ribosomal proteins from *Saccharomyces cerevisiae*. Europ. J. Biochem. **72**, 167 to 174 (1977)

[104] Horak, J., and D. Schiffmann: Acidic phosphoproteins of HeLa and rat 60 S ribosomal subunits. FEBS-Lett. **82**, 82 to 84 (1977)

[105] Horak, J., and D. Schiffmann: Acidic phosphoproteins of the 60 S ribosomal subunits of HeLa cells. Europ. J. Biochem. **79**, 375–380 (1977)

[106] Cenatiempo, Y., A. J. Cozzone, A. Genot, and J.-P. Reboud: In vitro phosphorylation of protein from free and membrane bound rat liver polysomes. FEBS-Lett. **79**, 165–169 (1977)

[107] DuVernay, V. H. jr., and J. A. Traugh: Two-step purification of the major phosphorylated protein in reticulocyte 40 S ribosomal subunits. Biochem. **17**, 2045 to 2049 (1978)

[108] Kudlicki, W., N. Grankowski, and E. Gasior: Isolation and properties of two protein kinases from yeast which phosphorylate casein and some ribosomal proteins. Europ. J. Biochem. **84**, 493–498 (1978)

[109] Genot, A., J.-P. Reboud, Y. Cenatiempo, and A. J. Cozzone: Endogeneous phosphorylation of ribosomal proteins from membrane-free rat liver polysomes. FEBS-Lett. **86**, 103–107 (1978)

[110] Genot, A., J.-P. Reboud, Y. Cenatiempo, and A. J. Cozzone: Differential phosphorylation of basic and acidic ribosomal proteins by protein kinases bound to membrane-free rat liver polysomes. FEBS-Lett. **99**, 261–264 (1979)

[111] Chihara-Nakashima, M., M. Hashimoto, and Y. Nishizuka: Intrinsic activity of guanosine 3′,5′-monophosphate-dependent protein kinase. II. Phosphorylation of ribosomal proteins. J. Biochem. **81**, 1863–1867 (1977)

[112] Coggins, J. R., E. A. Hooper, and R. N. Perham: Use of dimethyl suberimidate and novel periodate-cleavable bis (imidoesters) to study the quaternary structure of the pyruvate dehydrogenase multienzyme complex of *Escherichia coli*. Biochem. **15**, 2527–2533 (1976)

[113] Westermann, P., B. Gross, and H. Bielka: Neighbourhoods of proteins S 2–S 3, S 3–S 3a, S 15–S 15a, and

S 5—S 25 within the small subunit of rat liver ribosomes. Acta Biol. Med. Germ. **39**, 1147—1152 (1980)

[114] WESTERMANN, P., W. HEUMANN, U.-A. BOMMER, H. BIELKA, O. NYGARD, and T. HULTIN: Cross-linking of initiation factor eIF-2 to proteins of the small subunit of rat liver ribosomes. FEBS-Lett. **97**, 101—104 (1979)

[115] SVOBODA, A. J., and E. H. MCCONKEY: Cross-linking of proteins to ribosomal RNA in HeLa cell polysomes by sodium periodate. Biochem. Biophys. Res. Commun. **81**, 1145—1152 (1978)

[116] BUISSON, M., A.-M. REBOUD, M.-J. MARION, and J.-P. REBOUD: Photo-induced protein-RNA cross-linking in mammalian 60 S ribosomal subunits. Europ. J. Biochem. **97**, 335—344 (1979)

[117] REYES, R., D. VAZQUEZ, and J. P. G. BALLESTA: Structure of yeast ribosomes. Proteins associated with the rRNA. Biochim. Biophys. Acta **521**, 229—234 (1978)

[118] NOLL, F., K. WAHN, and H. BIELKA: Studies on the structures of animal ribosomes. IV. Electron microscopic projection of ribosome-antibody complexes. Acta Biol. Med. Germ. **31**, 807—812 (1973)

[119] NOLL, F., U.-A. BOMMER, G. LUTSCH, H. THEISE, and H. BIELKA: Localization of rat liver ribosomal protein S 2 and its involvement in initiation factor eIF-2 binding to the 40 S ribosomal subunit. FEBS-Lett. **87**, 129—131 (1978)

[120] LUTSCH, G., F. NOLL, H. THEISE, and H. BIELKA: Localization of ribosomal protein S 2 in rat liver ribosomes by immune electron microscopy. Acta Biol. Med. Germ. **36**, 287—290 (1977)

[121] LUTSCH, G., F. NOLL, H. THEISE, G. ENZMANN, and H. BIELKA: Localization of proteins S 1, S 2, S 16, and S 23 on the surface of small subunits of rat liver ribosomes by immune electron microscopy. Molec. Gen. Genetics **176**, 281—291 (1979)

[122] LUTSCH, G., unpublished results

[123] WESTERMANN, P., W. HEUMANN, and H. BIELKA: On the stoichiometry of proteins in the small ribosomal subunit of hepatoma ascites cells. FEBS-Lett. **62**, 132—135 (1976)

[124] NOLL, F., H. THEISE, and H. BIELKA: Studies on proteins of animal ribosomes. XVIII. Reaction of ribosomes and ribosomal proteins with antibodies against ribosomal proteins. Acta Biol. Med. Germ. **33**, 547—553 (1974)

[125] THEISE, H., F. NOLL, and H. BIELKA: Studies on proteins of animal ribosomes. XXVIII. Preparation and antigenic properties of 40 S subunit proteins of rat liver ribosomes. Acta Biol. Med. Germ. **37**, 1353—1362 (1978)

[126] BEHLKE, J., H. THEISE, F. NOLL, and H. BIELKA: Size and shape of isolated proteins of the small ribosomal subunit of rat liver. FEBS-Lett. **106**, 223—225 (1979)

[127] REYES, R., D. VAZQUEZ, and J. P. G. BALLESTA: Peptidyl transferase center of rat liver ribosome cores. Europ. J. Biochem. **73**, 25—31 (1977)

[128] RICHTER, D., and W. MÖLLER: Properties and function of ethanol potassium chloride extractable proteins from 60 S ribosomes and their interchangeability with the bacterial proteins L 7/L 12. Hoppe-Seylers Z. Physiol. Chem. **356**, 9—10 (1975)

[129] AGTHOVEN, A. J. van, J. A. MAASSEN, and W. MÖLLER: Structure and phosphorylation of an acidic protein from 60 S ribosomes and its involvement in elongation factor 2 dependent GTP hydrolysis. Biochem. Biophys. Res. Commun. **77**, 989—998 (1977)

[130] MÖLLER, W., L. I. SLOBIN, and K. AMONS: Isolation and characterization of two acidic proteins of 60 S ribosomes from *Artemia salina* cysts. Proc. Natl. Acad. Sci. U.S. **72**, 4744—4748 (1975)

[131] HOWARD, G. A.: Chicken liver ribosomes: Characterization of cross-reaction and inhibition of some functions by antibodies prepared against. *E. coli* ribosomal proteins L 7 and L 12. J. Mol. Biol. **106**, 623—637 (1976)

[132] TIRYAKI, D., U. ÜCER, and E. BERMEK: The effect of sulfhydryl reagents upon the activity of 40 S ribosomal subunits. Experientia **32**, 1270—1272 (1976)

[133] ÜCER, U., H. MÖNKEMEYER, and E. BERMEK: Characterization of N-ethylmaleimide-reactive proteins from human tonsillar ribosomes by two-dimensional polyacrylamide gel electrophoresis. Biochim. Biophys. Acta **402**, 206—213 (1975)

[134] McAllister, H. C., and R. S. Schweet: Involvement of sulfhydryl groups in the binding of tRNA to reticulocyte ribosomes. J. Mol. Biol. **34**, 519–525 (1968)

[135] Bermek, E., W. Krämer, H. Mönkemeyer, and H. Matthaei: Mechanism in protein synthesis. XII. Sites of action of showdomycin in a cell-free polyphenylalanine synthesizing system from human lymphatic tissue: ribosomes and elongation factor TF II. Biochem. Biophys. Res. Commun. **40**, 1311–1318 (1970)

[136] Bermek, E., H. Mönkemeyer, and R. Berg: The role of SH-groups of human ribosomal subunits in polypeptide synthesis. Biochem. Biophys. Res. Commun. **45**, 1294–1299 (1971)

[137] Cheng, T. C., and H. C. McAllister: A protein fraction of the 60 S subunit from reticulocyte ribosomes associated with acceptor-site binding activity. J. Mol. Biol. **78**, 123–134 (1973)

[138] Mönkemeyer, H., and E. Bermek: Inactivation of human ribosomes by N-ethylmaleimide.
Hoppe-Seylers Z. Physiol. Chem. **355**, 26 to 32 (1974)

[139] Terao, K., and K. Ogata: The different effects of N-ethylmaleimide and iodoacetamide on the activity of rat liver 60 S subunits for peptide bond elongation. J. Biochem. **84**, 1119–1123 (1978)

[140] Stockmar, R., D. Ringer, and H. C. McAllister: Stimulation by sulfhydryl-reactive reagents of peptidyl transferase and donor site binding activities in a reticulocyte system. FEBS-Lett. **23**, 51–55 (1972)

[141] Carrasco, L., and D. Vazquez: The involvement of sulfhydryl groups in the peptidyl transferase centre of eukaryotic ribosomes. Europ. J. Biochem. **50**, 317 to 323 (1975)

[142] Steinert, P. M., B. S. Baliga, and H. N. Munro: Available sulphydryl groups of mammalian ribosomes in different functional states. J. Mol. Biol. **88**, 895–911 (1974)

[143] Cheng, T.-C., and H. C. McAllister: Poly(U) and transfer RNA binding sites on reticulocytes ribosomes. Biochim. Biophys. Acta **319**, 401–409 (1973)

[144] Westermann, P.: Ribosomale Proteine aus Eukaryonten. Beiträge zur Isolierung und ihren Eigenschaften sowie zur Topographie und funktionellen Bedeutung im Ribosom. Promotion B, Akademie der Wissenschaften der DDR, Berlin, 1977

[145] Stahl, J., K. Dressler, and H. Bielka: Studies on proteins of animal ribosomes. XVII. Affinity labeling of rat liver ribosomes by N-bromoacetylpuromycin. FEBS-Lett. **47**, 167–170 (1974)

[146] Stahl, J., H. Böhm, V. A. Pozdnjakov, and A. S. Girshovich: Photoaffinity labeling of rat liver ribosomes by phenylalanine-tRNA N-acylated by 2-nitro-4-azidobenzoic acid (NAB). FEBS-Lett. **102**, 273–276 (1979)

[147] Czernilofsky, A. P., E. Collatz, A. M. Gressner, I. G. Wool, and E. Küchler: Identification of the tRNA-binding sites on rat liver ribosomes by affinity labeling. Molec. Gen. Genet. **153**, 231–235 (1977)

[148] Perez-Gonsalbez, M., D. Vazquez, and J. P. G. Ballesta: Affinity labeling of yeast ribosomal peptidyl transferase. Molec. Gen. Genetics **163**, 29–34 (1978)

[149] Tolan, D. R. S., J. W. B. Hershey, and R. R. Traut: Cross-linking of eukaryotic initiation factor eIF-3 to the 40 S ribosomal subunit of rabbit reticulocytes. Abstr. Internat. Congress Biochem. 1979, Toronto, p. 129

[150] Peterson, D. T., W. C. Merrick, and B. Safer: Binding and release of radiolabeled eukaryotic initiation factors 2 and 3 during 80 S initiation complex formation. J. Biol. Chem. **254**, 2509–2516 (1979)

[151] Bommer, U.-A., F. Noll, G. Lutsch, and H. Bielka: Immunochemical detection of proteins in the small subunit of rat liver ribosomes involved in binding of the ternary initiation complex. FEBS-Lett. **111**, 171–174 (1980)

[152] Boublik, M., H. Trachsel, and T. Staehelin: Biomolecular interaction of ribosomes by electron microscopy. Steenbock Symposium on Ribosomes. Madison (Wis., USA), Poster-Abstr. A 5 (1979)

[153] Emanuilov, I., D. D. Sabatini, J. A. Lake, and Ch. Freienstein: Localisation of eukaryotic initiation factor 3 on native small ribosomal subunits. Proc. Natl. Acad. Sci. U.S. **75**, 1389–1393 (1978)

[154] Barbacid, M., M. Fresno, and D. Vazquez: Inhibitors of polypeptide elonga-

tion on yeast polysomes. J. Antibiotics **28**, 453–462 (1975)

[155] BOERSMA, D., S. M. MCGILL, J. W. MOLLENKAMP, and D. J. ROUFA: Emetine resistance in Chinese hamster cells is linked genetically with an altered 40 S ribosomal subunit protein S 20. Proc. Natl. Acad. Sci. U.S. **76**, 413–419 (1979)

[156] REICHENBERGER, V. E. jr., and C. T. CASKEY: Emetine-resistant Chinese hamster cells. The identification of an electrophoretically altered protein of the 40 S ribosomal subunit. J. Biol. Chem. **254**, 6207–6210 (1979)

[157] BERGUERET, J., M. PERROT, and M. CROUZET: Ribosomal proteins in the fungus *Podospora anserina*: Evidence for an electrophoretically altered 60 S protein in a cycloheximide resistant mutant. Molec. Gen. Genetics **156**, 141–144 (1977)

[158] CODDINGTON, A., and R. FLURI: Characterization of the ribosomal proteins from *Schizosaccharomyces pombe* by two-dimensional polyacrylamide gel electrophoresis. Molec. Gen. Genetics **158**, 93–100 (1977)

[159] CARRASCO, L., C. FERNANDEZ-PUENTES, and D. VAZQUEZ: Effects of ricin on the ribosomal sites involved in the interaction of elongation factors. Europ. J. Biochem. **54**, 499–503 (1975)

[160] FERNANDEZ-PUENTES, C., L. CARRASCO, and D. VAZQUEZ: Site of action of ricin on the ribosome. Biochem. **15**, 4364 to 4369 (1976)

[161] HEDBLOM, M. L., D. B. COWLEY, and L. L. HOUSTON: The binding of ricin and its polypeptide chains to rat liver ribosomes and ribosomal subunits. Fed. Proc. **35**, 1356 (1976)

[162] COWLEY, D. B., M. L. HEDBLOM, and L. L. HOUSTON: Protection and rescue of ribosomes from the action of ricin A chain. Biochem. **18**, 2648–2654 (1979)

[163] HOWARD, G. A., and H. P. SCHWEBLI: Eukaryote ribosomes posses a binding site for concanavalin A. Proc. Natl. Acad. Sci. U.S. **74**, 818–821 (1977)

[164] ERDMANN, V. A.: Structure and function of 5 S and 5.8 S RNA. Progr. Nucleic Acid Res. Mol. Biol. **18**, 45–90 (1976)

[165] AZAD, A. A., and B. G. LANE: A possible role of 5 S rRNA in the reversible association of ribosomal subunits. Proc. Austral. Biochem. Soc. **9**, 41 (1976)

[166] SHINE, J., and L. DALGARNO: Identical 3′-terminal octanucleotide sequence in 18 S ribosomal ribonucleic acid from different eukaryotes. A proposed role for this sequence in the recognition of terminator codons. Biochem. J. **141**, 609 to 615 (1974)

[167] DALGARNO, L., and J. SHINE: Conserved terminal sequence in 18 S rRNA may represent terminator codons. Nature New Biol. **245**, 261–262 (1973)

[168] AZAD, A. A.: Possible involvement of the m_2^6Am$_2^6$A hairpin structure at the 3′ end of 18 S RNA in the initiation of protein synthesis — Hypothesis. Proc. Austral. Biochem. Soc. **12**, 78 (1979)

[169] VLASSOV, V. V., and P. WESTERMANN: Studies on the modification of ribosomes from rat liver with alkylating derivatives of tRNA. Mol. Biol. (russ.) **10**, 670–674 (1976)

[170] NIKA, H., and T. HULTIN: Disulfide interaction in situ between two neighboring proteins in mammalian 60 S ribosomal subunits. Isolation of the contact region of the larger protein. Biochim. Biophys. Acta **579**, 10–19 (1979)

[171] BÖHM, H., J. STAHL, and H. BIELKA: Photoaffinity labeling of rat liver ribosomes by N-(2-Nitro-4-azidobenzoyl)puromycin. Acta Biol. Med. Germ. **38**, 1447 to 1452 (1979)

X. Interactions of ribosomes with membranes

Contents

1. Activity of free and membrane-bound polysomes

Biochemical and electron microscopic studies have clearly shown that ribosomes involved in protein synthesis are present in the cell either free in the cytoplasm (free ribosomes) or attached to membranes mainly of the endoplasmic reticulum (bound ribosomes). As already pointed out in Chapter III, the relative proportion of free and membrane-bound ribosomes and polysomes, respectively, depends on the type of tissues and their physiological states.

While in the morphological approach this proportion is estimated by counting of the particles in thin tissue sections, in the biochemical approaches free and bound ribosomal particles are separated from tissue homogenates. For the isolation of free and bound ribosomes mainly the following techniques are used: Differential centrifugation, centrifugation in continous or discontinous sucrose gradients, and flotation. Differences in the estimations of the amount of free and membrane-bound ribosomes and their properties, found by various authors, may be due at least partially to different techniques used for their preparation (for critical valuations see [3, 4]).

It is now commonly accepted as a rule that in terms of their biological function, different types of proteins are synthesized on free and membrane-bound polysomes. On polysomes attached to the membranes of the endoplasmic reticulum preferentially two classes of proteins are formed, namely proteins destined for the export (secretory proteins) or for transport to other cellular compartments and, secondly, at least some of the membrane proteins. Thus, association of polysomes with membranes results in segregation or compartmentalization of newly synthesized proteins. On free polysomes, on the other hand, mainly so-called soluble tissue proteins are made (for further details see reviews [1—4]).

The biological activity of free and bound polysomes has been studied repeatedly by various authors with variable results. Some failed to find significant differences in the amino acid incorporation activity of free and bound polysomes from rat liver [5—7], rat spleen [8] or kidney [9]. In agreement with these results is the finding that Met-$tRNA_f$ binding and the peptidyl-transferase function do not differ in ribosomes of free and bound polysomes of the rat liver [10]. On the other hand, free polysomes from rat liver were found more efficient in amino acid incorporation [11, 12], while other authors [13—18] reported that free polysomes are less active than bound polysomes. These conflicting results are difficult to interpret at present.

Differences in the functional properties of free and membrane-bound polysomes were described also with regard to various requirements for amino acid incorporation as, e.g., microsomal wash fractions [19, 20], poly(U) [19], KCl [19, 21], divalent cations [22] or glutathione [23].

Furthermore, protein synthetic activities of bound polysomes are more sensitive to cycloheximide [17, 24], fusidic acid, aurin tricarboxylic acid and emetine [24] than those of free polysomes.

The possibility that components of the membranes of the endoplasmic reticulum may influence the activity of polysomes attached to them is suggested by the observation that the Arrhenius activation energy for polypeptide synthesis on free polysomes is 25.5 kcal $\times$ mol^{-1}, while for polysomes of the microsomal fraction (bound polysomes) a value of only 16 kcal $\times$ mol^{-1} has been determined [25].

Another suggestion for differences in the activity of free and membrane-bound polysomes is the finding that separate, non-interchangeable AA-tRNA pools may exist for protein synthesis on both classes of polysomes [26].

The numerous facts that free and membrane-bound polysomes show various significant differences raise the question as to the mechanism by which ribosomes are selected for translation of mRNA molecules on membranes while others are active as free polysomes in the cytoplasm. In this context the question whether the ribosomal subunits of free and bound polysomes originate from the same or from different pools is of interest.

2. Dynamics of ribosomal subunits between free and membrane-bound polysomes

a) *RNA labeling experiments*

The question as to whether the ribosomal particles of free and bound polysomes, respectively, originate from the same or from different pools has been repeatedly studied by in vivo incorporation of precursors into the RNA moiety of ribosomes of both classes of polysomes [13, 27—35]. Despite many problems in interpreting the results of such experiments, and thus some contradictory conclusions, most authors have found that the rates of labeling of the structural RNA of ribosomal particles of both classes of polysomes are very similar [27, 28, 29, 31, 33, 35]. So, the majority of the experimental results suggest that the ribosomes of free and membrane-bound polysomes are in dynamic equilibrium and that they originate from the same precursor pool of free native ribosomal subunits.

Turnover studies revealed that the half-life of the RNA of free and membrane-bound ribosomes is the same [36, 37]; for rat liver values of close to 5 [36] and 6 days [37], respectively, have been determined.

b) *Reassociation experiments*

Another technique to study the exchangeability of ribosomal subunits between free and bound polysomes are experiments on the in vitro attachment of ribosomal particles of both classes of polysomes to membranes of the endoplasmic reticulum [23, 32,

38—49]. Most authors found no differences in the binding to the membranes, using either polysomes [23, 41, 42], ribosomes [47] or ribosomal subunits [39, 40, 46, 49] derived from the free or the bound fraction. From studies with ribosomal subunits it follows that stripped membranes of the endoplasmic reticulum bind mainly 60 S ribosomal subunits and that the 40 S subunit is bound efficiently only in the presence of the large ribosomal subunit [32, 39, 49]. Upon release of nascent polypeptide chains from ribosomes in vitro it could be shown that small ribosomal subunits of the membrane-bound fraction could exchange with small subunits derived from free polysomes while a comparable exchange could not be demonstrated for the large subunits [40]. The small subunit exchange requires a macromolecular fraction of the cell sap and ATP or GTP [40]. For further details see also Section 4 b of this chapter.

3. Chemical properties of free and membrane-bound ribosomes

The question whether structural differences exist between free and membrane-bound ribosomal particles which could be responsible for selecting particular ribosomes for their attachment to membranes has been studied by various groups. In these experiments, mainly some properties of the RNA moieties and the number of proteins have been determined.

a) *Ribosomal RNA*

No significant differences in the base compositions of the high molecular weight rRNA species of rat liver [29] and rat spleen [8] and of the 5 S RNA of reticulocytes [50], respectively, of free and bound ribosomes were found. Also finger print analysis of fragments of the rRNA of HeLa cells obtained by pancreatic or T_1 ribonuclease did not reveal major differences between free and membrane-bound ribosomes [51].

b) *Ribosomal proteins*

The results of the analysis of the proteins from free and membrane-bound ribosomes are somewhat contradictory and difficult to interpret, and the possibility that some of the discrepant findings arose artefactually by the preparation procedures cannot definetely be ruled out.

While some authors could not detect major qualitative differences in the electrophoretic pattern of the proteins of free and bound ribosomes of rat liver [40, 52—54] or mouse liver [55], other reports suggest that some differences might exist [56—62]. A limited number of specific proteins were found in the large subunit of free ribosomes only, not present in bound polysomes of rabbit reticulocytes [58, 59], HeLa cells [60], and rat liver [61, 62]. On the other hand, some additional proteins were found by means of two-dimensional polyacrylamide gel electrophoresis only in membrane-bound polysomes of reticulocytes [58, 59], HeLa cells [60], and rat liver [61], which belong to the large ribosomal subunit [59—61] or also to the small one [59]. It has been suggested [60] that especially those proteins which are absent or additionally present, respectively, in the large ribosomal subunit are involved in regulating the binding of the 60 S ribosomal particles to the membranes of the endoplasmic reticulum.

The involvement of ribosomal proteins in membrane binding of ribosomal particles was demonstrated, at least in the sense that they are in contact with the membranes, by the finding that proteins in free polysomes are more accessible to in vivo [63] or in vitro phosphorylation [64] as well as to iodination [65] than those of membrane-bound polysomes.

4. Components and mechanisms involved in binding of ribosomes to membranes

a) *The role of ribosomal subunits*

Electron microscopic studies [66—69] (see Fig. 28) as well as biochemical approaches [32, 39, 49, 67, 70] have clearly shown that ribosomes are attached to the membranes by their large subunit. In reconstitution experiments it could be demonstrated [32, 39, 49, 70] that large ribosomal subunits bind more efficiently to the membranes and that the small subunit is bound significantly only in the presence of the 60 S subunit [32, 39]. In accordance with these findings are results of so-called stripping experiments. Treatment of the rough microsomal fraction with EDTA at low concentrations releases only small ribosomal subunits; at higher concentrations of the chelating agent, increasing amounts also of the large subunit are detached from the membranes (for details see Section 4c of this chapter).

The apparent dissociation constant for binding of large ribosomal subunits of rat liver to membranes of the rough endoplasmic reticulum was estimated to be in the order of 2×10^{-8} "M" [71]. Since the intracellular concentration of ribosomes is higher than 10^{-6} "M", it was concluded from these studies that the number of membrane-bound ribosomes is determined by the number of corresponding binding sites on the membrane. For ribosomal monomers of rat liver a binding constant of close to $8 \times 10^{7} \cdot \text{mol}^{-1}$ for their attachment to stripped rough membranes of the endoplasmic reticulum and $\sim 5 \times 10^{-8}$ mol binding sites per gram of membrane protein have been estimated [70].

From differences in the protein pattern of the large subunit of bound and free ribosomes [58—61] (see also Section 3b of this chapter) and the reduced accessibility of proteins in bound ribosomes in comparison to free against phosphorylation [63, 64] or iodination [65] it is suggested that at least a few specific proteins of the large ribosomal subunit may be directly involved in its interaction with membranes.

b) *The role of membrane components*

The specific role of various constituents of membranes for the organization of ribosomal binding sites results from different experimental studies.

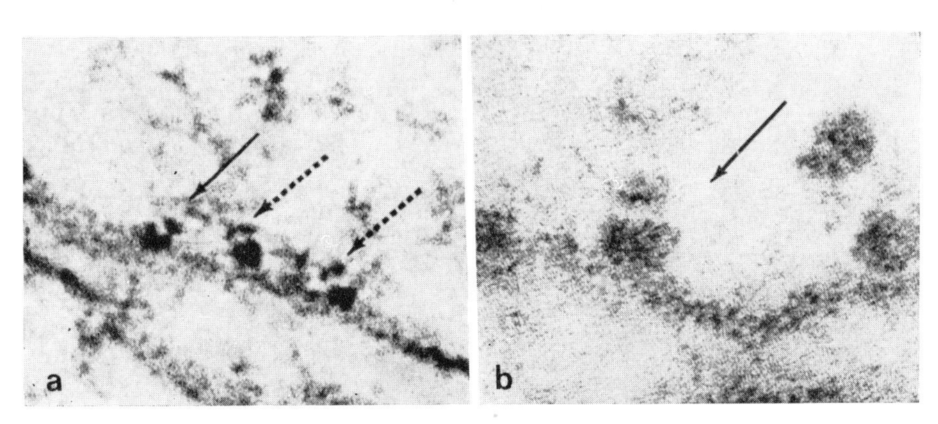

Fig. 28. Selected electron micrographs of thin sections of mouse liver cells, showing ribosomes attached to membranes of the endoplasmic reticulum by their large subunit (arrows). a) $\times$ 200000; dotted arrows: bipartite substructure in small subunits; b) $\times$ 410000.

From N. T. Florendo [68]. With permission of The Rockefeller University Press (New York)

In reconstitution studies mainly with rat liver systems it has been demonstrated that smooth membranes of the endoplasmic reticulum bind ribosomal particles with the same affinity as rough membranes [71] but accept less ribosomes than stripped rough membranes [41, 42, 70—75]. Furthermore, the in vitro binding of polysomes to smooth membranes of the endoplasmic reticulum of rat liver was found to be more sensitive to ionic strength than the binding to stripped rough membranes [42]. Erythrocyte ghosts were found to be incapable of binding ribosomes specifically [70], whereas reticulocyte cell membranes accept free ribosomes and polysomes [76]. In the latter case it was interestingly noted that the binding capacity declines with increasing cell maturation.

The capacity of the membranes of the endoplasmic reticulum to bind ribosomal particles is destroyed by trypsin [70, 74, 77—80], chymotrypsin [70, 79], papain [79], and by neuraminidase [81], but not by carboxypeptidase [79] or phospholipase C [70, 82]. The release of ribosomal particles from the rough endoplasmic reticulum by treatment with proteolytic enzymes and the inability of membranes treated with protein destroying enzymes to rebind ribosomes, respectively, point to the involvement of proteins in the formation of membrane "ribosome receptors". From the results of the experiments with neuraminidase and phospholipase C, respectively, it has been concluded that glycoproteins might be involved in the organization of membrane binding sites for ribosomes and that the ionic head groups of the membrane phospholipids are not solely responsible for the attachment of ribosomes to the membranes.

The in vitro binding of ribosomal particles to membranes was found to be enhanced by spermine [83].

The suggestion that specific proteins of the membranes are involved in ribosome binding has further been confirmed also by other findings. In membranes of the rough endoplasmic reticulum, which are characterized by their high number of ribosomal binding sites [70, 71], a protein of molecular weight 36000 was found, only barely detectable in the membranes of the smooth endoplasmic reticulum [84], and in polysomal preparations from the microsomal fraction a specific membrane protein with a high affinity to ribosomes could be detected [85].

A more detailed study of the membrane proteins destroyed by proteolytic enzymes, whereby the binding ability of the membranes for ribosomes is abolished, revealed mainly three proteins with molecular weights of 120000, 93000, and 36000, respectively [79]. Other authors [86—88] have reported on two integral proteins on the surface of rat liver rough endoplasmic reticulum related to ribosome binding as revealed by cross-linking experiments and called ribophorins I and II (molecular weights of 65000 and 63000 respectively).

The involvement of specific membrane proteins in ribosome binding is furthermore supported by the finding that the activity of the membrane-bound enzyme protein disulphide-isomerase (E. C. 5.3.4.1) is masked by the binding of polysomes, but activated by removal of the polysomes from the membranes [89, 90]. Also the finding that some antigenic determinants of rough endoplasmic membranes, blocked by attached ribosomes, become available to antibodies after stripping off the ribosomes without a concomitant loss of membrane proteins [91] suggests that specific proteins of the endoplasmic reticulum are involved in ribosome binding.

c) The role of ionic forces and the nascent peptide chain

Polysomes and ribosomes, respectively, can specifically be released from the membranes of the endoplasmic reticulum by various procedures and agents without destroying the structure of the membranes or of the ribosomal particles.

One of the most effective substance is

the chelating agent EDTA [67, 92—99], which brings about, depending on its concentration, the release at first of small subunits followed by the large one or, at higher concentrations, results in the partial release of both ribosomal subunits simultaneously in equal proportions [67, 92, 95, 97—99]. The fact that a certain portion of membrane-bound polysomes is refractory to EDTA suggests that besides divalent cation bridges also other forces and components are involved in binding of ribosomes to membranes.

Ribosomal particles can furthermore be released from their attachment to membranes by high concentrations of monovalent cations, especially K^+, and by the translation inhibitor puromycin [46, 93, 98—104]. While the treatment with either of both reagents results in a partial removal of bound ribosomal particles only [93, 99, 100, 102, 103, 105], the simultaneous application of high KCl and puromycin releases all membrane-bound ribosomes [93, 98, 99, 101—105]. These findings suggest that ribosomes may be bound to the membranes by ionic forces and by the nascent peptide chain. Ribosomes that can be released by KCl alone represent the so-called loosely bound fraction, whereas the KCl resistant ribosomes that can be removed only in the presence of puromycin — the so-called tightly bound ribosomes — are obviously anchored also by the nascent peptide chain to the membranes [93, 98, 99, 101—103]. The mechanism of binding of those ribosomes that can be removed by high KCl from the membranes and the concerted cooperation with the anchoring action of the nascent peptide chain is still unknown. The "loose" ribosomal fraction, which can be released from the membranes also by EDTA, and the "tight" fraction, which is resistant to EDTA [93, 94] have different densities, namely $1.55\ g \cdot cm^{-3}$ (heavy class) and $1.49\ g \cdot cm^{-3}$ (light class), respectively [94].

Treatment of the rough microsomal membrane fraction with high KCl and puromycin simultaneously brings about the release of ribosomal subunits [99, 102, 103] and parts of the mRNA [99, 103] (see also Section 4d of this chapter), while the nascent peptide chains, released from the ribosomes by puromycin remain associated with the membranes [102] (see also Section 5 of this chapter). Detachment of ribosomes from the membranes in media lacking Mg^{++} results in inactive subunits, while Mg^{++} preserves the biological activity of the particles [99].

The release of polysomes from the membranes by Na-deoxycholate and other detergents — procedures which are very commonly used for the preparation of "released, membrane-bound" polysomes — will not be discussed here, because these procedures disrupt the membrane structures.

In extracts obtained from free monosomes or polysomes of rabbit reticulocytes by treatment with 500 mM KCl, and in the ribosome free cytosolic fraction a so-called detachment factor has been found, and it was proposed that this factor functions in the release of ribosomes from the membranes after termination of polypeptide synthesis [106].

d) The role of messenger-RNA

The finding that part of the bound fraction of ribosomes can be detached from the membranes of the rough endoplasmic reticulum by RNase and that the majority of bound ribosomes is organized in polysomes has led to the suggestion that also mRNA may be involved in directing and binding of ribosomes to the membranes.

The results of studies on the action of ribonucleases on membrane-bound ribosomal particles and, therefore, also the interpretations are somewhat controversial, obviously mainly due to differences in the preparative and analytical procedures.

Most authors have reported that treatment of the rough microsomal fraction releases polysomes and other different types of ribosomal particles from the membranes [46, 74, 93, 95, 97, 107], obviously predominantly particles of the loosely bound ribosome fraction, while the ribosomes of the tightly bound fraction seem to be more resistant against RNase [93, 107]. In another study it could be shown very interestingly that treatment of microsomal membranes with RNase did not release bound ribosomes from the membranes, but bound ribosomes were no longer found to be homogeneously distributed on the membranes; instead large, tightly packed aggregates on the microsomal surfaces were analyzed [108]. From this experiment it has been concluded that RNase treatment leads to a lateral displacement of ribosomes at their attached binding sites [108].

Other studies relate in more detail to the involvement of mRNA in ribosome-membrane-interactions, following up the fate of the mRNA of bound polysomes after treatment of the rough microsomal fraction with RNase. Also here, the results are different, especially with regard to the significance of the 3′ poly(A) tail of the mRNA. On the one hand, poly(A) containing fragments of the mRNA were found to be extensively released from the membranes by mild RNase treatment in low salt [99]. From this finding it has been concluded that the 3′-end of mRNA is exposed on the outer membrane surface and, furthermore, that the mRNA molecule does not contribute significantly to the maintenance of the interactions between polysomes and the membranes of the endoplasmic reticulum [99]. This suggestion has been confirmed by the finding that poly(A) segments of membrane-bound mRNA molecules are accessible to in situ hybridization with poly(U) [99]. Other experiments, however, demonstrate that the poly(A) containing segment of mRNAs of membrane-bound polysomes is resistant to RNase [104, 109], while non-poly(A) segments are digested [104]. From the latter findings it has been concluded that the site of attachment of the mRNA to the membranes is located at or near the poly(A) containing 3′-end of the molecule [104]. This interpretation does not seem to be contradictory to the finding [99] that poly(A) sequences of mRNA molecules are destroyed by RNAse-treatment. Thus, the conclusion that parts of the mRNA molecules are involved in the binding of ribosomal particles to membranes [46, 95, 104] seems to be justified. This is supported by the finding that parts of the mRNA molecules remain firmly bound to the membranes after removal of most of the ribosomal particles by EDTA [109] or treatment with KCl and puromycin [104].

Differences in the poly(A) tail between free and membrane-bound polysomal mRNAs could not be detected [110, 111].

On experimental results with translation inhibitors, which also contribute to the enlightenment of the role of mRNA for ribosome-membrane-interaction, see Section 4e of this chapter.

It seems now well established that mRNA molecules coding for secretory and some other proteins (see Section 1 of this chapter) play a more indirect, but very important role for the attachment of polysomes to membranes, namely via their 5′-end sequences coding for the so-called signal sequences of these proteins, by which the ribosomes become bound to specific receptors of the membranes (for further details see Section 3 of Chapter XI).

e) The action of inhibitors of protein synthesis on ribosome-membrane interactions

Ribosomal particles originally bound to membranes of the endoplasmic reticulum can be detached from them by treatment with puromycin [46, 93, 98–102, 104, 105], which releases the nascent peptide chains from the ribosome [112, 113]. Simultaneously with the ribosomal subunits, also part of the mRNA is released from the membranes [99, 103].

Cycloheximide, which mainly inhibits the translocation reaction of the elongation step of protein synthesis [112, 113] (for details of this reaction see Section 2b of Chapter XI) protects by 50% the attachment of mRNA in polysomes to membranes but rather increases the formation of small free polysomes [114].

Also NaF, which inhibits the formation of translation initiation complexes, causes a significant release of ribosomes from membranes [115] and aurine tricarbocylic acid, which also inhibits the initiation complex formation [112, 113] were shown to inhibit the binding of ribosomes to microsomal membranes [116]. On the other hand, actinomycin D, an inhibitor of RNA synthesis, led to a conversion of large into small polysomes and ribosomes only, without changing the proportion of bound to free ribosomes [100].

From these results and others it has been concluded that the association of ribosomes with membranes follows their entry into polysomes, originally formed in the cytoplasm [46, 100, 101], which is in accordance with the "signal hypothesis" (see Chapter XI, Section 3).

5. Vectorial discharge of polypeptide chains from membrane-bound polysomes

As described before ribosomes attached to the membranes of the endoplasmic reticulum may be anchored to the membranes also by the nascent polypeptid chains.

Polypeptide chains of secretory proteins after natural termination of their synthesis on bound polysomes or nascent chains released from membrane-bound ribosomes by puromycin were found to remain associated with the membranes [44, 80, 117—124]. Non-secretory proteins, however, synthesized on membrane-bound polysomes in brain and muscle, e.g., are directly released to the cytosolic fraction [120, 121]. It has therefore been concluded [121] that a significant difference in the attachment of ribosomes to the membranes of the endoplasmic reticulum of secretory and non-secretory tissues exists, which results in a tissue-specific difference in the vectorial discharge of nascent proteins.

The release of the polypeptide chains of secretory proteins from the microsomal fraction by the detergents and/or sonication demonstrates that they are embedded within the membranes and the intravesicular compartment of the microsomal fractions, respectively. (Sonication releases the soluble intravesicular contents of the microsomes only, while treatment with detergents liberates also polypeptide chains embedded in the membranes or attached to them due to the destroying action of detergents [117, 118, 124, 125]).

These results clearly suggest that secretory proteins synthesized on membrane-bound ribosomes are discharged to membrane structures and to the intraluminal space of the endoplasmic reticulum [44, 103, 117—123]. This suggestion has been confirmed by the finding that nascent polypeptide chains are partially protected from digestion by proteolytic enzymes by both the ribosomal particle [126, 127] and by the membrane [128]. The protected C-terminal segment of the polypeptide buried in the large ribosomal subunit has a length of about 30 to 40 amino acid residues [126, 127] while in membrane bound ribosomes the polypeptide segment resistant to hydrolysis by proteases is longer [128]. Mainly from these experiments it has been concluded that the nascent polypeptide chain growths in a tunnel of the large ribosomal subunit and is then furthermore transferred into a transient tunnel of the membranes of the endoplasmic reticulum at the region on which the ribosome is attached to the membrane. In this way the nascent polypeptide chains of proteins synthesized on membrane-bound polysomes are discharged unidirectionally either into or across the membranes. The special signal by which ribosomes are directed to membranes are special N-terminal sequences of the growing peptide chain, consisting preferentially of hydrophobic amino acids, which are coded by corresponding 5'-terminal sequences of the mRNA molecule. As soon as these peptide segments of secretory and some other proteins (see Section 1 of this chapter) emerge from the large ribosomal subunits of polysomes formed in the cytoplasm, ribosomes with their mRNAs become attached to special binding sites on the membranes. Such special signal nucleotide sequences in the mRNA molecules and corresponding amino acid sequences in the polypeptides, respectively, synthesized on these mRNAs, were especially found for secretory proteins such as insulin, immunoglobulins, α-lactalbumin, melittin, and the parathyroid hormone (see Section 3 of Chapter XI). These proteins are originally synthesized as larger molecules, so-called preproteins, than are the final products secreted in vivo, because the N-terminal signal sequences are proteolytically removed as a cotranslational event. From these findings the so-called signal-hypothesis [129, 130] has been proposed (see Chapter XI, Section 3).

In most cases the vectorial transport of the peptide chain and the transformation of the preprotein to the final protein are dependent on the translation process on the ribosome; that means that only the growing peptide chain but not the final product can growth into the membrane. By synchronization experiments it was found that the vectorial transport into the membrane starts after about 70 to 80 amino acids have been connected [131].

In summary it can be suggested that the functional binding of ribosomes to the membranes of the endoplasmic reticulum only occurs during protein synthesis initiated on polysomal structures on special mRNA molecules in the cytoplasm. Once that these polysomes are directed to the membranes via the signal sequences of the corresponding polypeptide chains, additional forces contribute to the maintenance of the ribosome-membrane interactions, such as divalent and monovalent ion bridges and, furthermore, the nascent polypeptide chain and probably also parts of the mRNA molecule.

6. References

[1] ROLLESTON, F. S.: Membrane-bound and free ribosomes. Sub-Cell. Biochem. **3**, 91–117 (1974)

[2] BUTOW, R. A., W. F. BENNET, D. B. FINKELSTEIN, and R. E. KELLEMS: In: Membrane Biogenesis, 155–199. Plenum Press, New York, 1975

[3] SHORE, G. C., and J. R. TATA: Functions for polyribosome-membrane interactions in protein synthesis. Biochim. Biophys. Acta **472**, 197–236 (1977)

[4] MCINTOSH, P. R., and K. O'TOOLE: The interaction of ribosomes and membranes in animal cells. Biochim. Biophys. Acta **457**, 171–212 (1976)

[5] BLOEMENDAL, H., W. S. BONT, M. DEVRIES, and E. L. BENEDETTI: Isolation and properties of polyribosomes and fragments of the endoplasmic reticulum from rat liver. Biochem. J. **103**, 177–182 (1967)

[6] TAKAGI, M., and K. OGATA: Direct evidence for albumin biosynthesis by membrane-bound polysomes in rat liver. Biochem. Biophys. Res. Commun. **33**, 55–60 (1968)

[7] VENKATESAN, N., and W. J. STEELE: Free and membrane-bound polysomes of rat liver: Separation in nearly quantitative yield and analysis of structure and function. Biochim. Biophys. Acta **287**, 526 to 537 (1972)

[8] TALAL, N., and H. F. KALTREIDER: Functional and structural studies of membrane-bound and free ribosomes from rat spleen. J. Biol. Chem. **243**, 6504–6510 (1968)

[9] PRIESTLEY, G. C., M. L. PRUYN, and R. A. MALT: Glycoprotein synthesis by membrane-bound ribosomes and smooth membranes in kidney. Biochim. Biophys. Acta **190**, 154–160 (1969)

[10] Innanen, V. T., and D. M. Nicholls: fMet-tRNA$_f^{Met}$ binding and peptidyltransferase function in free and bound ribosomes from normal and puromycin aminonucleoside-treated rats. Chem.-Biol. Interact. **11**, 431—439 (1975)

[11] Ragnotti, G., G. R. Lawford, and P. N. Campbell: Biosynthesis of microsomal nicotinamide-adenine dinucleotide phosphate-cytochrom c reductase by membrane-bound and free polysomes from rat liver. Biochem. J. **112**, 139—147 (1969)

[12] MacDonald, R., and A. Korner: Growth hormone stimulation of protein synthetic activity of membrane-bound ribosomes. FEBS-Lett. **13**, 62—64 (1971)

[13] Hallinan, T., and H. N. Munro: Protein synthesis and ribonucleic acid turnover in rat-liver microsome subfractions. Biochim. Biophys. Acta **108**, 285—296 (1965)

[14] Tata, J. R., and H. G. William-Ashman: Effects of growth hormone and triiodothyronine on amino acid incorporation by microsomal subfractions from rat liver. Europ. J. Biochem. **2**, 366—374 (1967)

[15] Reid, I. M., D. S. R. Sarma, and H. Sidransky: Actinomycin D and hepatic polyribosomal disaggregation. Evidence for different half-lives of mRNA of free and membrane-bound polyribosomes. Lab. Invest. **25**, 141—148 (1971)

[16] Pain, V. M., J. Lanoix, J. J. M. Bergeron, and M. J. Clemens: Effect of diabetes on the ultrastructure of the hepatocyte and on the distribution and activity of ribosomes in the free and membrane-bound populations. Biochim. Biophys. Acta **353**, 487—498 (1974)

[17] Koffer, A.: Protein-synthesising activity of free and membrane-bound ribosomes in vitro and their differential sensitivity to protein synthesis inhibitors. FEBS-Lett. **46**, 326—332 (1974)

[18] Kraft-Creech, J., and E.-R. Lochmann: Protein synthesis on membrane-bound and free ribosomes in growing yeast. Biochim. Biophys. Acta **521**, 426 to 434 (1978)

[19] Shafritz, D. A., and K. J. Isselbacher: Liver protein synthesis: Differences in the properties of membrane-bound and free ribosomes. Biochem. Biophys. Res. Commun. **46**, 1721—1727 (1972)

[20] Uenoyama, K., and T. Ono: Specificities in messenger RNA and ribosomes from free and bound polyribosomes. Biochem. Biophys. Res. Commun. **49**, 713—719 (1972)

[21] Kubota, K., H. Yamaki, and T. Nishimura: Functional interaction of free polyribosomes with the membrane of the endoplasmic reticulum in a cell-free protein-synthesizing system from plasmacytoma X5563. Biochem. Biophys. Res. Commun. **52**, 489—495 (1973)

[22] Vernie, L. N., W. S. Bont, and P. Emmelot: Differences in Mg^{2+} and Ca^{2+} dependence of amino acid incorporation by free and membrane-bound polyribosomes isolated from liver and the effect of the hepatocarcinogen dimethylnitrosamine. Biochim. Biophys. Acta **281**, 253—262 (1972)

[23] Nolan, R. D., and N. H. Munro: Role of the endoplasmic reticulum membrane in the sulfhydryl requirement for protein synthesis. Biochim. Biophys. Acta **272**, 473—480 (1972)

[24] Glatzer, R. I., and A. C. Sartorelli: The differential sensitivity of free and membrane-bound polyribosomes to inhibitors of protein synthesis. Biochem. Biophys. Res. Commun. **46**, 1418—1424 (1972)

[25] Towers, N. R., J. K. Raison, G. M. Kellerman, and A. W. Linnane: Effects of temperature-induced phase changes in membranes on protein synthesis by bound ribosomes. Biochim. Biophys. Acta **287**, 301—311 (1972)

[26] Ilan, J., and M. Singer: Sampling of the leucine pool from the growing peptide chain: Differences in leucine specific activity of peptidyl-transfer RNA from free and membrane-bound polysomes. J. Mol. Biol. **91**, 39—51 (1975)

[27] Moulé, Y., and G. Delhumeau de Ongay: Relations métaboliques entre les ribosomes libres et liés du foie de rat. Biochim. Biophys. Acta **91**, 113—121 (1964)

[28] Bouvet, C., and Y. Moulé: Activité métabolique des acides ribonucleiques microsomiaux et solubles du foie de rat. Exptl. Cell Res. **33**, 330—343 (1964)

[29] Loeb, J. N., R. R. Howell, and G. M. Tomkins: Free and membrane-bound

ribosomes in rat liver. J. Biol. Chem. **242**, 2069–2074 (1967)

[30] Murty, C. N., and T. Hallinan: RNA turnover in free and endoplasmic reticulum-bound ribosomes from rat liver. Biochim. Biophys. Acta **157**, 414–416 (1969)

[31] Tanaka, T., M. Takagi, and K. Ogata: Studies on the metabolism of RNA of free and membrane-bound polysomes from rat liver. Biochim. Biophys. Acta **224**, 507 to 517 (1970)

[32] Baglioni, C., I. Bleiburg, and M. Zauderer: Assembly of membrane-bound polysomes. Nature New Biol. **232**, 8–12 (1971)

[33] Hayashi, T. T., and D. Kazmierowski: Changes in the free and membrane-bound ribosomes in the rat liver with starvation. Biochemistry **11**, 2371–2378 (1972)

[34] Hulse, J. L., and F. O. Wettstein: Two separable cytoplasmic pools of native ribosomal subunits in chick embryo tissue culture cells. Biochim. Biophys. Acta **269**, 265–275 (1972)

[35] Mechler, B., and P. Vassalli: Membrane-bound ribosomes of myeloma cells. II. Kinetic studies on the entry of newly made ribosomal subunits into the free and the membrane bound ribosomal particles. J. Cell Biol. **67**, 16–24 (1975)

[36] Izawa, M., and S. Ichii: RNA turnover in bound and free ribosomes from rat liver J. Biochem. **74**, 623–626 (1973)

[37] Mishra, R. K., J. F. Wheldrake, and L. A. W. Feetham: RNA turnover in endoplasmic reticulum-bound and free ribosomes. Biochim. Biophys. Acta **281**, 393–395 (1972)

[38] Shires, T. K., L. M. Narurkar, and C. Pitot: Polysome interaction in vitro with smooth microsomal membranes from rat liver. Biochem. Biophys. Res. Commun. **45**, 1212–1218 (1971)

[39] Rolleston, F. S.: The binding of ribosomal subunits to endoplasmic reticulum membranes. Biochem. J. **129**, 721–731 (1972)

[40] Borgese, D., G. Blobel, and D. D. Sabatini: In vitro exchange of ribosomal subunits between free and membrane-bound ribosomes. J. Mol. Biol. **74**, 415 to 438 (1973)

[41] Süss R., G. Blobel, and H. C. Pitot: Rat liver and hepatoma polysome-membrane interaction in vitro. Biochem. Biophys. Res. Commun. **23**, 299–304 (1966)

[42] Rolleston, F. S., and D. Mak: The binding of polyribosomes to smooth and rough endoplasmic-reticulum membranes. Biochem. J. **131**, 851–853 (1973)

[43] Burke, G. T., and C. M. Redman: The distribution of radioactive peptides synthesized by polysomes and ribosomal subunits combined in vitro with microsomal membranes. Biochim. Biophys. Acta **299**, 312–324 (1973)

[44] Shires, T. K., T. Ekren, L. M. Narurkar, and H. C. Pitot: Protein synthesis on rat liver polysome-membrane complexes formed in vitro and disposition of the discharged chains. Nature New Biol. **242**, 198–201 (1973)

[45] Shires, T. K., C. M. McLaughlin, and H. C. Pitot: The selectivity and stoichiometry of membrane binding sites for polyribosomes, ribosomes and ribosomal subunits in vitro. Biochem. J. **146**, 513 to 526 (1975)

[46] Mechler, B., and P. Vasalli: Membrane-bound ribosomes of myeloma cells. III. The role of the messenger-RNA and the nascent polypeptide chain in the binding of ribosomes to membranes. J. Cell. Biol. **67**, 25–37 (1975)

[47] Khawaja, J. A., and D. B. Lindholm: Synthesis and the puromycin-mediated release of nascent peptide chains by reconstituted rough endoplasmic reticulum frome rat liver and brain cortex. FEBS-Lett. **78**, 255–260 (1977)

[48] Feigenbaum, A. M., N. DeGroot, and A. A. Hochberg: The in vitro reassembly of rough endoplasmic reticulum: Ribosome binding capacity. Mol. Biol. Rep. **4**, 111–115 (1978)

[49] Ekren, T., T. Shires, and H. C. Pitot: Determining the affinity in vitro of hepatical ribosomal subunits for derivatives of the rough endoplasmic reticulum. Biochem. Biophys. Res. Commun. **54**, 283 to 289 (1973)

[50] Christman, J. K., and J. Goldstein: An apparent identity of 5 S RNA from free and membrane-bound reticulocyte ribosomes. Biochim. Biophys. Acta **217**, 346–355 (1970)

[51] KHAN, M. S. N., and B. E. H. MADEN: Comparison between the ribosomal ribonucleic acids from free and membrane-bound ribosomal fractions of HeLa cells. Biochem. J. **155**, 197–200 (1976)

[52] HANNA, N., G. BELLEMARE, and C. GODIN: Free and membrane-bound ribosomes. I. Separation by two-dimensional gel electrophoresis of proteins from rat liver monosomes. Biochim. Biophys. Acta **331**, 141–145 (1973)

[53] HANNA, N., and C. GODIN: Free and membrane-bound ribosomes. III. Analysis by two-dimensional gel electrophoresis of proteins from liver ribosomal subunits of rats with different dietary intakes of phenylalanine. Biochim. Biophys. Acta **374**, 342–349 (1974)

[54] LEWIS, J. A., and D. D. SABATINI: Proteins of rat liver free and membrane-bound ribosomes. Modification of two large subunit proteins by a factor detached from ribosomes at high ionic strength. Biochem. Biophys. Acta **478**, 331–349 (1977)

[55] HOFFMAN, W. L., and J. ILAN: Analysis by two-dimensional polyacrylamide gel electrophoresis of the in vivo phosphorylation of ribosomal proteins derived from free and membrane-bound polysomes. Mol. Biol. Rep. **2**, 219–224 (1975)

[56] FRIDLENDER, B. R., and F. O. WETTSTEIN: Differences in the ribosomal protein of free and membrane bound polysomes of chick embryo cells. Biochem. Biophys. Res. Commun. **39**, 247–253 (1970)

[57] BURKA, E. R., and I. BULOVA: Heterogeneity of reticulocyte ribosomes. Biochem. Biophys. Res. Commun. **42**, 801 to 805 (1971)

[58] FEHLMANN, M., G. BELLEMARE, and C. GODIN: Free and membrane-bound ribosomes. II. Two-dimensional gel electrophoresis of proteins from free and membrane-bound rabbit reticulocyte ribosomes. Biochim. Biophys. Acta **378**, 119 to 124 (1975)

[59] FEHLMANN, M., G. BELLEMARE, and C. GODIN: Free and membrane-bound rabbit reticulocyte ribosomes. Proteins from the large and the small subunits. FEBS-Lett. **59**, 8–12 (1975)

[60] MCCONKEY, E. H., and E. J. HAUBER: Evidence for heterogeneity of ribosomes within HeLa cells. J. Biol. Chem. **250**, 1311–1318 (1975)

[61] RAMSEY, J. C., and W. J. STEELE: Differences in size, structure and function of free and membrane-bound polyribosomes of rat liver. Biochem. J. **168**, 1–8 (1977)

[62] MADJAR, J.-J., M. ARPIN, M. J. MARION, and J.-P. REBOUD: Comparison of the protein content of free and membrane-bound rat liver polysomes and their derived subunits. Mol. Biol. Rep. **3**, 289 to 296 (1977)

[63] BARELA, T. D., and D. E. KIZER: In vivo phosphorylation of free and membrane-bound ribosomal protein. Biochim. Biophys. Acta **335**, 218–225 (1974)

[64] CENATIEMPO, Y., A. J. COZZONE, A. GENOT, and J.-P. REBOUD: In vitro phosphorylation of proteins from free and membrane-bound rat liver polysomes. FEBS-Lett. **79**, 165–169 (1977)

[65] LEWIS, J. A., and D. D. SABATINI: Accessibility of proteins in rat liver-free and membrane-bound ribosomes to lactoperoxidase-catalyzed iodination. J. Biol. Chem. **252**, 5547–5555 (1977)

[66] SHELTON, E., and E. I. KUPF: Substructure and configuration of ribosomes isolated from mammalian cells. J. Mol. Biol. **22**, 23–31 (1966)

[67] SABATINI, D. D., Y. TASHIRO, and G. E. PALADE: On the attachment of ribosomes to microsomal membranes. J. Mol. Biol. **19**, 503–524 (1966)

[68] FLORENDO, N. T.: Ribosome substructure in intact mouse liver cells. J. Cell Biol. **41**, 335–339 (1969)

[69] UNWIN, P. N. T.: Three-dimensional model of membrane-bound ribosomes obtained by electron microscopy. Nature **269**, 118–122 (1977)

[70] BORGESE, N., W. MOK, G. KREIBICH, and D. D. SABATINI: Ribosomal-membrane interaction: In vitro binding of ribosomes to microsomal membranes. J. Mol. Biol. **88**, 559–580 (1974)

[71] ROLLESTON, F. S., and T. Y. LAM: Dissociation constant of 60 S ribosomal subunit binding to endoplasmic reticulum membranes. Biochem. Biophys. Res. Commun. **59**, 467–473 (1974)

[72] KHAWAJA, J. A., and A. RAINA: Effect of spermine and magnesium on the attachment of free ribosomes to endoplasmic reticulum membranes in vitro. Biochem. Biophys. Res. Commun. **41**, 512–518 (1970)

[73] SHIRES, T. K., L. NARURKAR, T. EKREN, and H. C. PITOT: The in vitro association of polysomes with microsomal membranes from different rat tissues. J. Cell. Biol. **47**, 190a (1970)

[74] SHIRES, T. K., L. NARURKAR, and H. C. PITOT: The association in vitro of polyribosomes with ribonuclease-treated derivatives of hepatic rough endoplasmic reticulum. Biochem. J. **125**, 67–79 (1971)

[75] RAGLAN, W. L., T. K. SHIRES, and H. C. PITOT: Polyribosomal attachment of rat liver and hepatoma endoplasmic in vitro. Biochem. J. **121**, 271–278 (1971)

[76] BURKA, E. R., and L. F. SCHICKLING: Attachment of reticulocyte ribosomes to erythroid cell membranes in vitro. Biochemistry **9**, 459–463 (1970)

[77] CHEFURKA, W., and Y. HAYASHI: The effect of trypsin on rough endoplasmic membranes. Biochem. Biophys. Res. Commun. **24**, 633–638 (1966)

[78] deleted

[79] JOTHY, S., J.-L. BILODEAU, and H. SIMPKINS: The role of membrane proteins and phospholipids in the interaction of ribosomes with endoplasmic reticulum membranes. Canad. J. Biochem. **53**, 1039 to 1045 (1975)

[80] SHIRES, T. K., and H. C. PITOT: Correlation of rat liver membrane binding of polysomes in vitro with function of the complexes formed. Biochem. Biophys. Res. Commun. **50**, 344–351 (1973)

[81] SCOTT-BURDEN, T., and A. O. HAWTRY: The effect of neuraminidase treatment of ribosome-free membranes on their ribosomal reattachment ability. Biochem. Biophys. Res. Commun. **54**, 1288–1295 (1973)

[82] JOTHY, S., H. SIMPKINS, and S. TAY: The role of membrane phospholipids in the interaction of ribosomes with endoplasmic reticulum membrane. Biochem. J. **132**, 637–640 (1973)

[83] KHAWAJA, J. A.: Interaction of ribosomes and ribosomal subparticles with endoplasmic reticulum membranes in vitro: Effect of spermine and magnesium. Biochim. Biophys. Acta **254**, 117–128 (1971)

[84] BAILEY, D. J., R. K. MURRAY, and F. S. ROLLESTON: Electrophoretic studies of the proteins of rat liver endoplasmic reticulum. Canad. J. Biochem. **52**, 1003–1012 (1974)

[85] OLSNESS, S.: The isolation of polysomes from rat liver. Conta mination by membrane proteins with high affinity to ribosomes. Biochim. Biophys. Acta **232**, 705–716 (1971)

[86] KREIBICH, G., B. L. ULRICH, and D. D. SABATINI: Proteins of rough microsomal membranes related to ribosome binding. I. Identification of ribophorins I and II, membrane proteins characteristic of rough microsomes. J. Cell. Biol. **77**, 464–487 (1978)

[87] KREIBICH, G., C. M. FREIENSTEIN, B. N. PEREYRA, B. L. ULRICH, and D. D. SABATINI: Protein of rough microsomal membranes related to ribosome binding. II. Cross-linking of bound ribosomes to specific membrane proteins exposed at the binding site. J. Cell. Biol. **77**, 488 to 506 (1978)

[88] KREIBICH, G., M. CZAKÓ-GRAHAM, R. GREBENAU, W .MOK, RODRIQUEZ-BOULAN, and D. D. SABATINI: Characterization of the ribosomal binding site in rat liver rough microsomes: Ribophorins I and II, two integral membrane proteins related to ribosome binding. J. Supramol. Struct. **8**, 279–302 (1978)

[89] DANI, H. M., J. A. FIELDER, and B. R. RABIN: Degranulation and sex-specific regranulation of reticular membranes from rat liver as studied using a spectrophotometric assay of protein disulphide isomerase. FEBS-Lett. **65**, 377–382 (1976)

[90] FIELDER, J. A., H. M. DANI, D. RIDGE, and B. R. RABIN: The binding of monoribosomes, oligoribosomes and polyribosomes to reticular membranes from rat liver. Biochem. J. **172**, 109–114 (1978)

[91] LIN, J.-C., and E. FARBER: Effect of ribosome stripping procedures on antigenicity and conformation of endoplasmic reticulum membrane. Biochem. Biophys. Res. Commun. **76**, 1247–1252 (1977)

[92] ATTARDI, B., B. CRAVIOTO, and G. ATTARDI: Membrane-bound ribosomes in

HeLa cells. I. Their proportion to total cell ribosomes and their association with messenger-RNA. J. Mol. Biol. **44**, 47–70 (1969)

[93] Rosbash, M., and S. Penman: Membrane-associated protein synthesis of mammalian cells. I. The two classes of membrane-associated ribosomes. J. Mol. Biol. **59**, 227–241 (1971)

[94] Rosbash, M., and S. Penman: Membrane-associated protein synthesis of mammalian cells. II. Isopycnic separation of membrane-bound polyribosomes. J. Mol. Biol. **59**, 243–253 (1971)

[95] Lee, S. Y., V. Krsmanovic, and G. Brawermann: Attachment of ribosomes to membranes during polysome formation in mouse sarcoma 180 cells. J. Cell. Biol. **49**, 683–691 (1971)

[96] Khawaja, J. A.: Gradual release of spermine and RNA from rat liver microsomes treated with EDTA. FEBS-Lett. **21**, 49–52 (1972)

[97] Faiferman, I., A. O. Pogo, J. Schwartz, and M. E. Kaighn: Isolation and characterization of membrane-bound polysomes from ascites tumor cells. Biochim. Biophys. Acta **312**, 492–501 (1973)

[98] Dobberstein, B., D. Volkmann, and D. Klämbt: The attachment of polyribosomes to membranes of the hypocotyl of *Phaseolus vulgaris*. Biochim. Biophys. Acta **374**, 187–196 (1974)

[99] Kruppa, J., and D. D. Sabatini: Release of polyA+ messenger RNA from rat liver rough microsomes upon dissassembly of bound polysomes. J. Cell. Biol. **74**, 414 to 427 (1977)

[100] Blobel, G., and V. R. Potter: Studies on free and membrane-bound ribosomes in rat liver. II. Interactions of ribosomes and membranes. J. Mol. Biol. **26**, 293 to 301 (1967)

[101] Chua, N. H., G. Blobel, P. Siekevitz, and G. E. Palade: Attachment of chloroplast polysomes to thylakoid membranes in *Chlamydomonas reinhardtii*. Proc. Natl. Acad. Sci. U.S. **70**, 1554–1558 (1973)

[102] Adelman, M. R., D. D. Sabatini, and G. Blobel: Ribosome-mebrane interactions. Nondestructive dissassembly of rat liver rough microsomes into ribosomal and membranous components. J. Cell. Biol. **56**, 206–229 (1973)

[103] Harrison, T. M., G. G. Brownlee, and C. Milstein: Studies in polysome-membran interactions in mouse myeloma cells. Euro. J. Biochem. **47**, 613–620 (1974)

[104] Lande, M., M. Adesnik, M. Sumida, Y. Tashiro, and D. D. Sabatini: Direct association of messenger RNA with microsomal membranes in human diploid fibroplasts. J. Cell. Biol. **65**, 513–528 (1975)

[105] Engelman, F., and L. Barajas: Ribosome-membrane association in fat body tissue from reproductively active femals of *Leucophaea maderae*. Exptl. Cell Res. **92**, 102–110 (1975)

[106] Blobel, G.: Extraction from free ribosome of a factor mediating ribosome detachment from rough microsomes. Biochem. Biophys. Res. Commun. **68**, 1–7 (1976)

[107] Tanaka, T., and K. Ogata: Two classes of membrane-bound ribosomes in rat liver cells and their albumin synthesizing activity. Biochem. Biophys. Res. Commun. **49**, 1069–1074 (1972)

[108] Ojakian, G. K., G. Kreibich, and D. D. Sabatini: Mobility of ribosomes bound to microsomal membranes. J. Cell. Biol. **72**, 530–551 (1977)

[109] Milcarek, Ch., and S. Penman: Membrane-bound polyribosomes in HeLa cells: Association of polyadenylic acid with membranes. J. Mol. Biol. **89**, 327 to 338 (1974)

[110] Baglioni, C., R. Pemberton, and T. Delovitch: Presence of polyadenylic acid sequences in RNA of membrane-bound polyribosomes. FEBS-Lett. **26**, 320–322 (1972)

[111] Poirée, J.-C., A. Becarevic, J. Chebath, S. Manté, G. Cartouzou, and S. Lissitzky: Polyadenylic acid sequences in ribonucleic acid of free and membrane-bound polyribosomes of cultured thyroid cells. Biochimie **55**, 1095 to 1099 (1975)

[112] Vazquez, D.: Inhibitors of protein synthesis. FEBS-Lett. **40**, S 63–S 84 (1974)

[113] Vazquez, D.: Translation inhibitors. Internat. Rev. Biochem. **18**, 169–223 (1978)

[114] Rosbash, M.: Formation of membrane-bound polyribosomes. J. Mol. Biol. **65**, 413–422 (1972)

[115] Bleiberg, I., M. Zauderer, and C. Baglioni: Reversible disaggregation by NaF

of membrane-bound polyribosomes of mouse myeloma cells in tissue culture. Biochim. Biophys. Acta **269**, 453—464 (1972)

[116] Bussolati, E., R. Bidoglia, and N. Borgese: The inhibitory effect of ATA on the attachment of ribosomes to microsomal membranes in vitro. Mol. Pharmacol. **14**, 220—224 (1978)

[117] Campbell, P. N., O. Greengard, and B. A. Kernot: Studies on the synthesis of serum albumin by the isolated microsome fraction from rat liver. Biochem. J. **74**, 107—117 (1960)

[118] Redman, C. M., P. Siekevitz, and G. E. Palade: Synthesis and transfer of amylase in pigeon pancreatic microsomes. J. Biol. Chem. **241**, 1150—1158 (1966)

[119] Redman, C. M., and D. D. Sabatini: Vectorial discharge of peptides released by puromycin from attached ribosomes. Proc. Natl. Acad. Sci. U. S. **56**, 608—615 (1966)

[120] Andrews, T. M., and J. R. Tata: Differences in vectorial release of nascent protein from membrane-bound ribosomes of secretory and non-secretory tissues. Biochem. Biophys. Res. Commun **32**, 1050 to 1056 (1968)

[121] Andrews, T. M., and J. R. Tata: Protein synthesis by membrane-bound and free ribosomes of secretory and non-secretory proteins. Biochem. J. **121**, 683 to 694 (1971)

[122] Redman, C. M., and M. G. Cherian: The secretory pathways of rat serum glycoproteins and albumin. Localization of newly formed proteins within the endoplasmic. reticulum. J. Cell. Biol. **52**, 231 to 245 (1972)

[123] Milstein, C., G. G. Brownlee, T. M. Harrison, and M. B. Mathews: A possible precursor of immunoglobulin light chains. Nature New Biol. **239**, 117—120 (1972)

[124] Sauer, L. A., and G. N. Burrow: The submicrosomal distribution of radioactive proteins released by puromycin from the bound ribosomes of rat liver microsomes labelled in vitro. Biochim. Biophys. Acta **277**, 179—187 (1972)

[125] Kreibich, G., P. Debey, and D. D. Sabatini: Selective release of content from microsomal vesicles without membrane disassembly. I. Permeability changes induced by low detergent concentrations. J. Cell. Biol. **58**, 436—462 (1973)

[126] Blobel, G., and D. D. Sabatini: Controlled proteolysis of nascent polypeptide in rat liver cell fractions. I. Location of the polypeptide within the ribosome. J. Cell Biol. **45**, 130—145 (1970)

[127] Malkin, I. I., and A. Rich: Partial resistance of nascent polypeptide chains to proteolytic digestion due to ribosomal shielding. J. Mol. Biol. **26**, 329—346 (1967)

[128] Sabatini, D. D., and G. Blobel: Controlled proteolysis of nascent polypeptides in rat liver cell fractions. II. Location of the polypeptides in rough microsomes. J. Cell Biol. **45**, 146—157 (1970)

[129] Blobel, G., and B. Dobberstein: Transfer of proteins across membranes. I. Presence of proteolytically processed and unprocessed nascent immunoglobulin light chains on membrane-bound ribosomes of murine myeloma. J. Cell. Biol. **67**, 835—851 (1975)

[130] Blobel, G., and B. Dobberstein: Transfer of proteins across membranes. II. Reconstitution of functional rough microsomes from heterologous components. J. Cell. Biol. **67**, 852—862 (1975)

[131] Rothman, J. E., and H. F. Lodish: Synchronised transmembrane insertion and glycolysation of a nascent membrane protein Nature **269**, 775—780 (1977)

XI. Function of eukaryotic ribosomes

Contents

Translation is a highly complex process that requires the interaction of a large number of macromolecular and low-molecular components in a well-ordered sequence of events organized by the ribosome. During this process the nucleotide sequence of the genetic message is decoded and translated into the amino acid sequence of specific proteins.

In this chapter the different functions of eukaryotic ribosomes during protein synthesis will be described. At first, some properties of the nonribosomal components involved will be summarized. In the second part a survey is given about the sequence of events of translation occurring at the ribosome and, finally, particular aspects of eukaryotic protein synthesis on membrane-bound ribosomes and some regulatory aspects at the translational level will be briefly discussed.

1. Nonribosomal components of the protein synthesizing system

a) *Messenger RNA*

In contrast to prokaryotes, where polycistronic mRNA predominates, eukaryotic messengers are monocistronic. Transcription of eukaryotic DNA templates into mRNA takes place mainly in the cell nucleus, where the so-called hnRNA is synthesized that represents the precursor of the finally translatable mRNA. The hnRNA fraction has on the average a 5 to 10-fold larger size than the mature mRNA and undergoes drastic alterations on the way from the nucleus to the cytoplasm that are summarized by the term "processing" (for reviews see [1–4]).

During processing within the nucleus more or less extended RNA sequences encoded at noncontiguous positions within the genome are spliced together [3–17]. The process of splicing is not only restricted

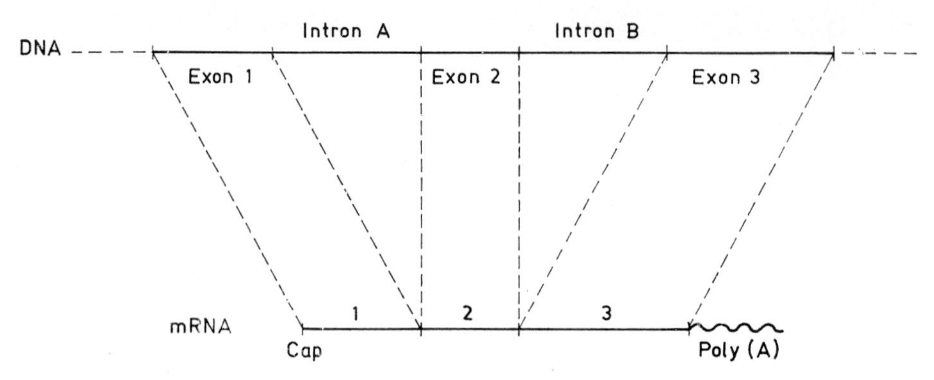

Fig. 29. Splicing mechanism of mRNA (according to [17])

to viruses where it was discovered in 1977 [5] but seems to be a general feature of eukaryotic mRNA maturation.

The basic mechanism of splicing can be demonstrated by the scheme given in Fig. 29: At first RNA polymerase synthesizes a complete primary transcript of a certain gene. Subsequently specific nucleases that were not yet isolated and in detail characterized, cut this RNA into distinct pieces. Some of these RNA segments are thereafter selectively spliced together by specific ligases the nature and mode of action of which is also not yet elucidated. As a consequence of this series of events noncontiguous stretches of the RNA become connected one with another.

According to a generally accepted concept [16, 17] those regions of the DNA that are expressed in the final mRNA are designated as "exons" and the intervening sequences that are cut out during the process are named "introns". The final mRNA molecule, therefore is the result of tied up exon-parts of the genetic message that originate from quite different parts of the gene. The splicing process was found operational during synthesis of various eukaryotic viral and cellular mRNAs. From sequence analyses it was concluded that all introns seem to start with the dinucleotide G—U and to end with A—G [17]. This at least goves some hint for the search for nucleases involved in cutting the primary transcript.

Most mRNA molecules can be isolated from the cell as mRNA protein complexes. The protein moieties seem to protect the mRNA and also to cooperate in protein biosynthesis. Different RNA protein complexes were found to exist in nuclei and in cytoplasm (for reviews see [18—21]).

Although only a limited number of eukaryotic mRNA molecules has been sequenced up to now, their most important structural features seem to be known (Fig. 30). At the 5′ end most eukaryotic cellular and viral mRNAs contain a so-called cap structure of 7-methyl-GTP connected with the following nucleotide by an unusual 5′-5′ triphosphate bond (Fig. 31) (for reviews see [22—25]). With regard to

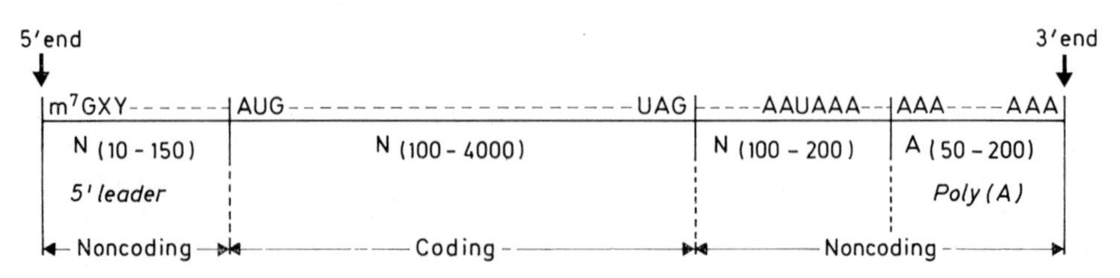

Fig. 30. Scheme of mRNA structure.

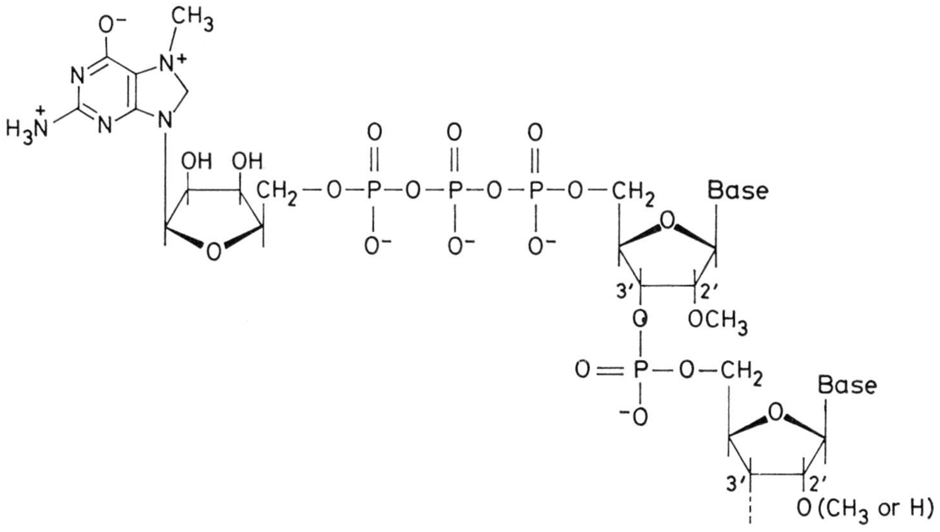

Fig. 31. Cap structure of mRNA.

the extent of methylation of the following nucleotides (X and Y) two different types of cap structures were found: "cap I" of the sequence $m^7G^{5'}pppX^mpYp$ ... and "cap II" of the sequence $m^7G^{5'}pppX^mpY^m$ p ... [26–28].

Cap structures are added to the 5′ end of messenger RNA precursors in connection with the transcription from DNA in the nucleus. The enzymes required for capping and methylation of the terminal GTP are located in the nucleus. In viruses these enzymes are constituents of the viral cores [25, 29].

The cap remains with the mRNA during the transport to the cytoplasm and its translation. In the cytoplasm of various cells nucleotide pyrophosphatases were found that are able to cleave specifically the pyrophosphate linkage of the cap [30–35].

The cap seems to have two main functions: it protects mRNA against degradation and facilitates its translation. The importance of the 5′ terminal m^7G-cap for efficient mRNA translation was demonstrated with a variety of different cellular and viral messengers in a number of translation systems [25, 36]. There is, however, no absolute cap requirement for mRNA translation because also decapped messengers were shown to be translated, although in most cases with lower efficiency. The molecular mechanism of interactions between the mRNA cap region and the components taking part in initiation of protein synthesis is still not understood. There is some evidence for certain factor proteins being involved in cap recognition [37–41] (see also Section 1c). Even prokaryotic messengers artificially capped at the 5′ end could be efficiently translated in a eukaryotic cell-free system [42, 43].

Adjacent to the cap structures eukaryotic messengers were found to consist of a noncoding region, the so-called 5′ terminal "leader" sequence of up to 150 nucleotides the function of which is not yet known. In prokaryotes, the 5′ terminal sequences of mRNA are important for the binding of the mRNA to ribosomes during initiation of protein synthesis [44]. The nucleotide sequence UGGAGGA ..... 5′ that is common to all prokaryotic messengers base-pairs with the complementary sequence ACCUCCU ..... near the 3′ terminus of the 16 S rRNA. In eukaryotes, only for a small number of mRNAs, in particular globin mRNAs, it has been shown that

Table 22. Nucleotide sequences of the 5′ noncoding region of eukaryotic mRNAs. The partial homologies are underlined. *Denotes that the identification of the nucleotide at that position is tentative. The arrow points to the most probable initiation codon.
(From BARALLE and BROWNLEE [45]).

a) Globin mRNA

β rabbit	m⁷GpppACACUUGCUUUUGACACAACUGUGUUUACUUGCAAUCCCCCAAAACAGACAGA↓AUGGUG
β human	m⁷GpppA*CAUUUGCUUCUGACACAACUGUGUUCACUAGCAACCUCAAACAGACACCAUGGUG
β_B mouse	m⁷GpppACA*UUUGCUUCUGACAUAGUUGUGUUGACUCACAACCCCAGAAACAGACAUCAUGGUG
β_A mouse	m⁷GpppACG*UUUGCUUCUGAUUCUGUUGUGUUGACUUGCAACCUCAGAAACAGACAUCAUGGUG
α mouse	m⁷GpppACUUCUGAUUCUGACAGACUCAGGAAGAAACCAUGGUG
α human	m⁷GpppA*CUCUUCUGGUCCCCACAGACUCAGAGAGAACCCACCAUGGUG
α rabbit	m⁷GpppACACUUCUGGUCCAGUCCGACUGAGAAGGAACCACCAUGGUG

b) Non-globin mRNA

Ovalbumin	m⁷GpppA*CAUACAGCUAGAAAGCUGUAUUGCCUUUAGCAGUCAAGCUCGAAAGACAACUCAGAGUUCACCAUGGGC
SV 40 VP 1	...AUGUUGCCUUACUUCUAGGCCUGUACGGAAGUGUUACUUCUGGCUCUAAACCUUAUGAAGAUGGCC
TYMV RNA	m⁷GpppGUAAUCAACUACCAAUUCCAGCUCUCUUUUGACAACUGGUCUUAUACCCACUUCCGUACACUUGCAACCCUCGUAAGACAAUUGCAAAUGAGUAAUGGCC
TYMV coat	m⁷GpppAAUAGCAAUCAGCCCCAACAUGGAA
ASV	m⁷GpppGCCAUUUUACCAUUCACCACAUUGGUGUGCACCUGGGUUGAUGGCUGGACCGUCGAUUCCCUAACGAUUGCGAACACCUGAAUGAAG
Reovirus s 54	m⁷GpppGCUAUUUUGCCUCUUCCCAGACGUUGUCGCAAUGGAG
Reovirus s 45	m⁷GpppGCUAAAGUCACGCCUGUCGUCGUCACUAUGGCU
Reovirus s 46	m⁷GpppGCUAUUCGCUGGUCAGUUAUGGCU
Reovirus m 52	m⁷GpppGCUAAUCUGCUGACCGUUACUCUGCAAAGAUGGGG
Reovirus m 44	m⁷GpppGCUAAAGUGACGGUGGUCAUGGCU
Reovirus m 30	m⁷GpppGCUAUUCGCGGUCAUGGCU
BMV coat protein	m⁷GpppGUAUUAAUAAUGUCG
VSV N	m⁷GpppAACAGUAAUCAAAUGUCU
VSV NS	m⁷GpppAACAGAUAUCAUGGAU
VSV G	...UUUCCUUGACACUAUGAAG
AlMV RNA$_4$	m⁷GpppG*UUUUUAUUUUUAAUUUUCUUUCAAAUACUUCCAUCAUGAGU
TMV RNA	m⁷GpppGUAUUUUUACAACAAUUACCAACAACAACAAACAACAAACAACAUUACAAUUACUAUUUACAAUUACAAUGGCA

some very short nucleotide sequences of the 5′ leader region can theoretically base-pair with parts of the 3′ terminal region of 18 S rRNA [45], whereas in a large number of other mRNAs a similar sequence complementarity is missing (Table 22). The only nucleotides common to all eukaryotic mRNAs at their 5′ terminal regions are the cap and the AUG codon that signalizes the beginning of the coding part of the mRNA, whereas the extension of the 5′ leader sequence, i.e. the distance between both positions, the cap and the AUG, is quite variant in the different mRNA molecules.

The coding part of eukaryotic mRNA starts with the initiation codon AUG and is terminated by one of the stop codons UAA, UAG or UGA. It amounts to between 30 and 75% of the messenger, and its translation follows the rules of the amino acid code.

Adjacent to the coding part in 3′ terminal direction another noncoding region of varying length follows [46—54], the role of which is still obscure. Most of the eukaryotic mRNAs contain within this 3′ noncoding part a sequence AAUAAA the function of which is not yet known.

At the 3′ end most eukaryotic mRNAs contain a more or less extended poly (A) sequence consisting of up to 200 adenylate residues (for reviews see [55, 56]). Polyadenylation of the mRNA may occur in the nucleus either directly after transcription or during or after processing and in the cytoplasm after binding to ribosomes. The main advantage of the poly(A) end seems to be the protection of the genetic message against uncontrolled degradation.

As to the function of eukaryotic mRNA during translation the first interaction of mRNA with the ribosome occurs in a cooperative process in which the 40 S ribosomal subunit and certain initiation factors are involved (see Section 2a).

The basic principle of mRNA recognition of prokaryotes operating by base-pairing of mRNA sequences with complementary sequences at the 16 S rRNA seems in eukaryotes to be substituted by a selection mechanism of the 7-methyl-GTP cap of the mRNA with one or several of the initiation factors in cooperation with the 40 S ribosomal subunit. Thereafter the 40 S subunit moves along the mRNA until the initiating AUG codon becomes positioned in the ribosomal P site opposite to the anticodon of the initiator-tRNA. Subsequently, the 60 S subunit joins to form the protein synthesizing 80 S ribosome.

Table 23. Nucleotide sequences 3′terminally adjacent to the AUG initiation codon of Reovirus mRNA protected by 40 S ribosomal subunits and of VSV mRNA protected by 80 S ribosomes (according to ref. [3]). The numbers in the columns designate the positions of the last phosphodiester bond following a G residue that is protected (column I) and the first phosphodiester bond following a G residue that is not protected by the ribosome (column II).

mRNA	Strain	Partial mRNA sequence	I	II
Reovirus	S 54	5′....AUG GAGGUGUGCUUGCCCAACG	13	20
	S 45	5′....AUG GCUUCCUCACUCAG	2	15
	S 46	5′....AUG GCU(CGCUGC)GCGUUCCUAUUCAAG	13	25
	m 52	5′....AUG GGGAACG(CU,CUUC)CUAAUCG	8	21
	m 44	5′....AUG GCUUCAUUCAAGGGAUUCUCCG	15	23
	m 30	5′....AUG GCUUACAUCGCAG	11	14
VSV	N	5′....AUG UCUGUUACAGUCAAG	11	16
	NS	5′....AUG GAUAAUCUCACAAAG	2	20

Studies on ribosome · mRNA complexes including ribonuclease treatment and sequence analyses of protected mRNA fragments show that in 40 S · mRNA complexes up to 60 nucleotides and in 80 S · mRNA complexes about 40 nucleotides [57—60] are covered by the ribosome. Sequence analyses show that about 13 nucleotides adjacent to the AUG within the coding part of the messenger are protected by the ribosome, as can be judged from the data summarized in Table 23. Thus, the ribosomal P site seems to be located about 4—4.5 nm distant from the end of the mRNA binding site at the small subunit [3, 61—64].

b) Transfer-RNA

Transfer RNAs play a key role in the decoding process of mRNA during protein synthesis. One region of the molecule reacts with a specific amino acid and another one interacts with the corresponding codon of the mRNA. Thus, transfer RNAs function as adaptor molecules between the nucleotide sequence of the mRNA and the amino acid sequence of the protein. Considerable information about the structure of tRNAs and of their role in the translation process has accumulated (for review see [68]). Here, only a short summary of the current knowledge about tRNA

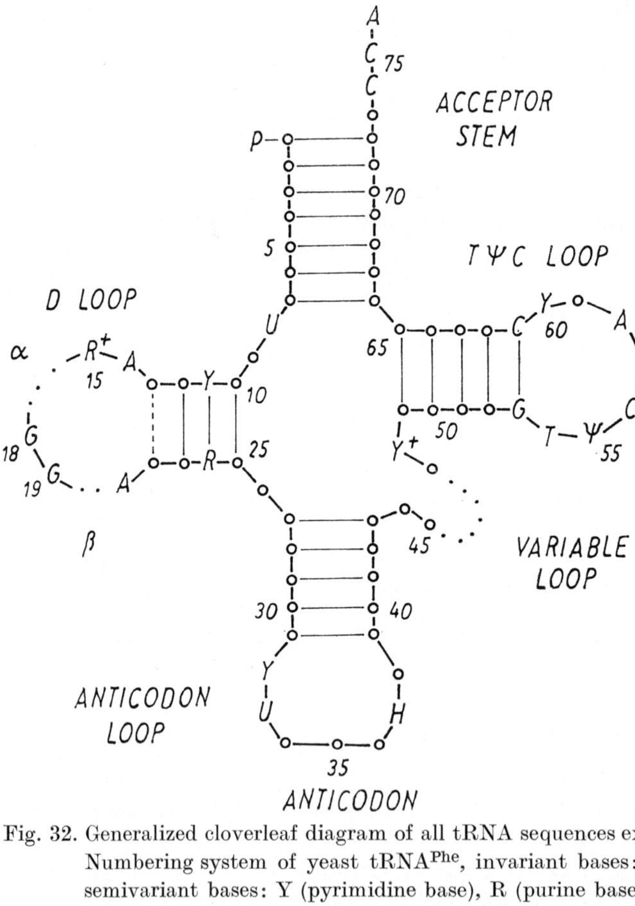

Fig. 32. Generalized cloverleaf diagram of all tRNA sequences except for initiator tRNAs. Numbering system of yeast $tRNA^{Phe}$, invariant bases: A, C, G, T, U and Ψ; semivariant bases: Y (pyrimidine base), R (purine base), H (hypermodified purine base). The dotted regions (α, β, variable loop) contain different numbers of nucleotides in various tRNA sequences.
From A. Rich and U. L. RajBhandary [77]. Reproduced with permission of the authors and Ann. Rev. Inc. (Palo Alto)

structure and function will be given. Functions of tRNAs others than those involved in ribosomal protein synthesis will not be considered (for reviews see [69—77]).

Structure of the transfer RNA molecule

Transfer RNAs are polyribonucleotides of an average molecular weight of about 25000 containing between 73 and 93 nucleotides and with a sedimentation coefficient of about 4 S. Since the elucidation of the primary structure of the first tRNA molecule (yeast $tRNA^{Ala}$) in 1965 [78] until the end of 1978 the nucleotide sequences of at least 150 tRNA molecules have been determined [79, 83]. All these sequences can be arranged in a two-dimensional cloverleaf diagram (Fig. 32), which reflects the actual secondary structure of all tRNA molecules (for reviews see [77, 80, 81]) characterized by the following common features:

1. Five main parts of the tRNA molecules can be distinguished: a) the acceptor arm containing the constant nucleotide sequence CCA at its 3′ end, b) the D arm which (normally) contains at least one dihydrouracil residue in its loop, c) the anticodon arm, d) the variable arm and e) the T arm containing the constant sequence TΨCG in its loop (this sequence is not present in eukaryotic initiator tRNA and in some bacterial tRNAs involved in functions other than ribosomal protein synthesis).
2. Certain positions are always occupied by either the same nucleotides (invariant bases) or the same type of nucleotides (semiinvariant bases) as indicated in Fig. 32. Most of these positions are located in the D arm and in the TΨCG loop.
3. The number of base pairs in the stem regions are constant (one exeption in the D stem). Generally, they are of the Watson-Crick type with occasional exceptions.
4. Mature tRNAs contain up to more than 16% modified nucleosides. Modifications occur during the maturation process of the tRNAs after transcription (for biosynthesis of tRNA see [69, 82]). The most frequent modification is methylation either of the ribose or of the base moieties of the nucleotides. More than 50 different base modifications in tRNAs have been

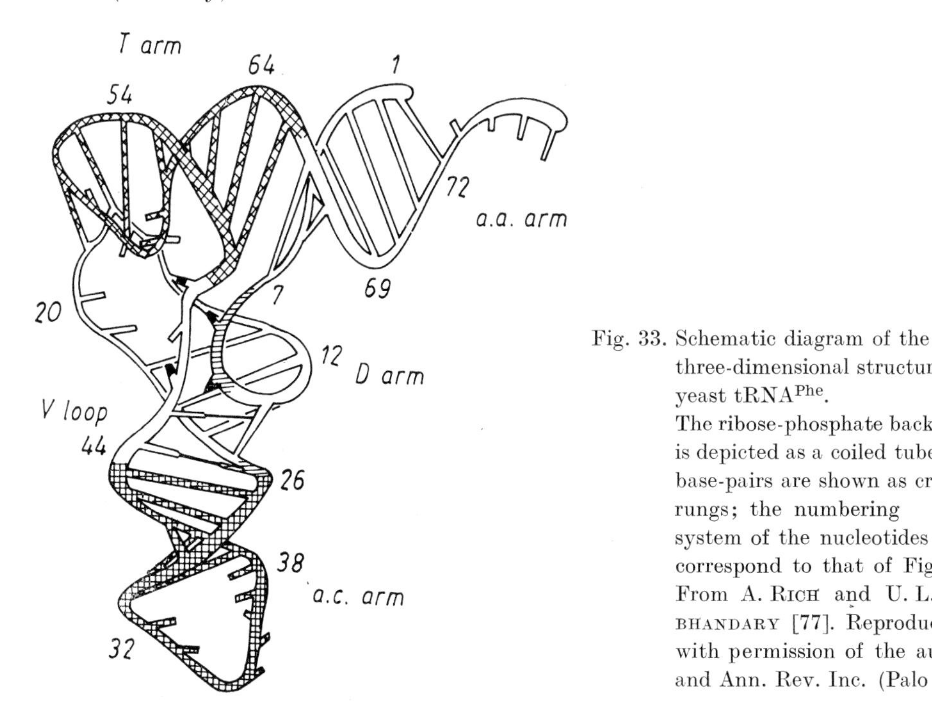

Fig. 33. Schematic diagram of the three-dimensional structure of yeast $tRNA^{Phe}$.
The ribose-phosphate backbone is depicted as a coiled tube; base-pairs are shown as cross-rungs; the numbering system of the nucleotides correspond to that of Fig. 32. From A. Rich and U. L. RajBhandary [77]. Reproduced with permission of the authors and Ann. Rev. Inc. (Palo Alto)

dentified so far. The functional role of some of these modifications is discussed in [81, 84].

X-ray crystallographic investigations led to the elucidation of the tertiary structure of $tRNA^{Phe}$ from yeast in the beginning of 1973 [85]. Since that time the model of the tertiary structure has been further refined [86, 87] (for reviews see [77, 80, 81, 88, 89]). The three-dimensional structure of $tRNA^{Phe}$ (Fig. 33) shows the following features:
The overall shape of the molecule is that of the letter L. The amino acid acceptor end is positioned at one and the anticodon at the other end of the L, both about 7 nm distant from each other. The D loop together with the TΨCG loop forms the corner of the L. The base-paired stem regions of the cloverleaf are maintained in the tertiary structure and are arranged in two nearly continuous double helices.

Three lines of evidences confirm the assumption that the tertiary structure found for $tRNA^{Phe}$ from yeast reflects the general tertiary structure of tRNAs:
1. Most of the tRNA sequences known so far can be arranged as predicted from $tRNA^{Phe}$ [77, 80, 81].
2. The majority of the data of physical and chemical investigations of different tRNAs in solution are in agreement with the data obtained by X-ray structure analysis of two crystal forms of $tRNA^{Phe}$ [77, 80, 81, 90, 91].
3. Recently, X-ray crystallographic investigation of yeast $tRNA_f^{Met}$ revealed the same three-dimensional structure with some slight deviations [92].

Aminoacylation of tRNA

The first step in protein biosynthesis occurring in the cytoplasm outside the ribosome is the activation of the amino acid and its binding to the cognate tRNA. These reactions are catalyzed by aminoacyl-tRNA synthetases (EC 6.1.1.) and can be summarized as follows:

(1) $\text{Amino acid}_n + \text{ATP} \rightleftharpoons \text{Aminoacyl}_n \sim \text{AMP} + \text{PP}_i$

(2) $\text{Aminoacyl}_n \sim \text{AMP} + \text{tRNA}_n \rightleftharpoons \text{Aminoacyl-tRNA}_n + \text{AMP}$

As the result of these reactions the amino acid is bound via an ester linkage to the 2′ or 3′ position of the terminal ribose of the tRNA molecule [93].

For each of the 20 amino acids there is one specific aminoacyl-tRNA synthetase in the cytoplasm. Studies on the structure, function and specificity of aminoacyl-tRNA synthetases have been extensively reviewed [93—98]. Here, only some of the results will be roughly summarized:

The enzymes have different molecular weights ranging from 46000 to about 290000, but most of them show a molecular weight of about 100000 [94, 96, 98]. In the cytoplasm of eukaryotic cells, several synthetases are organized in high molecular weight complexes (for reviews see [94, 99]). Subunit compositions of synthetases of the following types are known: α_1, α_2, α_4, $\alpha\beta$, $\alpha_2\beta_2$ [96]. The amino acid compositions of several aminoacyl-tRNA synthetases have been determined. Some of them contain repeated sequences with hydrophobic regions [94—96]; the total sequence has been determined in one case [100]. The three-dimensional structure of two aminoacyl-tRNA synthetases has been investigated by X-ray crystallographic analysis [101—104].

The high translational fidelity of protein synthesis (about three errors per 10000 amino acids [105]) requires the correct recognition of the amino acid as well as its cognate tRNAs by the aminoacyl-tRNA synthetase. The specific amino acid recognition by the enzyme seems to be achieved by a proof-reading mechanism [93, 94, 106—108].

Besides their cognate amino acid most synthetases are able to recognize several tRNA molecules (isoacceptor tRNAs) out of about 60 different tRNA molecules that all are of similar shape. The molecular mechanism of these highly specific interactions is not yet clearly understood. In the interaction with the enzyme all parts of the tRNA molecule but the D loop and the TΨCG arm are involved [80, 81, 94—96]. From the knowledge of the tertiary structure of yeast $tRNA^{Phe}$ it was concluded that the synthetase interacts mainly with the inner part of the L [80, 94, 95, 109, 110]. Different models have been proposed in order to explain the high specificity of tRNA recognition (see e.g. [112]). None of these models, however, is able to explain all experimental results obtained by various approaches [113]. Thus, one crucial question of the high translational fidelity at that step of protein synthesis remains to be clarified (for recent reviews see [98, 114, 115]).

Interaction with translation factors

The binding of aminoacyl-tRNA to the ribosome is mediated by at least two protein synthesis factors, eIF-2 and eEF-Tu. Initiation factor eIF-2 specifically binds initiator-tRNA (Met-$tRNA_f$), while the elongation factor eEF-Tu preferentially binds AA-tRNA but not initiator-tRNA and uncharged tRNA [116].

In the case of bacterial EF-Tu the presence of the α-amino group of the amino acid and an intact 3′-CCA end of the tRNA are required for the recognition by the factor (reviewed in [93]). Furthermore, the acceptor stem and the TΨCG stem only seem to be involved in the interaction [111].

In contrast to the other tRNA molecules the 5′ terminal nucleotide of bacterial initiator-tRNA is not involved in base-pairing. This seems to be of importance for the discrimination of initiator-tRNA by EF-Tu [93]. At present there are no indications that the constant sequence TΨCG that is replaced in eukaryotic initiator-tRNA by AUCG or AΨCG respectively, [79], is involved in the discrimination of initiator-tRNA by eukaryotic EF-Tu.

Interaction with mRNA

The codon-anticodon interaction can be considered as one of the most important steps of the whole translation process. It occurs by base pairing between the three bases of the codon on one side and the three anticodon bases of the tRNA on the other one. In the first two nucleotide positions of a codon, pairing obeys the classical pairing rules, whereas for the explanation of the weaker base interaction of the third position the wobble hypothesis was predicted in 1966 [117]. The present knowledge concerning this question, derived from a multitude of experimental data, has been reviewed recently [118—120].

Another problem of codon-anticodon interaction is the relative low energetical specificity of the interaction between corresponding trinucleotides, which cannot account alone for the high translational fidelity during protein synthesis [105]. The introduction of kinetic concepts [106, 121—124] helped to explain these differences (for review see [119]). Presumably proof-reading plays a role in codon-anticodon recognition [121]. Furthermore, rRNA-tRNA interactions possibly take part in enhancing the affinity of the correct aminoacyl-tRNA (see below) [123, 124]. Although the above mentioned hypotheses may contribute to answer some of the open questions, more experimental work has to be done to clarify the details of codon-anticodon recognition [119].

Interaction with the ribosome

During protein synthesis the aminoacyl-tRNA molecules interact with other components of the translation machinery in an ordered sequence of events. Although not yet all details of these interactions are known at present, the following statements can be made: From prokaryotic systems it is obvious that after correct codon-anticodon interaction a conformational rearrangement in the tRNA molecule occurs leading to an exposure of the TΨCG sequence at the corner of the L-shapped molecule [125—127], which is then able to base-pair with the constant CGAA sequence of the (prokaryotic) ribosomal 5 S RNA (for reviews see [116, 128, 129]). A model, how this rearrangement of the TΨCG sequence, normally involved in tertiary base pairing (Fig. 33), could occur, has been developed recently [81, 87]. The interaction of the tRNA with the ribosomal 5 S RNA seems to enhance the affinity of the correct aminoacyl-tRNA to the ribosome [123, 124]. Furthermore, 5 S RNA is assumed to be active in translocating the tRNA within the ribosome [116].

In eukaryotic ribosomes there are two different low molecular weight RNA molecules, 5 S RNA and 5.8 S RNA. From sequence data (cf. Chapter VI. 4) it is obvious that animal 5 S RNA contains a constant sequence complementary to the GAUC-sequence of initiator-tRNA, whereas 5.8 S RNA contains a constant sequence complementary to the GTΨC-sequence of non-initiator-tRNAs [130]. Whether these sequences are really involved in tRNA binding to the ribosome during protein synthesis remains to be proven.

During peptide bond formation two tRNA molecules (aminoacyl-tRNA and peptidyl-tRNA) interact with the peptidyltransferase center of the ribosome by their 3′ amino acid acceptor ends. Simultaneously, their anticodons are base-paired with two adjoining codons at the mRNA. Thus, both tRNA molecules come close together with their both ends, which can be achieved if the corners of the L are distant from each other. Assuming this positioning the prediction can be made that during translocation mainly the corner part of the molecule has to move by a rotatory operation around an assumed axis between the anticodon and the 3′ end [131]. This model of movement of tRNA, however, remains to be proven experimentally. In any case, during translocation the tRNA molecule as well as mRNA undergo slight conformational changes [81]. Crystallographic investigations demonstrated that tRNA molecules show a relative high flexibility in their anticodon region and in their acceptor ends rather than in the central parts [87]. This is presumably one of the properties, which enables tRNA to take part in the variety of interactions occurring during proteins synthesis.

c) *Initiation factors*

The assembly of the eukaryotic ribosomal initiation complex containing initiator tRNA (Met-tRNA$_f$), mRNA and both ribosomal subunits has been subjected to extensive investigations during the past decade (for reviews see [132—136]). Although not all details of this process are understood at present, at least 9 different protein factors (eukaryotic initiation factors = eIF) are known to be engaged in initiation of eukaryotic protein synthesis. The fact that in prokaryotes three proteins seem to suffice to start polypeptide synthesis points to the higher complexity of eukaryotic initiation [134].

Main progress in purification and characterization of the different initiation factors was achieved by Andersons group in Bethesda and by Staehelins laboratory in Basel, both working with the reticulocyte system. They purified all 9 initiation factors to homogeneity, characterized them physically and biologically and determined the amino acid composition of most of them [137—149] (for summary see Table 24). More recently, also other groups purified and characterized these factors from reticulocytes [151—156]. (Procedures for preparation and functional analyses of initiation factors are collected in [157].) From Krebs II ascites cells seven initiation factors showing the same properties as the corresponding factors from reticulocytes were purified [158].

Factor eIF-1 has been isolated from

Table 24. Eukaryotic initiation factors

Nomenclature according to [150]	Molecular weight	Functions	Other designations and references
eIF-1	15000	Binding of natural mRNAs to 40 S subunits. Inititation complex stabilization	IF-E1 [139] IF-M_β [239]
eIF-2	subunit α: 32000–38000 subunit β: 47000–52000 subunit γ: 50000–54000	GTP dependent template independent binding of Met-$tRNA_f$ to 40 S ribosomal subunit	IF-E2 [139, 152, 188] IF-MP [142, 143, 175, 177, 179] IF-I [161] EIF-1 [162–164] EIF-2 [171] EIF-3 [167] IF-2 [169, 170] aIF-2 [176] C3β [181] IF-L3 [165]
eIF-2A	50000–96000	Template dependent GTP independent binding of Met-$tRNA_f$ to 40 S ribosomal subunits	IF-M1 [144, 216] IF-1 [170, 217] aIF-2A [176] B.F. [215]
eIF-3	500000–750000 (complex of 7–11 different polypeptides	Prevents association of ribosomal subunits. Stabilization of 40 S initiation complexes	IF-E3 [139, 151] IF-M5 [145] IF-M3 [225] IF-3 [226, 228]
eIF-4A	48000–53000	Binding of natural mRNAs to 40 S initiation complexes	IF-E4 [139, 153, 240] IF-EMC [238–240] IF-M4 [145]
eIF-4B	80000–82000	Binding of natural mRNAs to 40 S initiation complexes. Cap recognition (?)	IF-E6 [139, 153] IF-M3 [146, 216]
eIF-4C	19000	Stabilization of 40 S initiation complexes. Stimulation of subunit joining (?)	IF-E7 [139] IF-M_α [239] IF-M2B_β [147]
eIF-4D	17000	Stabilization of 40 S initiation complex	IF-M2B_α [147]
eIF-5	125000–160000	Release of eIF-2 and eIF-3 from 40 S initiation complex. Binding of the 60 S ribosomal subunit to the 40 S initiation complex	IF-E5 [139] IF-M2A [148, 216, 247] IF-II [161] IF-A2A [248]

reticulocytes [137, 140, 154—156] and from ascites cells [158, 159]. It is a small protein, the precise role of which in peptide chain initiation is not yet exactly understood. It is possibly involved in binding of natural mRNA and in stabilizing the 40 S initiation complex [138—140, 160].

Factor eIF-2 is the most extensively studied initiation factor that has been prepared from a variety of eukaryotic cells [137, 140—142, 152, 158, 161—185]. While some authors have identified two subunits only [172, 174] most investigators found that eIF-2 consists of three unequal subunits (Table 24). Subunit α may be phosphorylated by specifically regulated protein kinases, by which the in vivo activity of eIF-2 is influenced. This phosphorylation step is of importance for the regulation of protein synthesis in reticulocytes (see in Section 4a). Another protein kinase phosphorylates the β-subunit of eIF-2 as well as distinct subunits of other initiation factors [186—189]. Also the γ-subunit of eIF-2 has been reported to be phosphorylatable [139]. The regulatory significance of these reactions, however, is not yet clear. First experiments on the functional role of the single subunits of eIF-2 point to different affinities to GTP and Met-tRNA$_f$ [194, 195].

The function of eIF-2 consists (a) in the formation of a ternary complex with initiator-tRNA and GTP, and (b) in the attachment of this ternary complex to the 40 S ribosomal subunit resulting in the formation of a quaternary initiation complex (I 2 in Fig. 35). The binding reactions are specific towards eukaryotic initiator-tRNA [143, 167, 169, 170, 178, 183, 190] and need neither GTP hydrolysis nor other components of the protein synthesizing system, such as natural or artificial templates [138, 143, 152, 160, 161, 165, 167—171]. The factor remains bound to the 40 S initiation complex until the 60 S subunit becomes bound to this complex [160, 191—193].

Several factors modulating the activity of eIF-2 have been prepared from different tissues [163, 164, 180, 196—207]. Their physiological significance, however, needs further investigation.

Recent results indicate that eIF-2 besides Met-tRNA$_f$ binds also mRNA [38, 208—214] and other RNAs, but with lower affinity [208, 209, 211]. The affinity of eIF-2 to various natural and artifical mRNAs differs significantly [211, 214]. The cap and other parts of the mRNA seem to be involved in this interaction [38, 213, 214]. Whether the differential mRNA binding activity of eIF-2 plays an essential role during protein synthesis initiation is not yet clear.

Another Met-tRNA$_f$ binding factor now called *eIF-2A* has been purified from reticulotytes [141, 144, 156, 215] as well as from other tissues [170, 176, 180, 216 to 220]. The only homogeneous preparation of this factor reported [144] consists of a single polypeptide chain of 65000 molecular weight. Other estimates deviate considerably (cf. Table 24). eIF-2A catalyzes the template-dependent but GTP-independent binding of Met-tRNA$_f$ to the small ribosomal subunit. The specificity towards initiator-tRNA is not as high as that of eIF-2 [221]. The simultaneous occurrence of both initiation factors in the same tissue [156, 170, 176, 222] led to the assumption of two different pathways of initiation complex formation operating in eukaryotic systems [222]. The eIF-2 mediated pathway, however, is favoured today, because translation of natural mRNA is absolutely dependent on eIF-2, but not on eIF-2A [222—224].

Factor eIF-3 is a high molecular weight complex consisting of 7 to 11 different protein components, the molecular weights of which range from about 30000 to 160000 [137, 140, 141, 145, 151, 155, 156, 158, 159, 184, 185, 225, 226,]. Some of them seem to be identical with other initiation factors [145], and in some cases additional minor bands can be observed by SDS-gel electrophoresis [145, 151]. The whole complex sediments with about 15 S and the total molecular weight has been determined to be 500000 to 750000 [137, 145, 151, 158, 225, 226, 230].

The following biological activities of eIF-3 have been detected so far: a) It binds to 40 S ribosomal subunits independently of other components of the translation system [151, 160, 192, 193] and can be found in native 40 S subunits [226, 227, 229, 230] and in all 40 S initiation complexes, but not in 80 S complexes [160, 192, 193]. b) It prevents reassociation of ribosomal subunits and even causes dissociation of ribosomes to a limited extent [138, 139, 226, 228]. In one case, however, dissociation activity could be separated from the complex factor [231, 232]. c) eIF-3 stabilizes the initiation complex containing 40 S subunits, eIF-2, GTP and Met-tRNA$_f$ [139, 160, 192, 193]. d) eIF-3 is essential for mRNA binding to 40 S subunits as well as for translation of natural mRNAs [138, 145, 160, 233—237]. Single subunits of eIF-3 may be phosphorylated in vivo and in vitro without affecting the factor activity [188, 189].

Initiation factors originally assumed to be involved in mRNA binding to 40 S subunits have been designated as eIF-4A, B, C, D. *Factor eIF-4A* is a single polypeptide of a molecular weight of about 50000. It has been purified from reticulocytes [137, 140, 141, 145, 153, 155, 156] and from ascites tumor cells [158, 159, 238, 239]. eIF-4A is required for translation of natural mRNAs but not of poly(U) [145, 239]. The factor promotes binding of mRNA but not of Met-tRNA$_f$ to 40 S subunits [137, 145, 160]. A stable interaction between eIF-4A and the ribosomal subunit, however, could not be detected [160]. The role of eIF-4A in mRNA binding is further confirmed by the observation that different amounts of eIF-4A are needed for optimal translation of different messages [238, 240, 241]. Thus, eIF-4A (as well as eIF-4B) seems to be responsible for the different efficiency of translation on different messengers (for review see [37]).

Factor eIF-4B also consists of a single polypeptide chain with a molecular weight of 80000. It has been obtained from several eukaryotic tissues [137, 140, 141, 146, 153, 155, 156, 158, 184, 185, 216]. Its mode of action seems to be similar to that of eIF-4A. It is also required for translation of natural mRNAs only [138, 160, 223, 224]. Moreover, eIF-4B apparently shows some mRNA discrimination activity [241 to 244], which was at first thought to be due to specific interactions between eIF-4B and the cap [245] (for reviews see [25, 37]). Recent studies, however, revealed that a minor polypeptide copurified with eIF-4B specifically binds to the cap [39, 246]. Nevertheless eIF-4B binds capped as well as uncapped mRNA [40, 244] and is necessary for mRNA attachment to 40 S ribosomal subunits in the course of initiation [138, 160].

Factors eIF-4C and eIF-4D are low molecular weight proteins (19000 and 17000, respectively), which initially copurified [216, 247]. Later on, they were separated one from another [147] and eIF-4C has now been purified in several laboratories [137, 140, 141, 147, 154—156, 158, 159, 239]. Its mode of action is not yet clear. Moderate stimulations (1.5- to 2-fold) by eIF-4C for Met-tRNA$_f$ and mRNA binding to 40 S subunits have been reported [138, 139, 147, 160]. eIF-4D, prepared in three laboratories only [141, 147, 154—156], seems slightly to stabilize Met-tRNA × 40 S complexes [193]. Moreover, it was found to stimulate methionyl-puromycin synthesis by the 80 S initiation complex [147, 160]. Further work is necessary to clarify the precise role of both factors.

Factor eIF-5 purified in several laboratories [137, 140, 141, 148, 154—156, 158, 161, 164, 168, 216, 248], consists of a large polypeptide chain of 125000—160000 molecular weight. It is responsible for the last step of initiation complex formation, the binding of the 60 S ribosomal subunit to the 40 S initiation complex [138, 148, 149, 160, 161, 168]. This "joining reaction" catalyzed by eIF-5 is accompanied by hydrolysis of the GTP first bound by eIF-2 [191, 249] together with the release of initiation factors eIF-2 and eIF-3 bound to the 40S initiation complex [160, 191 to 193]. There are recent indications that GTP-hydrolysis is necessary for the release of the initiation factors that occurs prior to 60 S subunit joining [192, 249].

There are some reports about further protein factors affecting initiation events in different ways [250—255]. These proteins cannot be correlated to the known initiation factors at present and their function in protein synthesis is still unclear.

Initiation factors can be found predominantly in the high salt wash fraction of ribosome preparations. This fraction is frequently used as starting material for factor preparations [157]. Furthermore, certain initiation factors are associated with native 40 S ribosomal subunits (see Chapter VIII), which, therefore, also often serve as source for initiation factors [256] (for review see [157]). Recently eIF-2 was described to be present in 16—20 S complexes [257].

d) Elongation factors

From the cytoplasm of eukaryotic cells two types of elongation factors have been isolated. Elongation factor 1, before named transferase I, catalyzes binding of amino-

Table 25. Eukaryotic elongation factor T

Material	Nomenclature of the authors	Molecular weight of the		References
		polymeric form	monomeric form(s)	
Reticulocytes	EF1H, EF1L	450000	50000 53000 30000	[268] [266]
Wheat embryos	EF1H, EF1L	180000—250000	62000	[269]
	EF1H, EF1A,B,C	240000—540000	A = 52000 B = 47000 C = 27000	[270—272]
Pig liver	EF1α,β,γ	β, γ = 90000	α = 53000 β = 55000 γ = 30000	[263] [273—278]
Calf brain	EF1H, EF1L	250000—1000000	50000—60000	[279]
Krebs II ascites cells	EF1H, EF1L	260000 230000 135000	47000	[280, 281]
	eEFTH, EF1H eEFTu, EF1α eEFTs, EF1β	β_2 = 52000	α = 47000 β = 26000	[267]
Artemia salina	EF1H, EF1L	200000	52000 47000	[282, 283]
	EF1H, EF1A,B,C	240000	A = 53000 B = 51000 C = 26000	[284, 285]
Silk gland	EF1H, EF1M, EF1a,b,c	H > 300000 M ≈ 150000	a = 51000 b = 26000 c = 46000	[261, 262]

acyl-tRNA to the ribosome and elongation factor 2, previously named transferase II, is involved in the translocation of peptidyl-tRNA from the ribosomal A site to the P site (reviews covering the literature until 1977 can be found in [116, 133, 258, 259]).

Whereas data obtained on EF-2 are unequivocal, the work with EF-1 has led to a number of conflicting results. There are recent tendencies to assimilate the nomenclature of eukaryotic elongation factors to the prokaryotic system [260 to 266].

Accordingly, aminoacyl-tRNA binding is catalyzed by eEF-T, that consists of two different factors, eEF-Tu and eEF-Ts, and translocation by eEF-G (e stands for eukaryotic). Correlations between the earlier and the novel system are given in Table 25. The proposed new nomenclature will be used here.

Elongation factor eEF-T occurs in the soluble fractions of various cell types in more or less aggregated forms with molecular weights over 100000 up to 1500000. The aggregates contain 3 different polypeptide chains. Disaggregation of the heavy forms of eEF-T can be accomplished by GTP under physiological conditions [283, 286, 287] or in vitro by specific proteolytic enzymes [288—292]. After rather puzzling results about number, molecular weights and the functional roles of the individual protein components [293] the data first obtained for pig liver EF-T [263, 273 to 278] proved true also for the majority of other tissues studied. Accordingly, eukaryotic EF-T consists of 3 different protein chains (see also Table 25). eEF-Tu has a molecular weight of 51000—53000 and was shown to be responsible for aminoacyl-tRNA binding to the ribosome; its function therefore corresponds with bacterial EF-Tu [294].

Eukaryotic EF-Ts was often found to consist of two components — EF-1β and γ [263, 276, 277], EF-1b and c [261, 262] or EF-1B and C [272, 284]. The molecular weights of the both components were estimated to 46000 to 53000 for the larger and to 23000 to 30000 for the smaller protein subunit. The proteins reveal a tendency to aggregate one with another. Functionally, the 30000 dalton component seems to correspond with the bacterial elongation factor EF-Ts [277], thus being responsible for recycling of EF-Tu during protein synthesis according to the following scheme (Fig. 34). The role of the 50000 dalton component of eEF-Ts remains to be clarified. Its function may be correlated to the inability of eEF-T to form stable complexes with GTP or GDP. However, it is certain that this protein compound is different from eEF-Tu with regard to amino acid composition and isoelectric points although their molecular weights are very similar.

Elongation factor eEF-G has been isolated from a large number of different tissues and purified to apparent homogeneity (for review see [116]). The molecular weights estimated range from 65000 to 112000. The amino acid composition of eEF-G is rather similar to bacterial EF-G but the number of cystein residues in the eukaryotic factor is considerably higher (22 to 6 Cys). The function of eEF-G is identical to that of EF-G, but the latter was not found to be able to substitute for eEF-G in eukaryotic protein synthesis [294].

During elongation, eEF-G at first forms a binary complex with GTP and then associates with the ribosome at the large subunit. eEF-G promotes translocation of the peptidyl-tRNA from the ribosomal A site to the P site. The subsequent hydrolysis of GTP to GDP causes conformational changes of eEF-G leading to the detachment of the factor from the ribosome [259, 295]. Consequently, eEF-G functions catalytically during protein synthesis.

A number of suitable procedures for preparation of elongation factors from various eukaryotic tissues was published recently [296—303].

e) Role of GTP during elongation

Two moles of GTP are required for the elongation of the peptide chain by one amino acid. One GTP is required for the binding of AA-tRNA to the ribosome promoted by eEF-Tu and one GTP is consumed after translocation catalyzed by eEF-G. During the first step, the AA-tRNA binding, GTP seems to alter the conformation of eEF-Tu in a manner that facilitates its interaction with AA-tRNA and the ribosome. After binding of the [AA-tRNA × eEF-Tu × GTP] complex to the ribosome, cleavage of GTP occurs in order to release the complex [eEF-Tu × GDP] from the ribosome.

A remarkable similarity is found in the mechanism of utilization of GTP energy for the translocation promoted by eEF-G. The affinity of eEF-G towards the ribosome is dependent on its association with GTP. Also the complex with a nonhydrolyzable GTP analogue, [eEF-G × Gpp(NH)p], can interact with ribosomes to form a stable ternary complex, whereas [EF-G × GDP] has practically no affinity to the ribosome. [eEF-G × GTP] promotes translocation as well as [eEF-G × Gpp(NH)p] of the peptidyl-tRNA from the ribosomal A site to the P site. The hydrolysis of GTP is necessary in order to release eEF-G from the ribosome after completion of the translocation step. Thus, the role of GTP in the above two reactions seems to be quite analogous. In both cases, the conformation as well as the reactivity of protein molecules are reversibly and qualitatively altered by the change of its nucleotide ligand. A single turnover reaction is accomplished utilizing the specific conformation induced by GTP, and the hydrolysis of GTP is required to shift the factor protein to the alternate conformation [295].

f) Termination factors

Relatively few information is available at present on factors catalyzing chain termination in eukaryotic protein biosynthesis.

Only one so-called release factor (RF) has been isolated from eukaryotes [304 to 310] in contrast to three release factors found in prokaryotes. The molecular

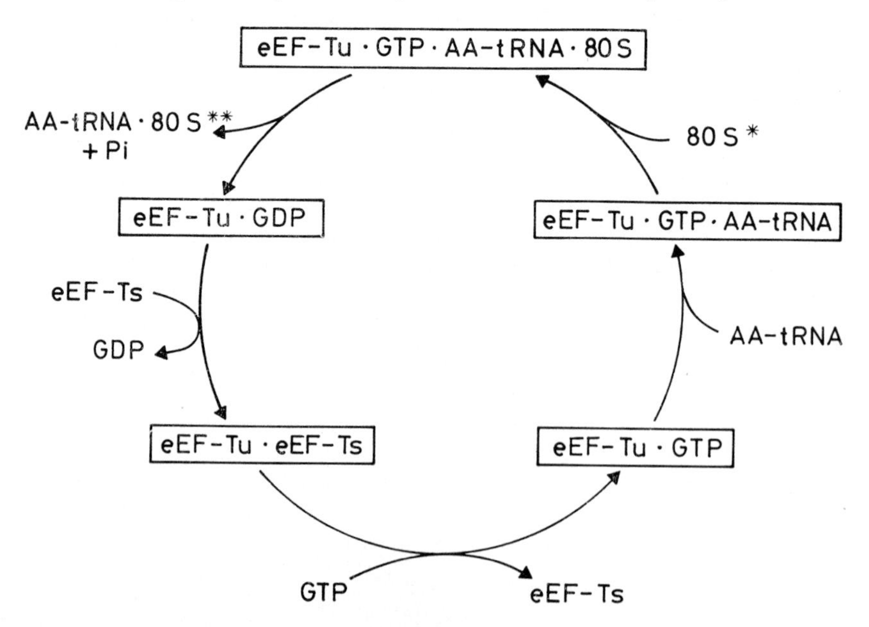

Fig. 34. Recycling of eEF-Tu during protein synthesis. *80 S ribosome of stage E 4 or I 4 in Fig. 36; **80 S ribosome of stage E 1 in Fig. 36.

weight of the eukaryotic RF was estimated to 105000 to 250000 after Sephadex G-200 filtration at varying salt concentrations and close to 50000 by sucrose gradient centrifugation. At increased ionic strength, RF was found to be separable into at least two species of 149000 and 43000 molecular weight. For RF from rabbit liver a molecular weight of 350000 was estimated [310]. The native 105000 component of reticulocyte release factor was found to consist of two subunits of 56500 molecular weight [307].

Regarding the function of RF, it is not known how the factor signals the end of the message to the peptidyltransferase centre. The action of RF depends on the hydrolysis of GTP to GDP and inorganic phosphate.

From rough microsomes of neonatal liver, a cAMP-activatable termination factor of 40000 molecular weight has been purified [311, 312] which catalyzes the release of the enzyme tyrosine aminotransferase from ribosomes.

g) Dissociation factors

There are several reports of a dissociation factor present in yeast [313, 314], rat liver [226, 230, 315], reticulocytes [316 to 319], ascites cells [320], muscle [321], *Xenopus laevis* [322] and mouse brain [323].

The factor, however, has not yet been extensively purified and was claimed to be closely related to initiation factor eIF-3 [321, 322] (see also Section 1c of this chapter).

The factor is preferably prepared from 40 S and 60 S subunits. If it is applied in several fold molar excess, the factor dissociates unprogrammed ribosomes into subunits at low magnesium concentrations. Increase of magnesium ions reverses the dissociating effects. However, there is a report that dissociation factor of yeast [314] is able to dissociate programmed ribosomes that still carry the peptidyl-tRNA at the 60 S subunit.

2. The process of eukaryotic protein synthesis

In ribosomal protein synthesis three phases can be distinguished: initiation, elongation and termination of the polypeptide chain. In the initiation steps the protein synthesizing complex consisting of the 80 S ribosome, mRNA, and initiator-tRNA is formed. During elongation the ribosome migrating along the mRNA synthesizes currently polypeptide bonds by using AA-tRNAs as substrate. In the cell several (2-100) elongating ribosomes can be found on one mRNA molecule, thus forming so-called polysomes. After completion the polypeptide chain is released during termination, and the ribosome dissociates under concomitant liberation of mRNA.

A schematic representation of the ribosome cycle during protein synthesis explained in the following text is given in Fig. 36.

a) Initiation

Initiation comprises all events leading to an 80 S initiation complex that consists of the 80 S ribosome, initiator-tRNA (Met-$tRNA_f$) bound to the ribosomal P site and messenger RNA. A tentative pathway for the assembly of the initiation complex is presented in Fig. 35 (for surveys see [138 to 140, 155, 160, 324, 325]).

The assembly starts with the free 40 S ribosomal subunit, to which eIF-3 is assumed to be bound at first [160, 192, 193]. This factor seems to prevent reassociation of ribosomal subunits and stabilizes the 40 S initiation complexes (see Section 1c of this chapter).

In the second step preformed ternary initiation complexes containing eIF-2, Met-$tRNA_f$, and GTP are bound to the 40 S subunits, thus forming the initiation complex I 2. This complex is stabilized by eIF-4C [139, 147, 160] and possibly also to some extent by eIF-4D [193].

A very important and, at least in some cases, rate limiting [139, 326] step during initiation is the binding of mRNA to the

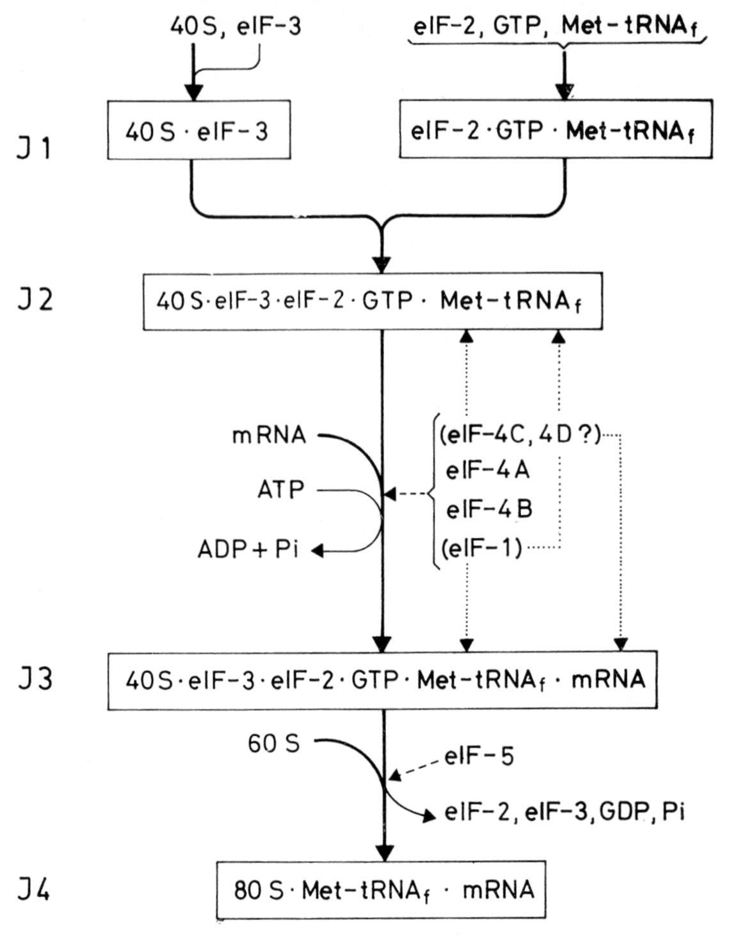

Fig. 35. Scheme of initiation of eukaryotic protein synthesis.
Stages of initiation complex formation (I 1—I 4) correspond to those given in Fig. 36. ⟶ reactions, - - -→ catalyzing effect, ·····→ stabilizing effect.

40 S subunit resulting in the relatively stable [139] initiation complex I 3. This step includes the selection of mRNA and the correct positioning of the first AUG-codon of the mRNA (for review see [37]). It requires ATP-hydrolysis [138, 160, 327], factors eIF-4A and eIF-4B, and to a lesser extent also eIF-1 and eIF-4C [138, 160]. eIF-1, -4A and 4B- are not required when AUG is used instead of a natural template. The resulting complex I 3 contains the 40 S subunit, Met-tRNA$_f$, mRNA, GTP, as well as factors eIF-2 and eIF-3 [138, 160, 191—193]. None of the other initiation factors could be found attached to any of the initiation complexes [160, 193].

The last step in initiation comprises the release of eIF-2 and eIF-3 connected with the hydrolysis of the bound GTP and with the attachment of the 60 S ribosomal subunit. These reactions are catalyzed by eIF-5 [138, 160, 191—193, 249] and seem to be stimulated by eIF-4C [193]. The resulting 80 S initiation complex (I 4) containing the 80 S ribosome, Met-tRNA$_f$ bound to the ribosomal P site, and mRNA is ready to accept a second aminoacyl-tRNA at its A site and thus to start with peptide chain elongation.

b) Elongation

Once the 80 S complex has been formed (stage I 4), protein synthesis can be defined as a repetitive sequence of the following three events (see Fig. 36).

i. Binding of aminoacyl-tRNA (stage I 4 (or E 4) → E 1).
ii. Transfer reaction by peptide bond formation (stage E 1 → E 2).
iii. Translocation of peptidyl-tRNA from the ribosomal A site[1]) to the P site[2]) (stage E 2 → E 3).

During each round of elongation the nascent peptide chain is prolonged by one amino acid.

In the initiated 80 S complex the ribosomal P site is occupied by the initiator-tRNA (stage I 4) which is a very similar situation as that reached later on after each completed round of elongation (stage E 4) when the peptidyl-tRNA is located in the P site. In all these cases a new codon triplet at the mRNA strand programs the ribosomal A site, that is thereby ready to accept a new aminoacyl-tRNA molecule with the complementary anticodon. This step is also called "decoding" because the mRNA codon directs the binding of the correct aminoacyl-tRNA to the A site.

Aminoacyl-tRNA binding is mediated by elongation factor eEF-Tu and GTP which form a ternary complex with aminoacyl-tRNA. On binding of this complex to the ribosome, GTP is hydrolyzed, and inorganic phosphate and eEF-Tu complexed with GDP are released. The complex eEF-Tu × GDP is recycled by eEF-Ts according to the scheme given in Fig. 34.

In the next step of elongation (stage E 1 → E 2) peptidyltransferase, an enzymatic activity of the large ribosomal subunit, catalyzes peptide bond formation. This reaction occurs between the nucleophilic NH_2 group of the A site-bound aminoacyl-tRNA and the P site-bound esterified carboxyl group of methionine of the initiator-tRNA (or later on, of the last amino acid of peptidyl-tRNA). As a result of this step, the peptide is anchored to the ribosome via the ribosomal A site and the discharged tRNA molecule is released from the P site of the ribosome.

The next step of elongation is the translocation of the peptidyl-tRNA from the ribosomal A site to the P site that is connected with the simultaneous movement of the mRNA codon to which peptidyl-tRNA is bound. Translocation is catalyzed by elongation factor eEF-G in presence of GTP. Upon hydrolysis of GTP, eEF-G is released from the ribosome (see Section 1 d, this chapter). After translocation, the peptidyl-tRNA is positioned in the ribosomal P site (stage E 4), and the ribosomal A site contains the codon to accept the next aminoacyl-tRNA molecule. Subsequently, a new round of elongation can be started.

c) Termination

The end of the mRNA to be read is signalized to the ribosome when the termination codons UAA, UAG, or UGA have entered the ribosomal A site. Peptide chain termination defines an event that results in the release of the polypeptide from its ribosome-bound tRNA.

In reticulocytes one larger release factor (RF) that may consist of several polypeptide chains catalyzes termination of protein synthesis [304]. RF binds to the ribosome only, a) when a termination codon is located in the A site and b) after eEF-G has left the ribosome. GTP is involved in the association of RF with the ribosome, and its hydrolysis is required as a subsequent rate-limiting step which was demonstrated by the existence of a ribosome-dependent GTPase activity of RF.

Although the precise mode of action is

[1]) A site designates the ribosomal locus by which aminoacyl-tRNA is accepted during elongation.

[2]) P site designates the ribosomal locus at which peptidyl-tRNA is positioned after translocation.

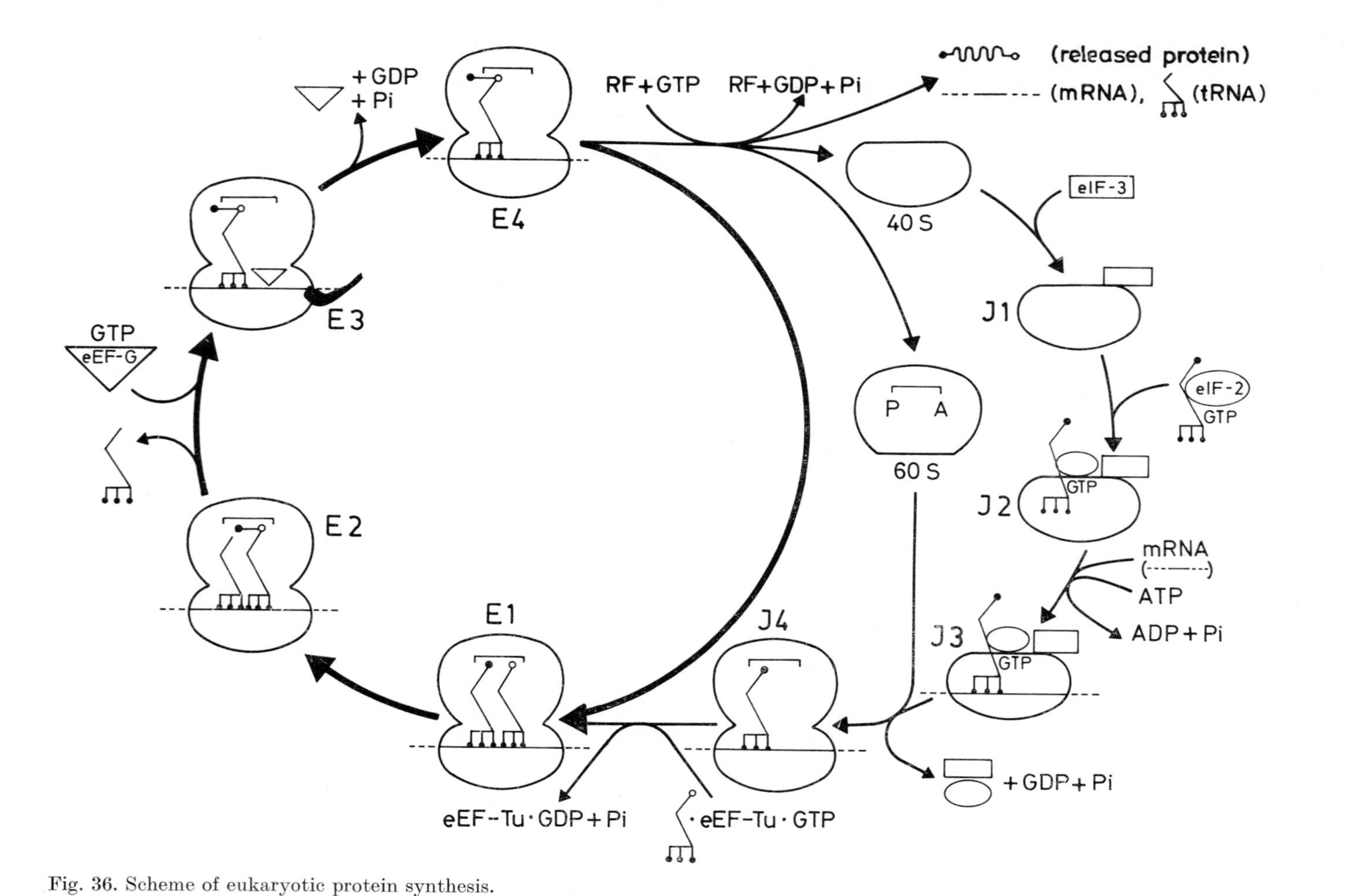

Fig. 36. Scheme of eukaryotic protein synthesis.

I 1—I 4: stages of initiation complex formation (cf. Fig. 35); E 1—E 4: stages of the repeating elongation cycle (thick arrows); for further details see text.

P: binding site for peptidyl-tRNA; A: binding site for AA-tRNA; eIF: eukaryotic initiation factor; eEF: eukaryotic elongation factor; RF: release factor; : Met-$tRNA_f$ (initiator-tRNA), elongator-AA-tRNA, deacylated tRNA, respectively.

not yet clear. RF seems to evoke esterase activity in the peptidyltransferase center that results in the hydrolysis of peptidyl-tRNA. It has still to be proved experimentally that a nucleophilic group of the factor participates in the attack and accepts transiently the nascent polypeptide chain in a covalent linkage. Thus, the role of RF might rather be to restrict the choice of peptidyltransferase for nucleophilic agents in order to promote hydrolysis of peptidyl-tRNA, and in this way to facilitate hydrolysis of the ester bond of the peptidyl-tRNA, thereby releasing the peptide.

After hydrolysis of GTP, RF is released, and the ribosome dissociates into its subunits. That enables them to reinitiate translation of a mRNA molecule under appropriate conditions.

3. Protein synthesis on membrane bound ribosomes

In the early sixtieth it became evident from biochemical and morphological experiments that most eukaryotic cells contain free and membrane-bound ribosomes (see Chapter III). Evidence accumulated that proteins for intracellular use are preferably synthesized on free ribosomes, and proteins for segregation and for membrane structure are preferably made on ribosomes associated with the endoplasmic reticulum (for review see [328]). Specific sites at the large ribosomal subunit on the one hand and at the endoplasmic reticulum on the other hand mediate the interaction between both structures. Besides, there is an independent selection principle by which the possible association or non-association of a given translating ribosome is directed. A possible mechanism for such a process was formulated in the so-called signal-hypothesis [329] (see also Chapter X), and in the following years largely verified and extended by a great number of experiments in different eukaryotic and even prokaryotic systems.

The basic implication of the hypothesis is that mRNAs for all proteins to be synthesized at membrane-bound ribosomes should contain special nucleotide sequences following the initiation codon that code for an extra sequence of amino acid residues, the so-called signal sequence. These NH_2-terminal amino acid sequences, when emerging from the large ribosomal subunit of — at first — free ribosomes, become attached to certain receptor sites at the microsomal membranes. In this way a transient tunnel through the large ribosomal subunit and the membrane seems to be formed. Proteins synthesized at membrane-bound ribosomes, therefore, can be expected to be made at first as larger precursors.

In the last few years, several authors have shown in fact that the primary translation products of various mRNAs, when synthesized in membrane-free systems, are larger by NH_2-terminal extensions of up to 30 amino acid residues than the proper authentic proteins. This was demonstrated for the light chain of certain immunoglobulins [330—344], proparathyroid hormone [345 to 351], promelittin [352—354], growth hormone [355—359], proinsulin [360—368], pancreatic enzymes [369, 370], serum albumin [371—375], alpha-lactalbumin [376, 377], lactogen [378—383], prolactin [384 to 388], corticotropin peptide [389], ribulose-1,5-biphosphate carboxylase [390, 391], egg white proteins [392—394] and viral glycoproteins [395].

The primary structures of a number of these preceding signals have been analyzed (Tab. 26). The originally expected sequence homologies between the different signals could not be confirmed. However, as obvious from the table, the majority of amino acids within the signals is of hydrophobic character resulting in hydrophobicity indices [396] of 2.0 to 3.0. Most signals contain strongly hydrophobic clusters (underlined in the table). Hydrophobicity as a common property of the signal sequences seems to direct the ribo-

Table 26. Amino acid sequences of NH_2-terminal signal peptides of various preproteins

Sequence of the signals*																										Synthesized protein	References
	−25					−20					−15					−10					−5				−1		
			Met	Asp	Met	Arg	Ala	Pro	Ala	Glu	Ile	Phe	Gly	Phe	Leu	Leu	Leu	Leu	Phe	Pro	Gly	Thr	Arg	Cys	Asp	Immunoglobulin MOPC41 (L-chain)	[342]
							Met	Ala	Trp	Ile	Ser	Leu	Ile	Leu	Ser	Leu	Leu	Ala	Leu	Ser	Ser	Gly	Ala	Ile	Ser	Immunoglobulin MOPC104E (L-chain)	[342]
						Met	×	Thr	×	Thr	Leu	Leu	Leu	Trp	Val	Leu	Leu	Leu	Trp	Val	Pro	×	×	Thr	×	Immunoglobulin MOPC321 (L-chain)	[343]
								×	Ala	Lys	Leu	Phe	Leu	Phe	Leu	Ala	Leu	Leu	Leu	Ala	Tyr	Val	Ala	Phe	Pro	Trypsinogen-2	in ref. [342]
		×	Leu	Lys	Met	×	Phe	Leu	Phe	Leu	Leu	Ala	Leu	Leu	Val	Leu	Trp	Glu	Pro	Lys	Pro	Ala	Gln	Ala	Phe	Proinsulin (rat)	in ref. [432]
		Met	Ala	Leu	×	Leu	×	×	Phe	×	Leu	Leu	Val	Leu	Leu	Val	Val	×	×	×	×	×	×	Ala	Val	Proinsulin (avian)	in ref. [342]
	Met	Met	Ser	Ala	Lys	Asp	Met	Val	Lys	Val	Met	Ile	Val	Met	Leu	Ala	Ile	Cys	Phe	Leu	Ala	Arg	Ser	Asp	Gly	Proparathyroid hormone	[351]
					Met	Lys	Phe	Leu	Val	×	Val	Ala	Leu	Val	Phe	Met	Val	Val	Tyr	Ile	×	Tyr	Ile	Tyr	Ala	Ala	Promelittin	[354]
							Met	Lys	Trp	Val	Thr	Phe	Leu	Leu	Leu	Leu	Phe	Ile	Ser	Gly	Ser	Ala	Phe	Ser	Arg	Proalbumin (rat)	[374, 375]
	Met	Pro	Gly	Ser	Arg	Thr	Ser	Leu	Leu	Leu	Ala	Phe	Ala	Leu	Leu	Cys	Leu	Pro	Trp	Leu	Gln	Glu	Ala	Gly	Ala	Lactogen	[382]
Met	Ala	Ala	Asp	Ser	Glu	Thr	Pro	Trp	Leu	Leu	Thr	Phe	Ser	Leu	Leu	Lys	Leu	Leu	Trp	Pro	Gln	Glu	Ala	Gly	Ala	Growth hormone	[359]
							Met	Arg	Ser	Leu	Leu	Ile	Leu	Val	Leu	Cys	Phe	Leu	Pro	Leu	Ala	Ala	Leu	Gly	Lys	Lysozyme	[392, 393]
	Met	Ala	Met	Ala	Gly	Val	Phe	Val	Leu	Phe	Ser	Phe	Val	Leu	Cys	Gly	Phe	Leu	Pro	Asp	Ala	Ala	Phe	Gly	Ala	Ovomocuid	[393, 394]
										Met	Lys	Cys	Leu	Leu	Tyr	Leu	Ala	Phe	Leu	Phe	Ile	His	Val	Asn	Cys	VSV glycoprotein	[395]
							Ser	×	Ser	Val	×	Leu	Val	Phe	Leu	Val	Leu	Val	Ser	Leu	×	Gly	Leu	Tyr	×	β-microglobulin	[399]

* Hydrophobic amino acid residues are underlined

somes carrying these nascent peptide chains to specific binding sites at the membrane and to facilitate penetration of this protein into the intravesicular lumen of the endoplasmic reticulum.

When the signal sequences of the proteins are transferred through the membrane, specific enzymes located within the membrane, called signal peptidases, accomplish the removal of the hydrophobic amino acid sequences, as obvious from a number of studies [377, 397, 398]. The enzyme(s) responsible and the mode of action, however, are still unknown.

Thus, the signal sequences represent metabolically short-lived peptide signals which trigger the attachment of the ribosomes to membranes and the transfer of secretory proteins across microsomal membranes.

There are more recent reports [400 to 402] that ovalbumin lacks a hydrophobic NH_2-terminal signal, although it is a typical protein for segregation. Instead, ovalbumin has a highly hydrophobic insert between the amino acid residues 229 and 276, which seems to compete with signals of other proteins located at their NH_2-terminus. The function and mode of interaction of ovalbumin with the endoplasmic reticulum are now under study.

Splitting off signal sequences from certain proteins is only one kind of posttranslational modification which occurs to many or all proteins to be segregated from the cell directly after their synthesis. Besides, a number of other posttranslational modifications are known that are necessary to convert the primary translational product into a protein of enzymic or structural function within or outside the cell.

4. Some regulatory aspects of protein biosynthesis in eukaryotes

The synthesis of proteins in the cell is regulated mainly at the transcriptional level, that means by the production of the corresponding mRNA, and by posttranscriptional modifications of the mRNA molecules. The spatial separation of the transcription and the translation process in eukaryotes, however, requires additional control mechanisms on the translational level. Basically, there are two possibilities concerning a) control of protein synthesis in general, and b) control of the synthesis of specific proteins.

Several examples of translational controls have been investigated in the past (for reviews see [3, 135, 403]). One type of translational control is obviously connected with the organization of mRNA in ribonucleoprotein particles (for reviews see [3, 18–21]). Here, three other well established examples of translational control acting predominantly at the initiation step of protein synthesis will be briefly described.

a) Hemin control of protein synthesis in reticulocytes

In rabbit reticulocyte lysates protein synthesis proceeds in the absence of hemin for several minutes only and then stops abruptly. This is accompanied by the disaggregation of polysomes and by a decrease of Met-$tRNA_f$ binding to 40 S ribosomal subunits. The inhibition of protein synthesis can be overcome by the addition of either hemin, or of initiation factor eIF-2.

Extensive investigations of this phenomenon led to the following conclusions (for reviews see [3, 135, 404, 405]): In the absence of hemin a high-molecular translational inhibitor (**h**emin-**c**ontrolled **i**nhibitor = HCI) is accumulated that seems to block at least one step of the initiation of protein synthesis [406–410]. This inhibitor is formed from a latent proinhibitor via various steps [411]. In the first stage a hemin-reversible translational inhibitor is made, which later on is converted into a form that cannot be reversed by hemin. Optimal conditions for the conversion of the proinhibitor into the inhibitor were described [411, 412].

Both forms of HCI have been purified and extensively characterized [413—417]. They act as a cAMP-independent protein kinase, which specifically phosphorylates the small subunit of the initiation factor eIF-2 and a 90000 molecular weight polypeptide [415—422]. This specific phosphorylation was observed in purified systems as well as in reticulocyte lysates [422, 423] and seems to be the basic effect of HCI in the control of protein biosynthesis [415, 423].

How initiation is affected by phosphorylation of the α-subunit of eIF-2 is still under discussion. Most of the investigators found the binding of the ternary initiation complex to the 40 S ribosomal subunit to be impaired by HCI action [404, 408, 415, 424, 425], but also the ternary complex formation [426, 427] and later steps of initiation [428] are considered to be affected. The observation that eIF-2 phosphorylated in the α-subunit acts only stoichiometrically but not catalytically, led to the conclusion that recycling of eIF-2 is impaired by the inhibitor rather than its direct function in initiation [191, 429]. A second explanation discussed more recently is that the interaction between eIF-2 and an additional protein factor responsible for maximal activity of eIF-2 ceases after phosphorylation of the small subunit of eIF-2 [203—206, 430, 431].

The mode of action of hemin in preventing the conversion of the proinhibitor to the inhibitor is also still a matter of controversy: From experiments with purified protein kinases [426, 427, 432—434] it was concluded that hemin action on the proinhibitor is mediated by a cAMP-dependent protein kinase. In contrast, other authors were not able to observe any participation of a cAMP-dependent protein kinase in this process [435] and it has been suggested that hemin acts directly on the reversible form of the inhibitor [416, 417].

Although not yet all details of the hemin control of protein synthesis are known, it seems to be certain that the basic principle of this translational control mechanism is also involved in a number of other translational control events. Thus, in reticulocytes and in their lysates, respectively, incubation at elevated temperatures [436—442], addition of oxidized glutathione [443, 444] or of dsRNA [415, 445—448] lead to the activation of a translational inhibitor which phosphorylates eIF-2 and causes a decrease in Met-$tRNA_f$ binding to 40 S ribosomal subunits. At least in the cases of elevated temperatures [439, 441] and of dsRNA [415, 450] it could be shown that the formed inhibitor is different from the hemin controlled inhibitor of protein synthesis. This kind of translational control seems also to be realized in several mammalian cells [451 to 456] as well as in *Artemia salina* and wheat germ [456, 457], indicating that it is widely distributed among eukaryotic cells.

Finally it should be noted that the degree of phosphorylation of eIF-2 within the cell is not only regulated by the activity of the corresponding protein kinase but also by the eIF-2 phosphatase activity which also might have a key position in regulatory events [458]. Furthermore, GTP [438, 445, 455], cAMP [438, 444, 445], hexose-6-phosphate [459, 460] and a protein factor [461] have been reported to prevent inhibition of protein synthesis caused by the translational inhibitor.

b) *Action of interferon on protein synthesis*

Interferons are glycoproteins produced by vertebrate cells after viral infection. They are able to interact with other cells and to induce an antiviral state in these cells (for reviews see [449, 462]).

One of the major targets of interferon action is protein synthesis which is affected in interferon-treated cells by at least three different mechanisms (for reviews see [3, 403, 463—467]). Two of them that act on the level of initiation are activated by dsRNA and ATP, while a third mechanism leading to an elongation block does not need dsRNA to be activated. The first two mechanisms of protein synthesis inhibition representing good examples for an inducible translational control system were extensively investigated in the past few years.

The first one includes the induction by interferon of a ribosome-associated cAMP-independent protein kinase that can be activated by dsRNA and is able to phosphorylate specifically two proteins of molecular weights of 67000 and 35000, respectively [467—471]. Recently, the latter one could be identified as the small subunit of eIF-2 [472—475], thus confirming the earlier assumption that initiation factors might be involved in this inhibition process [468, 476, 477]. Actually, phosphorylation of eIF-2 leads to a decrease of Met-tRNA$_f$ binding to 40 S subunits [467, 475, 478] and exogenous eIF-2 partially overcomes dsRNA dependent inhibition of translation [467, 478].

The second mechanism of interferon-induced inhibition of protein synthesis leading to an enhanced mRNA degradation in the presence of dsRNA and ATP is mediated by an endonuclease [478—484]. This enzyme is present in interferon-treated cells as well as in untreated cells [472, 475, 482, 484] and in reticulocyte lysates [472, 485, 486]. Interferon itself induces an enzymatic activity different from the endonuclease [474, 475, 482, 486, 487], which only in the presence of dsRNA is able to synthesize a low molecular weight translational inhibitor [468, 483, 486—489] using ATP as substrate. This low molecular weight inhibitor that has been identified as the trimer, tetramer or pentamer of the oligonucleotide pppA2′p5′A2′p5′A [490, 491], is the substance which activates the above mentioned endonuclease [472, 475, 482—485]. In the interaction with the nuclease the (2′—5′) oligoisoadenylate itself is degraded [482]. Thus, a continuous synthesis of oligoisoadenylate is required to maintain a constant rate of mRNA degradation. Although the endonuclease itself is not absolutely specific, differential rates of inactivation were observed for different mRNAs [472, 481—483]. Furthermore, it was reported that mRNA linked to dsRNA is more rapidly degraded than free mRNA, indicating that this could be the mechanism by which the specificity of the endonuclease for (the replicative intermediate form of) viral RNA is achieved [492].

The mechanism by which mRNA translation is impaired in extracts from interferon-treated cells even in the absence of dsRNA concerns elongation, because incomplete polypeptide chains are formed [493, 494]. This inhibition which seems to occur in the presence of ATP only [467] can be overcome by the addition of exogeneous eukaryotic tRNA [493—498]. Only very few minor species of tRNA are sufficient to restore translation [496—500], and the species required seem to depend on the template to be translated [496, 497, 500]. Extracts from interferon-treated cells are not deficient in these tRNA-species [500, 501], but need higher amounts of them for mRNA translation. Therefore, an increased inactivation of these tRNAs has been discussed to be the basic defect [467, 495, 502].

Furthermore, mRNA methylation required to form intact cap structures on mRNA was found to be impaired in interferon treated cells [503, 504]. A macromolecular inhibitor of this process was detected which is inactivated during incubation [505].

The different effects of interferon that all lead to an inhibition of protein synthesis were observed mainly in extracts from mouse L-cells or mouse Ehrlich ascites tumor cells, respectively. Some of the investigations were also extended to extracts from other mammalian cells [492, 498, 504, 506—509] as well as from avian cells [487, 508], indicating that these phenomena are not restricted to murine cells. The following questions remain open: Do these control mechanisms also occur in vivo, how do they (or at least some of them) lead to specific impairment of viral protein synthesis, are there additional control mechanisms and how do they act together? Concerning the last question a "multiphase antiviral state" hypothesis has been developed [467]. It is based on the observation that the three first mentioned phenomena need different concentrations of dsRNA: The tRNA-reversible elongation block occurs even in the absence of dsRNA, the interferon-induced protein kinase needs low dsRNA concentration to be activated, and the oligo-isoadenylate-synthetase is active also at high dsRNA concentrations. Thus, the cell should be able to control protein synthesis at different states of viral replication. On the other hand, kinetic investigations on interferon action revealed that during interferon treatment, oligo-isoadenylate-synthetase activity raises prior to eIF-2 kinase activity [506, 510]. Further investigations in cell-free systems as well as in intact cells are necessary to clarify these questions.

c) *Shut-off of host protein synthesis after viral infection*

Viral infection of animal cells leads in many cases to a rapid inhibition of host cell protein and RNA synthesis, whereas synthesis of viral RNA and protein persists. This "shut-off phenomenon" can be observed with several groups of viruses, especially with those that kill their host cells [511, 512]. The best studied examples are infections by picornavirus (polioviruses, encephalomyocarditis virus, and mengovirus) (for reviews see [3, 135, 403, 511, 512]). Very early after viral infection total protein synthesis of the cells decreases, and particularly cellular protein synthesis is strongly inhibited, whereas viral proteins are synthesized. The inhibition of host mRNA translation is not due to an enhanced mRNA degradation, but seems to occur at the level of polypeptide chain initiation [513]. Cell-free extracts derived from virus-infected cells are able to translate both, cellular and viral mRNAs (514 to 517] indicating that no stable inhibitor is formed and that the translational apparatus is not changed irreversibly.

Several models have been developed in order to explain the specific discrimination of host cell mRNAs versus picornavirus RNA (reviewed in [3, 511, 512]). An earlier assumption that dsRNA formed during virus replication directly inhibits initiation of host cell protein synthesis seems unlikely because (1) dsRNA in vitro inhibits both, host cell and viral protein synthesis [518], (2) shut-off occurs early after infection before any dsRNA has been synthesized, and (3) virus mutants defective in dsRNA synthesis are able to shut off host cell protein synthesis [519].

Another model supposes a direct competition between viral and cellular mRNAs as the reason for the shut-off phenomenon. In fact, under certain conditions picornavirus RNA is able to compete out cellular mRNA in cell-free translation systems [514—517, 520]. The various mRNAs compete for an initiation factor [242, 517, 521] which has been identified as eIF-4B [242 to 244]. Different affinities of host and viral mRNA to an initiation factor, however, cannot be the only explanation for the shut-off of host cell protein synthesis, because total protein synthesis is decreased and shut-off occurs before larger amounts of viral RNA are present in the cell.

Furthermore, several virus mutants defective in shut-off map in the coat protein region of the viral genome [519] indicating that a coat protein might be involved in this translational control system. These observations favour the so-called membrane-leakage model [512, 522] which predicts that after infection a viral coat protein becomes attached to the cellular membrane and causes an alteration in the membrane leading to an increased influx of sodium ions. The increased concentration of sodium in the cytoplasm inhibits cellular protein synthesis, and stimulates synthesis of viral proteins. In fact, pulse labeling experiments have shown that the ion influx is altered in cells after viral infection [522]. Furthermore, in a cell-free system it could be demonstrated that viral protein synthesis is stimulated at enhanced sodium chloride concentrations, whereas translation of cellular messages is drastically inhibited [522]. Similar results were obtained with whole cells upon change of ionic conditions [523, 524]. Although this model would explain many of the present experimental data, some questions remain still unanswered [512], and the shut-off phenomenon representing an example of a specific translational control mechanism is not yet clearly understood at present.

5. References

[1] PERRY, R.: Processing of RNA. Ann. Rev. Biochem. **45**, 605–629 (1976)

[2] CHAN, L., S. E. HARRIS, J. M. ROSEN, A. R. MEANS, and B. W. O'MALLEY: Processing of nuclear heterogeneous RNA: Recent developments. Life Sci. **20**, 1–16 (1977)

[3] REVEL, M., and Y. GRONER: Post-transcriptional and translational controls of gene expression in eukaryotes. Ann. Rev. Biochem. **47**, 1079–1126 (1978)

[4] ARNSTEIN, H. R. V.: The current state of eukaryotic mRNA. Nature **277**, 519–520 (1979)

[5] CHOW, L. T., R. E. GELINAS, T. R. BROKER, and R. J. ROBERTS: An amazing sequence arrangement at the 5′-end of adenovirus 2 mRNA. Cell **12**, 1–8 (1977)

[6] ROBERTS, R. J., D. F. KLESSIG, J. MANLEY, and B. S. ZAIN: The spliced messenger RNAs of adenovirus-2. FEBS-Symp. **51**, 245–254 (1979)

[7] DARNELL, J. E., jr.: Implications of RNA · RNA splicing in evolution of eukaryotic cells. Science **202**, 1257–1260 (1978)

[8] HASELTINE, W. A., A. M. MAXAM, and W. GILBERT: Rous sarcoma virus genome is terminally redundant: The 5′ sequence. Proc. Natl. Acad. Sci. U.S. **74**, 989–993 (1977)

[9] BERGET, S. M., C. MOORE, and P. A. SHARP: Spliced segments at the 5′ terminus of adenovirus 2 late mRNA. Proc. Natl. Acad. Sci. U.S. **74**, 3171–3175 (1977)

[10] ALONI, Y., R. DHAR, O. LAUB, M. HOROWITZ, and G. KHOURY: Novel mechanism for RNA maturation: The leader sequences of simian virus 40 mRNA are not transcribed adjacent to the coding sequences. Proc. Natl. Acad. Sci. U.S. **74**, 3686–3690 (1977)

[11] LAVI, S., and Y. GRONER: 5′-terminal sequences and coding region of late simian virus 40 mRNAs are derived from noncontiguous segments of the viral genome. Proc. Natl. Acad. Sci U.S. **74**, 5323–5327 (1977)

[12] TONEGAWA, S., A. M. MAXAM, R. TIZARD, O. BERNARD, and W. GILBERT: Sequence of a mouse germ-line gene for a variable region of an immunoglobulin light chain. Proc. Natl. Acad. Sci. U.S. **75**, 1485–1489 (1978)

[13] COLONNO, R. J., and A. K. BANERJEE: Complete nucleotide sequence of the leader RNA synthesized in vitro by vesicular stomatitis virus. Cell **15**, 93–101 (1978)

[14] KLESSIG, D. F.: Two adenovirus mRNAs have a common 5′ terminal leader sequence encoded at least 10 kb upstream from their main coding regions. Cell **12**, 9–21 (1977)

[15] BLANCHARD, J. M., J. WEBER, W. JELINEK, and J. E. DARNELL: In vitro RNA-RNA splicing in adenovirus-2 mRNA formation. Proc. Natl. Acad. Sci. U.S. **75**, 5344–5348 (1978)

[16] GILBERT, W.: Why genes in pieces? Nature **271**, 501 (1978)

[17] CRICK, F.: Split genes and RNA splicing. Science **204**, 264–271 (1979)

[18] SHAFRITZ, D. A.: Messenger RNA and its translation. In: Molecular Mechanisms of Protein Biosynthesis (H. WEISSBACH and S. PESTKA, eds.) 555–601, Academic Press, N.Y, San Francisco, London, 1977

[19] PREOBRAZHENSKY, A. A., and A. S. SPIRIN: Informosomes and their protein components: The present state of knowledge.

Progr. Nucl. Acid Res. Mol. Biol. **21**, 1 to 38 (1978)

[20] Spirin, A. S.: Messenger ribonucleoproteins (informosomes) and RNA-binding proteins. Mol. Biol. Rep. **5**, 53—57 (1979)

[21] Stevenin, J., and M. Jacob: Structure of pre-mRNP. Models and pitfalls. Mol. Biol. Rep. **5**, 29—35 (1979)

[22] Shatkin, A. J.: Capping of eucaryotic mRNAs. Cell **9**, 645—653 (1976)

[23] Busch, H., D. Henning, F. W. Hirsch, M. Rao, T. S. Ro-choi, W. H. Spohn, and B. C. Wu: Structural aspects of low molecular weight RNA and the implications of the 5′ cap for messenger RNA and protein synthesis. The Mol. Biol. of the Mammalian Genet. Apparatus **1**, 165 to 194 (1977)

[24] Shatkin, A. J., Y. Furuichi, M. Kozak, and N. Sonenberg: 5′-terminal caps in eukaryotic mRNAs. FEBS-Symp. **51**, 297—306 (1979)

[25] Filipowicz, W.: Functions of the 5′-terminal m^{7}G cap in eukaryotic mRNA. FEBS-Lett. **96**, 1—11 (1978)

[26] Perry, R. P., and D. E. Kelley: Kinetics of formation of 5′-terminal cap in mRNA. Cell **8**, 433—442 (1976)

[27] Dottin, R. P., A. M. Weiner, and H. F. Lodish: 5′-terminal nucleotide sequences of the messenger RNAs of *Dictyostelium discoideum*. Cell **8**, 233—244 (1976)

[28] Lockard, R. E.: Different cap 1:cap 2 ratios in rabbit α and β globin mRNA. Nature **275**, 153—154 (1978)

[29] Schibler, U., and R. P. Perry: Characterization of the 5′ termini of hnRNA in mouse L cells: Implications for processing and cap formation. Cell **9**, 121 to 130 (1976)

[30] Flavell, A. J., A. Cowie, S. Legon, and R. Kamen: Multiple 5′ terminal cap structures in late polyoma virus RNA. Cell **16**, 357—371 (1979)

[31] Abraham, K. A., and J. R. Lillehaug: Enzymatic cleavage of m^{7}GDP from eucaryotic mRNA. FEBS-Lett. **71**, 49—52 (1976)

[32] Shinshi, H., M. Miwa, T. Sugimura, K. Shimotohno, and K. Miura: Enzyme cleaving the 5′-terminal methylated blocked structure of messenger RNA. FEBS-Lett. **65**, 254—257 (1976)

[33] Walczewska, Z., M. M. Bretner, H. Sierakowska, E. Szczesna, W. Filipowicz, and A. J. Shatkin: Removal of 5′ terminal M7G from eukaryotic mRNAs by potato nucleotide pyrophosphatase and its effect on translation. Nucl. Acids Res. **4**, 3065—3081 (1977)

[34] Nuss, D. L., and Y. Furuichi: Characterization of the m^{7}G(5′)pppN-pyrophosphatase activity from HeLa cells. J. Biol. Chem. **252**, 2815—2821 (1977)

[35] Shimotohno, K., Y. Kodama, J. Hashimoto, and K.-J. Miura: Importance of 5′-terminally blocking structure to stabilize mRNA in eukaryotic protein synthesis. Proc. Natl. Acad. Sci. U.S. **74**, 2734 to 2738 (1977)

[36] Abraham, K. A., and A. Pihl: Translation of enzymically decapped mRNA. Europ. J. Biochem. **77**, 589—593 (1977)

[37] Kozak, M.: How do eucaryotic ribosomes select initiation regions in mRNA? Cell **15**, 1109—1123 (1978)

[38] Kaempfer, R., H. Rosen, and R. Israeli: Translational control: recognition of the methylated 5′end and an internal sequence in eukaryotic mRNA by the initiation factor that binds methionyl-tRNA$_f^{Met}$. Proc. Natl. Acad. Sci. U.S. **75**, 650—654 (1978)

[39] Sonenberg, N., M. A. Morgan, W. C. Merrick, and A. J. Shatkin: Polypeptide in eukaryotic initiation factors that crosslinks specifically to 5′-terminal cap in mRNA. Proc. Natl. Acad. Sci. U.S. **75**, 4843—4847 (1978)

[40] Padilla, M., D. Canaan, Y. Groner, J. A. Weinstein, M. Bar-Joseph, W. Merrick, and D. A. Shafritz: Initiation factor eIF-4B (IF-M$_3$) dependent recognition and translation of capped versus uncapped eucaryotic mRNAs. J. Biol. Chem. **253**, 5939—5945 (1978)

[41] Clemens, M. J.: Why do messenger wear caps? Nature **279**, 673—674 (1979)

[42] Paterson, B. M., and M. Rosenberg: Efficient translation of prokaryotic mRNAs in a eukaryotic cell-free system requires addition of a cap structure. Nature **279**, 692—696 (1979)

[43] Rosenberg, M., and B. M. Paterson: Efficient cap-dependent translation of polycistronic prokaryotic mRNAs is resitricted to the first gene in the operon. Nature **279**, 696—701 (1979)

[44] SHINE, J., and L. DALGARNO: Determinant of cistron specificity in bacterial ribosomes. Nature **254**, 34–38 (1975)

[45] BARALLE, F. E., and G. G. BROWNLEE: AUG is the only recognisable signal sequence in the 5′ non-coding regions of eukaryotic mRNA. Nature **274**, 84–87 (1978)

[46] HAGENBÜCHLE, O., M. SANTER, J. A. STEITZ, and R. J. MANS: Conservation of the primary structure at the 3′end of 18 S rRNA from eucaryotic cells. Cell **13**, 551 to 563 (1978)

[47] PROUDFOOT, N. J., and G. G. BROWNLEE: 3′ non-coding region sequences in eucaryotic messenger RNA. Nature **263**, 211–214 (1976)

[48] PROUDFOOT, N. J.: Complete 3′ non-coding region sequence of rabbit and human β-globin mRNAs. Cell **10**, 559 to 570 (1977)

[49] PROUDFOOT, N. J., SH. GILLAM, M. SMITH, and J. I. LONGLEY: Nucleotide sequence of the 3′ terminal region of rabbit α-globin mRNA: Comparison with human α-globin mRNA. Cell **11**, 807 to 818 (1977)

[50] HAMLYN, P. H., S. GILLAM, M. SMITH, and C. MILSTEIN: Sequence analysis of 3′ non-coding region of mouse immunoglobulin light chain mRNA. Nucl. Acids Res. **4**, 1123–1134 (1977)

[51] FRASER, N., and E. ZIFF: RNA structures near poly(A) of adenovirus-2 late mRNAs. J. Mol. Biol. **124**, 27–51 (1978)

[52] SCHWARZ, D. E., P. C. ZAMECNIK, and H. L. WEITH: Rous sarcoma virus genome is terminally redundant: The 3′ sequence. Proc. Natl. Acad. Sci. U.S. **74**, 994–998 (1977)

[53] POON, R., Y. KAN, and H. W. BOYER: Sequence of the 3′-noncoding and adjacent coding regions of human γ-globin mRNA. Nucl. Acids Res. **5**, 4625–4630 (1978)

[54] KRONENBERG, H. M., B. E. ROBERTS, and A. EFSTRATIADIS: The 3′ noncoding region of β-globin mRNA is not essential for in vitro translation. Nucl. Acids Res. **6**, 153–166 (1979)

[55] BRAWERMAN, G.:Characteristics and significance of the polyadenylate sequence in mammalian messenger RNA. Progr. Nucl. Acid Res. Mol. Biol. **17**, 117–148 (1976)

[56] MARBAIX, G., G. HUEZ, H. SOREQ, D. GALLWITZ, E. WEINBERG, R. DEVOS, E. HUBERT, and Y. CLEUTER: Role of polyadenylate segment in stability of eukaryotic messenger RNAs. FEBS Symp. **51**, 427–436 (1978)

[57] KOZAK, M., and A. J. SHATKIN: Migration of 40 S ribosomal subunits in messenger RNA in the presence of edeine. J. Biol. Chem. **253**, 6568–6577 (1978)

[58] KOZAK, M.: Migration of 40 S ribosomal subunits on messenger RNA when initiation is perturbed by lowering magnesium or adding drugs. J. Biol. Chem. **254**, 4731–4738 (1979)

[59] LEGON, S., H. D. ROBERTSON, and W. PRENSKY: The binding of ^{125}J-labelled rabbit globin messenger RNA to reticulocyte ribosomes. J. Mol. Biol. **106**, 23–36 (1976)

[60] LEGON, S.: Characterization of the ribosome-protected regions of ^{125}J-labelled rabbit globin messenger RNA. J. Mol. Biol. **106**, 37–53 (1976)

[61] KOZAK, M., and A. J. SHATKIN: Characterization of ribosome-protected fragments from reovirus messenger RNA. J. Biol. Chem. **251**, 4259–4266 (1976)

[62] CANCEDDA, R., and A. J. SHATKIN: Ribosome-protected fragments from sindbis 42-S and 26-S RNAs. Europ. J. Biochem. **94**, 41–50 (1979)

[63] KOZAK, M., and A. J. SHATKIN: Sequences of two 5′-terminal ribosome-protected fragments from reovirus mRNAs. J. Mol. Biol. **112**, 75–96 (1977)

[64] KOZAK, M., and A. J. SHATKIN: Sequences and properties of two ribosome binding sites from the small size class of reo virus messenger RNA. J. Biol. Chem. **252**, 6895–6908 (1977)

[65] LAZAREWITZ, S. G., and H. D. ROBERTSON: Initiator regions from the small size class of reovirus messenger RNA protected by rabbit reticulocyte ribosomes. J. Biol. Chem. **252**, 7842–7849 (1977)

[66] ROSE, J. K.: Nucleotide sequences of ribosome recognition sites in messenger RNAs of vesicular stomatitis virus. Proc. Natl. Acad. Sci. U.S. **74**, 3672–3676 (1977)

[67] ROSE, J. K.: Complete sequences of the ribosome recognition sites in vesicular stomatitis virus mRNAs: Recognition by the 40 S and 60 S complexes. Cell **14**, 345—353 (1978)
[68] OFENGAND, J.: tRNA and aminoacyl-tRNA synthetases. In: Molecular Mechanisms of Protein Biosynthesis. (Eds.: H. WEISSBACH and S. PESTKA) Academic Press, New York, San Francisco, London, 1977, 7—79.
[69] MORGAN, S. D., and D. SÖLL: Regulation of the biosynthesis of aminoacid-tRNA ligases and of tRNA. Progr. Nucl. Acid. Res. Mol. Biol. **21**, 181—207 (1978)
[70] SOFFER, R. L.: Aminoacyl-tRNA transferases. Adv. Enzymol. **40**, 91—139 (1974)
[71] WATERS, L. C., and B. C. MULLIN: Transfer RNA in RNA tumor viruses. Progr. Nucl. Acid Res. Mol. Biol. **20**, 131—160 (1977)
[72] CLARK, B. F. C.: Correlation of biological activities with structural features of transfer RNA. Progr. Nucl. Acid Res. Mol. Biol. **20**, 1—19 (1977)
[73] UMBARGER, H. E.: Amino acid biosynthesis and its regulation. Ann. Rev. Biochem. **47**, 533—606 (1978)
[74] LITTAUER, U. Z., and H. INOUYE: Regulation of tRNA. Ann. Rev. Biochem. **42**, 439—470 (1973)
[75] BRENCHLEY, J. E., and L. S. WILLIAMS: Transfer RNA involvement in the regulation of enzyme synthesis. Ann. Rev. Microbiol. **29**, 251—274 (1975)
[76] OSTERMAN, L. A.: Participation of tRNA in regulation of protein biosynthesis at the translational level in eukaryotes. Biochimie **61**, 323—342 (1979)
[77] RICH, A., and U. L. RAJBHANDARY: Transfer RNA: Molecular structure, sequence, and properties. Ann. Rev. Biochem. **46**, 805—860 (1976)
[78] HOLLEY, R. W., J. APGAR, G. A. EVERETT, J. T. MADISON, M. MARQUISEE, S. H. MERRILL, J. R. PENSWICK, and A. ZAMIR: Structure of a ribonucleic acid. Science **147**, 1462—1465 (1965)
[79] GAUSS, D. H., F. GRÜTER, and M. SPRINZL: Compilation of tRNA sequences. Nucl. Acids Res. **6**, r1—r19 (1979)
[80] KIM, S. H.: Three-dimensional structure of transfer RNA. Progr. Nucl. Acid Res. Mol. Biol. **17**, 181—216 (1976)
[81] KIM, S. H.: Three-dimensional structure of transfer RNA and its functional implications. Adv. Enzymol. **46**, 279—315 (1978)
[82] SMITH, J. D.: Transcription and processing of transfer RNA precursors. Progr. Nucl. Acid Res. Mol. Biol. **16**, 25—73 (1976)
[83] BARCISZEWSKI, J., and A. J. RAFALSKI: Atlas of primary structures of tRNAs. Polish Scientific Publisher, Poznan, in press
[84] MCCLOSKEY, J. A., and S. NISHIMURA: Modified nucleosides in transfer RNA. Accounts Chem. Res. **10**, 403—410 (1977)
[85] KIM, S. H., G. J. QUIGLEY, F. L. SUDDATH, A. MCPHERSON, D. SNEDEN, J. J. KIM, J. WEINZIERL, and A. RICH: Three-dimensional structure of yeast phenylalanine transfer RNA: Folding of the polynucleotide chain. Science **179**, 285—288 (1973)
[86] SUSSMAN, J. L., S. R. HOLBROOK, R. W. WARRANT, G. M. CHURCH, and S. H. KIM: Crystal structure of yeast phenylalanine transfer RNA. I. Crystallographic refinement. J. Mol. Biol. **123**, 607—630 (1978)
[87] HOLBROOK, S. R., J. L. SUSSMAN, R. W. WARRANT, and S. H. KIM: Crystal structure of yeast phenylalanine transfer RNA. II. Structural features and functional implications. J. Mol. Biol. **123**, 631—660 (1978)
[88] RICH, A.: Three-dimensional structure and biological function of transfer RNA. Accounts Chem. Res. **10**, 388—396 (1977)
[89] KIM, S. H., and G. J. QUIGLEY: Determination of a transfer RNA structure by crystallographic method. Meth. Enzymol. **59**, 3—21 (1979)
[90] REID, B. R., and R. E. HURD: Application of high-resolution nuclear magnetic resonance spectroscopy in the study of base pairing and the solution structure of transfer RNA. Accounts Chem. Res. **10**, 396—402 (1977)
[91] KEARNS, D. R.: High-resolution nuclear magnetic resonance investigation of the structure of tRNA in solution. Progr. Nucl. Acid Res. Mol. Biol. **18**, 91—149 (1976)
[92] SCHEVITZ, R. W., A. D. PODJARNY, N. KRISHNAMACHARI, J. J. HUGHES, P. B. SIGLER, and J. L. SUSSMAN: Crystal struc-

ture of a eukaryotic initiator tRNA. Nature **278**, 188–190 (1979)

[93] SPRINZL, M., and F. CRAMER: The –C–C–A end of tRNA and its role in protein biosynthesis. Progr. Nucl. Acid Res. Mol. Biol. **22**, 1–69 (1979)

[94] HOLLER, E.: Proteinbiosnythese: Die codonspezifische Aktivierung der Aminosäuren. Angew. Chem. **90**, 682–690 (1978)

[95] SÖLL, D., and P. R. SCHIMMEL: Aminoacyl-tRNA synthetases. Enzymes **10**, 489–538 (1974)

[96] KISSELEV, L. L., and O. O. FAVOROVA: Aminoacyl-tRNA synthetases: Some recent results and achievements. Adv. Enzymol. **40**, 141–238 (1974)

[97] NEIDHARDT, F. C., J. PARKER, and W. G. MCKEEVER: Function and regulation of aminoacyl-tRNA synthetases in prokaryotic and eukaryotic cells. Ann. Rev. Microbiol. **29**, 215–250 (1975)

[98] SCHIMMEL, P. R., and D. SÖLL: Aminoacyl-tRNA synthetases: General features and recognition of tRNAs. Ann. Rev. Biochem. **48**, 601–648 (1979)

[99] USSERY, M. A., W. K. TANAKA, and B. HARDESTY: Subcellular distribution of aminoacyl-tRNA synthetases in various eukaryotic cells. Europ. J. Biochem. **72**, 491–500 (1977)

[100] WINTER, G. P., and B. S. HARTLEY: The amino acid sequence of tryptophanyl tRNA synthetase from Bacillus stearothermophilus. FEBS-Lett. **80**, 340–342 (1977)

[101] IRWIN, M. J. , J. NYBORG, B. R. REID, and D. M. BLOW: The crystal structure of tyrosyl-transfer RNA synthetase at 2.7 Å resolution. J. Mol. Biol. **105**, 577–586 (1976)

[102] BLOW, D. M., M. J. IRWIN, and J. NYBORG: The peptide chain of tyrosyl tRNA synthetase: No evidence for a super-secondary structure of four α-helices. Biochem. Biophys. Res. Commun. **76**, 728 to 734 (1977)

[103] MONTEILHET, C., and D. M. FLOW: Binding of tyrosine, adenosine triphosphate and analogues to crystalline tyrosyl transfer RNA synthetase. J. Mol. Biol. **122**, 407 to 417 (1978)

[104] ZELWER, C., J. L. RISLER, and C. MONTEILHET: A low-resolution model of crystalline methionyl-transfer RNA synthetase from *Escherichia coli*. J. Mol. Biol. **102**, 93–101 (1976)

[105] LOFTFIELD, R. B., and D. VANDERJAGT: The frequency of errors in protein biosynthesis. Biochem. J. **128**, 1353–1356 (1972)

[106] HOPFIELD, J. J.: Kinetic proofreading: A new mechanism for reducing errors in biosynthetic processes requiring high specificity. Proc. Natl. Acad. Sci. U.S. **71**, 4135–4139 (1974)

[107] YAMANE, T., and J. J. HOPFIELD: Experimental evidence for kinetic proofreading in the aminoacylation of tRNA synthetase. Proc. Natl. Acad. Sci. U.S. **74**, 2246–2250 (1977)

[108] VON DER HAAR, F.: Enzyme specificity resulting from proofreading events. FEBS-Lett. **79**, 225–228 (1977)

[109] RICH, A.: Transfer RNA and protein synthesis. Biochimie **56**, 1441–1449 (1974)

[110] RICH, A., and P. R. SCHIMMEL: Structural organization of complexes of transfer RNAs with aminoacyl transfer RNA synthetases. Nucl. Acids Res. **4**, 1649 to 1665 (1977)

[111] JEKOWSKI, E., D. L. MILLER, and P. R. SCHIMMEL: Isolation, Characterization and structural implications of a nuclease-digested complex of aminoacyl transfer RNA and Escherichia coli elongation factor Tu. J. Mol. Biol. **114**, 451–458, (1977)

[112] LESTIENNE, P.: The specificity of aminoacylation: a tRNA-tRNA interaction model. J. Theoret. Biol. **73**, 159–180 (1978)

[113] SCHIMMEL, P. R.: Approaches to understanding the mechanism of specific protein-transfer RNA interactions. Accounts Chem. Res. **10**, 411–418 (1977)

[114] SCHIMMEL, P. R.: Understanding the recognition of tRNAs by aminoacyl-tRNA synthetases. Adv. Enzymol. **49**, 187–222 (1979)

[115] SCHIMMEL, P. R.: Recent results on how aminoacyl-tRNA synthetases recognize specific tRNAs. Mol. Cell. Biochem. **25**, 3–14 (1979)

[116] BERMEK, E.: Mechanisms in polypeptide chain elongation on ribosomes. Progr. Nucl. Acid Res. Mol. Biol. **21**, 63–100 (1978)

[117] CRICK, F. H. C.: Codon-anticodon pairing: The wobble hypothesis. J. Mol. Biol. **19**, 548–555 (1966)

[118] MITRA, S. K.: Recognition between codon and anticodon. Trends Biochem. Sci. **3**, 153–156 (1978)

[119] MITRA, S. K., and J. NINIO: Recognition between codon and anticodon. The limits of our knowledge. FEBS Symp. **51**, 437 to 444 (1978)

[120] JUKES, T. H.: The amino acid code. Adv. Enzymol. **47**, 375–432 (1978)

[121] THOMPSON, R. C., and P. J. STONE: Proofreading of the codon-anticodon interaction on ribosomes. Proc. Natl. Acad. Sci. U. S. **74**, 198–202 (1977)

[122] NINIO, J.: A semi-quantitative treatment of missense and nonsense suppression in the strA and ram ribosomal mutant of *Escherichia coli*. Evaluation of some molecular parameters of translation in vivo. J. Mol. Biol. **84**, 297–313 (1974)

[123] KURLAND, C. G., R. RIGLER, M. EHRENBERG, and C. BLOMBERG: Allosteric mechanism for codon-dependent tRNA selection on ribosomes. Proc. Natl. Acad. Sci. U.S. **72**, 4248–4251 (1975)

[124] KURLAND, C. G.: Aspects of ribosome structure and function. In: Molecular Mechanisms of Protein Biosynthesis. (Eds.: H. WEISSBACH and S. PESTKA). Academic Press, New York, San Francisco, London, 1977, 81–116

[125] SCHWARZ, U., R. LÜHRMANN, and H. G. GASSEN: On the mRNA induced conformational change of aa-tRNA exposing the T Ψ CG-sequence for binding to the 50 S ribosomal subunit. Biochem. Biophys. Res. Commun. **56**, 807–814 (1974)

[126] SCHWARZ, U., H. M. MENZEL, and H. G. GASSEN: Codon-dependent rearrangement of the three-dimensional structure of phenylalanyl tRNA, exposing the TΨCG sequence for binding to the 50 S ribosomal subunit. Biochem. **15**, 2484 to 2490 (1976)

[127] SCHWARZ, U., and H. G. GASSEN: Codon-dependent rearrangement of the tertiary structure of $tRNA^{Phe}$ from yeast. FEBS-Lett. **78**, 267–270 (1977)

[128] ERDMANN, V. A.: Structure and function of 5 S and 5.8 S RNA. Progr. Nucl. Acid Res. Mol. Biol. **18**, 45–90 (1976)

[129] KURLAND, C. G.: Structure and function of the bacterial ribosome. Ann. Rev. Biochem. **46**, 173–200 (1977)

[130] COX, R. A.: Structure and function of prokaryotic and eukaryotic ribosomes. Progr. Biophys. Molec. Biol. **32**, 193–231 (1977)

[131] RICH, A.: How transfer RNA may move inside the ribosome. In: Ribosomes. (Eds.: NOMURA, M., A. TISSIERES and P. LENGYEL). Cold Spring Harbor Lab. 1974, 871–884

[132] OCHOA, S., and R. MAZUMDER: Polypeptide chain initiation. Enzymes **10**, 1 to 51 (1974)

[133] WEISSBACH, H., and S. OCHOA: Soluble factors required for eukaryotic protein synthesis. Ann. Rev. Biochem. **45**, 191 to 216 (1976)

[134] GRUNBERG-MANAGO, M., and F. GROS: Initiation mechanism of protein synthesis. Progr. Nucl. Acid Res. Mol. Biol. **20**, 209–284 (1977)

[135] REVEL, M.: Initiation of messenger RNA translation into protein and some aspects of its regulation.In: Molecular Mechanism of Protein Biosynthesis. (H. Weissbach and S. Pestka, eds.). Academic Press, New York, San Francisco, London, 1977, 245 to 321

[136] BIELKA, H., and J. STAHL: Structure and function of eukaryotic ribosomes. Int. Rev. Biochem. **18**, 79–168 (1978)

[137] SCHREIER, M. H., B. ERNI, and T. STAEHELIN: Initiation of mammalian protein synthesis. I. Purification and Characterization of seven initiation factors. J. Mol. Biol. **116**, 727–753 (1977)

[138] TRACHSEL, H., B. ERNI, M. H. SCHREIER, and T. STAEHELIN: Initiation of mammalian protein synthesis. II. The assembly of the initiation complex with purified initiation factors. J. Mol. Biol. **116**, 755–767 (1977)

[139] ERNI, B.: Initiation of mammalian protein biosynthesis. Purification and characterization of initiation factors. Doctoral Dissertation, Swiss Federal Institute of Technology, Zürich (1976)

[140] STAEHELIN, T., B. ERNI, and M. H. SCHREIER: Purification and characterization of seven initiation factors for mammalian protein synthesis. Meth. Enzymol. **60**, 136–165 (1979)

[141] MERRICK, W. C.: Purification of protein synthesis initiation factors from rabbit-reticulocytes. Meth. Enzymol. **60**, 101 to 108 (1979)
[142] SAFER, B., W. F. ANDERSON, and W. C. MERRICK: Purification and physical properties of homogeneous initiation factor MP from rabbit reticulocytes. J. Biol. Chem. **250**, 9067—9075 (1975)
[143] SAFER, B., S. L. ADAMS, W. F. ANDERSON, and W. C. MERRICK: Binding of Met-tRNA$_f$ and GTP to homogeneous initiation factor MP. J. Biol. Chem. **250**, 9076—9082 (1975)
[144] MERRICK, W. C., and W. F. ANDERSON: Purification and characterization of homogeneous protein synthesis initiation factor M 1 from rabbit reticulocytes. J. Biol. Chem. **250**, 1197—1206 (1975)
[145] SAFER, B., S. L. ADAMS, W. M. KEMPER, K. W. BERRY, M. LLOYD, and W. C. MERRICK: Purification and characterization of two initiation factors required for maximal activity of a highly fractionated globin mRNA translation system. Proc. Natl. Acad. Aci. U.S. **73**, 2584—2588 (1976)
[146] PRICHARD, P. M., and W. F. ANDERSON: Preparation of rabbit reticulocyte initiation factor IF-M 3. Meth. Enzymol. **30**, 136—141 (1974)
[147] KEMPER, W. M., K. W. BERRY, and W. C. MERRICK: Purification and properties of rabbit reticulocyte protein synthesis initiation factor M2Bα and M2Bβ. J. Biol. Chem. **251**, 5551—5557 (1976)
[148] MERRICK, W. C., W. M. KEMPER, and W. F. ANDERSON: Purification and characterization of homogeneous initiation factor M2A from rabbit reticulocytes. J. Biol. Chem. **250**, 5556—5562 (1975)
[149] NOMBELA, C., N. A. NOMBELA, S. OCHOA, W. C. MERRICK, and W. F. ANDERSON: Nature of eukaryotic protein required for joining of 40 S and 60 S ribosome subunits. Biochem. Biophys. Res. Commun. **63**, 409—416 (1975)
[150] ANDERSON, W. F., L. BOSCH, W. E. COHN, H. LODISH, W. C. MERRICK, H. WEISSBACH, H. G. WITTMANN, and I. G. WOOL: International Symposium on Protein Synthesis. FEBS-Lett. **76**, 1—10, (1977)
[151] BENNE, R., and J. W. B. HERSHEY: Purification and characterization of initiation factor IF-E 3 from rabbit reticulocytes Proc. Natl. Acad. Sci. U.S. **73**, 3005—3009 (1976)
[152] BENNE, R., C. WONG, M. LUEDI, and J. W. B. HERSHEY: Purification and characterization of initiation factor IF-E 2 from rabbit reticulocytes. J. Biol. Chem. **251**, 7675—7681 (1976)
[153] BENNE, R., M. LUEDI, and J. W. B. HERSHEY: Purification and characterization of initiation factors IF-E 4 and IF-E 6 from rabbit reticulocytes. J. Biol. Chem. **252**, 5798—5803 (1977)
[154] BENNE, R., M. L. BROWN-LUEDI, and J. W. B. HERSHEY: Purification and characterization of protein synthesis initiation factors eIF-1, eIF-4C, eIF-4D, and eIF-5 from rabbit reticulocytes. J. Biol. Chem. **253**, 3070—3077 (1978)
[155] BENNE, R., M. L. BROWN-LUEDI, and J. W. B. HERSHEY: Protein synthesis initiation factors from rabbit reticulocytes: purification, characterization and radiochemical labeling. Meth. Enzymol. **60** 15—35 (1979)
[156] VOORMA, H. O., A. THOMAS, H. GOUMANS, H. AMESZ, and C. VAN DER MAST: Isolation and purification of initiation factors of protein synthesis from rabbi-reticulocyte lysate. Meth. Enzymol. **60**, 124—135 (1979)
[157] MOLDAVE, K., and L. GROSSMANN (eds.): Nucleic Acids and Protein Synthesis. Meth. Enzymol. **60** (1979)
[158] TRACHSEL, H., B. ERNI, M. H. SCHREIER, L. BRAUN, and T. STAEHELIN: Purification of seven protein synthesis initiation factors from Krebs II ascites cells. Biochim. Biophys. Acta **561**, 484—490 (1979)
[159] DAHL, H. H. M., and G. E. BLAIR: Purification of four eukaryotic initiation factors required for natural mRNA translation. Meth. Enzymol. **60**, 87—101 (1979)
[160] BENNE, R., and J. W. B. HERSHEY: The mechanism of action of protein synthesis initiation factors from rabbit reticulocytes. J. Biol. Chem. **253**, 3078—3087 (1978)
[161] CASHION, L. M., and W. M. STANLEY, jr.: Two eukaryotic initiation factors (IF-I and IF-II) of protein synthesis that are required to form an initiation complex

with rabbit reticulocyte ribosomes. Proc. Natl. Acad. Sci. U.S. **71**, 436–440 (1974)

[162] Majumdar, A., S. Reynolds, and N. K. Gupta: Protein synthesis in rabbit reticulocytes. XIII: Lack of mRNA (poly r(A)) binding activity in highly purified EIF-1. Biochem. Biophys. Res. Commun. **67**, 689–695 (1975)

[163] Dasgupta, A., A. Majumdar, A. D. George, and N. K. Gupta: Protein synthesis in rabbit reticulocytes. XV: Isolation of ribosomal protein factor (Co-EIF-1) which stimulates Met-tRNA$_f^{Met}$ binding to EIF-1. Biochem. Biophys. Res. Commun. **71**, 1234–1241 (1976)

[164] Majumdar, A., A. Dasgupta, B. Chatterjee, H. K. Das, and N. K. Gupta: Purification and properties of rabbit reticulocyte protein synthesis initiation factors EIF-1, EIF-2, and EIF-3. Meth. Enzymol. **60**, 35–52 (1979)

[165] Levin, D. H., D. Kyner, and G. Acs: Protein synthesis initiation in eukaryotes. Characterization of ribosomal factors from mouse fibroblasts. J. Biol. Chem. **248**, 6416–6425 (1973)

[166] Smith, K. E., and E. C. Henshaw: Binding of Met-tRNA$_f$ to native and derived 40 S ribosomal subunits. Biochem. **14**, 1060–1067 (1975)

[167] Ranu, R. S., and I. G. Wool: Preparation and characterization of eukaryotic initiation factor eIF-3: Formation of binary (EIF-3 · Met-tRNA$_f$) and ternary (EIF-3 · Met-tRNA$_f$ · GTP) complexes. J. Biol. Chem. **251**, 1926–1935 (1976)

[168] Van der Mast, C., A. Thomas, H. Goumans, H. Amesz, and H. O. Voorma: Initiation of protein synthesis in eukaryotes: Binding to sepharose-heparin and partial purification of initiation factors from Krebs II ascites cells. Europ. J. Biochem. **75**, 455–464 (1977)

[169] Bommer, U.-A., A. Henske, and H. Bielka: Preparation and properties of a Met-tRNA$_f$ binding factor from rat liver and rat hepatoma. Acta Biol. Med. Germ. **37**, 1363–1376 (1978)

[170] Herrera, F., I. Sadnik, G. Gough, and K. Moldave: Studies on native ribosomal subunits from rat liver. Purification and characterization of three eukaryotic binding factors specific for initiator tRNA. Biochem. **16**, 4664–4671 (1977)

[171] Harbitz, I., and J. G. Hauge: Purification and properties of a methionine formylmethionyl-tRNA-binding initiation factor from pig liver. Arch. Biochem. Biophys. **176**, 766–778 (1976)

[172] Harbitz, I., and J. G. Hauge: Purification and properties of eIF-2 from pig liver. Meth. Enzymol. **60**, 240–246 (1979)

[173] Stringer, E. A., A. Chaudhuri, and U. Maitra: Association of a GDP binding activity with initiation factor eIF-2 from calf liver. Biochem. Biophys. Res. Commun. **76**, 586–592 (1977)

[174] Stringer, E. A., A. Chaudhuri, and U. Maitra: Purified eukaryotic initiation factor 2 from calf liver consists of two polypeptide chains of 48000 and 38000 daltons. J. Biol. Chem. **254**, 6845–6848 (1979)

[175] Gilmartin, M. E., and I. M. Freedberg: Mammalian epidermal protein synthesis: Initiation factors. J. Invest. Dermatol. **67**, 240–245 (1976)

[176] Hejtmancik, J. E., and J. P. Comstock: Partial purification and characterization of two hen oviduct protein synthesis initiation factors capable of initiation complex formation. Biochem. **17**, 1396–1403 (1978)

[177] Filipowicz, W., J. M. Sierra, and S. Ochoa: Polypeptide chain initiation in eukaryotes: Initiation factor MP in *Artemia salina* embryos. Proc. Natl. Acad. Sci. U.S. **72**, 3947–3951 (1975)

[178] Ochiai-Yanagi, S., and R. Mazumder: Role of GTP in eukaryotic polypeptide-chain initiation. Purification and properties of a factor from *Artemia salina* embryos which interacts with initiator transfer RNA and guanine nucleotides. Europ. J. Biochem. **68**, 395–402 (1976)

[179] Treadwell, B. V., and W. G. Robinson: Isolation of protein synthesis initiation factor MP from the high-speed supernatant fraction of wheat germ. Biochem. Biophys. Res. Commun. **65**, 176–183 (1975)

[180] Treadwell, B. V., L. Mauser, and W. G. Robinson: Initiation factors for protein synthesis from wheat germ. Meth. Enzymol. **60**, 181–193 (1979)

[181] Giesen, M., R. Roman, S. N. Seal, and A. Marcus: Formation of an 80 S me-

thionyl-tRNA initiation complex with soluble factors from wheat germ. J. Biol. Chem. **251**, 6075–6081 (1976)

[182] Czarnecka, E., U. Cichocka, U. Szybiak, J. Zawielak, and A. B. Legocki: Studies of Met-$tRNA_f^{Met}$ binding factor from wheat germ. In: Translation of Natural and Synthetic Polynucleotides. (A. B. Legocki, ed.). University of Agriculture, Poznan, 1977, 127–132

[183] Spremulli, L. L., B. J. Walthall, S. R. Lax, and J. M. Ravel: Purification and properties of a Met-$tRNA_f$ binding factor from wheat germ. Arch. Biochem. Biophys. **178**, 565–575 (1977)

[184] Spremulli, L. L., B. J. Walthall, S. R. Lax, and J. M. Ravel: Partial purification of the factors required for the initiation of protein synthesis in wheat germ. J. Biol. Chem. **254**, 143–148 (1979)

[185] Walthall, B. J., L. L. Spremulli, S. R. Lax, and J. M. Ravel: Isolation and purification of protein synthesis initiation factors from wheat germ. Meth. Enzymol. **60**, 193–204 (1979)

[186] Traugh, J. A., S. M. Tahara, S. B. Sharp, B. Safer, and W. C. Merrick: Factors involved in initiation of haemoglobin synthesis can be phosphorylated in vitro. Nature **263**, 163–165 (1976)

[187] Tahara, S. M., J. A. Traugh, S. B. Sharp, T. S. Lundak, B. Safer, and W. C. Merrick: Effect of hemin on site specific phosphorylation of eukaryotic initiation factor 2. Proc. Natl. Acad. Sci. U.S. **75**, 789–793 (1978)

[188] Issinger, O.-G., R. Benne, J. W. B. Hershey, and R. R. Traut: Phosphorylation in vitro of eukaryotic initiation factors IF-E 2 and IF-E 3 by protein kinases. J. Biol. Chem. **251**, 6471–6474 (1976)

[189] Benne, R., J. Edman, R. R. Traut, and J. W. B. Hershey: Phosphorylation of eukaryotic protein synthesis initiation factors. Proc. Natl. Acad. Sci. U.S. **75**, 108–112 (1978)

[190] Ranu, R. S., and I. G. Wool: Discrimination between eukaryotic and prokaryotic and formylated and non-formylated initiator tRNAs by eukaryotic initiation factor EIF-3. Nature **257**, 616 to 618 (1975)

[191] Trachsel, H., and T. Staehelin: Binding and release of eukaryotic initiation factor eIF-2 and GTP during protein synthesis initiation. Proc. Natl. Acad. Sci. U.S. **75**, 204–208 (1978)

[192] Safer, B., D. T. Peterson, and W. C. Merrick: The effect of hemin controlled repressor on initiation factor functions during sequential formation of the 80 S initiation complex. In: Translation of Natural and Synthetic Polynucleotides. (A. B. Legocki ed.). University of Agriculture, Poznan, 1977, 24–31

[193] Peterson, D. T., W. C. Merrick, and B. Safer: Binding and release of radiolabeled eukaryotic initiation factors 2 and 3 during 80 S initiation complex formation. J. Biol. Chem. **254**, 2509–2516 (1979)

[194] Barrieux, A., and M. G. Rosenfeld: Characterization of GTP-dependent Met-$tRNA_f$ binding protein. J. Biol. Chem. **252**, 3843–3847 (1977)

[195] Barrieux, A., and M. G. Rosenfeld: Subunit function of the GTP-dependent Met-tRNA binding protein. Meth. Enzymol. **60**, 265–275 (1979)

[196] Reynolds, S. H., A. Dasgupta, S. Palmieri, A. Majumdar, and N. K. Gupta: Protein synthesis initiation in eukaryotic cells: A comparative study of the mechanism of peptide chain initiation using cell-free systems from mouse ascites tumor cells and rabbit reticulocytes. Arch. Biochem. Biophys. **184**, 328–335 (1977)

[197] Majumdar, A., R. Roy, A. Das, A. Dasgupta, and N. K. Gupta: Protein synthesis in rabbit reticulocytes. XIX. eIF-2 promotes dissociation of Met-$tRNA_f$ · EIF-1 · GTP complex and Met-$tRNA_f$ binding to 40 S ribosomes. Biochem. Biophys. Res. Commun. **78**, 161–169, (1979)

[198] Das, A., and N. K. Gupta: Protein synthesis in rabbit reticulocytes. XX. A supernatant factor (TDI) inhibits ternary complex (Met-$tRNA_f$ · EIF-1 · GTP) dissociation and Met-$tRNA_f$ binding to 40 S ribosomes. Biochem. Biophys. Res. Commun. **78**, 1433–1441 (1977)

[199] Dasgupta, A., A. Das, R. Roy, R. Ralston, A. Majumdar, and N. K. Gupta: Protein synthesis in rabbit reticulocytes. XXI. Purification and properties of a protein factor (Co-EIF-1) which stimulates

Met-tRNA$_f$ binding to EIF-1. J. Biol. Chem. **253**, 6054–6059 (1978)

[200] Dasgupta, A., A. Das, R. Roy, R. Ralston, A. Majumdar, and N. K. Gupta: Purification and properties of rabbit reticulocyte initiation factor Co-EIF-1. Meth. Enzymol. **60**, 53–61 (1979)

[201] Dasgupta, A., R. Roy, S. Palmieri, A. Das, R. Ralston, and N. K. Gupta: Protein synthesis in rabbit reticulocytes. 22. A heat stable dialysable factor (EIF-1^+) modulates Met-tRNA$_f$ binding to EIF-1. Biochem. Biophys. Res. Commun. **82**, 1019–1027 (1978)

[202] Malathi, V. G., and R. Mazumder: An *Artemia salina* factor which stimulates the activity of highly purified initiation factor eIF-2 from *A. salina* and reticulocytes. FEBS-Lett. **86**, 155–159 (1978)

[203] Haro, C. de, A. Datta, and S. Ochoa: Mode of action of the hemin-controlled inhibitor of protein synthesis. Proc. Natl. Acad. Sci. U.S. **75**, 243–247 (1978)

[204] Haro, C., de, and S. Ochoa: Further studies on the mode of action of the heme-controlled translational inibitor. Proc. Natl. Acad. Sci. U.S. **76**, 1741–1745 (1979)

[205] Ranu, R. S., and I. M. London: Regulation of protein synthesis in rabbit reticulocyte lysates: Additional initiation factor required for formation of ternary complex (eIF-2 · GTP · Met-tRNA$_f$) and demonstration of inhibitory effect of heme-regulated protein kinase. Proc. Natl. Acad. Sci. U.S. **76**, 1079–1083 (1979)

[206] Haro, C. de, and S. Ochoa: Further studies on the mode of action of the heme-controlled translational inhibitor: Stimulating protein acts at level of binary complex formation. Proc. Natl. Acad. Sci. U.S. **76**, 2163–2164 (1979)

[207] Osterhout, J. J., J. Phillips-Minton: and J. M. Ravel: Isolation and characterization of two factors from wheat germ that stimulate the binding of Met-tRNA$_f$ and GTP to eIF-2. Fed. Proc. **38**, 327 (1979)

[208] Hellerman, J. G., and D. A. Shafritz: Interaction of poly(A) and mRNA with eukaryotic initiator Met-tRNA$_f$ binding factor: Identification of this activity on reticulocyte ribonucleic acid protein particles. Proc. Natl. Acad. Sci. U.S. **72**, 1021 to 1025 (1975)

[209] Barrieux, A., and M. G. Rosenfeld: Comparison of mRNA binding by Met-tRNA$_f$ binding protein and mRNA-associated proteins. J. Biol. Chem. **252**, 392 to 398 (1977)

[210] Barrieux, A., and M. G. Rosenfeld: mRNA-induced dissociation of initiation factor 2. J. Biol. Chem. **253**, 6311–6314 (1978)

[211] Kaempfer, R., R. Hollender, W. R. Abrams, and R. Israeli: Specific binding of messenger RNA and methionyl-tRNA$_f^{Met}$ by the same initiation factor for eukaryotic protein synthesis. Proc. Natl. Acad. Sci. U.S. **75**, 209–213 (1978)

[212] Kaempfer, R.: RNA-affinity chromatography: its use in purification of eukaryotic initiation factor 2. Meth. Enzymol. **60**, 247–255 (1979)

[213] Kaempfer, R.: Binding of messenger RNA in inititation of eukaryotic translation. Meth. Enzymol. **60**, 380–392 (1979)

[214] Kaempfer, R., R. Hollender, H. Soreq, and U. Nudel: Recognition of messenger RNA in eukaryotic protein synthesis: Equilibrium studies of the interaction between messenger RNA and the initiation factor that binds methionyl-tRNA$_f$. Europ, J. Biochem. **94**, 591–600 (1979)

[215] Cimadevilla, J. M., and B. Hardesty: Isolation and partial characterization of a 40 S ribosomal subunit-transfer RNA binding factor from rabbit reticulocytes. J. Biol. Chem. **250**, 4389–4397 (1975)

[216] Picciano, D. J., P. M. Rrichard, W. C. Merrick, D. A. Shafritz, H. Graf, R. G. Crystal: and W. F. Anderson: Isolation of protein synthesis initiation factors from rabbit liver. J. Biol. Chem. **248**, 204–214 (1973)

[217] McCuiston, J., R. Parker, and K. Moldave: Partial purification and characterization of a binding factor specific for initiator tRNA and ribosomal 40 S subunit and of an aminoacyl-tRNA hydrolase specific for 40 S bound Met-tRNA$_f$ from rat liver. Arch. Biochem. Biophys. **172**, 387–398 (1976)

[218] Leader, D. P., and I. G. Wool: Isolation, purification, and assay of an initiation factor from rat liver cytosol that

promotes binding of aminoacyl-tRNA to 40 S ribosomal subunits. Meth. Enzymol. **30**, 180—186 (1974)

[219] Eich, F., and J. Drews: Isolation and characterization of a peptide chain initiation factor from Krebs II ascites tumor cells. Biochim. Biophys. Acta **340**, 334—338 (1974)

[220] Zasloff, M., and S. Ochoa: Purification of eukaryotic initiation factor 1 (EIF-1) from *Artemia salina* embryos. Meth. Enzymol. **30**, 197—206 (1974)

[221] Cimadevilla, J. M., and B. Hardesty: Aminoacyl-tRNA specificity of a 40 S ribosomal subunit binding factor from rabbit reticulocytes. Biochem. Biophys. Res. Commun. **63**, 16—23 (1975)

[222] Adams, S. L., B. Safer, W. F. Anderson, and W. C. Merrick: Eukaryotic initiation complex formation. Evidence for two distinct pathways. J. Biol. Chem. **250**, 9083—9089 (1975)

[223] Filipowicz, W., J. M. Sierra, C. Nombela, S. Ochoa, W. C. Merrick, and W. F. Anderson: Polypeptide chain initiation in eukaryotes: Initiation factor requirements for translation of natural messenger. Proc. Natl. Acad. Sci. U.S. **73**, 44—48 (1976)

[224] Nombela, C., N. A. Nombela, S. Ochoa, B. Safer, W. F. Anderson, and W. C. Merrick: Polypeptide chain intitiation in eukaryotes: Mechanism of formation of initiation complex. Proc. Natl. Acad. Sci. U.S. **73**, 298—301 (1976)

[225] Strycharz, W. A. M. Ranki, and H. H. M. Dahl: A high-molecular weight protein component required for natural messenger translation in ascites tumor cells. Europ. J. Biochem. **48**, 303—310 (1974)

[226] Thompson, H. A., I. Sadnik, J. Scheinbuks, and K. Moldave: Studies on native ribosomal subunits from rat liver. Purification and characterization of a ribosome dissociation factor. Biochem. **16**, 2221—2230 (1977)

[227] Freienstein, C., and G. Blobel: Nonribosomal proteins associated with eukaryotic native small ribosomal subunits. Proc. Natl. Acad. Sci. U.S. **72**, 3392 to 3396 (1975)

[228] Thompson, H. A., I. Sadnik, and K. Moldave: A novel method for the rapid quantitative determination of ribosome dissociation factor (IF-3) activity. An assay based on the coupling of the dissociation and peptidyltransferase reactions. Biochem. Biophys. Res. Commun. **73**, 532—538 (1976)

[229] Emanuilov, I., D. D. Sabatini, J. A. Lake, and C. Freienstein: Localization of eukaryotic initiation factor 3 on native small ribosomal subunits. Proc. Natl. Acad. Sci. U.S. **75**, 1389—1393 (1978)

[230] Moldave, K., H. A. Thompson, and I. Sadnik: Preparation of a ribosome dissociation factor from native ribosomal subunits of rat liver. Meth. Enzymol. **60**, 290 to 297 (1979)

[231] Russell, D. W., and L. L. Spremulli: Identification of a wheat germ ribosome dissociation factor distinct from initiation factor eIF-3. J. Biol. Chem. **253**, 6647—6649 (1978)

[232] Russell, D. W., and L. L. Spremulli: Purification and properties of a wheat germ ribosome dissociation factor. Fed. Proc. **38**, 328 (1979)

[233] Kennedy, D. S., and S. M. Heywood: The role of muscle and reticulocyte initiation factor 3 on the translation of myosin and globin messenger RNA in a wheat germ cell-free system. FEBS-Lett. **72**, 314—318 (1974)

[234] Gilmore-Hebert, M. A., and S. M. Heywood: Translation of tubulin messenger ribonucleic acid. Biochim. Biophys. Acta **454**, 55—66 (1976)

[235] Heywood, S. M., and D. S. Kennedy: mRNA affinity column fractionation of eukaryotic initiation factor and the translation of myosin mRNA. Arch. Biochem. Biophys. **192**, 270—281 (1979)

[236] Ilan, J., and J. Ilan: Requirement for homologous rabbit reticulocyte initiation factor 3 for initiation of α- and β- globin mRNA translation in a crude protozoal cell-free system. J. Biol. Chem. **251**, 5718—5725 (1876)

[237] Ilan, J., and J. Ilan: Unwinding protein specific for mRNA translation fractionated together with rabbit reticulocyte initiation factor 3 complex. Proc. Natl. Acad. Sci. U.S. **74**, 2325—2329 (1977)

[238] Wigle, D. T., and A. E. Smith: Specificity in initiation of protein synthesis

in a fractionated mammalian cell-free system. Nature New Biol. **242**, 136–140 (1973)

[239] DAHL, H.-H. M., E. TRUELSEN, and G. E. BLAIR: The purification and properties of two low-molecular weight proteins required for the initiation of translation in ascites tumour cells. Europ. J. Biochem. **77**, 209–216 (1977)

[240] BLAIR, G. E., H.-H. M. DAHL, E. TRUELSEN, and J. C. LELONG: Functional identity of a mouse ascites and a rabbit reticulocyte initiation factor required for natural mRNA translation. Nature **265**, 651–653 (1977)

[241] KABAT, D., and M. R. CHAPPELL: Competition between globin messenger ribonucleic acid for a discriminating initiation factor. J. Biol. Chem. **252**, 2684 to 2690 (1977)

[242] GOLINI, F., S. S. THACH, C. H. BIRGE, B. SAFER, W. C. MERRICK, and R. E. THACH: Competition between cellular and viral mRNAs in vitro is regulated by a messenger discriminatory initiation factor Proc. Natl. Acad. Sci. U.S. **73**, 3040 to 3044 (1976)

[243] ROSE, J. K., H. TRACHSEL, K. LEONG, and D. BALTIMORE: Inhibition of translation by poliovirus: Inactivation of specific initiation factor. Proc. Natl. Acad. Sci. U.S. **75**, 2732–2736 (1978)

[244] BAGLIONI, C., M. SIMILI, and D. A. SHAFRITZ: Initiation activity of EMC virus RNA binding to initiation factor eIF-4 B and shut off of host cell protein synthesis. Nature **275**, 240–243 (1978)

[245] SHAFRITZ, D. A., J. A. WEINSTEIN, B. SAFER, W. C. MERRICK, L. A. WEBER, E. D. HICKEY, and C. BAGLIONI: Evidence for role of m^7G^5-phosphate group in recognition of eukaryotic mRNA by initiation factor IF-M_3. Nature **261**, 291–294 (1976)

[246] BERGMANN, J. E., H. TRACHSEL, N. SONENBERG, A. J. SHATKIN, and H. F. LODISH: Characterization of rabbit reticulocyte factor(s) that stimulates the translation of mRNAs lacking 5′-terminal 7-methylguanosine. J. Biol. Chem. **254**, 1440–1443 (1979)

[247] SHAFRITZ, D. A., P. M. PRICHARD, J. M. GILBERT, W. C. MERRICK, and W. F. ANDERSON: Separation of reticulocyte initiation factor M 2 activity into two components. Proc. Natl. Acad. Sci. U.S. **69**, 983–987 (1972)

[248] HEJTMANCIK, J., and J. P. COMSTOCK: Isolation and purification of hen oviduct synthesis initiation factors A 2 A and A 2 B. Biochem. **15**, 3804–3811 (1976)

[249] MERRICK, W. C.,: Evidence that a single GTP is used in the formation of 80 S initiation complexes. J. Biol. Chem. **254**, 3708–3711 (1979)

[250] CIMADEVILLA, J. M., J. MORRISEY, and B. HARDESTY: A functional interaction between methionyl-transfer RNA hydrolase and a transfer RNA binding factor. J. Mol. Biol. **83**, 437–446 (1974)

[251] NYGARD, O., and T. HULTIN: Accumulation of $tRNA_f^{Met}$ on 80 S ribosomes in vitro under the influence of a Met-tRNA deacylase from rat-liver microsomes. Europ. J. Biochem. **72**, 537–542 (1977)

[252] ANDERSEN, K. B.: Methionyl-$tRNA_f^{Met}$ deacylase: Purification, characterization, and effects on translational initiation complexes. Europ. J. Biochem. **96**, 109 to 118 (1979)

[253] POHLREICH, P., O. KŘIŽ, Z. TUHACKOVA, Z. DUŠEK, and J. HRADEC: Purification and some properties of a protein factor binding and deacylating initiator transfer ribonucleic acid. Biochem. J. **177**, 707 to 719 (1979)

[254] ANDERSEN, K. B., G. BOLCSFOLDI, and M. H. VAUGHAN: Destabilization of eukaryotic 40 S translational initiation complex by ApUpG and partial characterization of a factor required for this activity. J. Biol. Chem. **253**, 6370–6378 (1978)

[255] ROBINSON, W. G., and L. MAUSER: Specific destruction of 40 S complexes containing AUG. Fed. Proc. **38**, 329 (1979)

[256] FRESNO, M., and D. VAZQUEZ: Initiation of protein synthesis in eukaryotic systems with native 40 S ribosomal subunits: Effects of translation inhibitors. Meth. Enzymol. **60**,‘566–577 (1979)

[257] AMESZ, H., T. HAUBRICH, and H. O. VOORMA: Postribosomal complexes containing eukaryot icinitiation factor eIF-2. Mol. Biol. Rep. **5**, 121–125 (1979)

[258] MILLER, D., and H. WEISSBACH: Factors involved in the transfer of aminoacyl-tRNA to the ribosome. In: Molecular

Mechanisms of Protein Biosynthesis. (Eds.: H. WEISSBACH and S. PESTKA). Acad. Press, New York, 1977, 323–373

[259] KAZIRO, Y.: The role of guanosine 5'-triphosphate in polypeptide chain elongation. Biochim. Biophys. Acta **505**, 95 to 127 (1978)

[260] PRATHER, N., J. M. RAVEL, B. HARDESTY, and W. SHIVE: Evidence for activities of rabbit reticulocyte elongation factor 1 analogous to bacterial factors EF-Ts and EF-Tu. Biochem. Biophys. Res. Commun. **57**, 578–583 (1974)

[261] EJIRI, S.-I., K. MURAKAMI, and T. KATSUMATA: Elongation factor 1 from the silk gland of silkworm. FEBS-Lett. **92**, 251 to 254 (1978)

[262] EJIRI, S.-I., Y. NAOKI, K. MURAKAMI, and T. KATSUMATA: Exchangeability of silk gland elongation factor 1b and pig liver elongation factor 1β in polypeptide chain elongation. FEBS-Lett. **95**, 277 to 280 (1978)

[263] NAGATA, S., K. MOTOYOSHI, and K. IWASAKI: Interaction of subunits of polypeptide chain elongation factor 1 from pig liver. Formation of EF-$1\alpha \cdot$ EF-$1\beta\gamma$ and EF-$1\alpha \cdot$ EF-$1\,\beta$ complexes. J. Biochem. **83**, 423–429 (1978)

[264] SLOBIN, L. I., and W. MÖLLER: Purification and properties of an elongation factor functionally analogous to bacterial elongation factor Ts from embryos of *Artemia salina*. Europ. J. Biochem. **84**, 69–77 (1978)

[265] ROOBOL, K., and W. MÖLLER: Transient interaction between elongation factor 1 from *Artemia salina* and the 80 S ribosome. FEBS-Lett. **96**, 377–380 (1978)

[266] SLOBIN, L. I.: Eucaryotic elongation factor Ts is an integral component of rabbit reticulocyte elongation factor 1. Europ. J. Biochem. **96**, 287–293 (1979)

[267] GRASMUK, H., R. D. NOLAN, and J. DREWS: The isolation and characterization of elongation factor eEF-Ts from Krebs-II-mouse-ascites-tumor cells and its role in the elongation process. Europ. J. Biochem. **92**, 479–490 (1978)

[268] KEMPER, W. M., W. C. MERRICK, B. REDFIELD, C. -K. LIU, and H. WEISSBACH: Purification and properties of rabbit reticulocyte elongation factor. Arch. Biochem. Biophys. **174**, 603–612 (1976)

[269] GOLINSKY, B., and A. B. LEGOCKI: Purification and some properties of elongation factor 1 from wheat germ. Biochim. Biophys. Acta **324**, 156–170 (1973)

[270] LANZANI, G. A., R. BOLLINI, and A. N. SOFFIENTINI: The heterogeneity of the elongation factor EF 1 from wheat embryos. Biochim. Biophys. Acta **335**, 275 to 283 (1974)

[271] BOLLINI, R., A. N. SOFFIENTINI, A. BERTANI, and G. A. LANZANI: Some molecular aspects of the elongation factor EF 1 from wheat embryos. Biochem. **13**, 5421 to 5425 (1974)

[272] LANZANI, G. A., E. CALDIROLI, L. A. MANZOCCHI, R. BOLLINI, and L. DE ALBERTI: The translational system from wheat embryos: some properties of the polypeptides associated in EF 1_H. FEBS Lett. **64**, 102 to 106 (1976)

[273] IWASAKI, K., K. MIZUMOTO, M. TANAKA, and Y. KAZIRO: A new protein factor required for polypeptide elongation in mammalian tissues. J. Biochem. **74**, 849 to 852 (1973)

[274] IWASAKI, K., K. MOTOYOSHI, S. NAGATA, and Y. KAZIRO: Purification and properties of a new polypeptide chain elongation factor, EF-1β, from pig liver. J. Biol. Chem. **251**, 1843–1845 (1976)

[275] NAGATA, S., K. IWASAKI, and Y. KAZIRO: Interaction of the low molecular weight form of elongation factor 1 with guanine nucleotides and aminoacyl-tRNA. Arch. Biochem. Biophys. **172**, 168–177 (1976)

[276] MOTOYOSHI, K., K. IWASAKI, and Y. KAZIRO: Purification and properties of polypeptide chain elongation factor-$1\beta\gamma$ from pig liver. J. Biochem. **82**, 145–155 (1977)

[277] MOTOYOSHI, K., and K. IWASAKI: Resolution of the polypeptide chain elongation factor $1\beta\gamma$ into subunits and some properties of the subunits. J. Biochem. **82**, 703–708 (1977)

[278] NAGATA, S., K. IWASAKI, and Y. KAZIRO: Purification and properties of polypeptide chain elongation factor-1α from pig liver. J. Biochem. **82**, 1633–1646 (1977)

[279] WEISSBACH, H., B. REDFIELD, and H. M. MOON: Further studies on the interactions of elongation factor 1 from animal tissues. Arch. Biochem. Biophys. **156**, 267–275 (1973)

[280] Drews, J., K. Bednarik, and H. Grasmuk: Elongation factor 1 from Krebs II mouse ascites cells. Purification, structure and enzymatic properties. Europ. J. Biochem. **41**, 217—227 (1974)

[281] Grasmuk, H., R. D. Nolan, and J. Drews: Functional identity of the monomeric and multiple forms of elongation factor 1 from Krebs-II-mouse-ascites tumour cells. Europ. J. Biochem. **67**, 421 to 431 (1976)

[282] Slobin, L. J., and W. Möller: Changes in form of elongation factor during developement of *Artemia salina*. Nature **258**, 452—454 (1975)

[283] Nombela, C., B. Redfield, S. Ochoa, and H. Weissbach: Elongation factor 1 from *Artemia salina*: Properties and disaggregation of the enzyme. Europ. J. Biochem. **65**, 395—402 (1976)

[284] Slobin, L. I., and W. Möller: Characterization of developmentally regulated forms of elongation factor 1 in *Artemia salina*. 1. Purification and structural properties of the enzymes. Europ. J. Biochem. **69**, 351—366 (1976)

[285] Slobin, L. I., and W. Möller: Characterization of developmentally regulated forms of EF 1 in *Artemia salina*. 2. Functional properties of the enzymes. Europ. J. Biochem. **69**, 367—375 (1976)

[286] Tarrago, A., J. E. Allende, B. Redfield, and H. Weissbach: The effect of guanosine nucleotides on the multiple forms of protein synthesis elongation factor 1 from wheat embryos. Arch. Biochem. Biophys. **159**, 353—361 (1973)

[287] Legocki, A. B., B. Redfield, and H. Weissbach: Interaction of the heavy and light forms of elongation factor 1 with guanine nucleotides and aminoacyl-tRNA. Arch. Biochem. Biophys. **161**, 709—712 (1974)

[288] Twardowski, T., B. Redfield, W. Kemper, W. C. Merrick, and H. Weissbach: Studies on the disaggregation of elongation factor 1 by elastase. Biochem. Biophys. Res. Commun. **71**, 272—279 (1976)

[289] Twardowski, T., J. M. Hill, and H. Weissbach: Disaggregation of elongation factor 1 by extracts of *Artemia salina*. Biochem. Biophys. Res. Commun. **71**, 826 to 833 (1976)

[290] Twardowski, T., J. M. Hill, and H. Weissbach: Studies on the disaggregation of EF-1 with carboxypeptidase A. Arch. Biochem. Biophys. **180**, 444—451 (1977)

[291] Kemper, W. M., W. C. Merrick, B. Redfield, C.-K. Liu, and H. Weissbach: Purification and properties of rabbit reticulocyte elongation factor. Arch. Biochem. Biophys. **174**, 603—612 (1976)

[292] Kolb, A. J., B. Redfield, T. Twardowski, and H. Weissbach: Binding of reticulocyte elongation factor 1 to ribosomes and nucleic acids. Biochim. Biophys. Acta **519**, 398—405 (1978)

[293] Grasmuk, H., R. D. Nolan, and J. Drews: A new concept of the function of EF 1 in peptide chain elongation. Europ. J. Biochem. **71**, 271—279 (1976)

[294] Grasmuk, H., R. D. Nolan, and J. Drews: Interchangeability of elongation factor-Tu and elongation factor 1 in aminoacyl-tRNA binding to 70 S and 80 S ribosomes. FEBS-Lett. **82**, 237—242 (1977)

[295] Kaziro, Y.: Polypeptide chain elongation in bacterial and mammalian systems. J. Biochem. **79**, 32p—33p (1976)

[296] Kemper, W. M., and W. C. Merrick: Preparation of protein synthesis elongation factors from rabbit reticulocytes Meth. Enzymol. **60**, 638—648 (1979)

[297] Nolan, R. D., H. Grasmuk, and J. Drews: Preparation of elongation factors from ascites cells. Meth. Enzymol. **60**, 649—657 (1979)

[298] Iwasaki, K., and Y. Kaziro: Polypeptide chain elongation factors from pig liver. Meth. Enzymol. **60**, 657—676 (1979)

[299] Skogerson, L.: Separation and characterization of yeast elongation factors. Meth. Enzymol. **60**, 676—684 (1979)

[300] Slobin, L. I., and W. Möller: Purification of elongation factor 1 from embryos of *Artemia salina*. Meth. Enzymol. **60**, 685—703 (1979)

[301] Legocki, A. B.: Elongation factor EF-2 from wheat germ: purification and properties. Meth. Enzymol. **60**, 703—712 (1979)

[302] Moretti, S., T. Staehelin, H. Trachsel, and J. Gordon: A new purification scheme for elongation factor 1 from rabbit reticulocytes and investigation of the homology of the subunits with those of initiation factor 2. Europ. J. Biochem. **97**, 609—614 (1979)

[303] TIBONI, O., G. DI PASQUALE, and O. CIFERRI: Purification of the elongation factors present in spinach chloroplasts. Europ. J. Biochem. **92**, 471–477 (1978)

[304] TATE, W. P., and C. T. CASKEY: The mechanism of peptide chain termination. Mol. Cell. Biochem. **5**, 115–126 (1974)

[305] GOLDSTEIN, J. L., A. L. BEAUDET, and C. T. CASKEY: Peptide chain termination with mammalian release factor. Proc. Natl. Acad. Sci. U.S. **67**, 99–106 (1970)

[306] BEAUDET, A. L., and C. T. CASKEY: Mammalian chain termination. II. Codon specificity and GTPase activity of release factor. Proc. Natl. Acad. Sci. U.S. **68**, 619 to 624 (1971)

[307] KONECKI, D. S., K. C. AUNE, W. TATE, and C. T. CASKEY: Characterization of reticulocyte release factor. J. Biol. Chem. **68**, **252**, 4514–4520 (1977)

[308] STEWART, J. W., F. SHERMAN, M. JACKSON, F. L. THOMAS, and N. SHIPMAN: Demonstration of the UAA ochre codon in baker's yeast by amino acid replacements in iso-1-cytochrome c. J. Mol. Biol. **68**, 83–96 (1972)

[309] STEWART, J. W., and F. SHERMAN: Demonstration of UAG as a nonsense codon in baker's yeast by amino-acid replacement in iso-1-cyto chrome c. J. Mol. Biol. **68**, 429–443 (1972)

[310] INNANEN, V. T., and D. M. NICHOLLS: Isolation of release factor from mammalian liver and the use of rat liver ribosomes in the release assay. Biochim. Biophys. Acta **324**, 533–544 (1973)

[311] CHONG-CHENG, C., and I. T. OLIVER: A translation control mechanism in mammalian protein synthesis modulated by cyclic adenosine monophosphate. Translational control of tyrosine aminotransferase synthesis in neonatal rat liver. Biochem. **11**, 2547–2553 (1972)

[312] DONOVAN, G., and I. T. OLIVER: Purification and properties of a microsomal cyclic adenosine monophosphate binding protein required for the release of tyrosine aminotransferase from polysomes. Biochem. **11**, 3904–3910 (1972)

[313] PÊTRE, J.: Specific dissociation of ribosomes from *Saccharomyces cerevisiae* by a protein factor. Europ. J. Biochem. **14**, 399–405 (1970)

[314] SURGUCHOV, A. P., E. S. FOMINYKCH, and L. V. LYZLOVA: Dissociability of the free and peptidyl-tRNA bound ribosomes. Mol. Biol. Rep. **4**, 79–81 (1978)

[315] LAWFORD, G. R., J. KAISER, and W. C. HEY: A factor capable of dissociating rat liver ribosomes. Can. J. Biochem. **49**, 1301–1306 (1971)

[316] LUBSEN, N. H., and B. D. DAVIS: A ribosome dissociation factor from rabbit reticulocytes. Proc. Natl. Acad. Sci. U.S. **69**, 353–357 (1972)

[317] MERRICK, W. C., N. H. LUBSEN, and W. F. ANDERSON: Ribosome dissociation factor from rabbit reticulocytes distinct from initiation factor M 3. Proc. Natl. Acad. Sci. U.S. **70**, 2220–2222 (1973)

[318] LUBSEN, N. H., and B. D. DAVIS: A ribosome dissociation factor on both native subunits in rabbit reticulocytes. Biochim. Biophys. Acta **335**, 196–200 (1974)

[319] KAEMPFER, R., and J. KAUFMANN: Translational control of hemoglobin synthesis by an initiation factor required for recycling of ribosomes and for their binding to messenger RNA. Proc. Natl. Acad. Sci. U.S. **69**, 3317–3321 (1972)

[320] NAKAYA, K., R. S. RANU, and I. G WOOL: Dissociation of eukaryotic ribosomes by purified initiation factor EIF-3. Biochem. Biophys. Res. Commun. **54**, 246–255 (1973)

[321] NAKAYA, K., R. S. RANU, and I. G. WOOL: Dissociation of skeletal muscle ribosomes from normal and diabetic animals by initiation factor EIF-3. Biochem. Biophys. Res. Commun. **59**, 237–245 (1974)

[322] DECROLY, M., and M. GOLDFINGER: A dissociation factor from embryos of *Xenopus laevis*. Biochim. Biophys. Acta **390**, 82–93 (1975)

[323] VAN DER SAAG, P. T., and S. BORDIN: A ribosome dissociation factor from neonatal mouse brain. J. Neurochem. **32**, 1857–1859 (1979)

[324] SAFER, B., R. JAGUS, and W. M. KEMPER: Analysis of initiation factor function in highly fractionated and unfractionated reticulocytes lysate systems. Meth. Enzymol. **60**, 61–87 (1979)

[325] MERRICK, W. C.: Assays for eucaryotic protein synthesis. Meth. Enzymol. **60**, 108–123 (1979)

[326] SAFER, B., W. M. KEMPER, and R. JAGUS: Identification of a 48 S preinitiation

complex in reticulocyte lysate. J. Biol. Chem. **253**, 3384–3386 (1978)

[327] KRAMER, G., D. KONECKI, J. M. CIMADEVILLA, and B. HARDESTY: ATP requirement for binding of ^{125}I-labeled globin-mRNA to *Artemia salina* ribosomes. Arch. Biochem. Biophys. **174**, 355–358 (1976)

[328] ROLLESTON, F. S.: Membrane-bound and free ribosomes. Sub-Cell. Biochem. **3**, 91–117 (1974)

[329] BLOBEL, G., and D. D. SABATINI: Ribosome-membrane interaction in eukaryotic cells. Biomembranes **2**, 193–195 (1971)

[330] MILSTEIN, C., G. G. BROWNLEE, T. M. HARRISON, and M. B. MATHEWS: A possible precursor of immunoglobulin light chains. Nature New Biol. **239**, 117–120 (1972)

[331] SWAN, D., H. AVIV, and P. LEDER: Purification and properties of biologically active mRNA for a myeolma light chain. Proc. Natl. Acad. Sci. U.S. **69**, 1967 to 1971 (1972)

[332] MACH, B., C. FAUST, and P. VASALLI: Purification of 14 S messenger RNA of immunoglobulin light chain that codes for a possible light-chain precursor. Proc. Natl. Acad. Sci. U.S. **70**, 451–455 (1973)

[333] SCHECHTER, I.: Biologically and chemically pure mRNA coding for mouse immunoglobulin L-chain prepared with the aid of antibodies and immobilized oligothymidine. Proc. Natl. Acad. Sci. U.S. **70**, 2256–2260 (1973)

[334] TONEGAWA, S., and I. BALD: Electrophoretically homogeneous myeloma light chain mRNA and its translation in vitro. Biochem. Biophys. Res. Commun. **51**, 81–87 (1973)

[335] SCHMECKPEPER, B. J., S. CORY, and J. M. ADAMS: Translation of immunoglobulin mRNAs in a wheat germ cell-free system. Mol. Biol. Rep. **1**, 355–363 (1974)

[336] SCHECHTER, I., D. J. MCKEAN, R. GUYER, and W. TERRY: Partial amino acid sequence of the precursor of immunoglobulin light chain programmed by mRNA in vitro. Science **188**, 160–162 (1974)

[337] BLOBEL, G., and B. DOBBERSTEIN: Transfer of proteins across membranes. I. Presence of proteolytically processed and unprocessed nascent immunoglobulin light chains on membrane-bound ribosomes of murine myeloma. J. Cell.. Biol. **67**, 835 to 851 (1975)

[338] BLOBEL, G., and B. DOBBERSTEIN: Transfer of proteins across membranes. II. Reconstitution of functional rough microsomes from heterologous components. J. Cell Biol. **67**, 852–862 (1975)

[339] BURSTEIN, Y., F. KANTOR, and I. SCHECHTER: Partial amino-acid sequence of the precursor of an immunoglobulin light chain containing NH_2-terminal pyroglutamic acid. Proc. Natl. Acad. Sci. U. S. 2604–2608 (1976)

[340] SCHECHTER, I., and Y. BURSTEIN: Marked hydrophobicity of the NH_2-terminal extra piece of immunoglobulin light-chain precursors: Possible physiological functions of the extra piece. Proc. Natl. Acad. Sci. U.S. **73**, 3273–3277 (1976)

[341] SCHECHTER, I., and Y. BURSTEIN: Identification of N-terminal methionine in the precursor of immunoglobulin light chain. Biochem. J. **153**, 543–550 (1976)

[342] BURSTEIN, Y., and I. SCHECHTER: Amino acid sequence of the NH_2-terminal extra piece segments of the precursors of mouse immunoglobulin λ_1-type and $\varkappa$-type light chains. Proc. Natl. Acad. Sci. U.S. **74**, 716–720 (1977)

[343] BURSTEIN, Y., R. ZEMELL, F. KANTOR, and I. SCHECHTER: Independent expression of the gene coding for the constant domain of immunoglobulin light chain: Evidence from sequence analyses of the precursor of the constant region polypeptide. Proc. Natl. Acad. Sci. U.S. **74**, 3157–3161 (1977)

[344] DOBBERSTEIN, B., and G. BLOBEL: Functional interaction of plant ribosomes with animal microsomal membranes. Biochem. Biophys. Res. Commun. **74**, 1675–1682 (1977)

[345] KEMPER, B., J. F. HABENER, J. T. POTTS jr., and A. RICH: Proparathyroid hormone: Identification of a biosynthetic precursor to parathyroid hormone. Proc. Natl. Acad. Sci. U.S. **69**, 643–647 (1972)

[346] KEMPER, B., J. F. HABENER, R. C. MULLIGAN, J. T. POTTS, jr., and A. RICH: Preproparathyroid hormone: A direct translation product of parathyroid messenger RNA. Proc. Natl. Acad. Sci. U.S. **71**, 3731 to 3735 (1974)

[347] HABENER, J. F., B. KEMPER, J. T. POTTS,

jr., and A. RICH: Parathyroid mRNA directs the synthesis of pre-proparathyroid hormone and proparathyroid hormone in the Krebs ascites cell-free system. Biochem. Biophys. Res. Commun. **67**, 1114 to 1121 (1975)

[348] HABENER, J. F., J. T. POTTS, jr., and A. RICH: Pre-proparathyroid hormone. Evidence for an early biosynthetic precursor of proparathyroid hormone. J. Biol. Chem. **251**, 3893–3899 (1976)

[349] KEMPER, B., J. F. HABENER, M. D. ERNST, J. D. POTTS, jr., and A. RICH: Pre-proparathyroid hormone: Analysis of radioactive tryptic peptides and amino acid sequence. Biochem. **15**, 15–19 (1976)

[350] KEMPER, B., J. F. HABENER, J. T. POTTS, jr., and A. RICH: Pre-proparathyroid hormone: Fidelity of the translation of parathyroid messenger RNA by extracts of wheat germ. Biochem. **15**, 20–25 (1976)

[351] HABENER, J. F., M. ROSENBLAT, B. KEMPER, H. M. KRONENBERG, A. RICH, and J. T. POTTS, jr.: Pre-proparathyroid hormone: Amino acid sequence, chemical synthesis, and some biological studies of the precursor region. Proc. Natl. Acad. Sci. U.S. **75**, 2616–2620 (1978)

[352] KREIL, G., and H. BACHMEYER: Biosynthesis of melittin, a toxic peptide from been venom. Detection of a possible precursor. Europ. J. Biochem. **20**, 344–350 (1971)

[353] SUCHANEK, G., I. KINDÅS-MÜGGE, G. KREIL, and M. H. SCHREIER: Translation of honeybee promelittin messenger RNA. Formation of a larger product in a mammalian cell-free system. Europ. J. Biochem. **60**, 309–315 (1975)

[354] SUCHANEK, G., G. KREIL, and M. A. HERMODSON: Amino acid sequence of honeybee prepromelittin synthesized in vitro. Proc. Natl. Acad. Sci. U.S. **75**, 701–704 (1978)

[355] STACHURA, M. E., and L. A. FROHMAN: Growth hormone: Independent release of big and small forms from rat pituitary in vitro. Science **187**, 447–449 (1975)

[356] SUSSMAN, P. M., R. J. TUSHINSKI, and F. C. BANCROFT: Pregrowth hormone: Product of the translation in vitro of messenger RNA coding for growth hormone. Proc. Natl. Acad. Sci. U.S. **73**, 29 to 33 (1976)

[357] WALLIS, M., and R. V. DAVIS: Studies on the chemistry of bovine and rat growth hormones. In: Growth hormone and related peptides. (A. PECILE and E. E. MÜLLER, eds.) Elsevier, New York, 1976, 1–14

[358] LINGAPPA, V. R., A. DEVILLERS-THIERY, and G. BLOBEL: Nascent prehormones are intermediates in the biosynthesis of authentic bovine pituitary growth hormone and prolactin. Proc. Natl. Acad. Sci. U.S. **74**, 2432–2436 (1977)

[359] SEEBURG, P. H., J. SHINE, J. A. MARTIAL, J. D. BAXTER, and H. M. GOODMAN: Nucleotide sequence and amplification in bacteria of structural gene for rat gıowth hormone. Nature **270**, 486–494 (1977)

[360] PERMUTT, M. A., J. BIESBROECK, R. CHYN, I. BOIME, E. SCZCESNA, and D. McWILLIAMS: In: Polypeptide Hormones: Molecular and Cellular Aspects, CIBA Found. Symp. No. 41, Elsevier, Amsterdam, 1976 97–116

[361] PERMUTT, M. A., and I. BOIME: Isolation of biologically active fish islet messenger RNA by oligo-d Tcellulose affinity chromatography. Diabetes **24**, Suppl. **2**, 405 (1975)

[362] WEBER, H. E.: Partial purification and translation of proinsulin messenger RNA. Diabetes **24**, Suppl. 2, 405 (1975)

[363] YIP, C., C.-L. HEW, and H. HSU: Translation of messenger ribonucleic acid from isolated pancreatic islets and human insulinomas. Proc. Natl. Acad. Sci. U.S. **72**, 4777–4779 (1975)

[364] CHAN, S. J., B. E. NOYES, K. L. AGARWAL, and D. F. STEINER: Construction and selection of recombinant plasmids containing full-lenght complementary DNAs corresponding to rat liver insulin I and II. Proc. Natl. Acad. Sci. U.S. **76**, 5036 to 5040 (1979)

[365] RAPOPORT, T. A., W. E. HÖHNE, D. KLATT, S. PREHN, and V. HAHN: Evidence for the synthesis of a precursor of carp proinsulin in a cell-free translation system. FEBS Lett. **69**, 32–36 (1976)

[366] LOMEDICO, P. T., and G. F. SAUNDERS: Preparation of pancreatic mRNA: cell-free translation of an insulin-immunoreactive polypeptide. Nucl. Acids Res. **3**, 381–386 (1976)

[367] SHIELDS, D., and G. BLOBEL: Cell-free synthesis of fish preproinsulin, and processing by heterologous mammalian microsomal membranes. Proc. Natl. Acad. Sci. U.S. **74**, 2059–2063 (1977)

[368] ULLRICH, A., J. SHINE, J. CHIRGWIN, R. PICTET, E. TISCHLER, W. J. RUTTER, and H. M. GOODMAN: Rat insulin genes: Constitution of plasmids containing the coding sequences. Science **196**, 1313–1319 (1977)

[369] DEVILLERS-THIERY, A., T. KINDT, G. SCHEELE, and G. BLOBEL: Homology in amino-terminal sequence of precursors to pancreatic secretory proteins. Proc. Natl. Acad. Sci. U.S. **72**, 5016–5020 (1975)

[370] SCHEELE, G., B. DOBBERSTEIN, and G. BLOBEL: Transfer of proteins across membranes. Biosynthesis in vitro of pretrypsinogen and trypsinogen by cell fractions of canine pancreas. Europ. J. Biochem. **82**, 593–599 (1978)

[371] YU, S., and C. REDMAN: In vitro synthesis of rat pre-proalbumin. Biochem. Biophys. Res. Commun. **76**, 469–476 (1977)

[372] STRAUSS, A. W., C. D. BENNETT, A. M. DONOHUE, J. A. RODKEY, and A. W. ALBERTS: Rat hepatoma 5123 TC albumin mRNA directs the synthesis of pre-proalbumin identical to rat liver pre-proalbumin. Biochem. Biophys. Res. Commun. **77**, 1224–1230 (1977)

[373] STRAUSS, A. W., A. M. DONOHUE, C. D. BENNETT, J. A. RODKEY, and A. W. ALBERTS: Rat liver pre-proalbumin: In vitro synthesis and partial amino acid sequence. Proc. Natl. Acad. Sci. U.S. **74**, 1358 to 1362 (1977)

[374] STRAUSS, A. W., C. D. BENNETT, A. M. DONOHUE, J. A. RODKEY, and A. W. ALBERTS: Rat liver pre-proalbumin: complete amino acid sequence of the prepiece. Analysis of the direct translation product of albumin messenger RNA. J. Biol. Chem. **252**, 6846–6855 (1977)

[375] STRAUSS, A. W., C. D. BENNETT, A. M. DONOHUE, J. A. RODKEY, I. BOIME, and A. W. ALBERTS: Conversion of rat preproalbumin to proalbumin in vitro by ascites membranes. Demonstration by NH_2-terminal sequence analysis. J. Biol. Chem. **253**, 6270–6274 (1978)

[376] CRAIG, R. K., P. A. BROWN, O. S. HARRISON, D. MCILREAVY, and P. N. CAMPBELL: Guinea-pig milk-protein synthesis. Isolation and characterization of messenger ribonucleic acids from lactating mammary gland and identification of caseins and pre-α-lactalbumin as translation products in heterologous cell-free systems. Biochem. J. **160**, 57–74 (1976)

[377] LINGAPPA, V. R., J. R. LINGAPPA, R. PRASAD, K. E. EBNER, and G. BLOBEL: Coupled cell-free synthesis, segregation, and core glycosylation of a secretory protein. Proc. Natl. Acad. Sci. U.S. **75**, 2338 to 2342 (1978)

[378] BOIME, I., S. BOGUSLAWSKI, and J. CAINE: The translation of a human placental lactogen mRNA fraction in heterologous cell-free systems: the synthesis of a possible precursor. Biochem. Biophys. Res. Commun. **62**, 103–109 (1975)

[379] COX, G. J., B. D. WEINTRAUB, S. W. ROSEN, and E. S. MAXWELL: Properties of biologically active messenger RNA from human placenta. J. Biol. Chem. **251**, 1723 to 1730 (1976)

[380] SZCZESNA, E., and I. BOIME: mRNA-dependent synthesis of authentic precursor to human placental lactogen: Conversion to its mature hormone form in ascites cell-free extracts. Proc. Natl. Acad. Sci. U.S. **73**, 1179–1183 (1976)

[381] BOIME, I., E. SZCZESNA, and D. L. SMITH: Membrane-dependent cleavage of the human placental lactogen precursor to its native form in ascites cell-free extracts. Europ. J. Biochem. **73**, 515–520 (1977)

[382] BIRKEN, S., D. L. SMITH, R. E. CANFIELD, and I. BOIME: Partial amino acid sequence of human placental lactogen precursor and its mature hormone form produced by membrane-associated enzyme activity. Biochem. Biophys. Res. Commun. **74**, 106–112 (1977)

[383] BIELINSKA, M., G. A. GRANT, and I. BOIME: Processing of placental peptide hormones synthesized in lysates containing membranes derived from tunicamycin-treated ascites tumor cells. J. Biol. Chem. **253**, 7117–7119 (1978)

[384] EVANS, G. A., and M. G. ROSENFELD: Cell-free synthesis of a prolactin precursor directed by mRNA from cultured rat pituitary cells. J. Biol. Chem. **251**, 2842–2847 (1976)

[385] DANNIS, P. S., and A. H. TASHJIAN: Thy-

rotropin-releasing hormone increases prolactin mRNA activity in the cytoplasm of GH-cells as measured by translation in a wheat germ cell-free system. Biochem. Biophys. Res. Commun. **70**, 1180–1189 (1976)

[386] Maurer, R. A., R. Stone, and J. Gorski: Cell-free synthesis of a large translation product of prolactin messenger RNA. J. Biol. Chem. **251**, 2801–2807 (1976)

[387] Jackson, R. C., and G. Blobel: Posttranslation cleavage of presecretory proteins with an extract of rough microsomes from dog pancreas containing signal peptidase activity. Proc. Natl. Acad. Sci. U.S. **74**, 5598–5602 (1977)

[388] Maurer, R. A., and D. J. McKean: Synthesis of preprolactin and conversion to prolactin in intact cells and a cell-free system. J. Biol. Chem. **253**, 6315–6318 (1978)

[389] Nakanishi, S., S. Tah, S. Hirata, S. Matsukura, H. Imura, and S. Numan: A large product of cell-free translation of messenger RNA coding for corticotropin. Proc. Natl. Acad. Sci. U.S. **73**, 4319 to 4323 (1976)

[390] Dobberstein, B., G. Blobel, and N.-H. Chua: In vitro synthesis and processing of a putative precursor for the small subunit of ribulose-1,5-bisphosphate carboxylase of *Chlamydomonas reinhardii*. Proc. Natl. Acad. Sci. U.S. **74**, 1082–1085 (1977)

[391] Chua, N.-H., and G. W. Schmidt: Posttranslational transport into intact chloroplasts of a precursor to the small subunit of ribulose-1,5-biphosphate carboxylase. Proc. Natl. Acad. Sci. U.S. **75**, 6110 to 6114 (1978)

[392] Palmiter, R. D., J. Gagnon, L. H. Ericson, and K. A. Walsh: Precursor of egg white lysozyme. Amino acid sequence of an NH_2-terminal extension. J. Biol. Chem. **252**, 6386–6393 (1977)

[393] Thibodeau, S. N., J. Gagnon, and R. D. Palmiter: Precursor forms of lysozyme and ovomucoid: sequence analysis. Fed. Proc. **36**, 656 (1977)

[394] Thibodeau, S. N., R. D. Palmiter, and K. A. Walsh: Precursor of egg white ovomucoid. Amino acid sequence of an NH_2-terminal extension. J. Biol. Chem. **253**, 9018–9023 (1978)

[395] Lingappa, V. R., F. N. Katz, H. F. Lodish, and G. Blobel: A signal for the insertion of a transmembrane glycoprotein. Similarities to the signals of secretory proteins in primary structure and function. J. Biol. Chem. **253**, 8667–8671 (1978)

[396] Segrest, J. P., and R. J. Feldman: Membrane proteins: Amino acid sequence and membrane penetration. J. Mol. Biol. **87**, 853–858 (1974)

[397] Shields, D., and G. Blobel: Efficient cleavage and segregation of nascent presecretory proteins in a reticulocyte lysate supplemented with microsomal membranes. J. Biol. Chem. **253**, 3753–3756 (1978)

[398] Warren, G., and B. Dobberstein: Protein transfer across microsomal membranes reassembled from separated membrane components. Nature **273**, 569 to 571 (1978)

[399] Lingappa, V. R., B. A. Cunningham, S. M. Jawinski, T. P. Hoppe, G. Blobel, and G. M. Edelman: Cell-free synthesis and segregation of β_2-microglobulin. Proc. Natl. Acad. Sci. U.S. **76**, 3651–3655 (1979)

[400] Palmiter, R. D., J. Gaguin, and K. A. Walsh: Ovalbumin: A secreted protein without a transient hydrophobic leader sequence. Proc. Natl. Acad. Sci. U.S. **75**, 94–98 (1978)

[401] Lingappa, V. R., D. Shields, S. L. C. Woo, and G. Blobel: Nascent chicken ovalbumin contains the functional equivalent of a signal sequence. J. Cell. Biol. **79**, 567–572 (1978)

[402] Lingappa, V. R., J. R. Lingappa, and G. Blobel: Chicken ovalbumin contains an internal signal sequence. Nature **281**, 117 to 121 (1979)

[403] Lodish, H. F.: Translational control of protein synthesis. Ann. Rev. Biochem. **45**, 39–72 (1976)

[404] London, I. M., M. J. Clemens, R. S. Ranu, D. H. Levin, L. F. Cherbas, and V. Ernst: The role of hemin in the regulation of protein synthesis in erythroid cells. Fed. Proc. **35**, 2218–2222 (1976)

[405] Ochoa, S.: Regulation of translation. In: Translation of Natural and Synthetic

Polynucleotides. (A. B. LEGOCKI, ed.). University of Agriculture, Poznan, 1977, 9–23

[406] MIZUNO, S., J. M. FISHER, and M. RABINOVITZ: Hemin control of globin synthesis: Action of an inhibitor formed in the absence of hemin on the reticulocyte cell-free system and its reversal by a ribosomal factor. Biochim. Biophys. Acta **272**, 638–650 (1972)

[407] BALKOW, K., S. MIZUNO, J. M. FISHER: and M. RABINOVITZ: Hemin control of globin synthesis: Effect of a translational repressor on Met-tRNA$_f$ binding to the small ribosomal subunit and its relation to the activity and availability of an initiation factor. Biochim. Biophys. Acta **324**, 397–409 (1973)

[408] CLEMENS, M. J.: Functional relationships between a reticulocyte polypeptide-chain-initiation factor (IF-MP) and the translational inhibitor involved in regulation of protein synthesis by haemin. Europ. J. Biochem. **66**, 413–422 (1976)

[409] RANU, R. S., D. H. LEVIN, J. DELAUNAY, V. ERNST, and I. M. LONDON: Regulation of protein synthesis in rabbit reticulocyte lysates: Characteristics of inhibition of protein synthesis by a translational inhibitor from heme-deficient lysates and its relationship to the initiation factor which binds Met-tRNA$_f$. Proc. Natl. Acad. Sci. U.S. **73**, 2720–2724 (1976)

[410] GROSS, M.: Isolation of two initiation factors that can partially reverse the inhibition of protein synthesis due to hemin deficiency or the hemin-controlled translational repressor in rabbit reticulocyte lysates. Arch. Biochem. Biophys. **180**, 121–129 (1977)

[411] GROSS, M.: Regulation of protein synthesis by hemin: Effect of dithiothreitol on the formation and activity of the hemin-controlled translational repressor. Biochem. Biophys. Res. Commun. **520**, 642–649 (1978)

[412] GROSS, M., and J. MENDELEWSKI: Control of protein synthesis by hemin. An association between the formation of the hemin-controlled translational repressor and the phosphorylation of a 100000 molecular weight protein. Biochim. Biophys. Acta **520**, 650–663 (1978)

[413] RANU, R. S., and I. M. LONDON: Regulation of protein synthesis in rabbit reticulocyte lysates: Purfication and initial characterization of the cyclic 3′:5′-AMP independent protein kinase of the heme-regulated translational inhibitor. Proc. Natl. Acad. Sci. U.S. **73**, 4349–4353 (1976)

[414] RANU, R. S., and I. M. LONDON: Regulation of protein synthesis in rabbit reticulocyte lysates: Preparation of efficient protein synthesis lysates and the purification and characterization of the heme-regulated translational inhibitory protein kinase. Meth. Enzymol. **60**, 459–484 (1979)

[415] FARRELL, P. J., K. BALKOW, T. HUNT, R. J. JACKSON, and H. TRACHSEL: Phosphorylation of initiation factor eIF-2 and the control of reticulocyte protein synthesis. Cell **11**, 187–200 (1977)

[416] TRACHSEL, H., R. S. RANU, and I. M. LONDON: Regulation of protein synthesis in rabbit reticulocyte lysates: Purification and characterization of heme-reversible translational inhibitor. Proc. Natl. Acad. Sci. U.S. **75**, 3654–3658 (1978)

[417] TRACHSEL, H., R. S. RANU, and I. M. LONDON: Purification of the heme-reversible form of the translational inhibitory protein kinase. Meth. Enzymol. **60**, 485 to 495 (1979)

[418] LEVIN, D. H., R. S. RANU, V. ERNST, and I. M. LONDON: Regulation of protein synthesis in reticulocyte lysates: Phosphorylation of methionyl-tRNA$_f$ binding factor by protein kinase activity of translational inhibitor isolated from heme-deficient lysates. Proc. Natl. Acad. Sci. U.S. **73**, 3112–3116 (1976)

[419] KRAMER, G., J. M. CIMADEVILLA, and B. HARDESTY: Specificity of the protein kinase acitvity with the hemin-controlled repressor of rabbit reticulocyte. Proc. Natl. Acad. Sci. U.S. **73**, 3078–3082 (1976)

[420] KRAMER, G., A. B. HENDERSON, P. PINPHANICHAKARN, M. H. WALLIS, and B. HARDESTY: Partial reaction of peptide initiation inhibited by phosphorylation of either initiation factor eIF-2 or 40 S ribosomal proteins. Proc. Natl. Acad. Sci. U.S. **74**, 1445–1449 (1977)

[421] GROSS, M., and J. MENDELEWSKI: Ad-

ditional evidence that the hemin-controlled translational repressor from rabbit reticulocytes is a protein kinase. Biochem. Biophys. Res. Commun. **74**, 559 to 569 (1977)

[422] Farrell, P. J., T. Hunt, and R. J. Jackson: Analysis of phosphorylation of protein synthesis initiation factor eIF-2 by two dimensional gel electrophoresis. Europ. J. Biochem. **89**, 517—521 (1978)

[423] Ernst, V., D. H. Levin, and I. M. London: In situ phosphorylation of the α subunit of eukaryotic initiation factor 2 in reticulocyte lysates inhibited by heme deficiency, double-stranded RNA, oxidized glutathione, or the heme-regulated protein kinase. Proc. Natl. Acad. Sci. U. S. **76**, 2118—2122 (1979)

[424] Pinphanichakarn, P., G. Kramer, and B. Hardesty: Partial reaction of peptide initiation inhibited by the reticulocyte hemin-controlled repressor. Biochem. Biophys. Res. Commun. **73**, 625—631 (1976)

[425] Gross, M.: Regulation of protein synthesis by hemin: Evidence that the hemin-controlled translational repressor inhibits the rate of formation of 40 S·Met-$tRNA_f$ complexes directly. J. Biol. Chem. **254**, 2378—2383 (1979)

[426] Datta, A., C. de Haro, J. M. Sierra, and S. Ochoa: Role of 3′:5′ cyclic-AMP-dependent protein kinase in regulation of protein synthesis in reticulocyte lysates. Proc. Natl. Acad. Sci. U.S. **74**, 1463 to 1467 (1977)

[427] Datta, A., C. de Haro, J. M. Sierra, and S. Ochoa: Mechanism of translational control by hemin in reticulocyte lysates. Proc. Natl. Acad. Sci. U.S. **74**, 3326—3329

[428] Gross, M.: Control of protein synthesis by hemin. Evidence that hemin-controlled translational repressor inhibits formation of 80 S initiation complexes from 48 S intermediate initiation complexes. J. Biol. Chem. **254**, 2370—2377 (1979)

[429] Cherbas, L., and I. M. London: On the mechanism of delayed inhibition of protein synthesis in heme-deficient rabbit reticulocyte lysates. Proc. Natl. Acad. Sci. U.S. **73**, 3506—3510 (1976)

[430] Haro, C. de, and S. Ochoa: Mode of action of the hemin-controlled inhibitor of protein synthesis: Studies with factors from reticulocytes. Proc. Natl. Acad. Sci. U.S. **75**, 2713—2716 (1978)

[431] Ranu, R. S., I. M. London, A. Das, A. Dasgupta, A. Majumdar, R. Ralston, R. Roy, and N. K. Gupta: Regulation of protein synthesis in rabbit reticulocyte lysates by the heme-regulated protein kinase: Inhibition of interaction of Met-$tRNA_f^{Met}$ binding factor with another initiation factor in formation of Met-$tRNA_f^{Met}$·40 S ribosomal subunit. Proc. Natl. Acad. Sci. U.S. **75**, 745—749 (1978)

[432] Levin, D. H., R. S. Ranu, V. Ernst, M. A. Fifer, and I. M. London: Association of a cyclic AMP-dependent protein kinase with a purified translational inhibitor isolated from hemin-deficient rabbit reticulocyte lysates. Proc. Natl. Acad. Sci. U.S. **72**, 4849—4853 (1975)

[433] Datta, A., C. de Haro, and S. Ochoa: Translational control by hemin is due to binding to cyclic AMP-dependent protein kinase. Proc. Natl. Acad. Sci. U.S. **75**, 1148—1152 (1978)

[434] Ochoa, S.: Regulation of protein synthesis. Europ. J. Cell Biol. **19**, 95—101 (1979)

[435] Grankowski, N., G. Kramer, and B. Hardesty: No effect of cAMP on protein synthesis in reticulocyte lysates. J. Biol. Chem. **254**, 3145—3147 (1979)

[436] Mizuno, S.: Temperature sensitivity of protein synthesis initiation in the reticulocyte lysate system. Reduced formation of the 40 S ribosomal subunit Met-$tRNA_f$ complex at an elevated temperature. Biochim. Biophys. Acta **414**, 273—282 (1975)

[437] Mizuno, S.: Temperature sensitivity of protein synthesis initiation in reticulocyte cell-free systems. A ribosomal factor which protects protein synthesis inactivation at elevated temperatures. Arch. Biochem. Biophys. **173**, 703—709 (1976)

[438] Mizuno, S.: Temperature sensitivity of protein synthesis initiation. Inactivation of a ribosomal factor by an inhibitor formed at elevated temperatures. Arch. Biochem. Biophys. **179**, 289—301 (1977)

[439] Bonanou-Tzedaki, S. A., K. E. Smith, B. A. Sheeran, and H. R. V. Arnstein: The high-temperature inactivation of rabbit reticulocyte lysates by a haemin-

independent mechanism. Europ. J. Biochem. **84**, 591–600 (1978)

[440] Bonanou-Tzedaki, S. A., K. E. Smith, B. A. Sheeran, and H. R. V. Arnstein: Reduced formation of initiation complexes between Met-tRNA$_f$ and 40 S ribosomal subunits in rabbit reticulocyte lysates incubated at elevated temperatures. Activity of the Met-tRNA$_f$ binding factor. Europ. J. Biochem. **84**, 601–610 (1978)

[441] Spieler, P. J., N. G. Ibrahim, and M. L. Freedman: Heat inhibition of reticulocyte protein synthesis. Evidence for a mechanism independent of the hemin controlled repressor. Biochim. Biophys. Acta **518**, 366–379 (1978)

[442] Henderson, A. B., A. H. Miller, and B. Hardesty: Multistep regulatory system for activation of a cyclic AMP-independent eukaryotic initiation factor 2 kinase. Proc. Natl. Acad. Sci. U.S. **76**, 2605–2609 (1979)

[443] Clemens, M. J., B. Safer, W. C. Merrick, W. F. Anderson, and I. M. London: Inhibition of protein synthesis in rabbit reticulocyte lysates by double-stranded RNA and oxidized glutathione: Indirect mode of action on polypeptide chain initiation. Proc. Natl. Acad. Sci. U.S. **72**, 1286–1290 (1975)

[444] Ernst, V., D. H. Levin, and I. M. London: Inhibition of protein synthesis initiation by oxidized glutathione: Activation of a protein kinase that phosphorylates the α-subunit of eukaryotic initiation factor 2. Proc. Natl. Acad. Sci. U.S. **75**, 4110–4114 (1978)

[445] Ernst, V., D. M. Levin, R. S. Ranu, and I. M. London: Control of protein synthesis in reticulocyte lysates: Effects of 3′5′ cyclic AMP, ATP, and GTP on inhibition induced by heme-deficiency, double-stranded RNA and a reticulocyte translational inhibitor. Proc. Natl. Acad. Sci. U.S. **73**, 1112–1116 (1976)

[446] Levin, D. H., and I. M. London: Regulation of protein synthesis: Activation by double-stranded RNA of a protein kinase that phosphorylates eukaryotic initiation factor 2. Proc. Natl. Acad. Sci. U.S. **75**, 1121–1125 (1978)

[447] Content, J., B. Lebleu, and E. de Clercq: Differential effects of various double-stranded RNAs on protein synthesis in rabbit reticulocyte lysates. Biochem. **17**, 88–94 (1978)

[448] Lenz, J. R., and C. Baglioni: Inhibition of protein synthesis by double-stranded RNA and phosphorylation of initiation factor eIF-2. J. Biol. Chem. **253**, 4219 to 4223 (1978)

[449] Stuart, W. E.: The Interferon System. Springer-Verlag, Wien, New York, 1979, 1–421

[450] Petryshyn, R., H. Trachsel, and I. M. London: Regulation of protein synthesis in reticulocyte lysates: Immune serum inhibits heme-regulated protein kinase activity and differentiates heme-regulated protein kinase from double-stranded RNA-induced protein kinase. Proc. Natl. Acad. Sci. U.S. **76**, 1575–1579 (1979)

[451] Cimadevilla, J. M., G. Kramer, P. Pinphanichakarn, D. Konecki, and B. Hardesty: Inhibition of peptide chain initiation by a nonhemin-regulated translational repressor from friend leukemia cells. Arch. Biochem. Biophys. **171**, 145 to 153 (1975)

[452] Pinphanichakarn, P., G. Kramer, and B. Hardesty: Partial purification and characterization of a translational inhibitor from friend leukemia cells. J. Biol. Chem. **252**, 2106–2112 (1977)

[453] Dabney, B. J. and A. L. Beaudet: Stimulation of protein synthesis by hemin in extracts of friend erythroleukemia cells. J. Biol. Chem. **253**, 7124–7126 (1978)

[454] Clemens, M. J., V. M. Pain, E. C. Henshaw, and I. M. London: Characterization of a macromolecular inhibitor of polypeptide chain initiation from Ehrlich ascites tumor cells. Biochem. Biophys. Res. Commun. **72**, 768–775 (1976)

[455] Delaunay, J., R. S. Ranu, D. H. Levin, V. Ernst, and I. M. London: Characterization of a rat liver factor that inhibits initiation of protein synthesis in rabbit reticulocyte lysates. Proc. Natl. Acad. Sci. U.S. **74**, 2264–2268 (1977)

[456] Sakamoto, Z., and T. Higashi: Studies on rat liver catalase: X. Effect of hemin. and an inhibitor on the translation of catalase messenger RNA. J. Biochem. **85**, 389–396 (1979)

[457] Sierra, J. M., C. de Haro, A. Datta, and S. Ochoa: Translational control by

protein kinases in *Artemia salina* and wheat germ. Proc. Natl. Acad. Sci. U.S. **74**, 4356—4359 (1977)

[458] SAFER, B., and R. JAGUS: Control of eIF-2 phosphatase activity in rabbit reticulocyte lysate. Proc. Natl. Acad. Sci. U.S. 1094—1098 (1979)

[459] ERNST, V., D. H. LEVIN, and I. M. LONDON: Evidence that glucose-6-phosphate regulates protein synthesis initiation in reticulocyte lysates. J. Biol. Chem. **253**, 7163—7172 (1978)

[460] WU, J. M., C. P. CHEUNG, A. R. BRUZEL, and R. J. SUHADOLNIK: The reversal of inhibition of protein synthesis by double-stranded RNA in lysed rabbit reticulocytes with fructose-6-phosphate. Biochem. Biophys. Res. Commun. **86**, 648—653 (1979)

[461] RALSTON, R. O., A. DAS, A. DASGUPTA, R. ROY, S. PALMIERI, and N. K. GUPTA: Protein synthesis in rabbit reticulocytes: Characteristics of a ribosomal factor that reverses inhibition of protein synthesis in heme-deficient lysates. Proc. Natl. Acad. Sci. U.S. **75**, 4858—4862 (1978)

[462] BARON, S., and F. DIANZANI (Eds.): The interferon system: A current review to 1978. Texas Rep. Biol. Med. **35**, 1—573 (1977)

[463] REVEL, M.: Interferon-induced translational regulation. Texas Rep. Biol. Med. **35**, 212—220 (1977)

[464] SEN, G. C., R. DESROSIERS, L. RATNER, S. SHAILA, G. E. BROWN, B. LEBLEU, E. SLATTERY, M. KAWAKITA, B. CABRER, M. TAIRA, and P. LENGYEL: Messenger RNA methylation, translation and degradation in extracts of interferon-treated cells. Texas Rep. Biol. Med. **35**, 221—229 (1977)

[465] HUNT, T.: Interferon, dsRNA and the pleiotypic effector. Nature **273**, 97—98 (1978)

[466] MORSER, J., and D. BURKE: Interferon and double-stranded RNA. Nature **277**, 435 (1979)

[467] REVEL, M., A. SCHMIDT, L. SHULMAN, A. ZILBERSTEIN, and A. KIMCHI: The regulation of protein synthesis by interferon. FEBS Symp. **51**, 415—426 (1978)

[468] ROBERTS, W. K., A. HOVANESSIAN, R. E. BROWN, M. J. CLEMENS, and I. M. KERR: Interferon-mediated protein kinase and low-molecular-weight inhibitor of protein synthesis. Nature **264**, 477—480 (1976)

[469] ZILBERSTEIN, A., P. FEDERMAN, L. SHULMAN, and M. REVEL: Specific phosphorylation in vitro of a protein associated with ribosomes of interferon-treated mouse L cells. FEBS Lett. **68**, 119—124 (1976)

[470] LEBLEU, B., G. C. SEN, S. SHAILA, B. CABRER, and P. LENGYEL: Interferon, double-stranded RNA, and protein phosphorylation. Proc. Natl. Acad. Sci. U.S. **73**, 3107—3111 (1976)

[471] SEN, G. C., H. TAIRA, and P. LENGYEL: Interferon, double-stranded RNA, and protein phosphorylation. Characteristics of a double-stranded RNA activated protein kinase system partially purified from interferon-treated Ehrlich ascites tumor cells. J. Biol. Chem. **253**, 5915—5921 (1978)

[472] FARRELL, P. J., G. C. SEN, M. F. DUBOIS, L. RATNER, E. SLATTERY, and P. LENGYEL: Interferon action: Two distinct pathways for inhibition of protein synthesis by double-stranded RNA. Proc. Natl. Acad. Sci. U.S. **75**, 5893—5897 (1978)

[473] SEN, G. C., P. J. FARRELL, M. F. DUBOIS, and P. LENGYEL: Characteristics of a double-stranded RNA activated protein kinase system from interferon-treated Ehrlich-ascites tumor cells. Fed. Proc. **37**, 1687 (1978)

[474] ZILBERSTEIN, A., A. KIMCHI, A. SCHMIDT, and M. REVEL: Isolation of two interferon-induced translational inhibitors: A protein kinase and an oligo-isoadenylate synthetase. Proc. Natl. Acad. Sci. U.S. **75**, 4734—4738 (1978)

[475] CHERNAJOVSKY, Y., A. KIMCHI, A. SCHMIDT, A. ZILBERSTEIN, and M. REVEL: Differential effects of two interferon-induced translational inhibitors on initiation of protein synthesis. Europ. J. Biochem. **96**, 35—41 (1979)

[476] OHTSUKI, K.: Possible role of initiation factor activity in the action of interferon. Texas Rep. Biol. Med. **35**, 270—281 (1977)

[477] COOPER, J. A., and P. J. FARRELL: Extracts of interferon-treated cells can inhibit reticulocyte lysate protein synthesis. Biochem. Biophys. Res. Commun. **77**, 124 to 131 (1977)

[478] LEWIS, J. A., E. FALCOFF, and R. FALCOFF: Dual action of double-stranded RNA in inhibiting protein synthesis in extracts of interferon-treated mouse L cells. Translation is impaired at the level of initiation and by mRNA degradation. Europ. J. Biochem. **86**, 497–509 (1978)

[479] BROWN, G. E., B. LEBLEU, M. KAWAKITA, S. SHAILA, G. C. SEN, and P. LENGYEL: Increased endonuclease activity in an extract from mouse Ehrlich ascites tumor cells which had been treated with a partially purified interferon preparation: Dependence on double-stranded RNA. Biochem. Biophys. Res. Commun. **69**, 114–122 (1976)

[480] SEN, G. C., B. LEBLEU, G. E. BROWN, M. KAWAKITA, E. SLATTERY, and P. LENGYEL: Interferon, double-stranded RNA and mRNA degradation. Nature **264**, 370 to 373 (1976)

[481] RATNER, L., G. C. SEN, G. E. BROWN, B. LEBLEU, M. KAWAKITA, B. CABRER, E. SLATTERY, and P. LENGYEL: Interferon, double-stranded RNA and RNA degradation. Characteristics of an endonuclease activity. Europ. J. Biochem. **79**, 565–577 (1977)

[482] SCHMIDT, A., A. ZILBERSTEIN, L. SHULMAN, P. FEDERMAN, H. BERISSI, and M. REVEL: Interferon action: Isolation of nuclease F, a translational inhibitor activated by interferon-induced (2′–5′) oligo-isoadenylate. FEBS Lett. **95**, 257 to 264 (1978)

[483] EPPSTEIN, D. A., and C. E. SAMUEL: Mechanism of interferon action: Properties of an interferon-mediated ribonucleolytic activity from mouse L_{929} cells. Virology **89**, 240–251 (1978)

[484] BAGLIONI, C., M. A. MINKS, and P. A. MARONEY: Interferon action may be mediated by activation of a nuclease by pppA2′p5′A2′p5′A. Nature **273**, 684–687 (1978)

[485] CLEMENS, M. J., and B. R. G. WILLIAMS: Inhibition of cell-free protein synthesis by pppA2′p5′A2′p5′A: a novel oligonucleotide synthesized by interferon-treated L cell extracts. Cell **13**, 565–572 (1978)

[486] HOVANESSIAN, A. G., R. E. BROWN, and I. M. KERR: Synthesis of low molecular weight inhibitor of protein synthesis with enzyme from interferon-treated cells. Nature **268**, 537–540 (1977)

[487] BALL, L. A., and C. N. WHITE: Oligonucleotide inhibitor of protein synthesis made in extracts of interferon-treated chick embryo cells: Comparison with the mouse low molecular weight inhibitor. Proc. Natl. Acad. Sci. U.S. **75**, 1167 to 1171 (1978)

[488] EPPSTEIN, D. A., and C. E. SAMUEL: Mechanism of interferon action: Partial purification and characterization of a low-molecular-weight interferon-mediated inhibitor of translation with nucleolytic activity. Biochem. Biophys. Res. Commun. **79**, 145–153 (1977)

[489] RATNER, L., R. C. WIEGAND, P. J. FARRELL, G. C. SEN, B. CABRER, and P. LENGYEL: Interferon, double-stranded RNA and RNA degradation: Fractionation of the endonuclease$_{INT}$ system into two macromolecular components: Role of a small molecule in nuclease activation. Biochem. Biophys. Res. Commun. **81**, 947–954 (1978)

[490] KERR, I. M., R. E. BROWN, and A. G. HOVANESSIAN: Nature of inhibitor of cell-free protein synthesis formed in response to interferon and double-stranded RNA. Nature **268**, 540–542 (1977)

[491] KERR, I. M., and R. E. BROWN: pppA2′p5′A2′p5′A: An inhibitor of protein synthesis synthesized with an enzyme fraction from interferon-treated cells. Proc. Natl. Acad. Sci. U.S. **75**, 256–260 (1978)

[492] NILSEN, T. W., and C. BAGLIONI: Mechanism for discrimination between viral and host mRNA in interferon-treated cells. Proc. Natl. Acad. Sci. U.S. **76**, 2600 to 2604 (1979)

[493] CONTENT, J., B. LEBLEU, U. NUDEL, A. ZILBERSTEIN, H. BERISSI, and M. REVEL: Blocks in elongation and initiation of protein synthesis induced by interferon treatment in mouse L cells. Europ. J. Biochem. **54**, 1–10 (1975)

[494] FALCOFF, R., E. FALCOFF, J. SANCEAU, and J. A. LEWIS: Influence of preincubation on the development of the inhibition of protein synthesis in extracts from interferon-treated mouse L cells. Action on tRNA. Virology **86**, 507–515 (1978)

[495] GUPTA, S. L., M. L. SOPORI, and P. LENGYEL: Release of the inhibition of mes-

senger RNA translation in extracts of interferon-treated Ehrlich ascites tumor cells by added transfer RNA. Biochem. Biophys. Res. Commun. **57**, 763–770 (1974)

[496] CONTENT, J., B. LEBLEU, A. ZILBERSTEIN, H. BERISSI, and M. REVEL: Mechanism of the interferon-induced block of mRNA translation in mouse L cells: Reversal of the block by transfer RNA. FEBS Lett. **41**, 125–130 (1974)

[497] FALCOFF, R., B. LEBLEU, J. SANCEAU, J. WEISSENBACH, G. DIRHEIMER, J. P. EBEL, and E. FALCOFF: The inhibition of translation of synthetic polyribonucleotide, and mengo RNA in extracts from interferon-treated L cells and its reversion by yeast tRNAs. Biochem. Biophys. Res. Commun. **68**, 1323–1331 (1976)

[498] MAYR, U., H. P. BERMAYER, G. WEIDINGER, C. JUNGWIRTH, H. J. GROSS, and G. BODO: Release of interferon-induced translation inhibition by tRNA in cell free extracts from mouse erythroleukemia cells. Europ. J. Biochem. **76**, 541–551 (1976)

[499] WEISSENBACH, J., G. DIRHEIMER, R. FALCOFF, J. SANCEAU, and E. FALCOFF: Yeast $tRNA^{Leu}$ (anticodon U–A–G) translates all six leucine codons in extracts from interferon treated cells. FEBS Lett. **82**, 71 to 76 (1977)

[500] ZILBERSTEIN, A., B. DUDOCK, H. BERISSI, and M. REVEL: Control of messenger RNA translation by minor species of leucyl-transfer RNA in extracts from interferon-treated L cells. J. Mol. Biol. **108**, 43–54 (1976)

[501] COLBY, C., E. PENHOET, and C. SAMUEL: A comparison of transfer RNA from untreated and interferon-treated murine cells. Virology **74**, 262–264 (1976)

[502] SEN, G. C., S. L. GUPTA, G. E. BROWN, B. LEBLEU, M. A. REBELLO, and P. LENGYEL: Interferon treatment of Ehrlich ascites tumor cells: effects on exogenous mRNA translation and tRNA inactivation in the cell extract. J. Virol. **17**, 191–203 (1976)

[503] SEN, G. C., B. LEBLEU, G. E. BROWN, M. A. REBELLO, Y. FURUICHI, M. MORGAN, A. J. SHATKIN, and P. LENGYEL: Inhibition of reovirus messenger RNA methylation in extracts of interferon treated Ehrlich ascites tumor cells. Biochem. Biophys. Res. Commun. **65**, 427–434 (1975)

[504] SHAILA, S., B. LEBLEU, G. E. BROWN, G. C. SEN, and P. LENGYEL: Characteristics of extracts from interferon-treated HeLa cells: presence of a protein kinase and endoribonuclease activated by double-stranded RNA and of an inhibitor of mRNA methylation. J. Gen. Virol. **37**, 535–546 (1977)

[505] SEN, G. C., S. SHAILA, B. LEBLEU, G. E. BROWN, R. C. DESROSIERS, and P. LENGYEL: Impairment of reovirus mRNA methylation in extracts of interferon-treated Ehrlich ascites tumor cells: further characteristics of the phenomenon. J. Virol. **21**, 69–83 (1977)

[506] BAGLIONI, C., P. A. MARONEY, and D. K. WEST: 2′5′oligo(A) polymerase activity and inhibition of viral RNA synthesis in interferon-treated HeLa cells. Biochem. **18**, 1765–1770 (1979)

[507] HOVANESSIAN, A., and I. M. KERR: Synthesis of an oligonucleotide inhibitor of protein synthesis in rabbit reticulocyte lysates analogous to that formed in extracts from interferon-treated cells. Europ. J. Biochem. **84**, 149–159 (1978)

[508] STARK, G. R., W. J. DOWER, R. T. SCHIMKE, R. E. BROWN, and I. M. KERR: 2-5 A synthetase: assay, distribution and variation with hormone status. Nature **278**, 471–473 (1979)

[509] WILLIAMS, B. R. G., and I. M. KERR: Inhibition of protein synthesis by 2′-5′ linked adenine oligonucleotides in intact cells. Nature **276**, 88–90 (1978)

[510] SAMUEL, C. E., D. A. FARRIS, and D. A. EPPSTEIN: Mechanism of interferon action: Kinetics of interferon action in mouse L_{929} cells: Translation inhibition, protein phosphorylation, and messenger RNA methylation and degradation. Virology **83**, 56–71 (1977)

[511] SMITH, A. E., and L. CARRASCO: Eukaryotic viral protein synthesis. Int. Rev. Biochem. **18**, 261–311 (1978)

[512] CARRASCO, L.: The inhibition of cell functions after viral infection. A proposed general mechanism. FEBS Lett. **76**, 11 to 15 (1977)

[513] KAUFMANN, Y., E. GOLDSTEIN, and S. PENMAN: Poliovirus-induced inhibition of polypeptide initiation in vitro on native

polyribosomes. Proc. Natl. Acad. Sci. U.S. **73**, 1834—1838 (1976)

[514] LAWRENCE, C., and R. E. THACH: Encephalomyocarditis virus infection of mouse plasmacytoma cells. I. Inhibition of cellular protein synthesis. J. Virol. **14**, 598—610 (1974)

[515] ABREU, S. L., and J. LUCAS-LENARD: Cellular protein synthesis shutoff by mengovirus: Translation of nonviral and viral mRNAs in extracts from uninfected and infected Ehrlich ascites tumor cells. J. Virol. **18**, 182—194 (1976)

[516] HACKETT, P. B., E. EGBERTS, and P. TRAUB: Translation of ascites and mengovirus RNA in fractionated cell-free systems from uninfected and mengovirus-infected Ehrlich-ascites-tumor cells. Europ. J. Biochem. **83**, 341—352 (1978)

[517] HACKETT, P. B., E. EGBERTS, and P. TRAUB: Selective translation of mengovirus RNA over host mRNA in homologous fractionated, cell-free translational systems from Ehrlich-ascites-tumor cells. Europ. J. Biochem. **83**, 353—361 (1978)

[518] CELMA, M. L., and E. EHRENFELD: Effect of poliovirus double-stranded RNA on viral and host-cell protein synthesis. Proc. Natl. Acad. Sci. U.S. **71**, 2440 to 2444 (1974)

[519] STEINER-PRYOR, A., and P. D. COOPER: Temperature-sensitive poliovirus mutants defective in repression of host protein synthesis are also defective in structural protein. J. Gen. Virol. **21**, 215—225 (1973)

[520] SVITKIN, YU. V., V. A. GINEVSKAYA, T. Y. UGAROVA, and V. I. AGOL: A cell-free model of the encephalomyocarditis virus-induced inhibition of host cell protein synthesis. Virology **87**, 199—203 (1978)

[521] HELENTJARIS, T., and E. EHRENFELD: Control of protein synthesis in extracts from poliovirus-infected cells. 1. mRNA discrimination by crude initiation factors. J. Virol. **26**, 510—521 (1978)

[522] CARRASCO, L., and A. E. SMITH: Sodium ions and the shut-off of host cell protein synthesis by picornaviruses. Nature **264**, 807—809 (1976)

[523] NUSS, D. L., H. OPPERMANN, and G. KOCH: Selective blockade of initiation of host protein synthesis in RNA-virus-infected cells. Proc. Natl. Acad. Sci. U.S. **72**, 1258—1262 (1975)

[524] GARRY, R. F., J. M. BISHOP, S. PARKER, K. WESTBROOK, G. LEWIS, and M. R. F. WAITE: Na^+ and K^+ concentrations and the regulation of protein synthesis in Sindbis virus-infected chick cells. Virology **96**, 108—120 (1979)

XII. Aphoristic synopsis and outlook

The Introduction (Chapter I) of this book was an attempt to summarize the main stages in the discovery of ribosomes and to outline some of their structural properties and biological functions. This chapter aims to integrate and comment some of the presently available data under various aspects and to outline some open questions which need further experimental studies.

As mentioned in the "Preface" and demonstrated in some chapters of this book, ribosomes and the translation system have developed considerably during evolution. This becomes clear when considering the number and properties of the components involved in the translation process in prokaryotes and eukaryotes. Although the final number of proteins in eukaryotic ribosomes is not yet known exactly, it is now certain that their total mass and number are about 50% higher than those of *E. coli* ribosomes. Furthermore, ribosomal proteins from prokaryotes and eukaryotes differ significantly in their electrophoretic and immunological properties, and, as far as analyzed, no identical amino acid sequences have been found in *E. coli* and rat liver ribosomal proteins. Furthermore, also the RNA moieties of prokaryotic and eukaryotic ribosomes are different in their molecular weights, base compositions and nucleotide sequences. Thus, ribosomal particles with essentially the same or at least very similar structural and functional features can be constructed from macromolecular components with different chemical properties. As far as the proteins are concerned, the only common property shared by prokaryotic and eukaryotic ribosomes is their basicity, which, however, seems to be one important prerequisite only for their interactions with ribosomal RNA. As to the molecular mechanisms of interactions between special domains of the macromolecular ribosomal components — in terms of recognition and specific binding — no definite data are available at present. In this context, also the elucidation of the biogenesis of ribosomal particles and the organization of their functional sites require further detailed studies. Thus, the ribosome is also a convenient model for the analysis of specific nucleic acid-protein interactions which play an important role in various steps of gene expression. Especially the so-called 7 S-complex of the large ribosomal subunit should be mentioned, which consists of one protein (L 5) and 5 S RNA only. Therefore, this RNP complex is a fairly simple model for studying RNA-protein interactions. Since this complex is found in eukaryotic ribosomes only, its structural and functional analysis could furthermore provide insights into some of the specificities of protein biosynthesis in eukaryotic cells.

Further evidence for peculiarities in eukaryotic protein synthesis in comparison to that in prokaryotes comes from observations that gene expression in eukaryotes seems to be regulatable also at the level of translation. Special structures of eukaryotic mRNA molecules, the compartmentation of eukaryotic cells, the higher numbers of ribosomal proteins as well as of initiation factors might be both the basis and explanation, for a more complex mechanism of translation and its regulation in eukaryotes; but this has yet to be confirmed in more detail.

Although data about the spatial arrangement of proteins and RNA and their functional role in ribosomes of eukaryotes are still very rare, a first attempt to correlate some of the experimental results is justified and attractive alike.

Chemical substitution, degradation by proteolytic enzymes and nucleases, enzymatic iodination and the reaction with antibodies have shown that all proteins and the RNA molecules are arranged with larger or smaller domains of their chains at the surface of the ribosome.

From experiments about the chemical substitution of proteins in ribosomal particles of different functional states as well as from affinity labeling, cross-linking and the application of antibodies, first clear-cut evidence for the functional role and localization of proteins in eukaryotic ribosomes is now available. On the basis of these data some preliminary and tentative conclusions can be drawn.

As to the function of ribosomal proteins, it is obvious that at least some of the active sites on the ribosome are organized by more than one protein. As expected, such proteins are with parts of their polypeptide chains closely neighboured as demonstrated, e.g., by immune electron microscopy and cross-linking. This could be proved, e.g., for proteins involved in mRNA and in eIF-2 and eIF-3 binding to the small ribosomal subunit.

On the other hand, there is strong evidence that some ribosomal proteins are involved in different reactions such as in the binding of eIF-2 and eIF-3, of eEF-T and eEF-G, of mRNA and eEF-T or of mRNA and eIF-2, respectively. These findings allow the conclusion that the corresponding active sites should be localized in close neighbourhood at the ribosome.

As a hypothesis, it can be assumed that the involvement of different proteins in one reaction, on the one hand, and the participation of the same proteins in different reactions of the translation process, on the other, is the basis for cooperative and anti-cooperative effects, respectively, by which the entire translation process is organized and channeled in the ribosomal structure.

The mutual stimulatory attachment of eIF-2 and eIF-3 to the small ribosomal subunit during the assembly of the initiation complex could be explained by cooperative effects of proteins involved in the binding of both initiation factors to the 40 S ribosomal subunit. The finding that some of these proteins are closely neighboured and that eIF-3 is attached to the small ribosomal subunit close to the proposed eIF-2 binding site supports this hypothesis.

On the other hand, the involvement of some proteins in different reactions may be responsible for anti-cooperative effects. This might be valid, e.g., for the finding that both elongation factors — eEF-T and eEF-G — cannot bind simultaneously to the ribosome during elongation.

Generalizing the data about the topography and function of ribosomal proteins, some conclusions as to the localization of functional sites on the ribosome can be drawn. By these procedures, part of the P site on the small ribosomal subunit of rat liver, which interacts with Met-tRNA$_f$ and eIF-2, could be mapped mainly close to the so-called cleft region between the head and the body of the 40S subunit, where also the mRNA seems to be located and to pass through the ribosome between both subunits.

One of the proteins involved in P site organization is very probably S 6. This protein is practically the only one of the small subunit phosphorylatable in vivo. The extent of its phosphorylation depends obviously on various functional states of a given tissue, e.g. on the growth rate. This supports the idea that phosphorylation and dephosphorylation, respectively, of protein S 6 might be of biological significance for the control of protein synthesis at the level of the first steps of initiation on the P site of the small ribosomal subunit.

Although the process of protein biosynthesis on the ribosome can formally be described satisfactorily in two-dimensional schemes, the molecular basis of the three-dimensional arrangement of the various steps of translation in the ribosomal structure is still obscure. To get more insights into the organization of these processes, the following problems seem to be important and thus require further experimental studies: How is the ribosome structure formed by specific interactions between special domains of its macromolecular components? Which components are involved and how do they interact and function in the organization of active sites on or in the ribosome? Where are the functional sites located on the ribosome and which is their spatial arrangement and cooperation as the basis for the ordered sequence of the various steps of the translation process? How does the ribosome specifically recognize and interact with nonribosomal components needed for decoding the genetic information and the synthesis of proteins? How are the various steps of the translation process supplied energetically and how does the peptidyltransferase center work? Especially the elucidation of the last two problems could be of interest with regard to the chemical synthesis of specific polypeptide chains by making use of biological principles. Furthermore, serious efforts have to be made in order to decide whether gene expression can be regulated also at the level of translation, and if so, how? This seems to be important especially for eukaryotes with regard to the regulation of development, growth and differentiation of cells and tissues.

XIII. Notes added in proof

Chapter III: Ribosomes within the cell

Ribosome crystals have been obtained from homogenates of chick embryos and were observed also in nuclei of chick embryos [1—3]. Furthermore, ribosome crystals were found attached to membranes of the endoplasmic reticulum and the nuclear envelope in lizard oocytes and chick embryo after cooling [4]. The crystalline ribosomes are bound to these membranes by their large subunit, and the long axis of the small subunit is oriented approximately parallel to the surface of the membranes [4]. More recently, the paracrystalline inclusions in the hippocampal region of the brain of humans under various pathological conditions — the so-called Hirana bodies — have been described as stacked sheets of ordered, crystalline arrays of membrane-bound ribosomes [5]. These structures are considered as inactive storage forms of ribosomes, which are most probably formed during periods of dormancy or degeneration [5].

References

[1] Barbieri, M.: Ribosome crystallization in homogenates and cell extracts of chick embryos. J. Supramolec. Structure **10**, 349 to 357 (1979)

[2] Barbieri, M.: The role of temperature in the crystallization of ribosomes in chick embryos. J. Supramolec. Structure **10**, 359—364 (1979)

[3] Barbieri, M.: Ribosome crystallization in nuclei of chick embryos. J. Supramolec. Structure **10**, 365—375 (1979)

[4] Unwin, P. N. T.: Attachment of ribosome crystals to intracellular membranes. J. Mol. Biol. **132**, 69—84 (1979)

[5] O'Brien, L., K. Shelley, J. Towfihgi, and A. McPherson: Crystalline ribosomes are present in brains from senile humans. Proc. Natl. Acad. Sci. U.S. **77**, 2260—2264 (1980)

Chapter V: Morphology of ribosomes and polysomes

In electron microscopic studies on ribosomes from the undifferentiated amoebae and differentiated spores of the cellular slime mold, *Dictyostelium discoideum,* no significant differences in the three-dimensional structure of monosomes and their 60 S and 40 S subunits of the two cell types could be detected [1], which are similar to corresponding particles of other eukaryotes.

Further electron microscopic investigations by the same authors on hybrid ribosomes formed from 40 S *Artemia salina* and 50 S *Escherichia coli* subunits, active in model assays for protein synthesis, showed that the 73 S hybrid particles strongly resemble homologous 70 S *E. coli* and 80 S *A. salina* monosomes. The morphological differences between the corresponding eukaryotic and prokaryotic ribosomes do not significantly affect the assembly and mutual orientation of 40 S *A. salina* and 50 S *E. coli* subunits within the hybrid monosomes [2].

References

[1] Boublik, M., and S. Ramagopal: Conformation of ribosomes from the vegetative amoebae and spores of *Dictyostelium discoideum.* Molec. Gen. Genetics **179**, 483 to 488 (1980)

[2] Boublik, M., R. M. Wydro, W. Hellmann, and F. Jenkins: Structure and functional *A. salina-E. coli* hybrid ribosomes by electron microscopy. J. Supramolec. Structure **10**, 397–404 (1979)

Chapter VI: Chemical components

1. Ribosomal proteins

a) Electrophoretic separation and number

Some of the more recent results are summarized in Table 27. The estimated number of ribosomal proteins may not reflect in all cases the real protein composition of the ribosomes due to different extent of characterization and to source dependent different difficulties in the preparation of ribosomes.

Table 27. Number of ribosomal proteins

	80 S particles	60 S subunits	40 S subunits	References
Human liver		27	21–22	[1]
Rat liver		45	34	[2]
Rabbit liver		45	34	[2]
Rabbit reticulocytes		45	34	[2]
Canine pancreas			31	[3]
Murine L5178 Y cells		38	31	[4]
Yeast		40	30	[5]
Yeast		44	29	[6]
Dictyostelium discoideum	78	42	34	[7]
Drosophila melanogaster	67			[8]
Drosophila melanogaster	78	35	31	[9]
Acheta domesticus, L.	76	38	30	[10]
Podospora anserina		41	28	[11]

The influence of proteolytic degradation on the number and position of protein spots in the 2-D pattern was demonstrated in an extensive study of ribosomal proteins from rat and rabbit liver and from rabbit reticulocytes using different electrophoretic systems [2]. The anomalous migration of ribosomal proteins of BHK cells with oxidized thiol groups was shown as a further possibility for the formation of artificial spots [12, 49] which is in agreement with earlier results (see ref. [78] in Chapter VI).

b) Preparation of single ribosomal proteins

Isolation of pure ribosomal proteins by carboxymethyl and phosphocellulose chromatography was described for 14 proteins from the small [14] and for 23 proteins from the large subunit of yeast [15]. Four pure proteins were obtained both from the 40 S subunits of *Artemia salina* and of wheat germ by carboxymethyl cellulose chromatography [16]; in this study a total number of 30 and 32 proteins was estimated in the 40 S subunits of *Artemia salina* and wheat germ, respectively [16]. Five pure proteins and a higher number of somewhat contaminated proteins were obtained from *Drosophila* ribosomes using LiCl gradient chromatography on carboxymethyl cellulose and gel filtration [17]. The acidic ribosomal phosphoproteins P 5/P 5′ from *Saccharomyces cerevisiae* have been isolated by preparative gel electrophoresis [18]. A method for the radioiodination of microgram quantities of ribosomal proteins extracted from polyacrylamide gels has been described [19].

c) *Molecular weights*

The results of more recent estimations of molecular weights of ribosomal proteins are summarized in Table 28.

d) *Amino acid composition and amino acid sequences*

Further progress was achieved in the characterization of yeast ribosomal proteins by the determination of the amino acid composition of 14 proteins of the small subunit [14] and of 23 proteins of the large subunit [15]. Complete sequences of proteins YP A 1 [20] and YP 55 [21] from *Saccharomyces cerevisiae,* of N-terminal regions of proteins YP 18, YP 41, YP 45, YP 58 [21] and of the 5 S RNA binding protein YL 3 and one of its CNBr fragments have been determined [22]. Furthermore, sequences of N-terminal regions of 11 rat liver proteins have been published [23].

e) *Posttranslational modifications, tissue and species specificities, ribosomal proteins of mitochondria and chloroplasts, and ribosomal mutants with altered proteins*

Eleven methylated proteins were identified in HeLa cell ribosomes; in one of them five different methylation sites have been found [24].

Only minor differences were found between rabbit liver and rabbit reticulocyte ribosomal proteins with respect to the extent of phosphorylation of S 6, the microheterogeneity of S 12 and the precise posi-

Table 28. Molecular weights of ribosomal proteins

Source of ribosomes	60 S subunits			40 S subunits				
	Range of molecular weights	M_n	Molecular mass	Range of molecular weights	M_n	Molecular mass	Method	References
Rat liver	10000–60000			10000–33000			a	[2]
Rabbit liver	10000–60000			10000–33000			a	[2]
Rabbit reticulocytes	10000–60000			10000–33000			a	[2]
Artemia salina				4800–36500		550000	b	[16]
Wheat germ				4300–43500		560000	b	[16]
Dictyostelium discoideum	13800–51000	23400	981900	13200–40900	21900	746000	a	[7]
Yeast				8000–37500			b	[14]
Yeast	8000–59000						b	[14]
Acheta domesticus L.	7000–77000	27600		8200–42000	21700		b	[11]
Human liver	14300–41800			16500–31000			b	[1]

a: SDS in the second dimension of 2-D electrophoresis
b: SDS electrophoresis of selected isolated ribosomal proteins

tion of S 24 in the 2-D map [2]. Ribosomes isolated from various tissues and developmental stages of the cricket, *Acheta domesticus,* have very similar protein composition [25]. A tissue-specific ribosomal protein component has been described for bound polysomes from rat kidney [26]. Quantitative variation of a 60 S ribosomal protein was found during the growth of the fungus *Podospora anserina* [13]. A comparison of the proteins of high salt washed ribosomes from young (4-day) and old (30-day) male *Drosophila melanogaster* revealed no detectable qualitative differences [8]. Changes of an acidic protein associated to ribosomes of *S. cerevisiae* were observed during the cell cycle [56].

Species specific differences were found between rat and rabbit for three proteins of the small and five proteins of the large ribosomal subunit from the livers [2]. Hybrid mammalian cells, obtained by fusion of 3T3 cells (mouse) and CHO cells (hamster) are characterized by hybrid ribosomes, containing ribosomal proteins, coded to about equal extents by both species [27].

The analysis of ribosomal proteins from 17 yeast species of the genera *Saccharomyces* and *Kluyveromyces* revealed differences in the protein patterns obtained by 1-D SDS electrophoresis [5]. Also for different strains of *Tetrahymena,* strain dependent variations of the protein composition of ribosomes have been found [28].

A 18000 D protein with a modified electrophoretic mobility from 40 S ribosomal subunits has been described for an emetine-resistant chinese hamster cell line [29]. Three mutants of *Podospora anserina* with modified ribosomal proteins S 15 and L 16 [30], and L 21 [31], respectively, were identified.

Altered chloroplast ribosomal proteins were found in three yellow mutants of *Chlamydomonas reinhardii* carrying the mutation probably in the chloroplast DNA [32]. Incomplete mitochondrial small ribosomal subunits with reduced amounts of proteins S 3, S 4, and S 15 were found in yeast mutants [33], and mitochondrial small ribosomal subunits deficient in proteins S 5 and S 9 were described for mutants of *Neurospora crassa* [34].

The acidic proteins L 41, L 44, L 45, and L 46 from the large ribosomal subunit of *Tetrahymena pyriformis* seem to be analogs of proteins L 7/L 12 from *E. coli* with regard to their electrophoretic mobility [52].

2. Phosphorylation of ribosomal proteins

Up to 10 ribosomal proteins from HeLa- or ascites-tumor cells can be phosphorylated in vitro by cAMP-dependent and cGMP-dependent protein kinases [35]. The individual proteins were phosphorylated with different rates by the two kinases, suggesting in vivo phosphorylation of S 6 and S 2 by the cAMP- and cGMP-dependent protein kinases, respectively, [35]. Furthermore, phosphorylation of ribosomal proteins by a cAMP-independent protein kinase from Soybean cotyledons was described [36].

Phosphorylation of protein S 6 was investigated in rat liver [37], rat brain [38], HeLa cells [39], Swiss mouse 3 T 3 cells [40], 3 T 3-L 1 adipocytes [41], chick embryo fibroblasts [42], hamster fibroblasts [43], virus infected Vero cells [44], rat thymocytes [45], *Tetrahymena pyriformis* [46], and *Mucor racemus* [47]. Phosphorylation of S 6 was found to be stimulated by cAMP [39], dibutyryl-cAMP [45], concanavalin A [45], prostaglandin [45], insulin [39, 41, 42], insulin like growth factor [42], epidermal growth factor [39, 41], antibodies to the insulin receptor [41], the combination of isoproterenol and 1-methyl-3-iso-butylxanthin [41], serum [40, 42], starvation [46], liver injury by carbon tetrachloride, and by extrahepatic cholestasis [37]. On the other hand, phosphorylation of cerebral ribosomal protein S 6 was found to be decreased in experimental hyperphenylalaninaemia [38].

Two metabolic pathways are probably involved in S 6 phosphorylation: one, which is insulin- and epidermal growth factor-dependent, and the other one, which is cAMP-dependent [39].

Differences in in-vitro interactions of poly(U) with 40 S subunits of rat liver ribosomes with low and highly phosphorylated protein S 6 have been described [48].

A ribosomal protein with a molecular weight of 65000 from 0.5 M KCl ribosomal extracts of mouse L cells has been found to function as a phosphate acceptor of interferon-induced protein kinase in vitro [50] and thus may participate in the regulation of eIF-2 activity.

Two strongly acidic proteins, L 44 and L 45, of the 60 S subunits of yeast ribosomes are phosphorylatable in vitro by a purified cAMP-independent protein kinase from yeast [51].

3. Immunological properties of ribosomal proteins

Ribosomal proteins of 80 S ribosomes and their subunits from a murine leukemic cell line have been studied by an immunochemical two-dimensional technique [53].

The proteins of ribosomes from various yeast species and their hybrids were compared by using immunochemical and gel electrophoretic techniques [5]. While antibodies against *S. cerevisiae* ribosomal proteins reacted also with those of some other yeast species, no significant cross-reactions were found with ribosomal proteins of *E. coli*, wheat germ or rat liver. On the other hand, homologies were described for *Chlamydomonas reinhardii* chloroplast and *E. coli* ribosomal proteins by using antibodies against *Chlamydomonas* chloroplast small ribosomal subunits [54]. Surprisingly, these antibodies were also reactive with the large ribosomal subunits of both sources.

Antibodies against purified acidic ribosomal proteins S 14 and S 7/8 from *Drosophila melanogaster* embryonic tissue were used to study the intracellular localization of these ribosomal proteins during oogenesis using indirect immuno fluorescence microscopy [55].

4. Ribonucleic acids

a) *High molecular weight ribosomal ribonucleic acids*

Investigations in this field are characterized mainly by a rapid progress in the collection of sequence data. In Table 29 the analyzed rRNA types and the sequenced regions are summarized. The data were obtained mostly by DNA sequencing techniques, but also direct RNA sequencing methods have been applied.

Homologies among rRNAs from various sources were tested also by using hybridization with DNA [62, 79—81]. Similar primary sequences have been found in rRNAs isolated from vegetative and developing cells of *Dictyostelium discoideum* [82].

Oligonucleotide patterns of rat and chicken 18 S rRNA [83], of 18 S and 28 S rRNA from several eukaryotic cell lines [84], of 18 S and 28 S rRNA from *Drosophila melanogaster* [85], and of 17 S and 25 S rRNA from *Saccharomyces cerevisiae* [86] were analyzed.

Oligoadenylation was found at the 3′ end of the 17 S rRNA of the large ribosomal subunit from hamster mitochondria [87].

Methylated residues in rRNA are obviously nonrandomly distributed, as described for hamster mitochondrial rRNA [88] and *Xenopus laevis* rRNA [89, 90]. Such methylated regions are comprised of conserved primary sequences and secondary structure domains, located at the subunit interface in the ribosome [88].

Complementary sequences were found near the 3′ ends of 18 S rRNA and 5 S rRNA of various species which are thought to be involved in subunit association [93].

Table 29. Sequence data of ribosomal RNAs

Source	RNA-type	Analyzed region(s)	References
Cytoplasm			
Rat liver	18 S	3′ end	[57]
Mouse sarcoma	18 S	3′ end	[57]
Rabbit reticulocytes	18 S	3′ end	[57]
Chicken reticulocytes	18 S	3′ end	[57)
Xenopus laevis	18 S, 28 S	18 S—28 S intergene	[58]
Xenopus laevis	18 S	complete	[59, 60]
Xenopus laevis	28 S	3′ end and 5′ end	[58, 61]
Xenopus laevis	28 S	surrounding of introns	[62]
Drosophila melanogaster	18 S	3′ end	[63]
Bombyx mori	18 S	3′ end	[64]
Yeast	25 S	3′ end	[65]
Saccharomyces cerevisiae	18 S	complete	[66, 67]
Barley embryos	18 S	3′ end	[57]
Tetrahymena	25 S	surrounding of introns	[68]
Chloroplasts			
Zea mays	23 S	complete	[69]
Zea mays	16 S	complete	[70]
Chlamydomonas	23 S	surrounding of introns	[71]
Euglena gracilis	16 S	3′ end, 16 S—23 S intergene	[72]
Mitochondria			
Yeast	15 S	complete	[73]
Yeast	21 S	surrounding of introns	[74]
Human	12 S, 16 S	complete	[75]
HeLa	12 S, 16 S	5′ end	[76]
Mouse	12 S, 16 S	complete	[77]
BHK-21	13 S	3′ end	[78]

The 3′ terminal sequences of yeast 26 S rRNA and *Xenopus laevis* 28 S rRNA can be arranged in secondary structure domains similar to those observed at the 3′ end of bacterial 23 S rRNA [91]. The region of *E. coli* 23 S rRNA comprising nucleotides 157—290 is homologous to sequences at the 5′ end of yeast and *Xenopus laevis* 28 S rRNA [92]. The secondary structure of the conserved 3′ end of 18 S rRNA was determined by limited nuclease digestion and psoralen cross-linking [94].

Molecular weights, the integrity and secondary structure features of the rRNA of mollusca were analyzed [97]. The 28 S rRNA of the insect *Rhodnius prolixus* contains a hidden break, which is protected against heat induced dissociation by a low molecular weight component [98].

Raman spectra of rat liver rRNAs were compared with spectra of intact ribosomes; the data point to differences in the secondary structure of isolated rRNA and rRNA in situ [99]. The dinucleotide $m_2^6Apm_2^6A$ which is a common sequence at the 3′ end of the small subunit rRNA shows stronger stacking interactions as the unmethylated ApA as measured by thermal denaturation ultraviolet absorption spectroscopy [100].

A secondary structure model of the 18 S rRNA from *Saccharomyces cerevisiae* was derived from sequence data and by comparison with a secondary structure model of *E. coli* 16 S rRNA [95]. Further secondary structure models were presented also for small subunit rRNAs of *E. coli*, *Z. mays* chloroplasts, *X. laevis* and *S. cerevisiae* cytoplasm, human and mouse mitochondria [96], and for the large subunit rRNAs from *E. coli*, *Z. mays* chloroplasts, and human and mouse mitochondrial ribosomes [69].

b) Low molecular weight ribosomal ribonucleic acids

5 S rRNA

5 S rRNAs from various sources have been sequenced; an updated collection of 5 S rRNA sequences is presented in [101]. Furthermore, the sequences of cytoplasmic 5 S rRNAs of wheat embryo [102], *Chlamydomonas reinhardii* [103], *Aspergillus nidulans* [104], *Neurospora crassa* [104], *Dictyostelium discoideum* [105], *Crithidia fasciculata* [106], *Schizosaccharomyces pombe* [107], *Bombyx mori* [108, 109], and *Lingula anatia* [108, 110] were estimated. The sequence of *Chlorella* 5 S rRNA (see ref. [624] in chapter VI) was revised [111].

By DNA sequencing were analyzed three 5 S rRNA genes of *Xenopus borealis* oocytes [112], the dominant sea urchin 5 S rRNA gene [113] and the 5 S rRNA gene of *Saccharomyces cerevisiae* [114].

The nucleotide sequence of chloroplast 5 S rRNA in flowering plants [115] and of the 5 S rRNA gene of tobacco chloroplasts [116, 117] was determined. A possible 5 S rRNA equivalent in hamster mitochondria containing 63 nucleotides was sequenced [118].

The secondary and tertiary structure of rat liver 5 S rRNA was studied by laser Raman spectroscopy [119] and by small- and wide-angle X-ray scattering [154]. The influence of magnesium ions [120] and temperature [121] on folding and unfolding of yeast 5 S rRNA was also characterized by ultraviolet absorption, circular dichroism, and 360 MHz proton nuclear magnetic resonance spectroscopy [122]. The number of base pairs in 5 S rRNA from *Saccharomyces carlsbergensis* was determined by infrared spectroscopy [123]. Psoralen induced cross-links in *Drosophila melanogaster* 5 S rRNA were found between U_{24} and U_{52} and U_{80} and U_{95}, respectively [124].

The 5 S rRNA-protein complex of *Saccharomyces cerevisiae* was studied by limited ribonuclease digestion [125]. A portion of the 5 S rRNA molecule composed of two base paired fragments (residues 1 to 12 and 79 to 121) was found to be able to bind yeast ribosomal protein YL 3 [125].

By affinity chromatography a ternary complex consisting of rat liver 5 S rRNA, 5.8 S rRNA and protein L 5 was formed [126].

Universal secondary structure models of eukaryotic ribosomal 5 S rRNA were constructed from comparative analysis of their sequences [111, 127—129, 144, 145].

Sequence data of 5 S rRNA were furthermore used for the construction of phylogenetic trees [108, 110, 130, 143].

5.8 S rRNA

Sequences of 5.8 S rRNA from *Drosophila melanogaster* [131], *Sciara coprophila* [132], wheat embryos [102], *Vicia faba* [133], *Chlamydomonas reinhardii* [103], and *Neurospora crassa* [134] were determined. The sequence of *Vicia faba* 5.8 S rRNA was obtained with a new improved direct RNA sequence method [133]. An updated collection of available sequences is published in [101].

The 5.8 S rRNA of *Xenopus laevis* was sequenced [58].

The sequences of the 2 S rRNA from *Drosophila melanogaster* [131] and *Sciara coprophila* [132] and the short spacer region between the coding region of 5.8 S and 2 S rRNA have been analyzed. Whereas the coding regions show a high degree of homology, there is no homology in the spacer sequences [63].

Methylation of 5.8 S rRNA takes place

of 5.8 S rRNA from whole cells and both the nuclear and the cytoplasmic fractions of rat liver, rat kidney cells in culture, and HeLa cells [135]. GmCp at position 77 was found methylated in 5.8 S rRNA from nuclear and cytoplasmic fractions, while Um at position 14 in rat liver is highly methylated in the cytoplasm and undermethylated in the nuclear fraction [135].

The number of base pairs in 5.8 S rRNA from *Saccharomyces carlsbergensis* was determined by infrared spectroscopy [123]. Carbodiimide reactive sites in 5.8 S rRNA of HeLa cells were identified [136] and are compatible with a secondary structure model proposed earlier (see ref. [642] in Chapter VI). Affinity chromatography showed the ability of 5.8 S rRNA to form a complex with 5 S rRNA and protein L 5 from rat liver [126].

Binding of 5.8 S rRNA to the high molecular weight rRNA of the large subunit involves probably both ends of the 5.8 S rRNA molecule [137]. The 3′ end of the 28 S rRNA of *Xenopus laevis* contains a binding site for 5.8 S rRNA [92].

A sequence homology of 54% between trout 5.8 S rRNA and a 5′ terminal sequence of *E. coli* 23 S rRNA was found containing two blocks of 8 completely homologous residues demonstrating the presence of 5.8 S rRNA-like sequences at the 5′ end of *E. coli* 23 S rRNA [138]. Furthermore, the 3′ end of *E. coli* 23 S rRNA contains a binding site for these 5.8 S rRNA-like sequences [92]. In eukaryotes, the 5.8 S rRNA sequence is located at the 5′ end of the 32 S pre-rRNA, proximal to the 28 S rRNA (see Chapter VII). This suggests that mutations altering processing of pre-rRNA might be responsible for the evolution of 5.8 S rRNA [92, 138].

4.5 S rRNA

Several variants of 4.5 S rRNA were found in chloroplasts of flowering plants [139]. They contain no modified nucleotides and in the cytoplasm as shown by the analysis are not phosphorylated at the 5′ terminus [139]. The nucleotide sequences of wheat chloroplast 4.5 S rRNA [140] and tobacco chloroplast 4.5 S rRNA [116, 117, 141] were determined. The location of the 4.5 S rRNA gene 3′ to the 23 S rRNA gene in tobacco chloroplast DNA suggests homology of the 4.5 S rRNA to the 3-terminus of prokaryotic 23 S rRNA which could be shown in the primary sequences [142]. Both tobacco and wheat chloroplast 4.5 S rRNA can form a secondary structure similar to that of the 3′ terminal region of 23 S rRNA [91]. Like 5.8 S rRNA also the 3′ end of chloroplast 4.5 S rRNA contains sequences complementary to the 5′ end of chloroplast 23 S rRNA suggesting basepairing of these regions [91].

5. Nuclease activities of ribosomal particles

Several more recent reports describe the identification of ribonuclease activities tightly bound to ribosomal particles from rat liver [146, 147], fibroblasts [148] and yeast [149–151]. On the other hand a procedure for the isolation of ribonuclease-free polysomes from human placenta has been published [152].

A deoxyribonuclease activity was found in cytoplasmic ribosomes from rye germs [153].

References

[1] GRESSNER, A. M.: Human liver ribosomal proteins: characterisation by two-dimensional electrophoresis and molecular weight determinations. Biochem. Med. **23**, 350–357 (1980)

[2] MADJAR, J. J., and R. R. TRAUT: Differences in electrophoretic behaviour of 8 ribosomal proteins from rat and rabbit tissues and evidence for proteolytic action on liver proteins. Molec. Gen. Genetics **179**, 89–101 (1980)

[3] Nesset, C. C. and S. R. Dickman: Effects of potassium chloride concentration on protein content and polyphenylalanine synthesizing capability of 40 S ribosomal subunits from canine pancreas. Biochem. **19**, 2731—2737 (1980)

[4] Longuet, M., M.-A. Auger-Buendia, and A. Tavitian: Studies on the distribution of ribosomal proteins in mammalian ribosomal subunits. Biochimie **61**, 1113 to 1123 (1979)

[5] Adoutte-Panvier, A., J. E. Davies, L. R. Gritz, and B. S. Littlewood: Studies of ribosomal proteins of yeast species and their hybrides. Gel electrophoresis and immunochemical cross-reactions. Molec. Gen. Genetics **179**, 273—282 (1980)

[6] Otaka, E. and S. Osawa: Yeast ribosomal proteins: V. Correlation of several nomenclatures and proposal of a standard nomenclature. Molec. Gen. Genetics **181**, 176—182 (1981)

[7] Ramagopal, S. and H. L. Ennis: Studies on ribosomal proteins in the cellular slime mold *Dictyostelium discoideum*. Resolution, nomenclature and molecular weights of proteins in the 40-S and 60-S ribosomal subunits. Europ. J. Biochem. **105**, 245 to 258 (1980)

[8] Schmidt, T. and G. T. Baker: Analysis of ribosomal proteins from adult *Drosophila melanogaster* in relation to age. Mechan. Ageing Develop. **11**, 105—112 (1979)

[9] Chooi, W. Y., L. M. Sabatini, M. Macklin, and D. Fraser: Group fractionation and determination of the number of ribosomal subunit proteins from *Drosophila melanogaster*. Biochem. **19**, 1425—1433 (1980)

[10] Bosselman, R. A. and M. S. Kaulenas: Ribosomal proteins of the cricket, *Acheta domesticus L.* Insect Biochem. **10**, 129 to 137 (1980)

[11] Coppin-Raynal, E.: Analytical chromatography of ribosomal proteins in the fungus, *Podospora anserina*. Analyt. Biochem. **109**, 395—398 (1980)

[12] Leader, D. P. and G. P. Mosson: The anomalous migration during two-dimensional gel electrophoresis of eukaryotic ribosomal proteins with oxidised thiol groups. Biochim. Biophys. Acta **622**, 360—364 (1980)

[13] Perrot, M. and J. Begueret: Quantitative variation of a 60 S ribosomal protein during growth of the fungus *Podospora anserina*. Biochimie **59**, 799—804 (1977)

[14] Higo, K. and E. Otaka: Isolation and characterization of fourteen ribosomal proteins from small subunits of yeast. Biochem. **18**, 4191—4196 (1979)

[15] Itoh, T., K. Higo, and E. Otaka: Isolation and characterization of twenty-three ribosomal proteins from large subunits of yeast. Biochem. **18**, 5787—5793 (1979)

[16] Shin, C.-Y. T., J. E. Toivonen, and G. R. Craven: Partial purification and characterization of the proteins from the 40-S ribosomes of *Artemia salina* and wheat germ. Europ. J. Biochem. **97**, 189—196 (1979)

[17] Chooi, W. Y.: Purification of *Drosophila* ribosomal proteins. Isolation of proteins S 8, S 13, S 14, S 16, S 19, S 20/L 24, S 21/L 26, S 24, S 25/S 27, S 26, S 29, L 4, L 10/11, L 12, L 13, L 16, L 18, L 19, L 27, 1, 7/8, 9, and 11. Biochem. **19**, 3469—3476 (1980)

[18] Zinker, S.: P 5/P 5′ — The acidic ribosomal phosphoproteins from *Saccharomyces cerevisiae*. Biochim. Biophys. Acta **606**, 76—82 (1980)

[19] Tolan, D. R., J. M. Lambert, G. Boileau, T. G. Fanning, and J. W. Kenny: Radioiodination of microgram quantities of ribosomal proteins from polyacrylamide gels. Analyt. Biochem. **103**, 101—109 (1980)

[20] Itoh, T.: Primary structures of yeast acidic ribosomal protein YP A 1. FEBS-Lett. **114**, 119—123 (1980)

[21] Itoh, T., K. Higo, E. Otaka, and S. Osawa: Studies on the primary structures of yeast ribosomal proteins. In: "Genetics and evolution of RNA polymerase, tRNA and ribosomes" (Eds. S. Osawa, H. Ozaki, H. Uchida, T. Yura), University of Tokyo Press, Tokyo 1980, p. 609—624

[22] Nazar, R. N., M. Yaguchi, G. E. Willick, C. F. Rollin, and C. Roy: The 5-S RNA binding protein from yeast (*Saccharomyces cerevisiae*) ribosomes. Evolution of the eukaryotic 5-S RNA binding protein. Europ. J. Biochem. **102**, 573 to 582 (1979)

[23] Wittmann-Liebold, B., A. W. Geiss-

LER, A. LIN, and I. G. WOOL: Sequence of the amino-terminal region of rat liver ribosomal proteins S 4, S 6, S 8, L 6, L 7a L 18, L 27, L 30, L 37, L 37a, and L 39. J. Supramolec. Structure **12**, 425—443 (1979)

[24] SCOLNIK, P. A. and G. L. ELICEIRI: Methylation sites in HeLa cell ribosomal proteins. Europ. J. Biochem. **101**, 93—101 (1979)

[25] BOSSELMAN, R. A. and M. S. KAULENAS: Analysis of the apparent tissue specific differences in *Acheta domesticus* ribosomal proteins. Insect Biochem. **10**, 139—146 (1980)

[26] HOFFMAN, W. L. and R. M. DOWBEN: Tissue-specific ribosomal component. Molec. Biol. Rep. **5**, 225—228 (1979)

[27] WEJKSNORA, P. L., and J. R. WARNER: Hybrid mammalian cells assemble hybrid ribosomes. Proc. Natl. Acad. Sci. U.S. **76**, 5554—5558 (1979)

[28] CUNY, M., M. MILLET, and D. H. HAYES: Strain dependent variation of the protein composition of *Tetrahymena* ribosomes. FEBS-Lett. **101**, 77—84 (1979)

[29] REICHENBERGER, V. E. jr., and C. T. CASKEY: Emetine-resistant chinese hamster cells. The identification of an electrophoretically altered protein of the 40 S ribosomal subunit. J. Biol. Chem. **254**, 6207—6210 (1979)

[30] CROUZET, M. and J. BEQUERET: Altered ribosomal proteins in mutants of the fungus *Podospora anserina*. J. Biol. Chem. **255**, 4996—4999 (1980)

[31] CROUZET, M. and J. BEQUERET: A new mutant form of the ribosomal protein L 21 in the fungus *Podospora anserina*. Identification of the structural gene for this protein. Molec. Gen. Genet. **180**, 177—183 (1980)

[32] GYURJAN, L., G. ERDÖS, N. P. YURINA, M. S. TURISCHEWA, and M. S. ODINTSOVA: Yellow mutations alter chloroplast ribosomal proteins in *Chlamydomonas reinhardii*. Acta Biochem. Biophys. **14**, 229—239 (1980)

[33] SOR, F. and G. FAYE: Mitochondrial and nuclear mutations that affect the biogenesis of the mitochondrial ribosomes of yeast. 2. Biochemistry. Molec. Gen. Genetics **177**, 47—56 (1979)

[34] COLLINS, R. A., H. BERTRAND, R. J. LAPOLLA, and A. M. LAMBOWITZ: Mitochondrial ribosome assembly in *Neurospora crassa* -mutants with defects in mitochondrial ribosome assembly. Molec. Gen. Genetics **177**, 73—84 (1979)

[35] ISSINGER, O.-G., H. BEIER, N. SPEICHERMANN, V. FLOKERZI, and F. HOFMANN: Comparison of phosphorylation of ribosomal proteins from HeLa and Krebs 2 ascites-tumour cells by cyclic AMP-dependent and cyclic GMP-dependent protein kinases. Biochem. J. **185**, 89—99 (1980)

[36] GOWDA, S. and D. T. N. PILLAY: Phosphorylation of ribosomal proteins by cyclic AMP independent protein kinase (ribosomal casein kinase) from soyabean cotyledons (Glycine-Max L.). Plant Cell Physiol. **21**, 1357—1365 (1980)

[37] GRESSNER, A. M.: Ribosomal protein modification in liver injury: Effect of carbon tetrachloride and extrahepatic cholestasis on protein phosphorylation. J. Clin. Chem. Clin. Biochem. **18**, 111—116 (1980)

[38] ROBERTS, S. and B. S. MORELOS: Cerebral ribosomal protein phosphorylation in experimental hyperphenylalaninaemia. Biochem. J. **190**, 405—419 (1980)

[39] LASTICK, S. M. and E. H. MCCONKEY: Control of ribosomal protein phosphorylation in HeLa cells. Biochem. Biophys. Res. Commun. **95**, 917—923 (1980)

[40] THOMAS, G., M. SIEGMANN, A.-M. KÜBLER, J. GORDON, and L. J. ASUA: Regulation of 40 S ribosomal protein S 6 phosphorylation in Swiss mouse 3 T 3 cells. Cell **19**, 1015—1023 (1980)

[41] SMITH, C., C. S. RUBIN, and O. M. ROSEN: Insulin-treated 3 T 3-L 1 adipocytes and cell-free extracts derived from them incorporate ^{32}P into ribosomal protein S 6. Proc. Natl. Acad. Sci. U.S. **77**, 2641 to 2645 (1980)

[42] HASELBACHER, G. K., R. E. HUMBEL, and G. THOMAS: Insulin-like growth factor: insulin or serum increase phosphorylation of ribosomal protein S 6 during transition of stationary chick embryo fibroblasts into early G 1 phase of the cell cycle. FEBS-Lett. **100**, 185—190 (1979)

[43] KENNEDY, I. M. and D. P. LEADER: The phosphorylation of protein S 6 in the newly-synthesized cytoplasmic ribosomes

in hamster fibroblasts. Molec. Biol. Rep. **6**, 199–208 (1980)

[44] Fenwick, M. L. and M. J. Walker: Phosphorylation of a ribosomal protein and of virus-specific proteins in cells infected with Herpes Simplex Virus. J. Gen. Virol. **45**, 397–405 (1979)

[45] Wettenhall, R. E. H. and G. J. Howlett: Phosphorylation of a specific ribosomal protein during stimulation of thymocytes by concanavalin A and prostaglandin E 1. J. Biol. Chem. **254**, 9317 to 9323 (1979)

[46] Kristiansen, K. and A. Krüger: Phosphorylation and degradation of ribosomes in starved *Tetrahymena pyriformis*. Exp. Cell Res. **118**, 159–169 (1979)

[47] Larsen, A. and P. S. Sypherd: Physiological control of phosphorylation of ribosomal protein S 6 in *Mucor racemus*. J. Bacteriol. **141**, 20–25 (1980)

[48] Gressner, A. M. and E. van der Leur: Interaction of synthetic polynucleotides with small rat liver ribosomal subunits possessing low and highly phosphorylated protein S 6. Biochim. Biophys. Acta **608**, 459–468 (1980)

[49] Leader, D. P.: Phosphorylated and other modified forms of eukaryotic ribosomal protein S 3 analysed by two-dimensional gel electrophoresis. Biochem. J. **189**, 241–245 (1980)

[50] Ohtsuki, K., M. Nakamura, T. Koike, N. Ishida, and S. Baron: A ribosomal protein mediates EIF-2 phosphorylation by interferon-induced kinase. Nature **287**, 65–67 (1980)

[51] Kudlicki, W., R. Szyszka, R. E. Palén, and E. Gasior: Evidence for a highly specific protein kinase phosphorylating two strongly acidic proteins of yeast 60 S ribosomal subunit. Biochim. Biophys. Acta **633**, 376–385 (1980)

[52] Sandermann, J., A. Krüger, and K. Kristiansen: Characterization of 60 S ribosomal proteins in *Tetrahymena pyriformis*. FEBS-Lett. **107**, 343–347 (1979)

[53] Ollier, M.-P., M.-A. Auger-Buendia, and L. Hartmann: Antigenicite des proteins ribosomiques d'une lignee cellulaire maligne chez la Souris, resultats preliminaires. C. R. Acad. Sc. Paris, Ser. D **289**, 189–192 (1979)

[54] Schneeman, R. and St. Surzycki: *E. coli* ribosomal proteins are cross reactive with antibody prepared against *Chlamydomonas reinhardi* chloroplast ribosomal subunit. Molec. Gen. Genetics **176**, 95–104 (1979)

[55] Burns, D. K. and W. Y. Chooi: Differential localization of ribosomal proteins S 14 and 7/8 in egg chambers of *Drosophila melanogaster*. Molec. Gen. Genetics **179**, 299–310 (1980)

[56] Sanchez-Madrid, F. and J. P. G. Ballesta: An acidic protein associated to ribosomes of *Saccharomyces cerevisiae*. Changes during cell cycle. Biochem. Biophys. Res. Commun. **91**, 643–650 (1979)

[57] Azad, A. A. and N. J. Deacon: The 3′-terminal primary structure of five eukaryotic 18 S rRNAs determined by the direct chemical method of sequencing. The highly conserved sequences include an invariant region complementary to eukaryotic 5 S rRNA. Nucl. Acids Res. **8**, 4365–4376 (1980)

[58] Hall, L. M. C. and B. E. H. Maden: Nucleotide sequence through the 18 S–28 S intergene region of a vertebrate ribosomal transcription unit. Nucl. Acids Res. **8**, 5993–6006 (1980)

[59] Salim, M. and B. E. H. Maden: Nucleotide sequence encoding the 5′ end of *Xenopus laevis* 18 S rRNA. Nucl. Acids Res. **8**, 2871–2884 (1980)

[60] Salim, M. and B. E. H. Maden: Nucleotide sequence of *Xenopus laevis* 18 S ribosomal RNA inferred from gene sequence. Nature **291**, 205–208 (1981)

[61] Sollner-Webb, B. and R. H. Reeder: The nucleotide sequence of the initiation and termination sites for ribosomal RNA transcription in *X. laevis*. Cell **18**, 485 to 499 (1979)

[62] Gourse, R. L. and S. A. Gerbi: Fine structure of ribosomal RNA. IV. Extraordinary evolutionary conservation in sequences that flank introns in rDNA. Nucl. Acids Res. **8**, 3623–3637 (1980)

[63] Jordan, B. R., Latil-Damotte, M., and R. Jourdan: Sequence of the 3′-terminal position of *Drosophila melanogaster* 18 S rRNA and of the adjoining spacer: Comparison with corresponding prokaryotic and eukaryotic sequences. FEBS-Lett. **117**, 227–231 (1980)

[64] Samols, D. R., O. Hagenbüchle, and

L. P. GAGE: Homology of the 3′ terminal sequences of the 18 S rRNA of *Bombyx mori* and the 16 S rRNA of *Escherichia coli*. Nucl. Acids Res. **7**, 1109–1119 (1979)

[65] VELDMAN, G. M., J. KLOOTWIJK, P. DE JONGE, R. J. LEER, and R. J. PLANTA: The transcription termination site of the ribosomal RNA operon in yeast. Nucl. Acids Res. **8**, 5179–5192 (1980)

[66] RUBTSOV, P. M., M. M. MUSAKHANOV, V. M. ZAKHARYEV, K. G. SKRYABIN, and A. A. BAYEV: The structure of the yeast ribosomal RNA genes. 1. The complete nucleotide sequence of the 18 S ribosomal RNA gene from *Saccharomyces cerevisiae*. Nucl. Acids Res. **8**, 5779–5794 (1980)

[67] SKRYABIN, K. G., V. M. ZACHARYEV, P. M. RUBTSOV, and A. A. BAYEV: Nucleotide sequence of the putative initiator site of transcription of the ribosomal operon of yeast. Dokl. Acad. Nauk USSR **247**, 1275–1277 (1979)

[68] WILD, M. A. and R. SOMMER: Sequence of a ribosomal RNA gene intron from *Tetrahymena*. Nature **283**, 693–694 (1980)

[69] GLOTZ, C., C. ZWIEB, R. BRIMACOMBE, K. EDWARDS, and H. KÖSSEL: Secondary structure of the large subunit ribosomal RNA from *Escherichia coli*, *Zea mays* chloroplast, and human and mouse mitochondrial ribosomes. Nucl. Acids Res. **9**, 3287–3306 (1981)

[70] SCHWARZ, Z. and H. KÖSSEL: The primary structure of 16 S rDNA from *Zea mays* chloroplast is homologous to *E. coli* 16 S rRNA. Nature **283**, 739–742 (1980)

[71] ALLET, B. and J. P. ROCHAIX: Structure analysis at the ends of the intervening DNA sequences in the chloroplast 23 S ribosomal genes of *C. reinhardii*. Cell **18**, 55–60 (1979)

[72] OROZCO, E. M. jr., K. E. RUSHLOW, J. R. DODD, and R. E. HALLICK: *Euglena gracilis* chloroplast ribosomal RNA transcription units. 2. Nucleotide sequence homology between the 16 S–23 S ribosomal RNA spacer and the 16 S ribosomal RNA leader regions. J. Biol. Chem. **255**, 10997–11003 (1980)

[73] SOR, F. and H. FUKUHARA: Nucleotide sequence of the gene for the mitochondrial 15 S ribosomal RNA of yeast. Compt. Rend. Acad. Sci. **291**, 933–936 (1980)

[74] DUJON, B.: Sequence of the intron and flanking exons of the mitochondrial 21 S rRNA gene of yeast strains having alleles at the ω and rib-1 loci. Cell **20**, 185–197 (1980)

[75] EPERON, I., S. ANDERSON, and D. NIERLICH: Distinctive sequence of human mitochondrial ribosomal RNA genes. Nature **286**, 460–467 (1980)

[76] CREWS, S. and G. ATTARDI: The sequences of the small ribosomal RNA gene and the phenylalanine tRNA gene are joined end to end in human mitochondrial DNA. Cell **19**, 775–784 (1980)

[77] VAN ETTEN, R. A., W. M. WALBERG, and D. A. CLAYTON: Precise genomic localization and nucleotide sequence of the two ribosomal RNA genes and three immediately adjacent novel transfer RNA genes in mouse mitochondrial DNA. Cell **22**, 157–170 (1980)

[78] BAER, R. and D. T. DUBIN: The 3′-terminal sequence of the small subunit ribosomal RNA from hamster mitochondria. Nucl. Acids Res. **8**, 4927–4941 (1980)

[79] BOHNERT, H. J., K. H. J. GORDON, and E. J. CROUSE: Homologies among ribosomal RNA and messenger RNA genes in chloroplasts, mitochondria and *E. coli*. Molec. Gen. Genetics **179**, 539–545 (1980)

[80] COX, R. A. and R. D. THOMPSON: Distribution of sequences common to the 25–28 S ribonucleic acid genes of *Xenopus laevis* and *Neurospora crassa*. Biochem. J. **187**, 75–90 (1980)

[81] GOURSE, R. L. and S. A. GERBI: Fine structure of ribosomal RNA. 3. Location of evolutionarily conserved regions within ribosomal DNA. J. Mol. Biol. **140**, 321 to 339 (1980)

[82] BATTS-YOUNG, B., H. F. LODISH, and A. JACOBSON: Similarity of the primary sequences of ribosomal RNAs isolated from vegetative and developing cells of *Dictyostelium discoideum*. Develop. Biol. **78**, 352–364 (1980)

[83] FUKE, M. and H. BUSCH: Comparison of nucleotide sequences of large T 1 ribonuclease fragments of 18 S ribosomal RNA of rat and chicken. Nucl. Acids Res. **7**, 1131–1135 (1979)

[84] ELADARI, M. E., A. HAMPE, and F. GALIBERT: Comparative studies of the primary structures of ribosomal RNAs of several

eukaryotic cell lines by the fingerprinting method. Biochimie **61**, 1097–1112 (1979)

[85] YAGURA, T., M. YAGURA, and M. MURAMATSU: *Drosophila melanogaster* has different ribosomal RNA sequences on X and Y chromosomes. J. Mol. Biol. **133**, 533–547 (1979)

[86] ELADARI, M.-E., A. HAMPE, and F. GALIBERT: Analysis of large specific T 1 oligonucleotides of 17 S and 25 S ribosomal RNAs from *Saccharomyces cerevisiae*. Biochimie **61**, 1073–1080 (1979)

[87] DUBIN, D. T., K. D. TIMKO, and R. J. BAER: The 3′ terminus of the large ribosomal subunit ("17 S") RNA from hamster mitochondria is ragged and oligoadenylated. Cell **23**, 271–278 (1981)

[88] BAER, R. J. and D. T. DUBIN: Methylated regions of hamster mitochondrial ribosomal RNA structural and functional correlates. Nucl. Acids Res. **9**, 323–337 (1981)

[89] BRAND, R. C. and S. A. GERBI: Fine structure of ribosomal RNA. 2. Distribution of methylated sequences within *Xenopus laevis* rRNA. Nucl. Acids Res. **7**, 1497–1511 (1979)

[90] MADEN, B. E. H.: Methylation map of *Xenopus laevis* ribosomal RNA. Nature **288**, 293–296 (1980)

[91] MACHATT, M. A., J.-P. EBEL, and C. BRANLANT: The 3′-terminal region of bacterial 23 S ribosomal RNA: structure and homology with the 3′-terminal region of eukaryotic 28 S rRNA and with chloroplast 4.5 S rRNA. Nucl. Acids Res. **9**, 1533–1549 (1981)

[92] COX, R. A. and J. M. KELLY: Comments on the evolution of 23–28 S rRNA: mature 23 S rRNA of prokaryotes appears homologous with the precursor of 25–28 S rRNA of eukaryotes. FEBS-Lett. **130**, 1–6 (1981)

[93] AZAD, A. A.: Intermolecular base-paired interaction between complementary sequences present near the 3′ ends of 5 S RNA and 18 S (16 S) rRNA might be involved in the reversible association of ribosomal subunits. Nucl. Acids Res. **7**, 1913–1929 (1979)

[94] DARZYNKIEWICZ, E., K. NAKASHIMA, and A. J. SHATKIN: Base-pairing in conserved 3′ end of 18 S rRNA as determined by psoralen photoreaction and RNase sensitivity. J. Biol. Chem. **255**, 4973–4975 (1980)

[95] BRIMACOMBE, R.: Secondary structure homology between *Escherichia coli* 16 S and *Saccharomyces cerevisiae* 18 S ribosomal RNA. Biochem. Internat. **1**, 162 to 171 (1980)

[96] ZWIEB, C., C. GLOTZ, and R. BRIMACOMBE: Secondary structure comparisons between small subunit ribosomal RNA molecules from six different species. Nucl. Acids Res. **9**, 3621–3640 (1981)

[97] CAMMARANO, P., P. LONDEI, F. MAZZEI, and A. FELSANI: Physicochemical characterization of the ribosomal RNA species of the mollusca. Molecular weight, integrity and secondary structure features of the RNA of the large and small ribosomal subunits. Biochem. J. **189**, 313–335 (1980)

[98] ISHIKAWA, H., W. G. FILHO, G. A. da SILVA PASSOS Jr., and F. L. DE LUCCA: Comparative studies on the thermal stability of animal ribosomal RNA's. 6. The 28 S ribosomal RNA of *Rhodnius prolixus* is heat-dissociable only after its purification. Comp. Biochem. Physiol. **68** B, 377–381 (1981)

[99] THOMAS, Jr., G. J., B. PRESCOTT, and M. G. HAMILTON: Raman spectra and conformational properties of ribosomes during various stages of disassembly. Biochem. **19**, 3604–3613 (1980)

[100] TAZAWA, I., T. KOIKE, and Y. INOUE: Stacking properties of a highly hydrophobic dinucleotide sequence, N^6,N^6-Dimethyladenylyl(3′-5′)N^6,N^6-dimethyladenosine, occurring in 16-18-S ribosomal RNA. Europ. J. Biochem. **109**, 33–38 (1980)

[101] ERDMANN, V. A.: Collection of published 5 S and 5.8 S RNA sequences and their precursors. Nucl. Acids Res. **9**, r25-r41 (1981)

[102] MACKAY, R. M., D. F. SPENCER, W. F. DOOLITTLE, and M. W. GRAY: Nucleotide sequences of wheat embryo cytosol 5 S and 5.8 S ribosomal nucleic acids. Europ. J. Biochem. **112**, 561–576 (1980)

[103] DARLIX, J.-L. and J.-D. ROCHAIX: Nucleotide sequence and structure of cytoplasmic 5 S RNA and 5.8 S RNA of *Chlamydomonas reinhardii*. Nucl. Acids Res. **9**, 1291–1299 (1981)

[104] PIECHULLA, B., U. HAHN, L. W. MC LAUGHLIN, and H. KÜNTZEL: Nucleotide sequence of 5 S ribosomal RNA from *Aspergillus nidulans* and *Neurospora crassa*. Nucl. Acids Res. **9**, 1445–1450 (1981)

[105] HORI, H., S. OSAWA, and M. IWABUCHI: The nucleotide sequence of 5 S rRNA from a cellular slime mold *Dictyostelium discoideum*. Nucl. Acids Res. **8**, 5535 to 5539 (1980)

[106] MACKAY, R. M., M. W. GRAY, and W. F. DOOLITTLE: Nucleotide sequence of *Crithidia fasciculata* cytosol 5 S ribosomal ribonucleic acid. Nucl. Acids Res. **8**, 4911–4917 (1980)

[107] KOMIYA, H., M. MIYAZAKI, and S. TAKEMURA: The nucleotide sequence of 5 S ribosomal RNA from *Schizosaccharomyces pombe*. J. Biochem. **89**, 1663–1666 (1981)

[108] KOMIYA, H., M. KAWAKAMI, N. SHIMIZU, and S. TAKEMURA: Nucleotide sequences and evolutionary aspects of 5 S ribosomal RNAs from *Lingula* and silkworm. Nucl. Acids Res. **8**, 5119–5123 (1980)

[109] KOMIYA, H., M. KAWAKAMI, and S. TAKEMURA: Nucleotide sequence of 5 S ribosomal RNA from the posterior silk glands of *Bombyx mori*. J. Biochem. **89**, 717–722 (1981)

[110] KOMIYA, H., N. SHIMIZU, M. KAWAKAMI, and S. TAKEMURA: Nucleotide sequence of 5 S ribosomal RNA from *Lingula anatina*. A study on the molecular evolution of 5 S ribosomal RNA from a living fossil. J. Biochem. **88**, 1449–1456 (1980)

[111] LUEHRSEN, K. R. and G. E. FOX: Secondary structure of eukaryotic cytoplasmic 5 S ribosomal RNA. Proc. Natl. Acad. Sci. U.S. **78**, 2150–2154 (1981)

[112] KORN, L. J. and D. D. BROWN: Nucleotide sequence of *Xenopus borealis* oocyte 5 S DNA: comparison of sequences that flank several related eucaryotic genes. Cell **15**, 1145–1156 (1978)

[113] LU, A. L., D. A. STEEGE, and D. W. STAFFORD: Nucleotide sequence of a ribosomal RNA gene in the sea urchin *Lytechinus variegatus*. Nucl. Acids Res. **8**, 1839–1853 (1980)

[114] VALENZUELA, P., G. I. BELL, A. VENEGOS, E. T. SEWELL, F. R. MASIARZ, L. J. DEGENNARO, F. WEINBERG, and W. J. RUTTER: Ribosomal RNA genes of *Saccharomyces cerevisiae*. 2. Physical map and nucleotide sequence of the 5 S ribosomal RNA gene and adjacent intergenic regions. J. Biol. Chem. **252**, 8126–8135 (1977)

[115] DYER, T. A. and C. M. BOWMAN: Nucleotide sequences of chloroplast 5 S ribosomal ribonucleic acid in flowering plants. Biochem. J. **183**, 595–604 (1979)

[116] TAKAIWA, F. and M. SUGIURA: Cloning and characterization of 4.5 S and 5 S RNA genes in tobacco chloroplasts. Gene **10**, 95–103 (1980)

[117] TAKAIWA, F. and M. SUGIURA: Nucleotide sequences of the 4.5 S and 5 S ribosomal RNA gene from tobacco chloroplasts. Molec. Gen. Genetics **180**, 1–4 (1980)

[118] BAER, R. J. and D. T. DUBIN: The sequence of a possible 5 S RNA equivalent in hamster mitochondria. Nucl. Acids Res. **8**, 3603–3610 (1980)

[119] FABIAN, H., S. BÖHM, H. WELFLE, P. REICH, and H. BIELKA: Laser Raman studies of rat liver ribosomal 5 S RNA. FEBS-Lett. **123**, 19–21 (1981)

[120] MARUYAMA, S. and S. SUGAI: Folding of yeast 5 S ribosomal RNA induced by magnesium binding. J. Biochem. **88**, 151 to 158 (1980)

[121] MARUYAMA, S., T. TATSUKI, and S. SUGAI: Equilibrium and kinetics of the thermal unfolding of yeast 5 S ribosomal RNA. J. Biochem. **86**, 1487–1494 (1979)

[122] LUOMA, G. A., P. D. BURNS, R. E. BRUCE, and A. G. MARSHALL: Melting of *Saccharomyces cerevisiae* 5 S ribonucleic acid: Ultraviolet absorption, circular dichroism, and 360 MHz proton nuclear magnetic resonance spectroscopy. Biochem. **19**, 5456–5462 (1980)

[123] STULZ, J., T. ACKERMANN, B. APPEL, and V. A. ERDMANN: Determination of base pairing in yeast 5 S and 5.8 S RNA by infrared spectroscopy. Nucl. Acids Res. **9**, 3851–3861 (1981)

[124] THOMPSON, J. F., M. R. WEGNEZ, and J. E. HEARST: Determination of the secondary structure of *Drosophila melanogaster* 5 S RNA by hydroxymethyltrimethylpsoralen crosslinking. J. Mol. Biol. **147**, 417–436 (1981)

[125] NAZAR, R. N.: The ribosomal protein binding site in *Saccharomyces cerevisiae* ribosomal 5 S RNA. A conserved protein binding site in 5 S RNA. J. Biol. Chem. **254**, 7724—7729 (1979)

[126] METSPALU, A., I. TOOTS, M. SAARMA, and R. VILLEMS: The ternary complex consisting of rat liver ribosomal 5 S RNA, 5.8 S RNA and protein L 5. FEBS-Lett. **119**, 81—84 (1980)

[127] BÖHM, S., H. FABIAN, and H. WELFLE: Universal secondary structures of prokaryotic and eukaryotic ribosomal 5 S RNA derived from comparative analysis of their sequences. Acta biol. med. germ. **40**, K19—K24 (1981)

[128] BÖHM, S., H. FABIAN, and H. WELFLE: Universal structural features of prokaryotic and eukaryotic ribosomal 5 S RNA derived from comparative analysis of their sequences. Acta biol. med. germ. **41**, 1—16 (1982)

[129] GARRETT, R. A., S. DOUTHWAITE, and H. F. NOLLER: Structure and role of 5 S RNA-protein complexes in protein biosynthesis. Trends Biochem. Sci. **5**, 137 to 139 (1981)

[130] KÜNTZEL, H., M. HEIDRICH, and B. PIECHULLA: Phylogenetic tree derived from bacterial, cytosol and organelle 5 S rRNA sequences. Nucl. Acids Res. **9**, 1451—1461 (1981)

[131] PAVLAKIS, G. N., B. R. JORDAN, R. M. WURST, and J. N. VOURNAKIS: Sequence and secondary structure of *Drosophila melanogaster* 5.8 S and 23 S rRNAs and of the processing site between them. Nucl. Acids Res. **7**, 2213—2237 (1979)

[132] JORDAN, B. R., M. LATIL-DAMOTTE, and R. JOURDAN: Coding and spacer sequences in the 5.8 S—2 S region of *Sciara coprophila* ribosomal DNA. Nucl. Acids Res. **8**, 3565—3573 (1980)

[133] TANAKA, Y., T. A. DYER, and G. G. BROWNLEE: An improved direct RNA sequence method; its application to *Vicia faba* 5.8 S ribosomal RNA. Nucl. Acids Res. **8**, 1259—1272 (1980)

[134] SELKER, E. and C. YANOFSKY: Nucleotide sequence and conserved features of the 5.8 S rRNA coding region of *Neurospora crassa*. Nucl. Acids Res. **6**, 2561—2567 (1979)

[135] NAZAR, R. N., T. O. SITZ, and K. D. SOMERS: Cytoplasmic methylation of mature 5.8 S ribosomal RNA. J. Mol. Biol. **142**, 117—121 (1980)

[136] KELLY, J. M. and B. E. H. MADEN: Chemical modification studies and the secondary structure of HeLa cell 5.8 S rRNA. Nucl. Acids Res. **8**, 4521—4534 (1980)

[137] NAZAR, R. N. and T. O. SITZ: Role of the 5′-terminal sequence in the RNA binding site of yeast 5.8 S rRNA. FEBS-Lett. **115**, 71—76 (1980)

[138] NAZAR, R. N.: A 5.8 S rRNA-like sequence in prokaryotic 23 S rRNA. FEBS-Lett. **119**, 212—214 (1980)

[139] BOWMAN, C. M. and T. A. DYER: 4.5 S ribonucleic acid, a novel ribosome component in the chloroplasts of flowering plants. Biochem. J. **183**, 605—613 (1979)

[140] WILDEMAN, A. G. and R. N. NAZAR: Nucleotide sequence of wheat chloroplastid 4.5 S ribonucleic acid. Sequence homologies in 4.5 S RNA species. J. Biol. Chem. **255**, 11896—11900 (1980)

[141] TAKAIWA, F. and M. SUGIURA: The nucleotide sequence of 4.5 S ribosomal RNA from tobacco chloroplasts. Nucl. Acids Res. **8**, 4125—4129 (1980)

[142] MACKAY, R. M.: The origin of plant chloroplast 4.5 S ribosomal RNA. FEBS-Lett. **123**, 17—19 (1981)

[143] OSAWA, S. and H. HORI: Molecular evolution of ribosomal components. In: "Ribosomes. Structure, function, and genetics". Eds.: G. CHAMBLISS, G. R. CRAVEN, J. DAVIES, K. DAVIS, L. KAHAN, and M. NOMURA, University Park Press, Baltimore 1979, 333—355

[144] SCHWARTZ, R. M. and M. O. DAYHOFF: 20 ribosomal and other RNAs. Atlas of Protein Sequence and Structure **5**, 293 to 299 (1976)

[145] SANKOFF, D., A.-M. MORIN, and R. J. CEDERGREEN: The evolution of 5 S RNA secondary structures. Canad. J. Biochem. **56**, 440—443 (1978)

[146] BRANSGROVE, A. B. and L. C. COSQUER: Ribosome bound ribonuclease; its preferential association with small polysomes. Biochem. Biophys. Res. Commun. **81**, 504—511 (1978)

[147] INGEBRETSEN, O. C., B. EKER, and A. PIHL: Purification and properties of a

ribonuclease from rat liver polysomes. FEBS-Lett. **25**, 217–220 (1972)

[148] JALKANEN, M. T., S. AHO, and E. KULONEN: Alkaline ribonuclease associated with polyribosomes in fibroblasts of experimental granulation tissue. Acta Chem. Scand. **B32**, 655–664 (1978)

[149] SCHULZ-HARDER, B., N. KÄUFER, and U. SWIDA: A ribonuclease from yeast associated with the 40 S ribosomal subunit. Biochim. Biophys. Acta **565**, 173–182 (1979)

[150] SCHULZ-HARDER, B. and C. KÜCHERER: The induction of a ribosomal ribonuclease in *Saccharomyces cerevisiae*. Z. Naturforsch. **35c**, 168–170 (1980)

[151] SWIDA, U., B. SCHULZ-HARDER, C. KÜCHERER, and N. KÄUFER: The occurence of two ribosomal ribonucleases depending on growth phase in yeast. Induction of ribonuclease in glucose-starved cells. Biochim. Biophys. Acta **652**, 129–138 (1981)

[152] KELLY, S., R. FOLMAN, A. HOCHBERG, and J. ILAN: Isolation of ribonuclease-free polysomes from human placenta. Biochim. Biophys. Acta **609**, 278–285 (1980)

[153] SIWECKA, M. A., M. RYTEL, and J. W. SZARKOWSKI: The presence of deoxyribonucleolytic activity in cytoplasmic ribosomes of rye (Secale cereale L) germs. Acta Biochim. Polon. **26**, 97–101 (1979)

[154] MÜLLER, J. J., H. WELFLE, G. DAMASCHUN, and H. BIELKA: A small-angle and wide-angle X-ray scattering study on the shape and secondary structure of native 5 S RNA from rat liver ribosomes. Biochim. Biophys. Acta **654**, 156–165 (1981)

Chapter VII: Biosynthesis of ribosomal components and biogenesis of ribosomal particles

1. Structural organization of ribosomal DNA

Rapid progress has been achieved in the characterization of rRNA genes. New and improved techniques have enabled the collection of sequence data of coding and noncoding regions of the rDNA. One of the aims of these studies is the identification of initiator and terminator sites of rRNA transcription and of recognition sites for processing enzymes.

The sequence of one kilobase pairs surrounding and upstream the transcription initiation site of cloned mouse rDNA [1, 2] was determined [2]. Few sequence similarities [2] were found in the region upstream the initiation site between mouse, *Xenopus laevis* [3] and *S. cerevisiae* [4, 5] rDNA. Further nucleotide sequences of the termination site [3, 6] and of the 18 S–28 S intergene region [7, 8] have been estimated in the rDNA of *Xenopus laevis*. About 900 nucleotides upstream the 5′ end of 18 S rRNA were found to be transcribed into 40 S pre-rRNA and excised during processing, whereas the termini of 28 S rRNA and 40 S pre-rRNA appear to coincide within ± 100 nucleotides [3]. The 5.8 S rDNA of *X. laevis* was located within the internal transcribed spacer and a secondary structure model of 5.8 S rRNA considering the flanking sequences has been proposed [8]. Analysis of the nucleotide sequence of *X. borealis* oocyte 5 S rDNA showed that a region of about 80 nucleotides is sufficient for promotor function [9].

Enriched rDNA of wheat and barley was cloned and stable full length ribosomal repeating units with length variation in the spacer region have been characterized [93].

Data on genetically determined changes of the quantity of rDNA in *Drosophila* were discussed recently [10].

The rRNA genes of *S. cerevisiae* consist of a set of repeating units of $5.8 \cdot 10^6$ dalton, and the coding regions of 18 S, 5.8 S, 25 S and 5 S rRNA were located on this repeating unit [11]. The polarity of the 35 S pre-rRNA is 5′ ... 18 S, 5.8 S, 25 S ... 3′ [11]. The 5.8 S rRNA coding region is separated from the 18 S and 25 S rRNA ones by transcribed spacer segments [11]. The rDNA of *S. cerevisiae* is characterized by a biphasic melting behaviour, which points to a considerable intramolecular heterogeneity in the base composition [12]. The sequence of a 419 base pair rDNA fragment of *S. cerevisiae* containing the transcription initiation site and its adjacent regions were determined [4]. The 5′ terminus of 35 S pre-rRNA of *S. cerevisiae* was mapped 670 nucleotides upstream the 17 S rRNA coding region [5]. The exact position of the 3′ end of the 26 S rRNA gene of yeast was mapped on the rDNA, and it has been found that 37 S pre-rRNA is 7 nucleotides longer at its 3′ end than the 26 S rRNA [13].

The coding regions of 17 S rRNA, 25 S rRNA and 35 S pre-rRNA and the promotor and terminator sites were located on the rRNA gene of *Tetrahymena pyriformis* by hybridization experiments [14] and by R-loop mapping [15]. A novel form of extra-chromosomal rDNA consisting of a 11 kb linear double-stranded DNA was found in conjugating *Tetrahymena* cells [16]. This rDNA contains a single rRNA gene and is an intermediate in rDNA amplification.

The sequences of a total spacer in *Euglena gracilis* chloroplast rDNA and flanking sequences coding for the 3′-terminal region of 16 S rRNA and for the 5′-terminal region of 23 S rRNA have been determined; two tRNA genes were found in the spacer region between the 16 S and 23 S rRNA gene [17, 18, 19]. The order of genes, which is repeated three times in *Euglena gracilis* chloroplast rDNA, is 16 S — tRNAs — 23 S — 5 S [17]. Only one set of rRNA genes was found in *Vicia faba* chloroplast DNA [20]. The rRNA region of mouse L-cell mitochondrial DNA was mapped by S 1 nuclease digestion of DNA × RNA hybrids [92].

Some of the DNA regions coding for the RNA of the cytoplasmic large ribosomal subunits of *Drosophila melanogaster* [21 to 25], *Drosophila hydei* [26], *Calliphora erythrocephala* [31], and *Tetrahymena* [15, 23, 27, 29, 30] are separated by an intervening sequence. Two insertions were found in the 26 S rRNA gene of *Physarum* [32]. Intervening sequences are also present in the mitochondrial large subunit rRNA gene of yeast [33, 34, 35], *Neurospora crassa* [36, 37], *Aspergillus nidulans* [38], and in the large subunit rRNA gene from chloroplasts of plants [39] and *Chlamydomonas reinhardii* [40]. These introns are transcribed and then removed during processing [27, 28, 34, 35, 40]. The large subunit rRNA is obtained by ligation [34, 35, 40]. Half of the rRNA genes on the X-chromosome of *D. melanogaster* are interrupted by intervening sequences, but these genes do not contribute significantly to the synthesis of 28 S rRNA [23, 24]. The length of insertion varies among the different 28 S rRNA genes of *D. melanogaster* [22, 25].

2. Synthesis and processing of ribosomal ribonucleic acids

a) *High molecular weight rRNA*

In yeast cells, processing of 37 S pre-rRNA to 18 S- and 29 S pre-rRNA, and conversion of the 29 S pre-rRNA to mature 26 S rRNA takes place in the nucleolus, whereas 18 S pre-rRNA is converted to mature 17 S rRNA in the cytoplasm [41]. A yeast mutant with a deficiency in cytoplasmic 49 S ribosomal subunits exhibits a slowed transport of 20 S pre-rRNA from the nucleus to the cytoplasm and a slowed cytoplasmic conversion into mature 17 S rRNA [94]. Possible recognition sites for processing enzymes were found in yeast pre-rRNA [42].

In *Tetrahymena* pre-rRNA, processing proceeds in the nucleus [43], which is characterized by a rapid excision of the 0.4 kb intervening sequence in the first step of the process. The accumulation of the excised RNA fragments can be observed in isolated nuclei [30] and nucleoli [27]. The excised intron is converted into a circular form by a post splicing process [28]. Correct termination of transcription of the rRNA gene in nucleoli isolated from *Tetrahymena* depends on a protein factor, which can be removed from the chromatin by salt-treatment [44].

The transcribed spacer sequence at the 3′ terminus of 45 S pre-rRNA from rat and mouse is gradually removed during the various processing steps [45]. In rat thymus cells partial degradation of newly synthesized 18 S rRNA was observed during processing [46]. Processing of preformed pre-rRNA in an in vitro-system from rat liver occurs in the absence of transcription but needs cytosolic proteins [47]. Also in vivo, upon inhibition of transcription by D-galactosamin, ribosomes were found to be formed and transported to the cytoplasm [48].

The first cleavage of pre-rRNA in *D. melanogaster* occurs at one of two different sites leading to two alternative pathways; 5 cleavage sites were mapped in the pre-rRNA [49].

The first example of simultaneous transcription and processing of rRNA in eukaryotes was observed by electron microscopic analysis of transcriptionally active rRNA genes of *Dictyostelium* [50]. Pre-rRNA of the 21 S rRNA from *S. cerevisiae* mitochondria contains sequences of mature rRNA, an intervening sequence, and an additional sequence at the 3′ end of the molecule; the intron is excised first, followed by the 3′ extra sequences [35].

In the mitochondrial 21 S rRNA gene of *S. cerevisiae* strains the intervening sequence has been localized 570 bp from the 3′ end of the gene; furthermore the splice point sequences were analyzed [34]. In mitochondria, a polyadenylated 20 S RNA species was identified as a precursor of 17 S rRNA [51].

Selective transcription of pre-rRNA in vitro was found to occur in isolated nucleoli of Novikoff hepatoma cells [52]. Electron microscopic studies of amplified rRNA genes in *X. laevis* oocyte nuclei revealed correct in vitro initiation and termination of transcription of pre-rRNA with about one-fifth of the in vivo frequency [53]. The synthesis of rRNA is not inhibited in rat liver under conditions of complete inhibition of protein synthesis [95].

Hypermethylated rRNA genes with low transcriptional activity were found in amplified chromosomal regions in a rat hepatoma cell line [54].

Regulation of the rRNA content was studied in human diploid fibroblasts [55] and during myoblast differentiation [56, 57]. The transcription of 45 S pre-rRNA was found to be the rate limiting step of ribosome formation [57]. Degradation of rRNA in growing and growth arrested *Tetrahymena thermophila* cells [58] seems to be rather independent of the cell cycle and to be regulated by growth conditions [58]. The ribosome content in the livers of mice protein-depleted for five days is reduced to about one-half of the normal value, and is restored to normal after one day of re-feeding [59].

b) 5 S rRNA

5 S rRNA from rat liver, rabbit reticulocytes, *S. cerevisiae* and *Euglena gracilis* contains 5′ triphosphates [60]. Correct in vivo transcription was observed with HeLa chromatin and endogenous RNA polymerase C [61], extracts of *X. laevis* oocyte nuclei and RNA polymerase III [62, 63], and yeast chromatin fractions and yeast RNA polymerase III [64]. Accurate transcription of cloned *Xenopus* 5 S rRNA genes depends on specific interactions of a protein factor [65] with a control region in the center of the 5 S rRNA gene [65—68]. This transcription factor is identical with the 5 S rRNA binding protein [69] found in the 5 S rRNA storage particles in amphibian oocytes [70, 71, 72].

5 S rRNA genes of several species have been characterized more recently. Rat liver 5 S rRNA genes were isolated [73]. The sequence of the *X. borealis* oocyte 5 S rRNA gene was determined and the flanking sequences of several related genes were compared [9]. Sequencing of 27 5 S rRNA gene copies of *D. melanogaster* revealed that in some of the gene copies a nucleotide was changed in the coding region, but the variant 5 S rRNAs coded by these gene copies have not yet been detected in vivo [74]. Furthermore, the nucleotide sequences of the 5 S rRNA gene and of the adjacent intergenic regions from *S. cerevisiae* were determined [75]. Different sequences were found in the upstream and downstream regions of the 5 S rRNA gene from *Torulopsis utilis* [90].

3. Biosynthesis of ribosomal proteins

Ribosomal proteins are coded by mRNAs of small size. mRNA from yeast coding ribosomal proteins sediments with 9 S [96]. The poly(A) rich mRNA from *X. laevis* oocytes coding for ribosomal proteins is enriched in a 10—16 S fraction [76]. Ribosomal protein genes in yeast seem to be clustered in a number of common transcriptional units [77]. cDNAs for mouse [78], *S. cerevisiae* [79, 80], and *D. melanogaster* [81] ribosomal proteins were cloned and isolated.

As analyzed for *D. melanogaster* embryos, the biosynthesis of ribosomal proteins begins in the pre-blastodermal stage, whereas rRNA synthesis and the formation of ribosomes are initiated in the blastodermal stage [82]. In cultured embryo fibroblasts, a rapid stimulation of ribosome synthesis was found by insulin [83]. Insulin seems to have a direct effect on the synthesis of ribosomal proteins but not on the synthesis of rRNA [84]. The total protein biosynthesis increases in mouse kidneys after contralateral nephrectomy during compensatory renal hypertrophy, but this process is not associated with a major increase in the synthetic rate of ribosomal proteins and rRNA [91]. In regenerating rat liver, however, the synthesis of ribosomal proteins is 2.5—3.6 times faster than in normal liver [85]. Newly synthesized ribosomal proteins were found to be transported into the nuclei of regenerating rat liver and to be degraded there when rRNA synthesis was inhibited by low doses of actinomycin D [86]. Infection of L-cells with Vesicular stomatitis virus inhibits the synthesis of ribosomal proteins [87].

The protein composition of ribosomal precursor particles in HeLa cells was analyzed and early-adding and late-adding groups of proteins were identified tentatively [88]. Binding of mitochondrial ribosomal proteins to a mitochondrial pre-rRNA containing a 2.3 kb intron was observed in *Neurospora* [37]; the stability of protein-RNA interaction is increased after removal of the intron [37]. Specific inhibition of mitochondrial ribosomal protein S 5 by chloramphenicol disturbed the assembly of the mitochondrial small ribosomal subunit [89].

References

[1] Grummt, I. and H. J. Gross: Structural organization of mouse rDNA. Comparison of transcribed and non-transcribed regions. Molec. Gen. Genetics **177**, 223—229 (1980)

[2] Urano, Y., R. Kominami, Y. Mishima, and M. Muramatsu: The nucleotide sequence of the putative transcription initiation site of a cloned ribosomal RNA gene of the mouse. Nucl. Acids Res. **8**, 6043—6058 (1980)

[3] Sollner-Webb, B. and R. H. Reeder: The nucleotide sequence of the initiation and termination sites for ribosomal RNA transcription in *X. laevis*. Cell **18**, 485—499 (1979)

[4] Bayev, A. A., O. I. Georgiev, A. A. Hadjiolov, M. B. Kermekchiev, N. Nikolaev, K. G. Skryabin, and V. M. Zakha-

RYEV: The structure of the yeast ribosomal genes. 2. The nucleotide sequence of the initiation site for ribosomal RNA transcription. Nucl. Acids Res. **8**, 4919 to 4926 (1980)

[5] KLEMENZ, R. and E. P. GEIDUSCHEK: The 5′ terminus of the precursor ribosomal RNA of *Saccharomyces cerevisiae*. Nucl. Acids Res. **8**, 2679—2689 (1980)

[6] MOSS, T., P. G. BOSELEY, and M. L. BIRNSTIEL: More ribosomal spacer sequences from *Xenopus laevis*. Nucl. Acids Res. **8**, 467—485 (1980)

[7] HALL, L. M. C. and B. E. H. MADEN: Nucleotide sequence through the 18 S to 28 S intergene region of a vertebrate ribosomal transcription unit. Nucl. Acids Res. **8**, 5993—6006 (1980)

[8] BOSELEY, P. G., A. TUYNS, and M. L. BIRNSTIEL: Mapping of the *Xenopus laevis* 5.8 S rDNA by restriction and DNA sequencing. Nucl. Acids Res. **5**, 1121—1137 (1978)

[9] KORN, L. J. and D. D. BROWN: Nucleotide sequence of *Xenopus borealis* oocyte 5 S DNA: comparison of sequences that flank several related eucaryotic genes. Cell **15**, 1145—1156 (1978)

[10] BASHKIROV, V. N.: Regulation of the number of ribosomal RNA genes in *Drosophila*. Genetika (russ.) **16**, 7—29 (1980)

[11] BELL, G. I., L. J. DEGENNARO, D. H. GELFAUD, R. J. BISHOP, P. VALENZUELA, and W. J. RUTTER: Ribosomal RNA genes of *Saccharomyces cerevisiae*. 1. Physical map of the repeating unit and location of the regions coding for 5 S, 5.8 S, 18 S, and 25 S ribosomal RNAs. J. Biol. Chem. **252**, 8118—8125 (1977)

[12] CRAMER, J. H. and R. H. ROWND: Denaturation mapping of the ribosomal DNA of *Saccharomyces cerevisiae*. Molec. Gen. Genetics **177**, 199—205 (1980)

[13] VELDMAN, G. M., J. KLOOTWIJK, P. de JONGE, R. J. LEER, and R. J. PLANTA: The transcription termination site of the ribosomal RNA operon in yeast. Nucl. Acids Res. **8**, 5179—5192 (1980)

[14] NILES, E. G. and R. K. JAIN: Physical map of the ribosomal ribonucleic acid gene from *Tetrahymena pyriformis*. Biochem. **20**, 905—909 (1981)

[15] CECH, T. R. and D. C. RIO: Localization of transcribed regions on extrachromosomal ribosomal RNA genes of *Tetrahymena thermophila* by R-loop mapping. Proc. Natl. Acad. Sci. U.S. **76**, 5051—5055 (1979)

[16] PAN, W.-C. and E. H. BLACKBURN: Single extrachromosomal ribosomal RNA gene copies are synthesized during amplification of the rDNA in *Tetrahymena*. Cell **23**, 459—466 (1981)

[17] OROZCO, E. M. jr., P. W. GRAY, and R. B. HALLICK: *Euglena gracilis* chloroplast ribosomal RNA transcription units. 1. The location of transfer RNA, 5 S, 16 S, and 23 S ribosomal RNA genes. J. Biol. Chem. **255**, 10991—10996 (1980)

[18] OROZCO, E. M. jr., K. E. RUSHLOW, J. R. DODD, and R. B. HALLICK: *Euglena gracilis* chloroplast ribosomal RNA transcription units. 2. Nucleotide sequence homology between the 16 S—23 S ribosomal RNA spacer and the 16 S ribosomal RNA leader regions. J. Biol. Chem. **235**, 10997—11003 (1980)

[19] GRAF, L., H. KÖSSEL, and E. STUTZ: Sequencing of 16 S—23 S spacer in a ribosomal RNA operon of *Euglena gracilis* chloroplast DNA reveals two tRNA genes. Nature **286**, 908—910 (1980)

[20] KOLLER, B. and H. DELIUS: *Vicia faba* chloroplast DNA has only one set of ribosomal RNA genes as shown by partial denaturation mapping and R-loop analysis. Molec. Gen. Genetics **178**, 261—269 (1980)

[21] GLOVER, D. M. and D. S. HOGNESS: A novel arrangement of the 18 S and 28 S sequences in a repeating unit of *Drosophila melanogaster* rDNA. Cell **10**, 167—176 (1977)

[22] INDIK, Z. K. and K. D. TARTOF: Long spacers among ribosomal genes of *Drosophila melanogaster*. Nature **284**, 477—479 (1980)

[23] JOLLY, D. J. and C. A. THOMAS, jr.: Nuclear RNA transcripts from *Drosophila melanogaster* ribosomal RNA genes containing introns. Nucl. Acids Res. **8**, 67—84 (1980)

[24] LONG, E. O. and I. B. DAWID: Expression of ribosomal DNA insertions in *Drosophila melanogaster*. Cell **18**, 1185—1196 (1979)

[25] WELLAUER, P. K. and I. B. DAWID: The structural organization of ribosomal DNA

in *Drosophila melanogaster*. Cell **10**, 193 to 212 (1977)

[26] Renkawitz-Pohl, R., K. H. Glätzer, and W. Kunz: Characterization of cloned ribosomal DNA from *Drosophila hydei*. Nucl. Acids Res. **8**, 4593–4611 (1980)

[27] Carin, M., B. F. Jensen, K. D. Jentsch, J. C. Leer, O. F. Nielsen, and O. Westergaard: In vitro splicing of the ribosomal RNA precursor in isolated nucleoli from *Tetrahymena*. Nucl. Acids Res. **8**, 5551 to 5566 (1980)

[28] Grabowski, P. J., A. J. Zaug, and T. R. Cech: The intervening sequence of the ribosomal RNA precursor is converted to a circular RNA in isolated nuclei of *Tetrahymena*. Cell **23**, 467–476 (1981)

[29] Wild, M. A. and J. G. Gall: An intervening sequence in the gene coding for 25 S ribosomal RNA of *Tetrahymena pigmentosa*. Cell **16**, 565–573 (1979)

[30] Zaug, A. J. and T. R. Cech: In vitro splicing of the ribosomal RNA precursor in nuclei of *Tetrahymena*. Cell **19**, 331–338 (1980)

[31] Beckingham, K. and R. White: The ribosomal DNA of *Calliphora erythrocephala*; an analysis of hybrid plasmids containing ribosomal DNA. J. Mol. Biol. **137**, 349–373 (1980)

[32] Gubler, U., T. Wyler, and R. Braun: The gene for the 26 S rRNA in Physarum contains two insertions. FEBS-Lett. **100**, 347–350 (1979)

[33] Bos, J. L., G. Heyting, P. Borst, A. C. Arnberg, and E. F. J. van Bruggen: An insert in the single gene for the large ribosomal RNA in yeast mitochondria. Nature **275**, 336–337 (1978)

[34] Bos, J. L., K. A. Osinga, G. van der Horst, N. B. Hecht, H. F. Tabak, G.-J. B. van Ommen, and P. Borst: Splice point sequence and transcripts of the intervening sequence in the mitochondrial 21 S ribosomal RNA gene of yeast. Cell **20**, 207–214 (1980)

[35] Merten, S., R. M. Synenki, J. Locker, T. Christianson, and M. Rabinowitz: Processing of precursors of 21 S ribosomal RNA from yeast mitochondria. Proc. Natl. Acad. Sci. U.S. **77**, 1417–1421 (1980)

[36] Heckman, J. E. and U. L. RajBhandary: Organization of tRNA and rRNA genes in *N. crassa* mitochondria: intervening sequence in the large rRNA gene and strand distribution of the RNA genes. Cell **17**, 583–595 (1979)

[37] Lapolla, R. J. and A. M. Lambowitz: Binding of mitochondrial ribosomal precursor RNA containing a 2.3-kilobase intron. J. Biol. Chem. **254**, 11746–11750 (1979)

[38] Lazarus, C. M., H. Lünsdorf, U. Hahn, P. P. Stephen, and H. Küntzel: Physical map of *Aspergillus nidulans* mitochondrial genes coding for ribosomal RNA: an intervening sequence in the large rRNA cistron. Molec. Gen. Genetics **177**, 389 to 397 (1980)

[39] MacKay, R. M.: The origin of plant chloroplast 4.5 S ribosomal RNA. FEBS-Lett. **123**, 17–19 (1981)

[40] Rochaix, J. D. and P. Malnoe: Anatomy of the chloroplast ribosomal DNA of *Chlamydomonas reinhardii*. Cell **15**, 661 to 670 (1978)

[41] Trapman, J. and R. J. Planta: Maturation of ribosomes in yeast. 1. Kinetic analysis by labelling of high molecular weight rRNA species. Biochim. Biophys. Acta **442**, 265–274 (1976)

[42] Veldman, G. M., R. C. Brand, J. Klootwijk, and R. J. Planta: Some characteristics of processing sites in ribosomal precursor RNA of yeast. Nucl. Acids Res. **8**, 2907–2920 (1980)

[43] Rodrigues-Pousada, C., M. L. Cyrne, and D. Hayes: Characterization of preribosomal ribonucleoprotein particles from *Tetrahymena pyriformis*. Europ. J. Biochem. **102**, 389–397 (1979)

[44] Leer, J. C., D. Tiryaki, and O. Westergaard: Termination of transcription in nucleoli from *Tetrahymena*. Proc. Natl. Acad. Sci. U.S. **76**, 5563–5566 (1979)

[45] Hamada, H., R. Kominami, and M. Muramatsu: 3′-Terminal processing of ribosomal RNA precursors in mammalian cells. Nucl. Acids Res. **8**, 889–903 (1980)

[46] Evtuschenko, V. I. and K. P. Hanson: Investigation of the maturation of ribosomal RNAs in the rat thymus cells. Biokhimiya **45**, 173–179 (1980)

[47] Schumm, D. E., M. A. Niemann, T. Palayoor, and T. E. Webb: In vivo equivalence of a cell-free system from rat liver for ribosomal RNA processing and transport. J. Biol. Chem. **254**, 12126–12130 (1979)

[48] Gajdardjieva, K. C., M. D. Dabeva, and A. A. Hadjiolov: Maturation and nucleocytoplasmic transport of rat liver ribosomal RNA upon D-galactosamine inhibition of transcription. Europ. J. Biochem. **104**, 451–458 (1980)

[49] Long, E. O. and I. B. Dawid: Alternative pathway in the processing of ribosomal RNA precursor in *Drosophila melanogaster*. J. Mol. Biol. **138**, 873–878 (1980)

[50] Grainer, R. M. and N. Maizels: Dictyostelium ribosomal RNA is processed during transcription. Cell. **20**, 619–623 (1980)

[51] Cleaves, G. R., T. Jones, and D. T. Dubin: Properties of a discrete high molecular weight poly-A-containing mitochondrial RNA. Arch. Biochem. Biophys. **175**, 303 to 311 (1976)

[52] Ballal, N. R., B. Samal, Y. C. Choi, and H. Busch: Studies on the specificity of preribosomal RNA transcription in nucleoli after selective deproteinization. Nucl. Acids Res. **7**, 919–924 (1979)

[53] Mc Knight, S. L., R. A. Hipskind, and R. Reeder: Ultrastructural analysis of ribosomal gene transcription in vitro. J. Biol. Chem. **255**, 7907–7911 (1980)

[54] Tantravahi, U., R. V. Guntaka, B. F. Erlanger, and O. J. Miller: Amplified ribosomal RNA genes in a rat hepatoma cell line are enriched in 5′-methylcytosine. Proc. Natl. Acad. Sci. U.S. **78**, 489–493 (1981)

[55] Wolf, S., M. Sameshima, S. A. Liebhaber, and D. Schlessinger: Regulation of ribosomal ribonucleic acid levels in growing ^{3}H-arrested and crisis-phase WI-38 human diploid fibroblasts. Biochem. **19**, 3484–3490 (1980)

[56] Krauter, K. S., R. Soeiro, and B. Nadal-Ginard: Transcriptional regulation of ribosomal RNA accumulation during L_6E_9 myoblast differentiation. J. Mol. Biol. **134**, 727–741 (1979)

[57] Krauter, K. S. and R. Soeiro: Uncoordinate regulation of ribosomal RNA and ribosomal protein synthesis during L_6E_9 myoblast differentiation. J. Mol. Biol. **142**, 145–159 (1980)

[58] Sutton, C. A. and R. L. Hallberg: Ribosomal biosynthesis in *Tetrahymena thermophila*. 3. Regulation of ribosomal RNA degradation in growing and growth arrested cells. J. Cell. Physiol. **101**, 349–358 (1979)

[59] Conde, R. D. and M. T. Franze-Fernandez: Increased transcription and decreased degradation control the recovery of liver ribosomes after a period of protein starvation. Biochem. J. **192**, 935–940 (1980)

[60] Soave, C., R. Nucca, E. Sala, A. Viotti, and E. Galante: 5-S RNA: investigation of the different extent of phosphorylation at 5′-terminus. Europ. J. Biochem. **32**, 392 to 400 (1974)

[61] Krause, B. and K. H. Seifart: Transcription of ribosomal 5 S RNA from HeLa chromatin by homologous and heterologous eukaryotic RNA polymerases. Biochem. Internat. **2**, 201–210 (1981)

[62] Korn, L. J., E. H. Birkenmeier, and D. D. Brown: Transcription initiation of *Xenopus* 5 S ribosomal RNA genes in vitro. Nucl. Acids Res. **7**, 947–968 (1979)

[63] Ng, S. Y., C. S. Parker, and R. G. Roeder: Transcription of cloned *Xenopus* 5 S RNA genes by *X. laevis* RNA polymerase III in reconstituted systems. Proc. Natl. Acad. Sci. U.S. **76**, 136–140 (1979)

[64] Tekamp, P. A., R. L. Garcea, and W. J. Rutter: Transcription and in vitro processing of yeast 5 S rRNA. J. Biol. Chem. **255**, 9501–9506 (1980)

[65] Engelke, D. R., S. Y. Ng, B. S. Shastry, and R. G. Roeder: Specific interaction of a purified transcription factor with an internal control region of 5 S RNA genes. Cell **19**, 717–728 (1980)

[66] Bogenhagen, D. F., S. Sakonju, and D. D. Brown: A control region in the center of the 5 S RNA gene directs specific initiation of transcription: 2. The 3′ border of the region. Cell **19**, 27–35 (1980)

[67] Sakonju, S., D. F. Bogenhagen, and D. D. Brown: A control region in the center of the 5 S RNA gene directs specific initiation of transcription: 1. The 5′ border of the region. Cell **19**, 13–25 (1980)

[68] Honda, B. M. and R. G. Roeder: Association of a 5 S gene transcription factor with 5 S RNA and altered levels of the factor during cell differentiation. Cell **22**, 119–126 (1980)

[69] Pelham, H. R. B. and D. D. Brown: A specific transcription factor that can bind either the 5 S RNA gene or 5 S RNA. Proc. Natl. Acad. Sci. U.S. **77**, 4170–4174 (1980)

[70] Picard, B., M. le Maire, M. Wegnez,

and H. Denis: Biochemical research on oogenesis. Composition of the 42-S storage particles of *Xenopus laevis* oocytes. Europ. J. Biochem. **109**, 359—368 (1980)

[71] Picard, B. and M. Wegnez: Isolation of a 7 S particle from *Xenopus laevis* oocytes: a 5 S RNA-protein complex. Proc. Natl. Acad. Sci. U.S. **76**, 241—245 (1979)

[72] Kloetzel, P.-M., W. Whitfield, and J. Sommerville: Analysis and reconstruction of an RNP particle which stores 5 S RNA and tRNA in amphibian oocytes. Nucl. Acids Res. **9**, 605—621 (1981)

[73] Arsenyan, S. G., T. A. Avdonina, A. Laving, M. Saarma, and L. L. Kisselev: Isolation of rat liver 5 S RNA genes. Gene **11**, 97—108 (1980)

[74] Tschudi, C. and V. Pirrotta: Sequence and heterogeneity in the 5 S RNA gene cluster of *Drosophila melanogaster*. Nucl. Acids Res. **8**, 441—451 (1980)

[75] Valenzuela, P., G. I. Bell, A. Venegos, E. T. Sewell, F. R. Masiarz, L. J. Degennaro, F. Weinberg, and W. J. Rutter: Ribosomal RNA genes of *Saccharomyces cerevisiae*. 2. Physical map and nucleotide sequence of the 5 S ribosomal RNA gene and adjacent intergenic regions. J. Biol. Chem. **252**, 8126—8135 (1977)

[76] Pierandrei-Amaldi, P. and E. Beccari: Messenger RNA for ribosomal proteins in *Xenopus laevis* oocytes. Europ. J. Biochem. **106**, 603—611 (1980)

[77] Mager, W. H., J. Retel, R. J. Planta, G. H. P. Bollen, V. C. H. F. de Regt, and H. Hoving: Transcriptional units for ribosomal proteins in yeast. Europ. J. Biochem. **78**, 575—583 (1977)

[78] Meyuhas, O. and R. P. Perry: Construction and identification of cDNA clones for mouse ribosomal proteins — Application for the study of r-protein gene expression. Gene **10**, 113—129 (1980)

[79] Fried, H. M. and J. R. Warner: Cloning of yeast gene for trichodermin resistance and ribosomal protein L 3. Proc. Natl. Acad. Sci. U.S. **78**, 238—242 (1981)

[80] Woolford, J. L., jr., Hereford, L. M., and M. Rosbash: Isolation of cloned DNA sequences containing ribosomal protein genes from *Saccharomyces cerevisiae*. Cell **18**, 1247—1259 (1979)

[81] Vaslet, C. A., P. O'Connell, M. Izquierdo, and M. Rosbash: Isolation and mapping of cloned ribosomal protein gene of *Drosophila melanogaster*. Nature **285**, 674—676 (1980)

[82] Santon, J. B. and M. Pellegrini: Expression of ribosomal proteins during *Drosophila* early development. Proc. Natl. Acad Sci. U.S. **77**, 5649—5653 (1980)

[83] DePhilip, R. M., D. E. Chadwick, R. A. Ignotz, W. E. Lynch, and I. Liebermann: Rapid stimulation by insulin of ribosome synthesis in cultured chick embryo fibroblasts. Biochem. **18**, 4812—4817 (1979)

[84] DePhilip, R. M., W. A. Rudert, and I. Liebermann: Preferential stimulation of ribosomal protein synthesis by insulin and in the absence of ribosomal and messenger ribonucleic acid formation. Biochem. **19**, 1662—1669 (1980)

[85] Nabeshima, Y. and K. Ogata: Stimulation of the synthesis of ribosomal proteins in regenerating rat liver with special reference to the increase in the amounts of effective mRNAs for ribosomal proteins. Europ. J. Biochem. **107**, 323—329 (1980)

[86] Tsurugi, K. and K. Ogata: Degradation of newly synthesized ribosomal proteins and histones in regenerating rat liver with and without treatment with a low dose of Actinomycin D. Europ. J. Biochem. **101**, 205—213 (1979)

[87] Jayne, M., F.-S. Wu, and J. M. Lucas-Lenard: Inhibition of synthesis of ribosomal proteins and of ribosome assembly after infection of L cells with Vesicular Stomatitis Virus. Biochim. Biophys. Acta **606**, 1—12 (1980)

[88] Lastick, S. M.: The assembly of ribosomes in HeLa cell nucleoli. Europ. J. Biochem. **113**, 175—182 (1980)

[89] Lambowitz, A. M., R. J. Lapolla, and R. A. Collius: Mitochondrial ribosome assembly in *Neurospora*. Twodimensional gelelectrophoretic analysis of mitochondrial ribosomal proteins. J. Cell. Biol. **82**, 17—31 (1979)

[90] Tabata, S.: Structure of the S 5 ribosomal RNA gene and its adjacent regions in *Torulopsis utilis*. Europ. J. Biochem. **110**, 107—114 (1980)

[91] Melvin, W. T., A. Kumar, and R. A. Malt: Synthesis and conservation of ribosomal proteins during compensatory renal

hypertrophy. Biochem. J. **188**, 229–235 (1980)

[92] NAGLEY, P. and D. A. CLAYTON: Transcriptional mapping of the ribosomal RNA region of mouse L-cell mitochondrial DNA. Nucl. Acids Res. **8**, 2947–2965 (1980)

[93] GERLACH, W. L. and J. R. BEDBROOK: Cloning and characterization of ribosomal RNA genes from wheat and barley. Nucl. Acids Res. **7**, 1869–1885 (1979)

[94] CARTER, C. J. and M. CANNON: Maturation of ribosomal precursor RNA in *Saccharomyces cerevisiae*. A mutant with a defect in both the transport and terminal processing of the 20 S species. J. Mol. Biol. **143**, 179–199 (1980)

[95] KARAGYOZOV, L. K., B. B. STOYANOVA, and A. A. HADJIOLOV: Effect of cycloheximide on the in vivo and in vitro synthesis of ribosomal RNA in rat liver. Biochim. Biophys. Acta **607**, 295–303 (1980)

[96] BOLLEN, G. H. P. M., W. H. MAGER, L. W. JENNESKENS, and R. J. PLANTA: Small-size mRNAs code for ribosomal proteins in yeast. Europ. J. Biochem. **105**, 75–80 (1980)

Chapter IX: Interactions, topography and function of ribosomal components

Proteins split from rat liver ribosomes by 50% ethanol or treatment with ethanol and KCl are able to stimulate poly(U)-dependent poly(Phe) synthesis of core particles [1]. Acidic proteins extracted from ribosomes of *Saccharomyces cerevisiae* by 0.4 M NH_4Cl and ethanol were shown to increase eEF-G dependent GTP hydrolysis [2, 3]. These proteins (L 44/45) reveal no immunological cross-reactivity with proteins of other eukaryotes or those of *E. coli*. The effects of KCl concentrations on the activity of 40 S ribosomal subunits from canine pancreas have also been investigated [4]. Large ribosomal subunits are stabilized against protein splitting with LiCl by bound tRNA [5].

UV-mediated cross-linking indicates that protein L 5 of rat liver ribosomes, which can be isolated in form of the so-called 7 S complex together with 5 S RNA from 60 S subunits, is in close contact to 5 S RNA also within the subunit [6, 7]. Protein L 5 binds tightly to immobilized 5 S RNA and the resulting complex interacts also with 5.8 S RNA [8]. The complex which has been characterized by hydrodynamic methods [9] can be degraded by different proteases resulting in the removal of an N-terminal 10000 molecular weight peptide of L 5 (H. WELFLE, personal communication). The GTP- and ATP-hydrolyzing activities of the rat liver ribosome 5 S RNA-L 5 protein complex are stimulated by pig liver elongation factor 2 plus AA-tRNA [10].

Besides ribosomal protein L 5, also proteins L 6, L 7, and L 19 can be bound to immobilized 5 S RNA of rat liver [11], and proteins L 6, L 7, L 19, L 35a, and S 9 could be attached to 5 S RNA of *E. coli* [12]. 5.8 S RNA of rat liver ribosomes binds three of these proteins (L 6, L 19, and S 9) and additionally L 8 and S 13 [13, 14].

Immobilized elongator-tRNA binds proteins L 6, L 35a, and S 15 of the total 80 S protein mixture, whereas initiator-tRNA binds L 6 and L 35a only [15]. By a similar technique it was demonstrated that 40 S ribosomal subunits interact with 5.8 S RNA and with complexes of 5 S RNA and ribosomal proteins [16].

The analysis of the three-dimensional arrangement of antigenic sites of ribosomal proteins on the surface of the small ribosomal subunit of rat liver by means of immune electron microscopy has been extended [17, 18] (see also Fig. 27 on page 172). These data are in good agreement with results of cross-linking experiments; using dimethyl suberimidate [19] and dimethyl 5.6-dihydroxy-4.7-dioxo-3.8-diazadecanbisimidate (Ref. [113] in Chapter IX and unpublished data of B. GROSS and P. WESTERMANN), the following protein dimers have been identified for the 40 S ribosomal subunit of rat liver:

S 2—S 3	S 4 —S 15a	S 11—S 15
S 2—S 4	S 4 —S 23/24	S 11—S 25
S 2—S 7	S 4 —S 28	S 13/16—S 17
S 2—S 15a	S 5 —S 7	S 13/16—S 25
S 3—S 3a	S 5 —S 13/16	S 13/16—S 27
S 3—S 3b	S 5 —S 20	S 14—S 15
S 3/3a—S 11	S 5 —S 25	S 14—S 17
S 3—S 8	S 5 —S 29	S 14—S 20
S 3—S 17	S 6 —S 8	S 15—S 15a
S 3—S 20	S 6 —S 23/24	S 15a—S 17
S 3a—S 13	S 7 —S 10	S 15a—S 20
S 3a—S 3b	S 7 —S 14	S 16—S 19
S 3b—S 4	S 7 —S 18	S 19—S 27
S 3b—S 11	S 7 —S 20	S 20—S 23/24
S 4—S 6	S 10—S 17	S 23—S 24
S 4—S 10	S 10—S 20	S 27—S 28
S 4—S 14		

Proteins which could be cross-linked within 60 S ribosomal subunits by oxidative formation of disulfide bridges (Ref. [65] in Chapter IX) or by treatment with dimethyl suberimidate and dimethyl dithiobispropionimidate [31] are:

L 3—L 5	L 7/7a—L 36
L 3—L 7/7a	L 8—L 35
L 3—L 8	L 8—L 36
L 4—L 6	L 13/13a—L 35
L 4—L 26	L 14—L 15
L 4—L 30	L 15—L 19
L 6—L 29	L 18a—L 27/27a
L 7/7a—L 15	L 27/27a—L 30.
L 7/7a—L 21/23/23a	

In a very recent study the topography of proteins of 40 S ribosomal subunits from rabbit reticulocytes using 2-iminothiolane as cross-linking reagent has been analyzed. Altogether 36 protein dimers were identified and a model of the 40 S subunit derived from these cross-linking data has been proposed [42].

The investigation of the quaternary initiation complex [eIF-2 × GTP × Met-tRNA$_f$ × 40 S ribosomal subunit] by similar techniques shows that Met-tRNA$_f$ can be cross-linked to eIF-2β and to ribosomal proteins S 3a and S 6 [20, 21], and that eIF-2α and eIF-2γ can be attached to ribosomal proteins S 3, S 3a, S 6, 13/16, and S 15 (Ref. [114] in Chapter IX) as well as to 18 S rRNA [22]. A model of the eIF-2 binding site on the 40 S ribosomal subunit which summarizes these results has recently been proposed [23].

By using diepoxibutane, the 18 S RNA could be cross-linked mainly to proteins S 3a, S 6, S 7, and to minor amounts also to proteins S 3, S 8, S 11, S 23/24, and S 25 within the small subunit of rat liver ribosomes [20]. When methyl-p-azidobenzoylaminoacetimidate was used [20], proteins S 3a, S 6, S 8, S 11, and in smaller amounts also S 2, S 3, S 4, S 9, S 16/18, S 23/24, S 25, and S 26 are covalently linked to 18 S rRNA (see also [24]).

The messenger binding site at the small ribosomal subunit has been investigated by UV-mediated cross-linking of 40 S subunit — poly(U) complexes, resulting in the attachment of poly(U) to 18 S rRNA [25] and to ribosomal proteins [26, 27]. Using a poly(U) analogue with an alkylating group at the 3′ end, ribosomal proteins S 3/3a were selectively labeled [28]. The 3′-terminal region of 18 S rRNA is highly conserved (cf. Chapter VI) and sequences complementary to 5 S RNA were found which allow interactions between both rRNAs within the 80 S ribosome [29, 30]. The peptides of the contact regions of proteins L 6 and L 29, which can be covalently linked by a disulfide bridge, were studied by degradation of the binary complex with different proteases [32, 33].

The P site of the peptidyltransferase center of *Drosophila* ribosomes could be labeled by mercurated C-A-C-C-A-[^{3}H]-Leu-Ac as an affinity reagent [34]. The activity of the peptidyltransferase of rabbit ribosomes was not diminished by nuclease or by trypsin degradation of the ribosomes, although the latter reduced protein synthesis and subunit reassociation activities considerably [35].

The results of studies on the binding of concanavalin A to ribosomal particles and ribosomal proteins are still contradictory [36—38].

The possible role of the 3′-end of the 18 S rRNA of the small subunit of eukaryotic ribosomes to interact with the 5′-noncoding region of the mRNA is still unsolved (see also Chapter XI and the corresponding Addendum). The only hints for such an event occurring during initiation of protein synthesis in eukaryotes, comparable to the Shine-Dalgarno concept as established for prokaryotes, are in vitro base-pairing interactions between both RNAs [39]. The comparison of corresponding sequences has revealed [40] that the number of potential binding sites for 18 S rRNA in 5′- or 3′-noncoding sequences of mRNAs is higher as compared to a random RNA sequence chain, but this finding can be explained also by clustering of purines and pyrimidines, common to noncoding sequences.

Nevertheless, a possible role of interactions of the 3′-region of the 18 S rRNA and the 5′-region of the mRNA — at least a more indirect functional one, may be mediated by proteins — is supported by the finding that the 3′-end of 18 S rRNA and 5′-sequences of the mRNA can be cross-linked within 40 S- and 80 S initiation complexes by psoralene [41]; thus these regions of both RNA species should be located in close proximity within this functional domain.

References

[1] MacConnel, W. P. and N. O. Kaplan: The role of ethanol extractable proteins from the 80 S rat liver ribosome. Biochem. Biophys. Res. Commun. **92**, 46—52 (1980)

[2] Sanchez-Madrid, F., R. Reyes, P. Conde, and J. P. G. Ballesta: Acidic ribosomal proteins from eukaryotic cells. Effect on ribosomal function. Europ. J. Biochem. **98**, 409—416 (1979)

[3] Sanchez-Madrid, F., P. Conde, D. Vazquez, and J. P. G. Ballesta: Acidic proteins from *Saccharomyces cerevisiae* ribosomes. Biochem. Biophys. Res. Commun. **87**, 281—291 (1979)

[4] Nesset, C. C. and S. R. Dickman: Effects of potassium chloride concentration on protein content and polyphenylalanine synthesizing capability of 40 S ribosomal subunits from canine pancreas. Biochemistry **19**, 2731—2737 (1980)

[5] Reboud, A.-M., S. Dubost, M. Buisson, and J.-P. Reboud: tRNA binding stabilizes rat liver 60 S ribosomal subunits during treatment with LiCl. J. Biol. Chem. **255**, 6954—6961 (1980)

[6] Terao, K., T. Uchiumi, and K. Ogata: Cross-linking of L 5 protein to 5 S RNA in rat liver 60 S subunits by ultraviolet irradiation. Biochim. Biophys. Acta **609**, 306 to 312 (1980)

[7] Marion, M.-J. and J.-P. Reboud: An argument for the existence of a natural complex between protein L 5 and 5 S RNA in rat liver 60 S ribosomal subunits. Biochim. Biophys. Acta **652**, 193—203 (1981)

[8] Metspalu, A., I. Toots, M. Saarma, and R. Villems: The ternary complex consisting of rat liver ribosomal 5 S RNA, 5.8 S RNA and protein L 5. FEBS-Lett. **119**, 81—84 (1980)

[9] Behlke, J., H. Welfle, I. Wendel, and H. Bielka: Physicochemical studies of the 7 S complex of rat liver ribosomes and its components. Acta Biol. Med. Germ. **39**, 33—40 (1980)

[10] Ogata, K., K. Terao, and T. Uchiumi: Stimulation by aminoacyl-tRNA of the GTPase and ATPase activities of rat liver 5 S RNA protein particles in the presence of EF-2. J. Biochem. **87**, 517—524 (1980)

[11] Ulbrich, N., K. Todokoro, E. J. Ackerman, and I. G. Wool: Characterization of the binding of rat liver ribosomal proteins L 6, L 7, and L 19 to 5 S ribosomal ribonucleic acid. J. Biol. Chem. **255**, 7712 to 7715 (1980)

[12] Ulbrich, N., A. Lin, K. Todokoro, and I. G. Wool: Identification by affinity

chromatography of the rat liver ribosomal proteins that bind to *Escherichia coli* 5 S ribosomal ribonucleic acid. J. Biol. Chem. **255**, 797—801 (1980)
[13] ULBRICH, N., A. LIN, and I. G. WOOL: Identification by affinity chromatography of the eukaryotic ribosomal proteins that bind to 5.8 S ribosomal ribonucleic acid. J. Biol. Chem. **254**, 8641—8645 (1979)
[14] TOOTS, I., A. METSPALU, A. LIND, M. SAARMA, and R. VILLEMS: Immobilized eukaryotic 5.8 S RNA binds *E. coli* and rat liver ribosomal proteins. FEBS-Lett. **104**, 193—196 (1979)
[15] ULBRICH, N., I. G. WOOL, E. ACKERMAN, and P. B. SIGLER: The identification by affinity chromatography of the rat liver ribosomal proteins that bind to elongator and initiator transfer ribonucleic acids. J. Biol. Chem. **255**, 7010—7016 (1980)
[16] VILLEMS, R., M. SAARMA, A. METSPALU, and I. TOOTS: New aspects of the eukaryotic ribosomal subunit interactions. FEBS-Lett. **107**, 66—68 (1979)
[17] BOMMER, U.-A., F. NOLL, G. LUTSCH, and H. BIELKA: Immunochemical detection of proteins in the small subunit of rat liver ribosomes involved in binding of the ternary initiation complex. FEBS-Lett. **111**, 171—174 (1980)
[18] LUTSCH, G., F. NOLL, H. THEISE, G. ENZMANN, and H. BIELKA: Three-dimensional mapping of proteins of rat liver ribosomes by immune electron microscopy. Studia Biophys. **79**, 125—126 (1980)
[19] TERAO, K., T. UCHIUMI, Y. KOBAYASHI, and K. OGATA: Identification of neighbouring protein pairs in the rat liver 40 S ribosomal subunits cross-linked with dimethyl suberimidate. Biochim. Biophys. Acta **621**, 72—82 (1980)
[20] WESTERMANN, P., O. NYGARD, and H. BIELKA: Cross-linking of Met-$tRNA_f$ to eIF-2β and to the ribosomal proteins S 3a and S 6 within the eukaryotic initiation complex, eIF-2 · GMPPCP · Met-$tRNA_f$ · small ribosomal subunit. Nucl. Acids Res. **9**, 2387—2396 (1981)
[21] NYGARD, O., P. WESTERMANN, and T. HULTIN: Met-$tRNA_f^{Met}$ is located in close proximity to the β subunit of eIF-2 in the eukaryotic initiation complex, eIF-2 · Met-$tRNA_f^{Met}$ · GDPCP. FEBS-Lett. **113**, 125 to 128 (1980)
[22] WESTERMANN, P., O. NYGARD, and H. BIELKA: The α and γ subunits of initiation factor eIF-2 can be cross-linked to 18 S ribosomal RNA within the quaternary initiation complex, eIF-2 · Met-$tRNA_f$ · GDPCP · small ribosomal subunit. Nucl. Acids Res. **8**, 3065—3071 (1980)
[23] NYGARD, O., P. WESTERMANN, and T. HULTIN: Identification of neighbouring components in the quaternary initiation complex, eIF-2 · GTP · Met-$tRNA_f$ · small ribosomal subunit. Acta Chem. Scand. **B 35**, 57—59 (1981)
[24] REBOUD, A.-M., M. BUISSON, S. DUBOST, and J.-P. REBOUD: Photoinduced protein-RNA cross-linking in mammalian 80 S ribosomes. Europ. J. Biochem. **106**, 33—40 (1980)
[25] REBOUD, A.-M., S. DUBOST, M. BUISSON, and J.-P. REBOUD: Covalent attachment of poly(U) template to 40 S mammalian ribosomal subunits. Biochem. Biophys. Res. Commun. **93**, 974—978 (1980)
[26] TERAO, K. and K. OGATA: Proteins of small subunits of rat liver ribosomes that interact with poly(U). I. Effects of preincubation of poly(U) with 40 S subunits on the interactions of 40 S subunits proteins with aurintricarboxylic acid and with N.N′-p-phenylenedimaleimide. J. Biochem. **86**, 597—603 (1979)
[27] TERAO, K. and K. OGATA: Proteins of the small subunit of rat liver ribosomes that interact with poly(U). II. Cross-links between poly(U) and ribosomal proteins in 40 S subunits induced by UV irradiation. J. Biochem. **86**, 605—617 (1979)
[28] STAHL, J. and N. D. KOBETS: Affinity labeling of proteins at the mRNA binding site of rat liver ribosomes by an analogue of octauridylate containing an alkylating group attached to the 3′-end. FEBS-Lett. **123**, 269—272 (1981)
[29] AZAD, A. A.: Intermolecular base-paired interactions between complementary sequences present near the 3′ ends of 5 S rRNA and 18 S (16 S) rRNA might be involved in the reversible association of ribosomal subunits. Nucl. Acids Res. **7**, 1913 to 1929 (1979)
[30] AZAD, A. A. and N. J. DEACON: The 3′-terminal primary structure of five eukaryotic 18.S rRNAs determined by the direct chemical method of sequencing. The highly

conserved sequences include an invariant region complementary to eukaryotic 5 S rRNA. Nucl. Acids Res. 8, 4365–4376 (1980)
[31] UCHIUMI, T., K. TERAO, and K. OGATA: Identification of neighbouring protein pairs in rat liver 60 S ribosomal subunits cross-linked with dimethyl suberimidate or dimethyl 3.3′-dithiobispropionimidate. J. Biochem. 88, 1033–1044 (1980)
[32] NIKA, H. and T. HULTIN: Disulfide interaction in situ between two neighbouring proteins in mammalian 60 S ribosomal subunits. Isolation of the contact region of the larger protein. Biochim. Biophys. Acta **579**, 10–19 (1979)
[33] NIKA, H. and T. HULTIN: Disulfide interaction in situ between the neighbouring proteins L 6 and L 29 in mammalian ribosomes. Structural shielding of proteins and contact region. Biochim. Biophys. Acta **624**, 142–152 (1980)
[34] FABIJANSKI, St. and M. PELLEGRINI: Affinity labeling of a reactive sulfhydryl residue at the peptidyl transferase P site in *Drosophila* ribosomes. Biochemistry **18**, 5674 to 5679 (1979)
[35] COX, R. A. and S. KOTECHA: Resistance of the peptidyltransferase centre of rabbit ribosomes to attack by nucleases and proteinases. Biochem. J. **190**, 199–214 (1980)
[36] MICHEL, S. and A. J. COZZONE: On the binding of concanavalin A to bacterial and mammalian ribosomes. Int. J. Biochem. **12**, 485–488 (1980)
[37] YOSHIDA, K.: The presence of ribosomal glycoproteins: agglutination of free and membran-bound ribosomes from wheat germ by concanavalin A. J. Biochem. **83**, 1609–1614 (1978)
[38] MICHEL, S., J. J. MADJAR, J.-P. REBOUD, and A. J. COZZONE: Interaction of concanavalin A with individual proteins from bacterial and mammalian ribosomes. Int. J. Biochem. **13**, 141–145 (1981)
[39] AZAD, A. A. and N. J. DEACON: Base-paired interaction in vitro between hen globin mRNA and eukaryotic rRNAs. Biochem. Biophys. Res. Commun. **86**, 568 to 576 (1979)
[40] WACHTER, R. DE: Do eukaryotic mRNA 5′ noncoding sequences base-pair with the 18 S ribosomal RNA 3′ terminus? Nucl. Acids. Res. **7**, 2045–2054 (1979)
[41] NAKASHIMA, K. E., E. DARZYNKIEWICZ, and A. J. SHATKIN: Proximity of mRNA 5′ region and 18 S rRNA in eukaryotic initiation complexes. Nature **286**, 5770, 226 to 230 (1980)
[42] TOLAN, D. R. and R. R. TRAUT: Protein topography of the 40 S ribosomal subunit from rabbit reticulocytes shown by cross-linking with 2-iminothiolane. J. Biol. Chem. **256**, 10129–10136 (1981)

Chapter X: Interactions of ribosomes with membranes

In agreement with results mentioned on page 187, no significant differences have been found in the peptidyltransferase reaction between free and membrane bound ribosomes as measured by the formation of peptidyl-puromycin [1]. Thus it can be concluded that binding of ribosomes to the membranes of the endoplasmic reticulum has no significant effect on the peptidyltransferase center of the large subunit.

As to the chemical composition of loose and tight membrane bound ribosomes (see page 188), no differences in the proteins between these classes of ribosomes from mouse liver were detected [2].

In ribosomal preparation isolated from membranes of the rough endoplasmic reticulum of rat liver by using non-ionic detergents in the presence of 25 mM KCl, a number of non-ribosomal proteins of molecular weights between 3600 and 166000 were found. These proteins are presumably of membranous origin and could thus be involved in binding of ribosomes to the membranes [3]. For rough microsomes from liver, intramembrane protein particles, arranged in definite structures, have been described, which seem to be involved in ribosome binding [4]. While protein-free liposomes are not able to bind ribosomes, artificial lipid vesicles with reincorporated proteins did bind ribosomes.

The specificity of the binding of ribosomes to membranes, using degranulated microsomes, has been demonstrated in vitro also by gel filtration experiments for ribosomes, isolated from the microsomal fraction with different "degranulating agents" [5].

The attachment of ribosomes through a unique site of the large subunit to membranes of the endoplasmic reticulum and the nuclear envelope has been demonstrated also for ribosomes organized in crystals [6]. In this study it has furthermore been shown that ribosomes in membrane bound complexes are organized such that the long axis of the small subunit is oriented approximately in parallel to the membrane surface and that the bond between the large ribosomal subunit and the membranes depends on the concentration of K^+, but not of Mg^{2+}.

Specific receptors for pre-proteins on the cytoplasmic side of membranes of the rough endoplasmic reticulum of dog pancreas, sensitive to proteases, have been demonstrated for carp preproinsulin, while smooth endoplasmic reticulum membranes and erythrocyte plasma membranes were not able to bind signal sequences of this pre-protein [7].

Membrane-associated protein components required for the transport of nascent proteins across the membranes of the endoplasmic reticulum have been isolated [8, 11]. These factors have been desribed as a protein fragment with an apparent molecular weight of 60000, representing a cytoplasmic domain of a larger membrane protein [8], and in other studies [9] as a protein complex consisting of 6 different polypeptides with molecular weights between 9000 and 72000. Conditions to remove and to restore the translocation activity of rough microsomal membranes have also been described [10, 11].

An enzyme responsible for processing pre-human placental lactogen has been demonstrated in ascites lysates, dog pancreas and rat liver. This enzyme is an endopeptidase that cleaves peptides on the COOH side of alanine residues [12]. The contranslational proteolytic event was inhibited by chymostatin, an inhibitor of proteases with chymotrypsin-like activity [12].

The sequence complexity, frequency distribution, and template activities of mRNA populations of free and membrane-bound polysomes from myeloma cells have been analyzed by mRNAxcDNA hybridization and cell-free translation [13, 14].

Current models about mechanisms of protein secretion across membranes [15], the assembly of proteins into membranes [16], and intracellular protein topogenesis [17] have been reviewed more repently.

References

[1] Kuehl, L. and W. Robison: Peptidyl-puromycin synthesis by free and membrane-bound ribosomes. Biochim. Biophys. Acta **563**, 454–465 (1979)

[2] Hoffman, W. L. and R. M. Dowben: Ribosome-membrane interactions: Characterization of ribosomal proteins from loose and tight bound ribosomes. Mol. Biol. Rep. **6**, 79–82 (1980)

[3] Aulinskas, T. H. and T. S. Burden: Hepatic membrane proteins involved in ribosome binding: Identification by three procedures. Hoppe Seyler's Z. Physiol. Chem. **360**, 709–720 (1979)

[4] Nagata, I.: Binding of ribosomes to intramembrane particles of rough microsomes: An electron-microscope study. Cell Structure and Function **5**, 1–11 (1980)

[5] Hawkins, H. C. and R. B. Freedman: A gel filtration approach to the study of ribosome-membrane interactions. Biochim. Biophys. Acta **558**, 85–98 (1979)

[6] Unwin, P. N. T.: Attachment of ribosome crystals to intracellular membranes. J. Mol. Biol. **132**, 69–84 (1979)

[7] Prehn, S., A. Tsamaloukas, and T. Rapoport: Demonstration of specific receptors of the rough endoplasmic membrane for the signal sequence of carp pre-

proinsulin. Europ. J. Biochem. **107**, 185 to 195 (1980)
[8] MEYER, D. I. and B. DOBBERSTEIN: Identification and characterization of a membrane component essential for the translocation of nascent proteins across the membrane of the endoplasmic reticulum. J. Cell Biol. **87**, 503—508 (1980)
[9] WALTER, P. and G. BLOBEL: Purification of a membrane-associated protein complex required for protein translocation across the endoplasmic reticulum. Proc. Natl. Acad. Sci. U.S. **77**, 7112—7116 (1980)
[10] WALTER, P., R. C. JACKSON, M. M. MARCUS, V. R. LINGAPPA, and G. BLOBEL: Tryptic dissection and reconstitution of translocation activity for nascent presecretory proteins across microsomal membranes. Proc. Natl. Acad. Sci. U.S. **76**, 1795—1799 (1979)
[11] MEYER, D. I. and B. DOBBERSTEIN: A membrane component essential for vectorial translocation of nascent proteins across the endoplasmic reticulum: Requirements for its extraction and reassociation with the membrane. J. Cell Biol. **87**, 498 to 502 (1980)
[12] STRAUSS, A. W., M. ZIMMERMAN, I. BOIME, B. ASHE, T. A. MUMFORD, and A. W. ALBERTS: Characterization of an endopeptidase involved in pre-protein processing. Proc. Natl. Acad. Sci. U.S. **76**, 4225 to 4229 (1979)
[13] MECHLER, B. and T. H. RABBITTS: Membrane-bound ribosomes of myeloma cells IV. mRNA complexity of free and membrane-bound polysomes. J. Cell Biol. **88**, 29—36 (1981)
[14] MECHLER, B.: Membrane-bound ribosomes of myeloma cells. V. Subcellular distribution of immunoglobulin mRNA molecules. J. Cell. Biol. **88**, 37—41 (1981)
[15] DAVIS, B. D. and P.-C. TAI: The mechanism of protein secretion across membranes. Nature **283**, 433—438 (1980)
[16] WICKNER, W.: Assembly of proteins into membranes. Science **210**, 861—868 (1980)
[17] BLOBEL, G.: Intracellular protein topogenesis. Proc. Natl. Acad. Sci. U.S. **77**, 1496—1500 (1980)

Chapter XI: Function of eukaryotic ribosomes

1. Nonribosomal components of the protein synthesizing system

a) *Messenger RNA*

Processing of pre-mRNA has been shown to be a general prerequisite for the formation of stable and functional active mRNAs in eukaryotes that occurs in the following sequence of events: Transcription — capping — cleavage — polyadenylation — methylation — splicing (for more recent reviews see [1—4]).

Considerable progress in understanding of the molecular basis of mRNA splicing was achieved very recently. In 1979 specific immunoprecipitable RNP complexes were found in human tissues, the RNA moieties of which consist of 100—200 nucleotides [5—7]. These small nuclear RNAs (snRNA), already described 10 years ago, are rich in uridine residues and contain so-called consensus sequences. For example, U_1-RNA as one of the most prominent snRNAs contains adjacent to the 5′ initial $m_3^{2,7,7}$GpppAmUm the hexanucleotide ACUUAC that accurately matches the 5′-end of the intron of hnRNA GUAAGU ... (3′). The consensus sequence then continues with CUG which exactly matches the 3′end of the intron CAG (3′). On the basis of these data the following mechanism during splicing seems to operate: The 5′end of U_1RNA base-pairs with both ends of the intron, the remainder of the intron then loops out and two exons are thereby aligned for ligation (for reviews see [8—11]).

In combined SV 40 — mouse globin gene experiments the existence of at least one functioning "splice junction" was shown to be a necessary prerequisite for the formation of stable mRNA [12].

Two recent reviews concern the structure and function of the 5′terminal cap of eukaryotic mRNAs [13, 14]. In addition to earlier reports the existence of cap structures was shown for mRNA in maize [15], hen ovalbumin mRNA [16] and chick lens mRNA [17, 18]. The cap binding protein (see Section 2a) is able to stimulate the translation of capped and also of uncapped mRNAs to which artificially the cap was covalently linked in 5′position.

Transfer of cap structures to mRNA, according to a series of recent studies (for summary see [19]), also takes place in vivo. Upon infection of a number of different eukaryotic cells with influenza virus, the synthesis of viral mRNA is started and primed by a 10—14 nucleotides long oligonucleotide that originates from corresponding host cell mRNAs. Thus, the 5′-terminal cap and part of the adjacent leader sequences of the influenza virus mRNA are donated by capped cellular mRNAs.

More recent studies demonstrate that translation of capped mRNA moieties can be inhibited by 7-methyl-GTP and related analogs [20—22].

From yeast ribosomes an mRNA decapping enzyme was isolated by a high-salt washing procedure [23].

Confirming the earlier statement, a close inspection of the available mRNA sequences [24] does not reveal conclusive indication of specific base-pairing mechanisms operating between the 5′noncoding segments of the mRNA and the 18 S rRNA 3′terminus during initiation of protein synthesis (see also Addendum to Chapter IX).

As to the coding part of mRNA there is strong evidence that AUG is the only initiation codon in eukaryotes whereas in prokaryotes also GUG is used [25].

The influence of the chain length of 3′terminal poly(A) tails on translation and stability of mRNA was studied in different systems [26—29]. Messenger RNAs containing longer stretches of poly(A) seem to be less stable than those with 20 to 25 adenylate residues at the 3′end [27]. An explanation for these surprising results was derived by microinjection experiments with *Physarum* in which it was demonstrated that mRNAs with long poly(A) tracts were relatively short-lived whereas the polysome-bound mRNA contained only short poly(A) tails of about 15 nucleotides [28, 29]. Removal of considerable parts of long poly(A) tails seems to take place before polysome formation.

Further studies on the protective role of 80 S and 40 S ribosomal subunits, respectively, against nuclease degradation of mRNA bound in initiation complexes confirm the conclusions drawn from Table 23 [30—32].

Studies with mRNA modified in various ways (circular RNA, fragmentation, decrease of secondary structure by chemical substitution and formation of artificial initiation codons) support the "scanning model" of initiation of protein synthesis, according to which after recognition of the normally capped 5′terminus the 40 S ribosomal subunit scans along the mRNA until the AUG initiation codon is reached [33 to 38]. There are also strong indications that the secondary structure of the mRNA influences the initiation velocity [39].

b) Transfer RNA

Comprehensive representations of the current knowledge of transfer RNA structure and function are given in [40—42]. The nucleotide sequences of more than 180 tRNA molecules were listed up in recent reviews [43—47]. Surveys on the structure, properties and function of modified nucleotides found in tRNA are given in [48—53].

Crystal structure analysis of various tRNAs [54—58] revealed that there are no significant differences between tRNA and the corresponding aminoacyl-tRNA [54] and no substantial deviations from the well-known three-dimensional structure of $tRNA^{Phe}$ from yeast (one exception in [57]).

Structural studies on aminoacyl-tRNA

synthetases, as sequencing, which revealed a wide distribution of repeating sequences and few sequence homologies only, X-ray crystallographic investigations on the tertiary structure as well as collections of all known molecular weights and subunit compositions of synthetases are summarized in [59—61].

The problem of translational fidelity which is in the range of one error per 10^3 to 10^4 amino acids and of the contribution of aminoacyl-tRNA synthetases to this fairly low error frequency has been reviewed recently [62—64]. For the recognition of the correct amino acid by the synthetase, a so-called double sieve editing mechanism was proposed [65, 66] that comprises sorting of amino acids according to size and chemical proof-reading [67] by a hydrolytic site of the enzyme. In a more hypothetical model, an editing mechanism was proposed for tRNA recognition by these enzymes [68]. Experimental data on tRNA-synthetase interactions, reviewed in [69], confirm the proposal that the "recognition elements" of the tRNA are located at the inner side of the L, mainly at the acceptor end, in the D stem and in some cases in the anticodon region. X-ray crystallographic investigations on tRNA-synthetase complexes [70] may possibly contribute to elucidate this problem.

The second step important for translational fidelity is the codon-anticodon interaction between tRNA and mRNA. The knowledge of this process was extended by the discovery of a simpler codon recognition system in mitochondria [71 to 73], operating with a smaller number of tRNA molecules. The results obtained support the "two out of three" hypothesis of codon reading, which predicts that in most cases two of the three codon letters are sufficient for discrimination of non-cognate aminoacyl-tRNAs [74, 75]. Data from in vitro-protein synthesis systems, however, favour the codon-anticodon interaction according to the wobble hypothesis [76]. Information on codon recognition and codon usage patterns originates from investigations on isoacceptor-tRNAs [77] and from sequence studies on DNA and mRNA, respectively [78, 79].

The dynamics of the tRNA molecule during the decoding process at the ribosome was considered in several recent articles [80—86]. A number of investigations, mainly in prokaryotic systems, indicate that codon-anticodon interactions alone do not suffice to account for the high accuracy of decoding. Besides, conformational changes in the tRNA molecule lead to the stabilization of the complex between the ribosome and the cognate tRNA.

Slow tritium exchange studies [87] revealed that both, the anticodon along with the adjacent stem region and the CCA-terminus with the adjacent stem of the tRNA are in close contact with the ribosome and, furthermore, that conformational changes in the T- and D-loop occur (for review see [88]). With the same technique it has been shown that the CGAA sequence of 5 S RNA is not accessible on the ribosome surface [89, 90]. Therefore, it seems to be doubtful that base pairing between the CGAA sequence of ribosomal 5 S RNA and the TΨCG sequence of tRNA is involved in ribosome — tRNA interaction. Further arguments for and against the CGAA-TΨCG base pairing theory are discussed in [81—83]. The involvement of the amino acid acceptor end of the tRNA in the peptidyltransferase center function [91] as well as conformational changes of the tRNA molecule that take place during the different events of the whole translation process have been reviewed recently [83].

2. The process of eukaryotic protein synthesis

Two more general reviews comprising the components and the sequence of events of the translation process in eukaryotic as well as in prokaryotic cells have been published recently [92, 93].

a) Initiation

The process of polypeptide chain initiation in eukaryotes was summarized in [94—98]. Some more information on the role of the individual factors is now available, however, substantial deviations from the initiation scheme given in Fig. 35 did not arise.

Initiation factors eIF-2 [99—101] and eIF-3 [102, 103] have been prepared from further sources, the procedures for eIF-2 [104], eIF-4 A and eIF-4 D [104] preparation were improved, the subcellular distribution of initiation factors was reevaluated [106, 107], and the variability of initiation factor preparations from rabbit reticulocytes was described [108]. Further physical and chemical properties of the subunits of eIF-2 have been reported [109].

Progress has been made in the understanding of the role of several factors in initiation, although not yet all details are established. Thus, the effect of eIF-1 in various model assay systems was investigated [110]. eIF-1 stimulates the binding of Met-tRNA$_f$ as well as of mRNA to the 40 S ribosomal subunit, and it can replace AUG in the Met-puromycin assay. The results are interpreted in terms of a coordinating function of eIF-1 in codon-anticodon interaction between Met-tRNA$_f$ and mRNA.

Further reports on mRNA binding activity of eIF-2 have been published [111 to 114]. By affinity chromatography it was shown that eIF-2 binds also to rRNA [104, 115]. The finding that Met-tRNA$_f$ is bound to the β-subunit of eIF-2 in the ternary initiation complex was further confirmed by cross-linking experiments [116]. Additional reports on factors stimulating eIF-2 activity were published [117, 118]. However, the assays used for measuring the stimulating activities of the ternary complex formation should be considered with caution [119]. The mode of action of the so-called Co-eIF-2 A described earlier, was further investigated [120, 121].

The multiple functions of eIF-3 described in Chapter XI (p. 213) were confirmed for reticulocyte eIF-3 [122]. eIF-3 from chicken embryonic muscle was reported to contain mRNA discriminatory activity [123].

For wheat germ the ribosome dissociating activity was separated from eIF-3 (cf Chapter XI, p. 213) and could be correlated to a small polypeptide [124, 125] (cf. Section 2d). Furthermore, also in rat liver a dissociation activity distinct from eIF-3 was found [126]. Investigations on the precise role of eIF-4 C revealed that this factor is accessary to eIF-3 in dissociating 80 S ribosomes, binds to 40 S subunits (presumably at an early stage of initiaton complex formation), prevents dimerisation of native 40 S subunits, and stabilizes the 40 S initiation complex [127, 128].

Concerning mRNA binding to 40 S ribosomal subunits during initiation, evidence has accumulated that confirms the ,,scanning model" (cf. Section 1a). ATP hydrolysis needed at this step seems to be important for the movement of the 40 S subunit along the 5′ terminal noncoding region of the mRNA [129]. The cap binding protein (m. w. 24000) (cf Chapter XI, p. 213) has been purified by affinity chromatography [130, 13] as well as by conventional techniques [132]. This protein is present in preparations of eIF-3 or eIF-4 B [131, 132], and it partially restores the translational activity of capped mRNAs [132, 133].

eIF-5 catalyzes the release of eIF-2 and eIF-3 [134] and presumably also of eIF-4 C [98, 128] from the 40 S $\times$ Met-tRNA $\times$ mRNA complex that occurs simultaneously with GTP-hydrolysis and precedes 60 S subunit joining [134].

b) Elongation

A survey on elongation and the factors involved has been published recently [93]. Purified eEF-Ts [135] was shown to be responsible for the formation of the aggregated forms of eEF-T in eukaryotes. eEF-T can be disaggregated by elastase treatment to defined polypeptides [136]. One mode of eEF-Ts action seems to be the transfer of the γ-phosphate from various nucleoside triphosphates onto GDP [137]. Specific interactions of eEF-G with the antibiotics modeccin [138], the diphtheria toxin causing ADP-ribosylation [139], as well as with the pokeweed antiviral protein and ricin [140] which all influence translocation by inhibition of eEF-G activity, have been described. Furthermore, eEF-G was shown to stimulate GTPase and ATPase activity of the rat liver 5 S RNA-protein complex [141].

c) Termination

A release factor was isolated from *Artemia salina* cysts [142] and a permanent increase of its activity during embryonic development of the cysts was observed [143].

d) Dissociation factors

The molecular weights of dissociation factors isolated from native subunits of rat liver ribosomes were determined as to 51000 [126] and from wheat germ as to 23000 [124]. The rat liver factor was found able to bind to the 40 S and to the 60 S ribosomal subunit [126].

Alterations of the intracellular distribution of dissociation factors between ribosomes and cytoplasm were reported for rat liver during growth of the animals [144]. In rat ascites hepatoma cells, association factor activity was detected, in contrast to the dissociation factor activity present in identical fractions from rat liver [145]. The process of dissociation and reassociation of ribosomes seems to be markedly influenced by polyamines [146, 147].

3. Some regulatory aspects of protein biosynthesis in eukaryotes

Recent developments in the field of protein synthesis regulation in eukaryotes have been reviewed by various authors [98, 148—151]. Storage of mRNA in eggs and embryos becomes a more and more attractive example for translational control in eukaryotes [152]. Here, the developments in the three examples for translational control that were described in Chapter XI will be considered only.

a) Hemin control of protein synthesis in reticulocytes

Recent advances in the analysis of the translational control mechanism in reticulocyte lysates were summarized in [98, 153—155]. The hemin controlled translational inhibitor (HCI) was shown to be activated by hemin deficiency, oxygen supply, elevated temperatures or pH, high pressure and by reaction with sulfhydryl-reagents [153, 155], indicating that changes in the reduction state or possibly other modifications are involved in the activation of the proinhibitor. There are some indications that one or several proteins are involved in the activation process [156]; the participation of a cAMP-dependent protein kinase, however, seems to be ruled out [157].

Phosphorylation of the α-subunit of eIF-2 by HCI action was again demonstrated in reticulocyte lysates [158, 159]. This well-established phenomenon stimulated systematic investigations about protein kinases in rabbit reticulocytes [160 to 162] as well as the phosphatase responsible for the dephosphorylation of eIF-2α [163 to 167]. First results on the phosphorylation state of eIF-2 in lysates [168] and on the regulation of eIF-2 phosphatase [169] were obtained.

The mode, how phosphorylation of eIF-2 influences the proteosynthetic activity in lysates is still obscure. In all reactions tested so far no differences between the activities of eIF-2 and of its phosphorylated form could be detected [170, 171]. There are some indications that the recycling of eIF-2 is also not impaired by phosphorylation [171, 172]. Further reports support the suggestion that phosphorylation inhibits the interaction of eIF-2 with one (or more) cofactor(s) that are normally necessary for maximal eIF-2 activity [118, 173, 174]. Two groups described a protein factor antagonizing the action of HCI without affecting the phosphorylation state of eIF-2 [175, 176]. Very recently an alternative to the phosphorylation-dephosphorylation mechanism of eIF-2 was proposed for HCI action [98, 177–179], according to which the oxidation-reduction state of SH-groups may be important for the activity of eIF-2.

The protein kinase from reticulocytes dependent on double-stranded RNA was purified and characterized [180–183]. This enzyme is different from HCI but phosphorylates the same site(s) of eIF-2α [184, 185]. Detection of inhibitors similar to HCI also in other cells of different species [186, 187] supports the assumption that phosphorylation of eIF-2 is a more general mechanism of protein synthesis regulation.

b) Interferon action on protein synthesis

Recent results in this field are summarized in [188–192]. Increased attention has been paid to the 2′–5′oligoadenylate (2–5 A) synthetase-endonuclease system of interferon action. The structural requirements of dsRNA for the activation of the 2–5 A synthetase [193–195] and the process of induction of this enzyme [196–199] were investigated; furthermore, the 2–5 A synthetase was purified [200 to 202], and new methods for detection of 2–5 A and of 2–5 A synthetase were developed [203–205]. Chemically synthesized 2–5 A is equally active as enzymatically synthesized 2–5 A [206]. Degradation of 2–5 A [207–209] also influences the regulation of the endonuclease, which is active in the presence of 2–5 A only. The specificity of the endonuclease is going to be investigated [210 to 211]. The 2–5 A system is not restricted to interferon-treated cells, but was also found to be operational in reticulocytes [212], in cells treated with glucocorticoids [213], and in cells unable to develop an antiviral state after interferon treatment [214]. The inhibition of protein synthesis can be induced by 2–5 A in permeabilized BHK cells [215] as well as in mitochondria [216]. Furthermore, 2–5 A inhibits DNA-synthesis [204] and cell growth [217].

The phenomenon of induction by interferon of a ribosome-associated dsRNA dependent protein kinase, phosphorylating the same tryptic peptide of eIF-2α as HCI [218] was described by several authors [219–223]. The inhibition of mengovirus-RNA translation caused by interferon can be overcome by adding purified eIF-2 [224] which indicates that eIF-2 is really involved in this process. Induction of protein kinase can also be triggered by endogenous interferon [225]. Attempts have been made to correlate both phenomena of interferon action on protein synthesis [193 to 195]. Furthermore, impairment of cap methylation by interferon treatment has been reported [226].

c) Shut-off of host cell protein synthesis after viral infection

The shut-off phenomenon of host cell protein synthesis after viral infection cannot be explained satisfactorily at present. Results obtained by several groups led to different explanations (for review see [150]). Few reports [227, 228] are in favour with the membrane leakage model. Others point to correlations between viral RNA synthesis and inhibition of protein synthesis [229, 230]. Enhanced phosphorylation of ribosomal proteins [230] and of a ribosome-associated viral protein [231], respectively, was observed in cells infected with vaccinia virus. The participation of an "anti shut-off protein" in the translational control was also proposed [232].

Reports on the inhibition of initiation complex formation by vaccinia viral cores [233, 234] are interesting in connection with the observation that 2—5 A synthetase activity is accumulated in the RNP-cores of vesicular stomatitis virus and of Moloney mouse leukemia virus [235]. Other authors working with vesicular stomatitis virus assume that mainly competition between cellular and viral RNA for ribosomes leads to the inhibition of host cell protein synthesis [236].

Reports concerning shut-off of host cell protein synthesis by poliovirus [132, 237 to 240], vaccinia virus [241] and Semliki forest virus [242] indicate that initiation factors are affected after viral infection. Thus, the addition of the cap binding protein (CBP) and of eIF-4 B [242] can restore the proteosynthetic activity of extracts from poliovirus-infected cells (cf. Section 2a) as well as from Semliki forest virus-infected cells [242]. In the latter case it was shown that the late viral RNA needs only small amounts of these factors to be efficiently translated [242]. Also, poliovirus RNA containing a protein (VPg) covalently linked to its 5′ terminus instead of a cap [243] is able to form initiation complexes [244], presumably in the absence of CBP. Thus, inactivation of initiation factor(s) seems to be one mechanism leading to an inhibition of cellular protein synthesis without affecting viral protein synthesis. The diversity of the results obtained with other viral systems (see above), however, indicates that different groups of viruses may use different mechanism to programme the translational machinery of the host cell and to get preferentially viral protein synthesized.

References

[1] Perry, R. P., U. Schibler, and L. Meyuhas: Processing of messenger RNA and the determination of its relative abundance. In: From Gene to Protein. Miami Winter Symp. Ser., Vol. **16**. (Eds.: Russell, T. R., K. Brew, H. Faber, and J. Schultz) Academic Press, New York, 1979, 187—206

[2] Darnell, J. E., jr.: Steps in processing of mRNA: implications for gene regulation. In: From Gene to Protein. Miami Winter Symp. Ser., Vol. 16. (Eds.: Russell, T. R., K. Brew, H. Faber, and J. Schultz) Academic Press, New York, 1979, 207 to 228

[3] Scherrer, K., M.-T. Imaizumi-Scherrer, C.-A. Reynaud, and A. Therwath: On pre-messenger RNA and transcription. A review. Mol. Biol. Rep. **5**, 5—15 (1979)

[4] Coutelle, C.: The precursor to animal cell messenger RNA. Biochem. J. **197**, 1—6 (1981)

[5] Lerner, M. R., and J. A. Steitz: Antibodies to small nuclear RNAs complexed with proteins are produced by patients with systemic lupus erythematosus. Proc. Natl. Acad. Sci. U.S. **76**, 5495—5499 (1979)

[6] Lerner, M. R., J. A. Boyle, S. M. Mount, S. L. Wolin, and J. A. Steitz: Are snRNPs involved in splicing? Nature **283**, 220—222 (1980)

[7] Rogers, J., and R. Wall: A mechanism for RNA splicing. Proc. Natl. Acad. Sci. U.S. **77**, 1877—1879 (1980)

[8] Roberts, R.: Small RNAs and splicing. Nature **283**, 132—134 (1980)

[9] Knowler, J. T., and A. F. Wilks: Ribonucleoprotein particles and the maturation of eukaryote mRNA. Trends Biochem. Sci. **5**, 268—271 (1980)

[10] Lewin, B.: Alternatives for splicing: recognizing the ends of introns. Cell **22**, 324—326 (1980)

[11] Aloni, J.: Splicing of viral mRNAs. Progr. Nucl. Acid Res. Mol. Biol. **25**, 1 to 31 (1981)

[12] Hamer, D. H., and P. Leder: Splicing and formation of stable RNA. Cell **18**, 1299 to 1305 (1979)

[13] Shatkin, A. J., N. Sonenberg, and Y. Furuichi: 5′-capping and eukaryotic mRNA function. In: From Gene to Protein. Miami Winter Symp. Ser., Vol. 16. (Eds.: Russell, T. R., K. Brew, H. Fa-

BER, and J. SCHULTZ) Academic Press, New York, 1979, 229–240

[14] BANERJEE, A.: 5′-terminal cap structure in eucaryotic messenger ribonucleic acids. Microbiol. Rev. **44**, 175–205 (1980)

[15] NICHOLS, J. L.: 'Cap' structures in maize poly(A)-containing RNA. Biochim. Biophys. Acta **563**, 490–498 (1979)

[16] MALEK, L. T., H.-J. BRETER, G. M. HELLMANN, K. H. FRIDERICI, F. M. ROTTMAN, and R. E. RHOADS: 5′-terminal structure of hen ovalbumin messenger ribonucleic acid. J. Biol. Chem. **254**, 10415 to 10420 (1979)

[17] LAVERS, G. C., G. GANG, and E. M. CICCONE: Differential influence of mRNA cap metabolism on protein synthesis in differentiating embryonic chick lens cells. In: From Gene to Protein. Miami Winter Symp. Ser., Vol. 16. (Eds.: RUSSELL, T. R., K. BREW, H. FABER, and J. SCHULTZ) Academic Press, New York, 1979, 605

[18] LAVERS, G. C.: Influence of m^7GpppN mRNA caps on lens protein synthesis and on mRNA binding to ribosomes in a homologous lens translation system derived from embryonic chick lens cells. Exp. Eye Res. **30**, 747–752 (1980)

[19] KRUG, R. M., M. BOULOY, and S. J. PLOTCH: The 5′-ends of influenza viral messenger RNAs are donated by capped cellular RNAs. Trends Biochem. Sci. **6**, 7–10 (1981)

[20] SASAVAGE, N. L., K. FRIDERICI, and F. M. ROTTMAN: Specific inibition of capped mRNA translation in vitro by $m^7G^{5'}pppp^{5'}G$ and $m^7G^{5'}pppp^{5'}m^7G$. Nucl. Acids Res. **6**, 3613–3624 (1979)

[21] MIURA, K.-I., Y. KODAMA, K. SHIMOTOHNA, T. FUKUI, M. IKEHARA, I. NAKAGAWA, and T. HATA: Inhibitory effect of methylated derivatives of guanylic acid for protein synthesis with reference to the functional structure of the 5′-cap in viral messenger RNA. Biochim. Biophys. Acta **564**, 264–274 (1979)

[22] CHU, L.-Y., and R. E. RHOADS: Inhibition of cell-free messenger ribonucleic acid translation by 7-methylguanosine 5′-triphosphate: Effect of messenger ribonucleic acid concentration. Biochem. **19**, 184–188 (1980)

[23] STEVENS, A.: An mRNA decapping enzyme from ribosomes of *Saccharomyces cerevisiae*. Biochem. Biophys. Res. Commun. **96**, 1150–1159 (1980)

[24] DE WACHTER, R.: Do eukaryotic mRNA sequences base-pair with the 18 S ribosomal RNA 3′terminus? Nucl. Acids Res. **7**, 2045–2054 (1979)

[25] SHERMAN, F., G. MCKNIGHT, and J. W. STEWART: AUG is the only initiation codon in eukaryotes. Biochim. Biophys. Acta **609**, 343–346 (1980)

[26] DESPHANDE, A. K., B. CHATTERJEE, and A. K. ROY: Translation and stability of rat liver messenger RNA for α_{2u}-globulin in *Xenopus oocytes*. The role of terminal poly(A). J. Biol. Chem. **18**, 8937–8942 (1979)

[27] PALATNIK, C. M., R. V. STORTI, A. K. CAPONE, and A. JACOBSON: Messenger RNA stability in *Dictyostelium discoideum*: Does poly(A) have a regulatory role? J. Mol. Biol. **141**, 99–118 (1980)

[28] ADAMS, D. S., D. NOONAN, and W. R. JEFFREY: Cytoplasmic polyadenylate processing events accompany the transfer of mRNA from the free mRNP particles to the polysomes in *Physarum*. Proc. Natl. Acad. Sci. U.S. **78**, 83–87 (1981)

[29] JEFFERY, W. R., D. S. ADAMS, and D. NOONAN: Cytoplasmic processing events in the polyadenylate region of *Physarum* messenger RNA. Mol. Biol Rep. **7**, 63 to 70 (1981)

[30] SCHROEDER, H. W. jr., Ch. D. LIARAKOS, R. C. GUPTA, K. RANDERATH, and B. W. O'MALLEY: Ribosome binding site analysis of ovalbumin messenger ribonucleic acid. Biochem. **18**, 5798–5808 (1979)

[31] BROWNING, K. S., D. W. LEUNG, and J. M. CLARK, jr.: Protection of satellite tobacco necrosis virus ribonucleic acid by wheat germ 40 S and 80 S ribosomes Biochem. **19**, 2276–2283 (1980)

[32] BROWNING, K. S., and J. M. CLARK, jr.: Translation initiation site of the coat protein messenger ribonucleic acid of the cowpea strain of tobacco mosaic virus. Biochem. **19**, 5922–5926 (1980)

[33] FILIPOWICZ, W., and A. L. HAENNI: Binding of ribosomes to 5′-terminal leader sequences of eukaryotic mRNAs. Proc. Natl. Acad. Sci. U.S. **76**, 3111–3115 (1979)

[34] KOZAK, M.: Inability of circular mRNA

to attach to eukaryotic ribosomes. Nature **280**, 82—85 (1979)

[35] KOZAK, M.: Influence of mRNA secondary structure on binding and migration of 40 S ribosomal subunits. Cell **19**, 79 to 90 (1980)

[36] KOZAK, M.: Evaluation of the "Scanning Model" for initiation of protein synthesis in eukaryotes. Cell **22**, 7—8 (1980)

[37] KOZAK, M.: Binding of wheat germ ribosomes to bisulfite-modified reovirus messenger RNA: Evidence for a scanning mechanism. J. Mol. Biol. **144**, 291—304 (1980)

[38] KOZAK, M.: Binding of wheat germ ribosomes to fragmented viral mRNA. J. Virology **35**, 748—756 (1980)

[39] PAVLAKIS, G. N., R. E. LOCKARD, N. VAMVAKOPOULOS, L. RIESER, U. L. RAJ BHANDARY, and J. N. VOURNAKIS: Secondary structure of mouse and rabbit α- and β-globin mRNAs: Differential accessibility of α and β initiator AUG codons towards nucleases. Cell **19**, 91—102 (1980)

[40] ALTMAN, S.: Transfer RNA. MIT Press, Cambridge, Massachuetts, 1978

[41] SCHIMMEL, P. R., D. SÖLL, and J. N. ABELSON (Eds.): Transfer RNA: Structure, properties, and recognition. Cold Spring Harbor Laboratory, 1979

[42] SÖLL, D., J. N. ABELSON, and P. R. SCHIMMEL (Eds.): Transfer RNA: Biological aspects. Cold Spring Harbor Laboratory, 1980

[43] SINGHAL, R. P., and P. A. M. FALLIS: Structure, function, and evolution of transfer RNAs (with appendix giving complete sequences of 178 tRNAs). Progr. Nucl. Acid Res. Mol. Biol. **23**, 227—290 (1979)

[44] GAUSS, D. H., F. GRÜTER, and M. SPRINZL: Compilation of tRNA sequences. In Ref. [41] pp. 520—537

[45] CELIS, J. E.: Collection of mutant tRNA sequences. In Ref. [41] pp. 539—545

[46] SPRINZL, M., F. GRÜTER, A. SPELZHAUS, and D. H. GAUSS: Compilation of tRNA sequences. Nucl. Acids Res. **8**, r 1—r 22 (1980)

[47] GAUSS, D. H., and M. SPRINZL: Compilation of tRNA sequences. Nucl. Acids Res. **9**, r 1—r 23 (1981)

[48] NISHIMURA, S.: Modified nucleosides in tRNA. In Ref. [41] pp. 59—79

[49] NISHIMURA, S.: Structures of modified nucleosides found in tRNA. In Ref. [41] pp. 547—549

[50] NISHIMURA, S.: Chromatographic mobilities of modified nucleotides. In Ref. [41] pp. 551—552

[51] DIRHEIMER, G., G. KEITH, A.-P. SIBLER, and R. P. MARTIN: The primary structure of tRNAs and their rare nucleosides. In Ref. [41] pp. 19—41

[52] SINGER, B., and M. KRÖGER: Participation of modified nucleosides in translation and transcription. Progr. Nucl. Acid Res. Mol. Biol. **23**, 151—194 (1979)

[53] AGRIS, P. F.: The modified nucleosides of transfer RNA: A bibliography of biochemical and biophysical studies from 1970 to 1979. Alan R. Liss Inc., New York, 1980

[54] RICH, A., G. J. QUIGLEY, M. M. TEETER, A. DECRUIX, and N. WOO: Recent progress in tRNA structural analysis. In Ref. [41] pp. 101—113

[55] WOO, N. H., B. A. ROE, and A. RICH: Three-dimensional structure of *Escherichia coli* initiator $tRNA_f^{Met}$. Nature **286**, 346—351 (1980)

[56] SCHEVITZ, R. W., A. D. PODJARNY, N. KRISHNAMACHARI, J. J. HUGHES, and P. B. SIGLER: A crystallographic analysis of yeast initiator tRNA. In Ref. [41] pp. 133—143

[57] WRIGHT, H. T., P. C. MANOR, K. BEURLING, R. L. KARPEL, and J. R. FRESCO: The structure of baker's yeast $tRNA^{Gly}$: A second tRNA conformation. In Ref. [41] pp. 145—160

[58] MORAS, D., M. B. COMARMOND, J. FISCHER, R. WEISS, J. C. THIERRY, J. P. EBEL, and R. GIEGE: Crystal structure of yeast $tRNA^{Asp}$. Nature **288**, 669—674 (1980)

[59] HARTLEY, B. S.: Structural studies of aminoacyl-tRNA synthetases. In Ref. [41] pp. 223—234

[60] SÖLL, D., and P. R. SCHIMMEL: Characteritics of aminoacyl-tRNA synthetases. In Ref. [41] pp. 553—563

[61] JOACHIMIAK, A., and J. BARCISZEWSKI: Amino acid: tRNA ligases (EC 6.1.1.) FEBS-Lett. **119**, 201—211 (1980)

[62] YARUS, M.: The relationship of the accuracy of aminoacyl-tRNA synthesis to that of translation. In Ref. [41] pp. 501—515

[63] YARUS, M.: Accuracy of translation. Progr. Nucl. Acid Res. Mol. Biol. **23**, 195–225 (1979)

[64] HOPFIELD, J. J., and T. YAMANE: The fidelity of protein synthesis. In: Ribosomes: Structure, Function and Genetics (Eds.: CHAMBLISS, G., G. R. CRAVEN, J. DAVIES, K. DAVIS, L. KAHAN, and M. NOMURA) University Park Press, Baltimore, 1980, 585–596

[65] FERSHT, A. R.: Editing mechanisms in the aminoacylation of tRNA. In Ref. [41] pp. 247–254

[66] FERSHT, A. R.: Enzymic editing mechanisms in protein synthesis and DNA replication. Trends Biochem. Sci. **5**, 262 to 265 (1980)

[67] CRAMER, F., F. von der HAAR, and G. L. IGLOI: Mechanism of aminoacyl-tRNA synthetases: Recognition and proofreading processes. In Ref. [41] pp. 267–279

[68] WRIGHT, H. T.: A new mechanism for the hydrolytic editing function of aminoacyl-tRNA synthetases. Kinetic specificity for the tRNA substrate. FEBS-Lett. **118**, 165–171 (1980)

[69] SCHIMMEL, P. R.: Similarities in the structural organization of complexes of tRNAs with aminoacyl-tRNA synthetases and the mechanism of recognition. In Ref. [41] pp. 297–310

[70] DIETRICH, A., R. GIEGÉ, M. B. COMARMOND, J. C. THEIRRY, and D. MORAS: Crystallographic studies on the aspartyl-tRNA synthetase-$tRNA^{Asp}$ system from yeast: The crystalline aminoacyl-tRNA synthetase. J. Mol. Biol. **138**, 129–135 (1980)

[71] HECKMAN, J. E., J. SARNOFF, B. ALZNER-DEWEERD, S. YIN, and U. L. RAJBHANDARY: Novel features in the gentic code and codon reading patterns in *Neurospora crassa* mitochondria based on sequences of 6 mitochondrial transfer RNAs. Proc. Natl. Acad. Sci. U.S. **77**, 3159–3163 (1980)

[72] BARRELL, B. G., S. ANDERSON, A. T. BANKIER, M. H. L. DEBRUIJN, E. CHEN, A. R. COULSON, J. DROUIN, I. C. EPERON, D. P. NIERLICH, B. A. ROE, F. SANGER, P. H. SCHREIER, A. J. H. SMITH, R. STADEN, and I. G. YOUNG: Different pattern of codon recognition by mammalian mitochondrial transfer RNAs. Proc. Natl. Acad. Sci. U.S. **77**, 3164–3166 (1980)

[73] BONITZ, S. G., R. BERLANI, G. CORUZZI, M. LI, G. MACINO, F. G. NOBREGA, M. P. NOBREGA, B. E. THALENFELD, and A. TZAGOLOFF: Codon recognition rules in yeast mitochondria. Proc. Natl. Acad. Sci. U.S. **77**, 3167–3170 (1980)

[74] LAGERKVIST, U.: Two out of three – An alternative method of codon reading. Proc. Natl. Acad. Sci. U.S. **75**, 1759 to 1762 (1978)

[75] LAGERKVIST, U.: Unorthodox codon reading and the evolution of the genetic code. Cell **23**, 305–306 (1981)

[76] GOLDMAN, E. and G. W. HATFIELD: Use of protein synthesis in vitro to study codon recognition by *Escherichia coli* $tRNA^{Leu}$ isoaccepting species. In Ref. [42] pp. 427–437

[77] HATFIELD, D., C. R. MATTHEWS, and M. RICE: Aminoacyl-tRNA populations in mammalian cells. Chromatographic profiles and patterns of codon recognition. Biochim. Biophys. Acta **564**, 414–423 (1979)

[78] GROSJEAN, H.: Codon usage in several organisms. In Ref. [42] pp. 565–569

[79] GRANTHAM, R., C. GAUTIER, M. GOUY, M. JACOBZONE, and R. MERCIER: Codon catalog usage is a genome strategy modulated for gene expressivity. Nucl. Acids Res. **9**, r43–r74 (1981)

[80] CANTOR, C. R.: tRNA-ribosome interactions. In Ref. [41] pp. 363–392

[81] MÖLLER, A., U. MANDERSCHMIED, R. LIPECKY, S. BERTRAM, M. SCHMITT, and H. G. GASSEN: Codon-induced structural transition in tRNA. In Ref. [41] pp. 459 to 471

[82] GASSEN, H. G.: Ligand-induced conformational changes in ribonucleic acids. Progr. Nucl. Acid Res. Mol. Biol. **24**, 57–86 (1980)

[83] CLARK, B. F. C.: Structure of tRNA during protein biosynthesis. In: Ribosomes: Structure, Function and Genetics (Eds.: CHAMBLISS, G., G. R. CRAVEN, J. DAVIES, J. DAVIS, L. KAHAN, and M. NOMURA) University Park Press, Baltimore, 1980, 413–444

[84] KURLAND, C. G.: On the accuracy of elongation. In Ribosomes: Structure, Function and Genetics (Eds.: CHAMBLISS, G.,

G. R. CRAVEN, J. DAVIES, K. DAVIS, L. KAHAN, and M. NOMURA) University Park Press, Baltimore, 1980, 597–614

[85] BUCKINGHAM, R. H., and C. G. KURLAND: Interactions between UGA-suppressor $tRNA^{Trp}$ and the ribosome: mechanism of tRNA selection. In Ref. [42] pp. 421 to 426

[86] LABUDA, D., and D. PÖRSCHKE: Multistep mechanism of codon recognition by transfer ribonucleic acid. Biochem. **19**, 3799–3805 (1980)

[87] FARBER, N. M., and C. R. CANTOR: Comparison of the structures of free and ribosome-bound $tRNA^{Phe}$ by using slow tritium exchange. Proc. Natl. Acad. Sci. U.S. **77**, 5135–5139 (1980)

[88] CANTOR, C. R.: Physical and chemical techniques for the study of RNA structure on the ribosome. In Ribosomes: Structure, Function and Genetics (Eds.: CHAMBLISS, G., G. R. CRAVEN, J. DAVIES, K. DAVIS, L. KAHAN, and M. NOMURA) University Park Press, Baltimore, 1980, 23 to 49

[89] FARBER, N. M., and C. R. CANTOR: A slow tritium exchange study of the solution structure of *Escherichia coli* 5 S ribosomal RNA. J. Mol. Biol. **146**, 223 to 239 (1981)

[90] FARBER, N. M., and C. R. CANTOR: Accessibility and structure of ribosomal RNA monitored by slow tritium exchange of ribosomes. J. Mol. Biol. **146**, 241–257 (1981)

[91] KRAYEVSKY, A. A. and M. K. KUKHANOVA: Peptidyltransferase center of ribosomes. Progr. Nucl. Acid Res. Mol. Biol. **23**, 1–51 (1979)

[92] HERSHEY, J. W. B.: The translational machinery: Components and mechanism. Cell Biology 4, 1–68 (1980)

[93] WEISSBACH, H.: Soluble factors in protein synthesis. In: Ribosomes: Structure, Function and Genetics (Eds.: CHAMBLISS, G., G. R. CRAVEN, J. DAVIES, K. DAVIS, L. KAHAN, and M. NOMURA) University Park Press, Baltimore, 1980, 347–411

[94] SAFER, B., and W. F. ANDERSON: The molecular mechanism of hemoglobin synthesis and its regulation in the reticulocyte. Crit. Rev. Biochem., CRC Press, Ohio (Cleveland), 1978, 261–290

[95] GRUNBERG-MANAGO, M.: Initiation of protein synthesis as seen in 1979. In: Ribosomes: Structure, Function and Genetics (Eds.: CHAMBLISS, G., G. R. CRAVEN, J. DAVIES, K. DAVIS, L. KAHAN, and M. NOMURA) University Park Press, Baltimore, 1980, 445–477

[96] HUNT, T.: The initiation of protein synthesis. Trends Biochem. Sci. **5**, 178–181 (1980)

[97] THOMAS, A., R. BENNE, and H. O. VOORMA: Initiation of eukaryotic protein synthesis. FEBS-Lett. **128**, 177–185 (1981)

[98] JAGUS, R., W. F. ANDERSON, and B. SAFER: The regulation of initiation of mammalian protein synthesis. Progr. Nucl. Acid Res. Mol. Biol. **25**, 127–185 (1981)

[99] BENNE, R., M. KASPERAITIS, H. O. VOORMA, E. CEGLARZ, and A. B. LEGOCKI: Initiation factor eIF-2 from wheat germ. Purification, functional comparison to eIF-2 from rabbit reticulocytes and phosphorylation of its subunits. Europ. J. Biochem. **104**, 109 to 117 (1980)

[100] JAKUBOWICZ, T., A. J. SVOBODA, and M. H. VAUGHAN: Purification and phosphorylation of human initiation factor eIF-2. Biochem. Biophys. Res. Commun. **97**, 1420–1428 (1980)

[101] KREUTZFELDT, C.: Initiation of protein synthesis in yeast: Binding of Met-$tRNA_f$. Z. Naturforsch. 36c, 142–148 (1981)

[102] CEGLARZ, E., H. GOUMANS, A. THOMAS, and R. BENNE: Purification and characterization of protein synthesis initiation factor eIF-3 from wheat germ. Biochim. Biophys. Acta **610**, 181–188 (1980)

[103] CHECKLEY, J. W., L. COOLEY, and J. M. RAVEL: Characterization of initiation factor eIF-3 from wheat germ. J. Biol. Chem. **256**, 1582–1586 (1981)

[104] NYGÅRD, O., P. WESTERMANN, and T. HULTIN: The use of rRNA-cellulose chromatography in the rapid isolation of homogeneous protein synthesis initiation factor eIF-2 from rat liver microsomes. Biochim. Biophys. Acta **608**, 196–200 (1980)

[105] van der MAST, C., and H. O. VOORMA: Purification of free eukaryotic initiation

factors eIF-4A and eIF-4D on cibacron blue F3G-A. Biochim. Biophys. Acta **607**, 512—519 (1980)

[106] THOMAS, A., H. GOUMANS, H. AMESZ, R. BENNE, and H. O. VOORMA: A comparison of the initiation factors of eukaryotic protein synthesis from ribosomes and from the postribosomal supernatant. Europ. J. Biochem. **98**, 329—337 (1979)

[107] CLEMENS, M. J., and V. M. PAIN: Association of initiation factor eIF-2 with a rapidly sedimenting fraction from Ehrlich ascites tumour cells. Biochem. J. **194**, 357—360 (1981)

[108] MEYER, L. J., M. L. BROWN-LUEDI, S. CORBETT, D. R. TOLAN, and J. W. B. HERSHEY: The purification and characterization of the multiple forms of protein synthesis eukaryotic initiation factors 2, 3, and 5 from rabbit reticulocytes. J. Biol. Chem. **256**, 351—356 (1981)

[109] LLOYD, M. A., J. C. OSBORNE (Jr.), B. SAFER, G. M. POWELL, and W. C. MERRICK: Characteristics of eukaryotic initiation factor 2 and its subunits. J. Biol. Chem. **255**, 1189—1193 (1980)

[110] THOMAS, A., W. SPAAN, H. van STEEG, H. O. VOORMA, and R. BENNE: Mode of action of protein synthesis initiation factor eIF-1 from rabbit reticulocytes. FEBS Lett. **116**, 67—71 (1980)

[111] SEGNI, G., H. ROSEN, and R. KAEMPFER: Competition between alpha- and beta-globin messenger ribonucleic acids for eukaryotic initiation factor 2. Biochem. **18**, 2847—2854 (1979)

[112] ROSEN, H., and R. KAEMPFER: Mutually exclusive binding of messenger RNA and initiator methionyl transfer RNA to eukaryotic initiation factor 2. Biochem. Biophys. Res. Commun. **91**, 449—455 (1979)

[113] MALATHI, V. G., and R. MAZUMDER: An *Artemia salina* factor which counteracts the mRNA-induced inhibition of initiator Met-tRNA binding to initiation factor eIF-2. Biochem. Biophys. Res. Commun. **89**, 585—590 (1979)

[114] KAEMPFER, R., J. van EMMELO, and W. FIERS: Specific binding of eukaryotic initiation factor 2 to satellite tabacco necrosis virus RNA at a 5′-terminal sequence comprising ribosome binding site. Proc. Natl. Acad. Sci. U.S. **78**, 1542—1546 (1981)

[115] VLASIK, T. N., S. P. DOMOGATSKY, T. A. BEZLEPKINA, and L. P. OVCHINNIKOV: RNA-binding activity of eukaryotic initiation factors of translation. FEBS Lett. **116**, 8—10 (1980)

[116] NYGÅRD, O., P. WESTERMANN, and T. HULTIN: Met-tRNA$_f^{Met}$ is located in close proximity to the β-subunit of eIF-2 in the eukaryotic initiation complex [eIF-2 · Met-tRNA^{Met} · GDPCP]. FEBS Lett. **113**, 125—128 (1980)

[117] CLEMENS, M. J., C. O. ECHETEBU, V. J. TILLERAY, and V. M. PAIN: Stimulation of initiation factor eIF-2 by a rat liver protein with GDPase activity. Biochem. Biophys. Res. Commun. **92**, 60—67 (1980)

[118] SIEKIERKA, J., K. I. MITSUI, and S. OCHOA: Mode of action of the heme-controlled translational inhibitor: Relationship of eukaryotic initiation factor 2-stimulating protein to translation restoring factor. Proc. Natl. Acad. Sci. U.S. **78**, 220—223 (1981)

[119] BENNE, R., H. AMESZ, J. W. B. HERSHEY, and H. O. VOORMA: The activity of eukaryotic initiation factor eIF-2 in ternary complex formation with GTP and Met-tRNA$_f$. J. Biol. Chem. **254**, 3201—3205 (1979)

[120] DASTIDAR, D. G., B. YAGHMAI, A. DAS, H. K. DAS, and N. K. GUPTA: Protein synthesis in rabbit reticulocytes. Demonstration of the requirements for eIF-2 and Co-eIF-2A for peptide chain initiation using immune sera. J. Biol. Chem. **255**, 365—368 (1980)

[121] GHOSH-DASTIDAR, P., D. GIBLIN, B. YAGMAI, A. DAS, H. K. DAS, L. J. PARKHURST, and N. K. GUPTA: Protein synthesis in rabbit reticulocytes. A study of the mechanism of interaction of fluorescently labeled Co-eIF-2A with eIF-2 using fluorescence polarization. J. Biol. Chem. **255**, 3826—3829 (1980)

[122] TRACHSEL, H., and T. STAEHELIN: Initiation of mammalian protein synthesis. The multiple functions of the initiation factor eIF-3. Biochim. Biophys. Acta **565**, 305—314 (1979)

[123] GETTE, W. R., and S. M. HEYWOOD: Translation of myosin heavy chain

messenger RNA in an eukaryotic initiation factor 3- and messenger-dependent muscle cell-free system. J. Biol. Chem. **254**, 9879–9885 (1979)

[124] Russell, D. W., and L. L. Spremulli: Purification and characterization of a ribosome dissociation factor (eukaryotic initiation factor 6) from wheat germ. J. Biol. Chem. **254**, 8796–8800 (1979)

[125] Russell, D. W., and L. L. Spremulli: Mechanism of action of the wheat germ ribosome dissociation factor: Interaction with the 60 S subunit. Arch. Biochem. Biophys. **201**, 518–526 (1980)

[126] Jones, R. L., I. Sadnik, H. A. Thompson, and K. Moldave: Studies on native ribosomal subunits from rat liver. Evidence for a low molecular weight ribosome dissociation factor. Arch. Biochem. Biophys. **199**, 277–285 (1980)

[127] Goumans, H., A. Thomas, A. Verhoeven, H. O. Voorma, and R. Benne: The role of eIF-4C in protein synthesis initiation complex formation. Biochim. Biophys. Acta **608**, 39–46 (1980)

[128] Thomas, A., H. Goumans, H. O. Voorma, and R. Benne: The mechanism of action of eukaryotic initiation factor 4C in protein synthesis. Europ. J. Biochem. **107**, 39–45 (1980)

[129] Kozak, M.: Role of ATP in binding and migration of 40 S ribosomal subunits. Cell 22, 459–467 (1980)

[130] Sonenberg, N., K. M. Rupprecht, S. M. Hecht, and A. J. Shatkin: Eukaryotic messenger RNA cap binding protein – Purification by affinity chromatography on sepharose-coupled m^7GDP. Proc. Natl. Acad. Sci. U.S. **76**, 4345–4349 (1979)

[131] deleted

[132] Trachsel, H., N. Sonenberg, A. J. Shatkin, J. K. Rose, K. Leong, J. E. Bergmann, J. Gordon, and D. Baltimore: Purification of a factor that restores translation of vesicular stomatitis virus mRNA in extracts from poliovirus-infected HeLa cells. Proc. Natl. Acad. Sci. U.S. **77**, 770–774 (1980)

[133] Sonenberg, N., H. Trachsel, S. Hecht, and A. J. Shatkin: Differential stimulation of capped mRNA translation in vitro by cap binding protein. Nature **285**, 331–333 (1980)

[134] Peterson, D. T., B. Safer, and W. C. Merrick: Role of eukaryotic initiation factor 5 in the formation of 80 S initiation complexes. J. Biol. Chem. **254**, 7730 to 7735 (1979)

[135] Roobol, K., I. Vianden, and W. Möller: Aggregation of eukaryotic elongation factor eEF-T_s and its isolation by means of hydrophobic adsorption chromatography. FEBS Lett. **111**, 136–140 (1980)

[136] Pulikowska, J., M. Barciszewska, J. Joachimiak, A. J. Rafalski, and T. Twardowski: Effect of elastase on elongation factor I from wheat germ. Biochem. Biophys. Res. Commun. **91**, 1011–1017 (1979)

[137] Roobol, K., and W. Möller: Protein synthesis in *Artemia salina*. Eukaryotic elongation factor is a transphosphorylase. Mol. Biol. Rep. **7**, 197–202 (1981)

[138] Olsnes, S., and A. K. Abraham: Elongation-factor-2-induced sensitization of ribosomes to modeccin. Evidence for specific binding of elongation factor 2 to ribosomes in the absence of nucleotides. Europ. J. Biochem. **93**, 447–452 (1979)

[139] Brown, B. A., and J. W. Bodley: Primary structure at the site in beef and wheat elongation factor 2 of ADP-ribosylation by diphteria toxin. FEBS Lett. **103**, 253–255 (1979)

[140] Gessner, S. L., and J. D. Irvin: Inhibition of elongation factor 2–dependent translocation by the pokeweed antiviral protein and ricin. J. Biol. Chem. **255**, 3251–3253 (1980)

[141] Ogata, K., K. Terao, and T. Uchiumi: Stimulation by aminoacyl-tRNA of the GTPase and ATPase activities of rat liver 5 S RNA protein particles in the presence of EF-2. J. Biochem. **87**, 517 to 523 (1980)

[142] Reddington, M. A., and W. P. Tate: A polypeptide chain release factor from the undeveloped cyst of the brine shrimp *Artemia salina*. FEBS Lett. **97**, 335–338 (1979)

[143] Reddington, M. A., I. L. Armstrong, and W. P. Tate: Isolation of a polypeptide chain release factor from undeveloped cyst of *Artemia salina*. Proc. Australian Biochem. Soc. **11**, 87 (1978)

[144] COMOLLI, R., A. SCHUBERT, and L. RIBONI: Changes in the activity and distribution of the ribosomal dissociation factor of rat liver during growth. Mechan. Ageing Develop. **11**, 199—207 (1979)

[145] COMOLLI, R., L. RIBONI, and A. SCHUBERT: Reassociation of eukaryotic ribosomal subunits by a factor from rat ascites hepatoma cytosol. Experientia **35**, 1305—1307 (1979)

[146] SPERRAZZA, J. M., D. W. RUSSELL, and L. L. SPREMULLI: Reversible dissociation of wheat germ ribosomal subunits: Cation-dependent equilibria and thermodynamic parameters. Biochemistry **19**, 1053—1058 (1980)

[147] COMOLLI, R., and L. RIBONI: Reassociation of eukaryotic ribosomal subunits and polyamine concentration in Yoshida ascites hepatoma and Ehrlich ascites carcinoma cells during growth. Cancer Biochem. Biophys. **5**, 25—31 (1981)

[148] OCHOA, S., and C. de HARO: Regulation of protein synthesis in eukaryotes. Ann. Rev. Biochem. **48**, 549—580 (1979)

[149] KRAMER, G., A. B. HENDERSON, N. GRANKOWSKI, and B. HARDESTY: Translational control in eukaryotes. In: Ribosomes: Structure, Function and Genetics (Eds.: CHAMBLISS, G., G. R. CRAVEN, J. DAVIES, K. DAVIS, L. KAHAN, and M. NOMURA) University Park Press, Baltimore, 1980

[150] KRAMER, G., and B. HARDESTY: Regulation of eukaryotic protein synthesis. Cell Biology 4, 69—105 (1980)

[151] AUSTIN, S. A., and M. J. CLEMENS: Control of the initiation of protein synthesis in mammalian cells. FEBS Lett. **110**, 1—7 (1980)

[152] RAFF, R. A.: Masked messenger RNA and the regulation of protein synthesis in eggs and embryos. Cell Biology, 4, 107—136 (1980)

[153] HUNT, T.: The control of protein synthesis in rabbit reticulocyte lysates. In: From Gene to Protein. Miami Winter Symp. Ser., Vol. 16. (Eds.: RUSSELL, T. R., K. BREW, H. FABER, and J. SCHULTZ) Academic Press, New York, 1979, 321—346

[154] HUNT, T.: Phosphorylation and the control of protein synthesis in reticulocyte. In: Recently discovered systems of enzyme regulation by reversible phosphorylation. Molec. Aspects Cell. Reg., Vol. 1 (Ed.: COHEN, P.) Amsterdam, Elsevier North-Holland, Biomed. Press, 1980

[155] GROSS, M.: The control of protein synthesis by hemin in rabbit reticulocytes. Molec. Cell. Biochem. **31**, 25—36 (1980)

[156] WALLIS, M. H., G. KRAMER, and B. HARDESTY: Partial purification and characterization of a 90.000-dalton peptide involved in activation of the eIF-2α protein kinase of the hemin-controlled translational repressor. Biochem. **19**, 798—804 (1980)

[157] LEVIN, D. H., V. ERNST, and I. M. LONDON: Effects of the catalytic subunit of cAMP-dependent protein kinase (type 2) from reticulocytes and bovine heart muscle on protein phosphorylation and protein synthesis in reticulocyte lysates. J. Biol. Chem. **254**, 7935—7941 (1979)

[158] ERNST, V., D. H. LEVIN, and I. M. LONDON: The in situ phosphorylation of the α-subunit of eIF-2 in reticulocyte lysates inhibited by heme deficiency, dsRNA, GSSG, or the heme-regulated inhibitor. In: From Gene to Protein. Miami Winter Symp. Ser., Vol. 16. (Eds.: RUSSELL, T. R., K. BREW, H. FABER, and J. SCHULTZ) Academic Press, New York, 1979, 588

[159] LEVIN, D. H., V. ERNST, A. LEROUX, R. PETRYSHYN, R. FAGARD, and I. M. LONDON: Regulation of eukaryotic protein synthesis by the phosphorylation of the initiation factor eIF-2. In: Protein Phosphorylation and Bioregulation. FMI-EMBO Workshop, Basel 1979, Karger, Basel, 1980, 128—141

[160] TRAUGH, J. A., G. M. HATHAWAY, P. T. TUAZON, S. M. TAHARA, G. A. FLOYD, R. W. DELGRANDE, and T. S. LUNDAK: Cyclic nucleotide-independent protein kinases from rabbit reticulocytes and phosphorylation of translational components. In: Modulation of Protein Function ICN-UCLA Symp. Mol. Cell. Biol., Vol. 13 (Eds.: ATKINSON, D. E. and C. F. FOX) Academic Press, New York, 1979, 233 to 245

[161] FLOYD, G. A., and J. A. TRAUGH: Heme deficiency and phosphorylation of ribosome-associated proteins. Europ. J. Biochem. **106**, 269—277 (1980)

[162] Tuazon, P. T., W. C. Merrick, and J. A. Traugh: Site-specific phosphorylation of initiation factor 2 by three cyclic nucleotide-independent protein kinases. J. Biol. Chem. **255**, 10954–10958 (1980)
[163] Mumby, M., and J. A. Traugh: Dephosphorylation of translational initiation factors and 40 S ribosomal subunits by phosphoprotein phosphatases from rabbit reticulocytes. Biochem. **18**, 4548 to 4556 (1979)
[164] Mumby, M., and J. A. Traugh: Multiple forms of phosphoprotein phosphatase from rabbit reticulocytes. Biochim. Biophys. Acta **611**, 342–350 (1980)
[165] Grankowski, N., D. Lehmusvirta, G. Kramer, and B. Hardesty: Partial purification and characterization of reticulocyte phosphatase with activity for phosphorylated peptide initiation factor 2. J. Biol. Chem. **255**, 310–317 (1980)
[166] Crouch, D., and B. Safer: Purification and properties of eIF-2 phosphatase. J. Biol. Chem. **255**, 7918–7924 (1980)
[167] Stewart, A. A., D. Crouch, P. Cohen, and B. Safer: Classification of an eIF-2 phosphatase as a type-2 protein phosphatase. FEBS Lett. **119**, 16–19 (1980)
[168] Jagus, R., and B. Safer: Control of eIF-2 phosphorylation in rabbit reticulocyte lysate. In: From Gene to Protein. Miami Winter Symp. Ser., Vol. 16. (Eds.: Russell, T. R., K. Brew, H. Faber, and J. Schultz) Academic Press, New York, 1979, 598
[169] Grankowski, N., D. Lehmusvirta, G. B. Stearns, G. Kramer, and B. Hardesty: The isolation and partial characterization of two substrate-specific protein activators of the reticulocyte phosphoprotein phosphatase. J. Biol. Chem. **255**, 5755 to 5762 (1980)
[170] Merrick, W. C.: Effect of phosphorylation of eIF-2. In: Modulation of protein function. ICN-UCLA Symp. Mol. Cell. Biol., Vol. 13. (Eds.: Atkinson, D. E. and C. F. Fox) Academic Press, New York 1979, 391–405
[171] Benne, R., M. Salimans, H. Goumans, H. Amesz, and H. O. Voorma: Regulation of protein synthesis in rabbit reticulocyte lysates: Phosphorylation of eIF-2 does inhibit its capacity to recycle. Europ. J. Biochem. **104**, 501–509 (1980)
[172] Safer, B., W. Kemper, and R. Jagus: The use of [^{14}C] eukaryotic initiation factor 2 to measure the endogenous pool size of eukaryotic initiation factor 2 in rabbit reticulocyte lysate. J. Biol. Chem. **254**, 8091–8094 (1979)
[173] Das, A., R. O. Ralston, M. Grace, R. Roy, P. Goshdastidar, H. K. Das, B. Yaghmai, S. Palmieri, and N. K. Gupta: Protein synthesis in rabbit reticulocytes: Mechanism of protein synthesis inhibition by heme-regulated inhibitor. Proc. Natl. Acad. Sci. U.S. **76**, 5076 to 5081 (1979)
[174] DeHaro, C., and S. Ochoa: Further studies on the mode of action of the heme-controlled translation inhibitor: Stimulating protein acts at level of binary complex formation. Proc. Natl. Acad. Sci. U.S. **76**, 2163–2164 (1979)
[175] Amesz, H., H. Goumans, T. Haubrich-Morree, H. O. Voorma, and R. Benne: Purification and characterization of a protein factor that reverses the inhibition of protein synthesis by the heme-regulated translational inhibitor in rabbit reticulocyte lysates. Europ. J. Biochem. **98**, 513–520 (1979)
[176] Ralston, R. O., A. Das, M. Grace, H. Das, and N. K. Gupta: Protein synthesis in rabbit reticulocytes: Characteristics of a postribosomal supernatant factor that reverses inhibition of protein synthesis in heme-deficient lysates and inhibition of ternary complex (Met-tRNA$_f$xeIF-2xGTP). Proc. Natl. Acad. Sci. U.S. **76**, 5490–5494 (1979)
[177] Safer, B., R. Jagus, and D. Crouch: Indirect inactivation of eukaryotic initiation factor 2 in reticulocyte lysate by selenite. J. Biol. Chem. **255**, 6913–6917 (1980)
[178] Jagus, R., and B. Safer: Activity of eukaryotic initiation factor 2 is modified by processes distinct from phosphorylation: Activities of eukaryotic initiation factor 2 and eukaryotic initiation factor 2α kinase in lysate gelfiltered under different conditions. J. Biol. Chem. **256**, 1317–1323 (1981)
[179] Jagus, R., and B. Safer: Activity of eukaryotic initiation factor 2 is modified by processes distinct from phosphorylation: Activity of eukaryotic initiation fac-

tor 2 in lysate is modified by oxidation-reduction state of its sulfhydryl groups. J. Biol. Chem. **256**, 1324—1329 (1981)

[180] RANU, R. S.: Regulation of protein synthesis in rabbit reticulocyte lysates: Purification and initial characterization of the double-stranded RNA activated protein kinase. Biochem. Biophys. Res. Commun. **97**, 252—262 (1980)

[181] PETRYSHYN, R., D. H. LEVIN, and I. M. LONDON: Purification and characterization of a latent precursor of a double-stranded RNA dependent protein kinase from reticulocyte lysates. Biochem. Biophys. Res. Commun. **94**, 1190—1198 (1980)

[182] LEVIN, D. H., R. PETRYSHYN, and I. M. LONDON: Characterization of double-stranded RNA activated protein kinase that phosphorylates α-subunit of eukaryotic initiation factor 2(eIF-2) in reticulocyte lysates. Proc. Natl. Acad. Sci. U.S. **77**, 832—836 (1980)

[183] GROSFELD, H., and S. OCHOA: Purification and properties of the double-stranded RNA activated eukaryotic initiation factor-2 kinase from rabbit reticulocytes. Proc. Natl. Acad. Sci. U.S. **77**, 6526—6530 (1980)

[184] RANU, R. S.: Regulation of protein synthesis in rabbit reticulocyte lysates: The heme-regulated protein kinase (HRI) and double-stranded RNA induced protein kinase (dRI) phosphorylate the same site(s) on initiation factor eIF-2. Biochem. Biophys. Res. Commun. **91**, 1437 to 1444 (1979)

[185] ERNST, V., D. H. LEVIN, A. LEROUX, and I. M. LONDON: Site-specific phosphorylation of the α-subunit of eukaryotic initiation factor eIF-2 by the heme-regulated and double-stranded RNA activated eIF-2α kinase from rabbit reticulocyte lysates. Proc. Natl. Acad. Sci. U.S. **77**, 1286—1290 (1980)

[186] RANU, R. S.: Isolation of a translational inhibitor from wheat germ with protein kinase activity that phosphorylates initiation factor eIF-2. Biochem. Biophys. Res. Commun. **97**, 1124—1132 (1980)

[187] RANU, R. S.: Regulation of protein synthesis in eukaryotes by the protein kinases that phosphorylate initiation factor eIF-2: evidence for a common mechanism of inhibition of protein synthesis. FEBS Lett. **112**, 211—215 (1980)

[188] BAGLIONI, C.: Interferon-induced enzymatic activities and their role in the antiviral state. Cell **17**, 255—264 (1979)

[189] CLEMENS, M. J.: Interferons and cellular regulation. Nature **282**, 364—365 (1979)

[190] YABROV, A. A.: Interferon and nonspecific resistance. Human Science Press, New York, London, 1980

[191] REVEL, M.: Molecular mechanisms involved in the antiviral effects of interferon. In: Interferon, Vol. 1 (Eds.: CANTELL, K., E. DEMAEYER, M. LANDY, M. REVEL, J. VILCEK). Academic Press, London, New York, Toronto, Sydney, San Francisco 1979

[192] WILLIAMS, B. R. G., and I. M. KERR: The 2—5 A (pppA2′p5′A2′p5′A) system in interferon-treated and control cells. Trends Biochem. Sci. **5**, 138—140 (1980)

[193] TORRENCE, P. F., and R. M. FRIEDMAN: Are double-stranded RNA-directed inhibition of protein synthesis in interferon-treated cells and interferon induction related phenomena? J. Biol. Chem. **254**, 1259—1267 (1979)

[194] MINKS, M. A., D. K. WEST, S. BENVIN, and C. BAGLIONI: Structural requirement of double-stranded RNA for the activation of 2′,5′-oligo(A) polymerase and protein kinase of interferon-treated HeLa cells. J. Biol. Chem. **254**, 10180—10183 (1979)

[195] MINKS, M. A., D. K. WEST, S. BENVIN, J. J. GREENE, P. O. P. TS'O, and C. BAGLIONI: Activation of 2′5′-oligo(A) polymerase and protein kinase of interferon-treated HeLa cells by 2′-O-methylated poly(inosinic acid) · poly(cytidylic acid). J. Biol. Chem. **255**, 6403—6407 (1980)

[196] EPPSTEIN, D. A., T. C. PETERSON, and C. E. SAMUEL: Mechanism of interferon action: Synthesis and activity of the interferon-mediated low-molecular weight oligonucleotide from murine and human cells. Virology **98**, 9—19 (1979)

[197] MINKS, M. A., S. BENVIN, and C. BAGLIONI: Mechanism of $pppA(2'p5'A)_n2'p5'A_{OH}$ synthesis in extracts of interferon-treated HeLa-cells. J. Biol. Chem. **255**, 5031 to 5035 (1980)

[198] BAGLIONI, C., and P. A. MARONEY: Mechanism of action of human interferons:

Induction of 2′,5′-oligo(A) polymerase. J. Biol. Chem. **255**, 8390–8393 (1980)

[199] Baglioni, C., and P. A. Maroney: Human leukocyte interferon induces 2′,5′-oligo(A) polymerase and protein kinase. Virology **101**, 540–544 (1980)

[200] Hovanessian, A. G., and I. M. Kerr: The (2′-5′)oligoadenylate (pppA2′-5′A2-5′A) synthetase and protein kinase(s) from interferon-treated cells. Europ. J. Biochem. **93**, 515–526 (1979)

[201] Samanta, H., J. P. Dougherty, and P. Lengyel: Synthesis of (2′-5′)$(A)_n$ from ATP: Characteristics of the reaction catalyzed by (2′-5′) $(A)_n$ synthetase purified from mouse Ehrlich ascites tumor cells treated with interferon. J. Biol. Chem. **255**, 9807–9813 (1980)

[202] Dougherty, J. P., H. Samanta, P. J. Farrell, and P. Lengyel: Interferon, double-stranded RNA, and RNA degradation: Isolation of homogeneous pppA-$(2'p5'A)_{n-1}$ synthetase from Ehrlich ascites tumor cells. J. Biol. Chem. **255**, 3813–3816 (1980)

[203] Justesen, J., D. Ferbus, and M. N. Thang: 2′5′oligoadenylate synthetase, an interferon induced enzyme: Direct assay methods for the products, 2′5′oligoadenylates and 2′5′cooligonucleotides. Nucl. Acids Res. **8**, 3073–3085 (1980)

[204] Knight, M., P. J. Cayley, R. H. Silverman, D. H. Wreschner, C. S. Gilbert, R. E. Brown, and I. M. Kerr: Radioimmune, radiobinding and HPLC analysis of 2-5A and related oligonucleotides from intact cells. Nature **288**, 189–192 (1980)

[205] Johnston, M. I., R. M. Friedman, and P. F. Torrence: Interferon-induced (2′-5′)oligoadenylate synthetase: Adsorption to and assay on adenosine 2′-5′diphosphate sepharose. Biochem. **19**, 5580–5585 (1980)

[206] Martin, E. M., N. J. M. Birsdall, R. E. Brown, and I. M. Kerr: Enzymic synthesis, characterization and nuclear magnetic resonance spectra of pppA2′p5′A2′p5′A and related oligonucleotides: Comparison with chemically synthesized material Europ. J. Biochem. **95**, 295–307 (1979)

[207] Williams, B. R. G., I. M. Kerr, C. S. Gilbert, C. N. White, and L. A. Ball: Synthesis and breakdown of pppA2′-p5′A2′p5′A and transient inhibition of protein synthesis in extracts from interferon-treated and control cells. Europ. J. Biochem. **92**, 455–462 (1978)

[208] Minks, M. A., S. Benvin, P. A. Maroney, and C. Baglioni: Metabolic stability of 2′5′oligo(A) and activity of 2′5′-oligo(A)-dependent endonuclease in extracts of control and interferon-treated HeLa cells. Nucl. Acids Res. **6**, 767–780 (1979)

[209] Schmidt, A., Y. Chernajovsky, L. Schulman, P. Federman, H. Berissi, and M. Revel: Interferon-induced phosphodiesterase degrading (2′-5′)oligoisoadenylate and the CCA terminus of transfer RNA. Proc. Natl. Acad. Sci. U.S. **76**, 4788–4792 (1979)

[210] Nilsen, T. W., S. G. Weissman, and C. Baglioni: Role of 2′,5′-oligo(adenylic acid) polymerase in the degradation of ribonucleic acid linked to double-stranded ribonucleic acid by extracts of interferon-treated cells. Biochem. **19**, 5574 to 5579 (1980)

[211] Wreschner, D. H., J. W. McCauley, J. J. Skehel, and I. M. Kerr: Interferon action—sequence specificity of the $ppp(A2'p)_nA$-dependent ribonuclease. Nature **289**, 414–417 (1981)

[212] Vaquero, C. M., and M. J. Clemens: Inhibition of protein synthesis and activation of a nuclease in rabbit reticulocyte lysates by the unusual oligonucleotide pppA2′p5′A2′p5′A. Europ. J. Biochem. **98**, 245–252 (1979)

[213] Krishnan, I., and C. Baglioni. Increased levels of (2′–5′) oligo(A) polymerase activity in human lymphoblastoid cells treated with glucocorticoids. Proc. Natl. Acad. Sci. U.S. **77**, 6506–6510 (1980)

[214] Wood, J. N., and A. G. Hovanessian: Interferon enhances 2–5A synthetase in embryonal carcinoma cells. Nature **282**, 74–76 (1979)

[215] Williams, B. R. G., R. R. Golgher, and I. M. Kerr: Activation of a nuclease by pppA2′p5′A2′p5′A in intact cells. FEBS Lett. **105**, 47–51 (1979)

[216] Kortsaris, A., T. Karemfillis, S. I. Koliais, and J. Taylor-Papadimitrou: Mitochondrial protein synthesis is inhibited by 2′,5′linked oligoadenylic acid triphos-

phate. Biochem. Biophys. Res. Commun. **96**, 1466–1471 (1980)
[217] Kimchi, A., H. Shure, and M. Revel: Regulation of lymphocyte mitogenesis by (2′–5′)oligoisoadenylate. Nature **282**, 849–851 (1979)
[218] Samuel, C. E.: Mechanism of interferon action: Phosphorylation of protein synthesis initiation factor eIF-2 in interferon-treated human cells by a ribosome-associated kinase possessing site specificity similar to hemin-regulated rabbit reticulocyte kinase. Proc. Natl. Acad. Sci. U.S. **76**, 600–604 (1979)
[219] Gupta, S. L.: Specific protein phosphorylation in interferon-treated uninfected and virus-infected mouse L_{929}-cells: Enhancement by double-stranded RNA. J. Virol. **29**, 301–311 (1979)
[220] Samuel, C. E.: Mechanism of interferon action: Kinetics of interferon action in mouse L_{929}-cells, phosphorylation of protein synthesis initiation factor eIF-2 and ribosome-associated protein P 1. Virology **93**, 281–285 (1979)
[221] West, D. K., and C. Baglioni: Induction by interferon in HeLa cells of a protein kinase activated by double-stranded RNA. Europ. J. Biochem. **101**, 461 to 468 (1979)
[222] Ohtsuki, K., and S. Baron: An interferon-induced ribosome-associated protein kinase which reduces the activity of initiation factor. J. Biochem. **85**, 1495 to 1502 (1979)
[223] Ohtsuki, K., M. Nakamura, T. Koike, N. Ishida, and S. Baron: A ribosomal protein mediates eIF-2 phosphorylation by interferon-induced kinase. Nature **287**, 65–67 (1980)
[224] Kaempfer, R., R. Israeli, H. Rosen, S. Knoller, A. Zilberstein, A. Schmidt, and M. Revel: Reversal of the interferon-induced block of protein synthesis by purified preparations of eukaryotic initiation factor 2. Virology **99**, 170–173 (1979)
[225] Hovanessian, A. G., F. Barresinoussi, and L. Montagnier: Interferon-mediated protein kinase in mouse cells treated with iododeoxyuridine and induced to express endogeneous retroviruses. J. Gen. Virol. **52**, 199–204 (1981)
[226] Desrosiers, R. C., and P. Lengyel: Impairement of reovirus mRNA "cap" methylation in interferon-treated mouse L_{929} cells. Biochim. Biophys. Acta **562**, 471–480 (1979)
[227] Contreras, A., and L. Carrasco: Selective inhibition of protein synthesis in virus-infected mammalian cells. J. Virol. **29**, 114–122 (1979)
[228] Garry, R. F., and M. R. F. Waite: Na^+ and K^+ concentrations and the regulation of the interferon system in chick cells. Virology **96**, 121–128 (1979)
[229] Ehrenfeld, E., and S. Manis: Inhibition of 80 S initiation complex formation by infection with poliovirus. J. Gen. Virol. **43**, 441–445 (1979)
[230] Schrom, M., and R. Bablanian: Inhibition of protein synthetis by vaccinia virus: Studies on the role of virus-induced RNA synthesis. J. Gen. Virol. **44**, 625 to 638 (1979)
[231] Sagot, J., and G. Beaud: Phosphorylation in vivo of a vaccinia-virus structural protein found associated with the ribosomes from infected cells. Europ. J. Biochem. **98**, 131–140 (1979)
[232] Beaud, G., and A. Dru: Protein synthesis in vaccinia virus-infected cells in the presence of amino acid analogues: A translational control mechanism. Virology **100**, 10–21 (1980)
[233] Person, A., F. Ben-Hamida, and G. Beaud: Inhibition of 40 S · Met-$tRNA_f^{Met}$ ribosomal initiation complex formation by vaccinia virus. Nature **287**, 355–357 (1980)
[234] Ghosh-Dastidar, P., B. B. Goswami, A. Das, P. Das, and N. K. Gupta: Vaccinia viral core inhibits Met-$tRNA_f$ · 40 S initiation complex formation with physiological mRNAs. Biochem. Biophys. Res. Commun. **99**, 946–953 (1981)
[235] Wallach, D. and M. Revel: An interferon-induced cellular enzyme is incorporated into virions. Nature **287**, 68 to 70 (1980)
[236] Lodish, H. F., and M. Porter: Translational control of protein synthesis after infection by vesicular stomatitis virus. J. Virol. **36**, 719–733 (1980)
[237] Jen, G., B. M. Detjen, and R. E. Thach: Shutoff of HeLa cell protein synthesis by encephalomyocarditis virus and polio-

virus: A comparative study. J. Virol. **35**, 150–156 (1980)

[238] Helentjaris, T., E. Ehrenfeld, M. L. Brown-Luedi, and J. W. B. Hershey: Alterations in initiation factor activity from poliovirus-infected HeLa cells. J. Biol. Chem. **254**, 10973–10978 (1979)

[239] Brown, B. A., and E. Ehrenfeld: Initiation factor preparations from poliovirus-infected cells restrict translation in reticulocyte lysates. Virology **103**, 327 to 339 (1980)

[240] Brown, D., J. Hansen, and E. Ehrenfeld: Specificity of initiation factor preparations from poliovirus-infected cells. J. Virol. **34**, 573–575 (1980)

[241] Schrom, M., and R. Bablanian: Inhibition of protein synthesis by vaccinia virus: Characterization of an inhibited cell free protein-synthesizing system from infected cells. Virology **99**, 319–328 (1979)

[242] Van Steeg, H., H. van Grinsven, F. van Mansfeld, H. O. Voorma, and R. Benne: Initiation of protein synthesis in neuroblastoma cells infected by Semliki forest virus: A decreased requirement of late viral mRNA for eIF-4 B and cap binding protein. FEBS Lett. **129**, 62 to 66 (1981)

[243] Kitamura, N., B. L. Semler, P. G. Rothberg, G. R. Larsen, C. J. Adler, A. J. Dorner, E. A. Emini, R. Hanecak, J. L. Lee, S. van der Werf, C. W. Anderson, and E. Wimmer: Primary structure, gene organization and polypeptide expression of poliovirus RNA. Nature **291**, 547–553 (1981)

[244] Golini, F., B. L. Semler, A. J. Dorner, and E. Wimmer: Protein-linked RNA of a poliovirus is competent to form an initiation complex of translation in vitro. Nature **287**, 600–603 (1980)

XIV. Subject index

XV. Reference index